World Geography

AFRICA

Regions | Physical Geography | Biogeography and Natural Resources
Human Geography | Economic Geography | Gazetteer

Second Edition

Volume 4

Editor
Joseph M. Castagno
Educational Reference Publishing, LLC

SALEM PRESS
A Division of EBSCO Information Services, Inc.
Ipswich, Massachusetts

GREY HOUSE PUBLISHING

Cover photo: Africa from outer space. Image by 1xpert.

For information contact Grey House Publishing/Salem Press, 4919 Route 22, PO Box 56, Amenia, NY 12501.

∞ The paper used in these volumes conforms to the American National Standard for Permanence of Paper for Printed Library Materials, Z39.48 1992 (R2009).

Publisher's Cataloging-In-Publication Data
(Prepared by The Donohue Group, Inc.)

Names: Castagno, Joseph M., editor.
Title: World geography / editor, Joseph M. Castagno, Educational Reference Publishing, LLC.
Description: Second edition. | Ipswich, Massachusetts : Salem Press, a division of EBSCO Information Services, Inc. ; Amenia, NY : Grey House Publishing, [2020] | Interest grade level: High school. | Includes bibliographical references and index. | Summary: A six-volume geographic encyclopedia of the world, continents and countries of each continent. In addition to physical geography, the set also addresses human geography including population distribution, physiography and hydrology, biogeography and natural resources, economic geography, and political geography. | Contents: Volume 1. South & Central America — Volume 2. Asia — Volume 3. Europe — Volume 4. Africa — Volume 5. North America & the Caribbean — Volume 6. Australia, Oceania & the Antarctic.
Identifiers: ISBN 9781642654257 (set) | ISBN 9781642654288 (v. 1) | ISBN 9781642654318 (v. 2) | ISBN 9781642654301 (v. 3) | ISBN 9781642654295 (v. 4) | ISBN 9781642654271 (v. 5) | ISBN 9781642654325 (v. 6)
Subjects: LCSH: Geography—Encyclopedias, Juvenile. | CYAC: Geography—Encyclopedias. | LCGFT: Encyclopedias.
Classification: LCC G133 .W88 2020 | DDC 910/.3—dc23

First Printing
PRINTED IN CANADA

World Geography

AFRICA

Regions | Physical Geography | Biogeography and Natural Resources
Human Geography | Economic Geography | Gazetteer

Second Edition

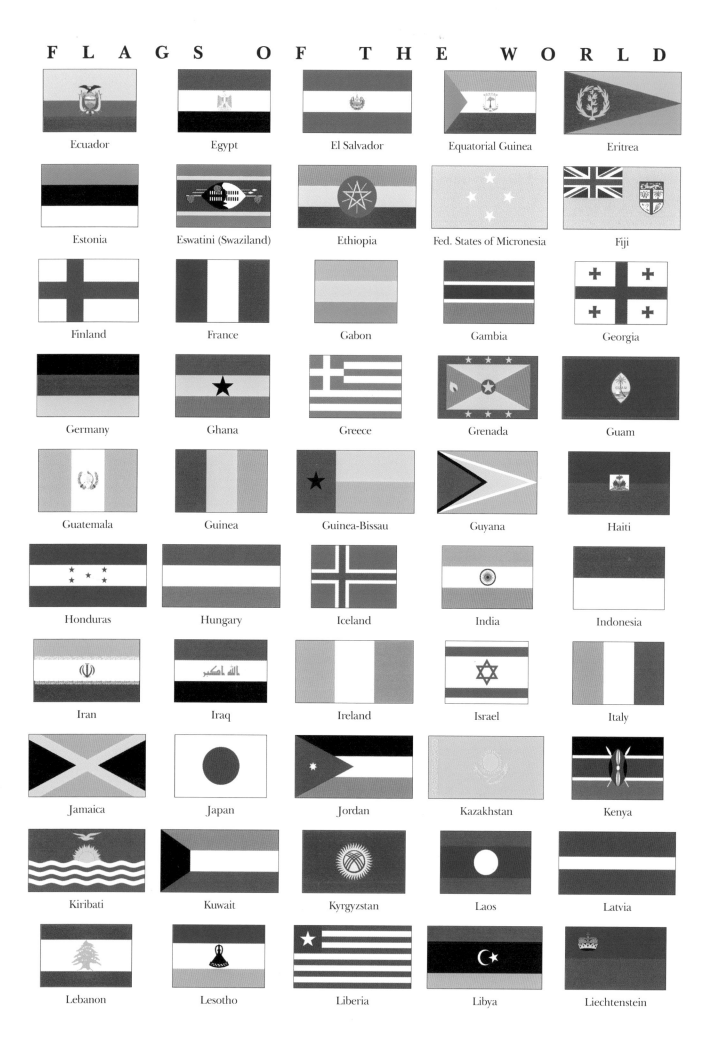

Ecuador

Egypt

El Salvador

Equatorial Guinea

Eritrea

Estonia

Eswatini (Swaziland)

Ethiopia

Fed. States of Micronesia

Fiji

Finland

France

Gabon

Gambia

Georgia

Germany

Ghana

Greece

Grenada

Guam

Guatemala

Guinea

Guinea-Bissau

Guyana

Haiti

Honduras

Hungary

Iceland

India

Indonesia

Iran

Iraq

Ireland

Israel

Italy

Jamaica

Japan

Jordan

Kazakhstan

Kenya

Kiribati

Kuwait

Kyrgyzstan

Laos

Latvia

Lebanon

Lesotho

Liberia

Libya

Liechtenstein

CONTENTS

THE CHALLENGE OF COVID-19

As *World Geography: Africa* goes to press in July 2020, the entire globe is grappling with the worst and most widespread pandemic in more than a century. The cause of the pandemic is a highly contagious viral condition known as Coronavirus Disease 2019, or COVID-19.

The first documented emergence of COVID-19 occurred in December 2019 as an outbreak of pneumonia in Wuhan City, in Hubai Province, China. By January 2020, Chinese health officials had reported tens of thousands of infections and dozens of deaths. That same month, COVID-19 cases were appearing across Asia, Europe, and North America, and spreading rapidly. On March 11, the World Health Organization (WHO) declared COVID-19 viral disease a pandemic.

The rapid spread of COVID-19 viral disease has strong geographical components. The virus emerged in Wuhan, a huge, densely populated city. It spread rapidly among a population living largely indoors during the cold-weather months. But the most significant geographical factor in the spread of the virus may well be the globalization of transportation. Every day, thousands of people fly to destinations near and far. Each traveler carries the potential to unknowingly spread disease.

To curtail COVID-19's spread, countries have closed their borders, air travel has been drastically reduced, and sweeping mitigation policies have been inaugurated. Some densely populated Asian countries, including South Korea, Singapore, and Taiwan, enforced these measures very early, and have, as of July 2020, managed to dampen the effect of the virus and limit the number of confirmed cases and deaths due to COVID-19. Other places, slower to act, such as China and Iran, have been very hard hit.

Many COVID-19 questions remain: Where will the disease strike next? Will the onset of warmer weather reduce the communicability of the virus? Will a vaccine be available soon? Will there be a second wave of infection? When will life be back to normal?

Geographers will continue to play a unique role in answering these questions, applying the tools and techniques of their discipline to achieving the fullest possible epidemiological understanding of the pandemic.

PUBLISHER'S NOTE

North Americans have long thought of the field of geography as little more than the study of the names and locations of places. This notion is not without a basis in fact: Through much of the twentieth century, geography courses forced students to memorize names of states, capitals, rivers, seas, mountains, and countries. Both students and educators eventually rebelled against that approach, geography courses gradually fell out of favor, and the future of geography as a discipline looked doubtful. Happily, however, the field has undergone a remarkable transformation, starting in the 1990s. Geography now has a bright and pivotal significance at all levels of education.

While learning the locations of places remains an important part of geography studies, educators recognize that place-name recognition is merely the beginning of geographic understanding. Geography now places much greater emphasis on understanding the characteristics of, and interconnections among, places. Modern students address such questions as how the weather in Brazil can affect the price of coffee in the United States, why global warming threatens island nations, and how preserving endangered plant and animal species can conflict with the economic development of poor nations.

World Geography, Second Edition, addresses these and many other questions. Designed and written to meet the needs of high school students, while being accessible to both middle school and undergraduate college students, these six volumes take an integrated approach to the study of geography, emphasizing the connections among world regions and peoples. The set's six volumes concentrate on major world regions: South and Central America; Asia; Europe; Africa, North America; and Australia, Oceania, and the Antarctic. Each volume begins with common overview information related to the geography, maps and mapmaking. The core essays in the volumes begin with an overview section to provide global context and then go on to examine important geographic aspects of the regions in that area of the world: its physical geography;

biogeography and natural resources; human geography (including its political geography); and economic geography. These essays range in length from three to ten pages. A gazetteer indicates major political, geographic, and manmade features throughout the region.

A robust appendix found in each volume provides further information:

- The Earth in Space (The Solar System, Earth's Moon, The Sun and the Earth, The Seasons);
- Earth's Interior (Earths Internal Structure, Plate Tectonics, Volcanoes, Geologic Time Scale);
- Earth's Surface (Internal Geological Processes, External Processes, Fluvial and Karst Processes, Glaciation, Desert Landforms, Ocean Margins);
- Earth's Climates (The Atmosphere, Global Climates; Cloud Formation, Storms);
- Earth's Biological Systems (Biomes);
- Natural Resources (Soils, Water);
- Exploration and Transportation (Exploration and Historical Trade Routes, Road Transportation, Railways, Air Transportation);
- Energy and Engineering (Energy Sources, Alternative Energies, Engineering Projects);
- Industry and Trade (Manufacturing, Globalization of Manufacturing and Trade, Modern World Trade Patterns);
- Political Geography (Forms of Government, Political Geography, Geopolitics, National Park Systems);
- Boundaries and Time Zones (International Boundaries, Global Time and Time Zones);
- Global Education (Themes and Standards in Geography Education);
- Global Data (The World Gazetteer of Oceans and Continents, The World's Oceans and Seas, Major Land Areas of the World, Major Islands of the World, Countries of the World (including population and pollution density), Past and Projected World Population Growth, 1950-2050, The World's Largest Countries by Area, The World's Smallest Countries by Area, The World's Largest Countries by Population,

The World's Smallest Countries by Population, The World's Most Densely Populated Countries, The World's Least Densely Populated Countries, The World's Most Populous Cities, Major Lakes of the World, Major Rivers of the World, The Highest Peaks in Each Continent, Major Deserts of the World, Highest Waterfalls of the World).

• A Glossary, General Bibliography, and Index complete the backmatter.

The regional divisions in the set make it possible to study specific countries or parts of the world. Pairing the specific regional information, organized by regions, physical geography, biogeography and natural resources, human geography, economic geography, and a gazetteer, with global information makes it possible for students to see the connections not only between countries and places within the region, but also between the regions and the entire global system, all within a single volume.

To make this set as easy as possible to use, all of its volumes are organized in a similar fashion, with six major divisions—Regions (organized into subregions by volume), Physical Geography, Biogeography and Natural Resources, Human Geography, Economic Geography, and Gazetteer. The number of subregions in each volume varies, depending upon the major world division being examined—Africa, for example, incudes the continent and offshore islands.

Physical geography considers a world region's physiography, hydrology, and climatology. Biogeography and natural resources explores renewable and nonrenewable resources, flora, and fauna. Human geography addresses the people, population distribution, culture regions, urbanization, and political geography of the area. Economic geography considers the region's agriculture, industries, engineering projects, transportation, trade, and communications.

Gazetteers include descriptive entries on hundreds of important places, especially those mentioned in the volume's essays. A typical entry gives the place name and location, indicating the category into which the place falls (mountain, river, city, country, lake, etc).

The entries also include statistics relevant to the categories of place (height of mountains, length of rivers, population of cities and countries).

A feature new to this edition is the discussion questions included throughout the volume. These questions are meant to foster discussion and further research into the topics related to the history, current issues, and future concerns related to physical, human, economic, and political geography.

Both a physical and a social science, geography is unique among social sciences in the demands it makes for visual support. For this reason, *World Geography* contains more than 100 maps, more than 1300 photographs, and scores of other graphical elements. In addition, essays are punctuated with more than 500 textual sidebars and tables, which amplify information in the essays and call attention to especially important or interesting points.

Both English and metric measures are used throughout this set. In most instances, English measures are given first, followed by their metric equivalents in parentheses. It should be noted that in cases of measures that are only estimates, such as the areas of deserts or average heights of mountain ranges, the metric figures are often rounded off to estimates that may not be exact equivalents of the English-measure estimates. In order to enhance clarity, units of measure are not abbreviated in the text, with these exceptions: kilometers are abbreviated as km. and square kilometers as sq. km. This exception has been made because of the frequency with which these measures appear.

Reference works such as this would be impossible without the expertise of a large team of contributing scholars. This project is no exception. Salem Press would like to thank the more than 175 people who wrote the signed essays and contributed entries to the gazetteers. A full list of contributors follows this note. We recognize the efforts of Dr. Ray Sumner, of California's Long Beach City College, for the expertise and insights that she brought to the previous edition of this book, and which have formed the strong foundation for this new edition. We also acknowledge the work of the editor of this current volume, Joseph Castagno, Educational Reference Publishing, LLC.

Introduction

When Henry Morton Stanley of the *New York Herald* shook the hand of David Livingstone on the shore of Central Africa's Lake Tanganyika in 1871, the moment represented the high point of geography to many people throughout the world. A Scottish missionary and explorer, Livingstone had been out of contact with the outside world for nearly two years, and European and American newspapers had buzzed with speculation about his disappearance. At that time, so little was known about the geography of the interior of Africa that Stanley's finding Livingstone was acclaimed as a brilliant triumph of explorations.

The field of geography in Stanley and Livingstone's time was—and to a large extent still is—synonymous with explorations. Stories of epic journeys, both historic and contemporary, continue to exert a powerful attraction on readers. Mountains, deserts, forest, caves, and glaciers still draw intrepid explorers, while even more armchair travelers are thrilled by accounts and pictures of these exploits and discoveries. We all love to travel—to the beach, into the mountains, to our great national parks, and to foreign countries. In the need and desire to explore our surroundings, we are all geographers.

Numerous geographical societies welcome both professional geographers and the general public into their membership, as they promote a greater knowledge and understanding of the earth. The National Geographic Society, founded in 1888 "for the increase and diffusion of geographical knowledge," has awarded more than 11,000 grants for scientific exploration and research. Each year, the society invests millions of dollars in expeditions and fieldwork related to environmental concerns and global geographic issues. The findings are recorded in the pages of the familiar yellow-bordered *National Geographic* magazine, now produced in 40 local-language editions in many countries around the world, publishing around 6.8 million copies monthly, with some 60 million readers. The magazine, along with the National Geographic International television network, reaches more than 135 million readers and viewers worldwide and has more than 85 million subscribers.

An even older geographical association is Great Britain's Royal Geographical Society, which grew out of the Geographical Society of London, founded in 1830 with the "sole object" of promoting "that most important and entertaining branch of knowledge—geography." Over the century that followed, the Royal Geographical Society focused on exploration of the continents of Africa and Antarctica. In the society's London headquarters adjacent to the Albert Hall, visitors can still view such historic artifacts as David Livingstone's cap and chair, as well as diaries, sketches, and maps covering the great period of the British Empire and beyond. Today the society assists more than five hundred field expeditions every year.

With the aid of satellites and remote-sensing instruments, we can now obtain images and data from almost anywhere on Earth. However, remote and inaccessible places still invite the intrepid to visit and explore them in person. Although the outlines of the continents have now been completed, and their interiors filled in with details of mountains and rivers, cities and political boundaries, remote places still exert a fascination on modern urbanites.

The enchantment of tales about strange sights and courageous journeys has been with us since the ancient voyages of Homer's *Ulysses,* Marco Polo's travels to China, and the nautical expeditions of Christopher Columbus, Ferdinand Magellan, and James Cook. While those great travelers are from the remote past, the age of exploration is far from over—a fact repeatedly demonstrated by the modern Norwegian navigator Thor Heyerdahl. Moreover, new journeys of discovery are still taking place. In 1993, after dragging a sled wearily across the frigid wastes of Antarctica for more than three months, Sir Ranulph Twisleton-Wykeham-Fiennes announced that the age of exploration is not dead. Six years later, in 1999, the long-missing body of British mountain climber George Mallory was found on the slopes of Mount Everest, near whose top he had mysteriously vanished in 1924. That discovery sparked a new wave of admiration and respect for explorers of such courage and endurance.

How many people have been enthralled by the bravery of Antarctic explorer Robert Falcon Scott and the noble sacrifice his injured colleague Lawrence Oates made in 1912, when he gave up his life in order not to slow down the rest of the expedition? There can be no doubt that the thrills and the dangers of exploring find resonance among many modern readers.

The struggle to survive in environments hostile to human beings reminds us of the power of our planet Earth. Significant books on this theme have included Jon Krakauer's *Into Thin Air* (1998), an account of a disastrous expedition climbing Mount Everest, and Sebastian Junger's *The Perfect Storm* (1997), the story of the worst gale of the twentieth century and its effect on a fishing fleet off the East Coast of North America. *Endurance* (1998), the epic of Sir Ernest Shackleton's survival and leadership for two years on the frozen Arctic, attracts the same people who avidly read *Undaunted Courage* (1996) the story of Meriwether Lewis and William Clark's epic exploration of the Louisiana Purchase territories in the early nineteenth century. In 1997 *Seven Years in Tibet* premiered, a popular film about the Austrian Heinrich Harrer, who lived in Tibet in the mid-twentieth century. The more urban people become, the greater their desire for adventurous, remote places, a least vicariously, to raise the human spirit.

There are, of course, also scientific achievements associated with modern exploration. In November 1999, the elevation of Mount Everest, the world tallest peak was raised by 7 feet (2.1 meters) to a new height of 29,035 feet (8,850 meters) above sea level; the previously accepted height had been based on surveys made during the 1950s. This new value was the result of Global Positioning System (GPS) technology enabling a more accurate measurement than had been possible with land-based earthbound surveying equipment. A team of climbers supported by the National Geographic Society and the Boston Museum of Science was equipped with GPS equipment, which enabled a fifty-minute re-

cording of data based on satellite signals. At the same time, the expedition was able to ascertain that Mount Everest is moving northeast, atop the Indo-Australian Plate, at a rate of approximately 2.4 inches (10 centimeters) per year.

In 2000, the International Hydrographic Organization named a "new" ocean, the Southern Ocean, which encompasses all the water surrounding Antarctica up to 60° south latitude. With an area of approximately 7.8 million square miles (20.3 million sq. km.), the Southern Ocean is about twice the size of the entire United States and ranks as the world's fourth largest ocean, after the Pacific, Atlantic, and Indian Oceans, but just ahead of the Arctic Ocean.

Despite the humanistic and scientific advantage of geographic knowledge, to many people today, geography is a subject where one merely memorized longs lists of facts dealing with "where" questions. (Where is Andorra? Where is Prince Edward Island? Where is Kalamazoo?) or "what" questions (What is the highest mountain in South America? What is the capital of Costa Rica?) This approach to the study of geography has been perpetuated by the annual National Geographic Bee, conducted in the United States each year for students in grades four through eight. Participants in the competition display an astonishing recall of facts but do not have the opportunity of showing any real geographic thought. To a geographer, such factual knowledge is simply a foundation for investigating and explaining the much more important questions dealing with "why"—"Why is the Sahara a desert?"

Geographers aim to understand why environments and societies occur where and as they do, and how they change. Geography must be seen as an integrative science; the collection of factual data and evidence, as in exploration, is the empirical foundation for deductive reasoning. This leads to the creation of a range of geographical methods, models, theories, and analytical approaches that serve to unify a very broad area of knowledge—the interaction between natural and human environments. Although geography as an academic discipline became established in nineteenth century Germany, there have always been geographers, in the sense of people curious about their world. Humans have always wanted to know about day and night, the shape of the earth, the nature of climates, differences in plants and animals, as well as what lies beyond the horizon. Today, as we hear about and actually experience the sweeping effects of globalization, we need more than ever to develop our geographic skills. Not only are we connected by economic ties to the countries of the world, but we must also appreciate the consequences of North America's high standard of living.

Political boundaries are artificial human inventions, but the natural world is one biosphere. As concern over global warming escalates, national leaders meet to seek a solution to the emission of greenhouse gases, rising ocean levels, and mass extinctions. Are we connected to our environment? At a time when the rate of species extinction is a hundred times above normal, and the human population is crowding in increasing numbers into huge urban centers, we have, nevertheless, taken time each year in April to celebrate Earth Day since 1970. We need now to realize that every day is Earth Day.

Geography languished in the United States in the 1960s, as social studies was taught with a history emphasis in schools. American students became alarmingly disadvantaged in geographic knowledge, compared with most other countries. Fortunately, members of the profession acted to restore geography to the curriculum. In 1984, the National Geographic Society undertook the challenge of restoring geography in the United States. The society turned to two organizations active in geographic education: The Association of American Geographers, the professional geographers' group with more than 10,000 members, mostly in higher education in the United States; and the National Council for Geographic Education that supports geography teaching at all levels—from kindergarten through university, with members that include U.S. and international teachers, professors, students, businesses, and others who support geography education. The council administers the Geographic Alliances, found in every state of the United Sates, with a national membership of about 120,000 schoolteachers. Together, they produced the "Guidelines in Geographic Education," which introduced the Five Themes of Geography, to

enhance the teaching of geography in schools. Using the themes of Location, Place, Human/Environment Interaction, Movement and Regions, teachers were able to plan and conduct lessons in which students encountered interesting real-world examples of the relevance and importance of geography. Continued research into geographic education led to the inclusion of geography in 1990 as one of the core subjects of the National Education Goals, along with English, mathematics, science, and history.

Another milestone was the publication in 1994 of "Geography for Life," the national Geography Standards. The earlier Five Themes were subsumed under the new Six Essential Elements: The World in Spatial Terms; Places and Regions; Physical Systems; Human Systems; Environment Systems; Environment and Society; and The Uses of Geography. Eighteen geography standards are included, describing what a geographically informed person knows and understands. States, schools, and individual teachers have welcomed the new prominence of geography, and enthusiastically adopted new approaches to introduce the geography standards to new learners. The rapid spread of computer technology, especially in the field of Geographical Information Science, has also meant a new importance for spatial analysis, a traditional area of geographical expertise. No longer is geography seen as an outdated mass of useless or arcane facts; instead, geography is now seen, again, to be an innovative an integrative science, which can contribute to solving complex problems associated with the human-environmental relationship in the twenty-first century.

Geographers may no longer travel across uncharted realms, but there is still much we long to explore, to learn, and seek to understand, even if it is only as "armchair" geographers. This reference work, *World Geography,* will help carry readers on their own journeys of exploration.

Ray Sumner
Long Beach City College

Joseph M. Castagno
Educational Reference Publishing, LLC

CONTRIBUTORS

Emily Alward
Henderson, Nevada Public Library

Earl P. Andresen
University of Texas at Arlington

Debra D. Andrist
St. Thomas University

Charles F. Bahmueller
Center for Civic Education

Timothy J. Bailey
Pittsburg State University

Irina Balakina
Writer/Editor, Educational Reference Publishing, LLC

David Barratt
Nottingham, England

Maryanne Barsotti
Warren, Michigan

Thomas F. Baucom
Jacksonville State University

Michelle Behr
Western New Mexico University

Alvin K. Benson
Brigham Young University

Cynthia Breslin Beres
Glendale, California

Nicholas Birns
New School University

Olwyn Mary Blouet
Virginia State University

Margaret F. Boorstein
C.W. Post College of Long Island University

Fred Buchstein
John Carroll University

Joseph P. Byrne
Belmont University

Laura M. Calkins
Palm Beach Gardens, Florida

Gary A. Campbell
Michigan Technological University

Byron D. Cannon
University of Utah

Steven D. Carey
University of Mobile

Roger V. Carlson
Jet Propulsion Laboratory

Robert S. Carmichael
University of Iowa

Joseph M. Castagno
Principal, Educational Reference Publishing, LLC

Habte Giorgis Churnet
University of Tennessee at Chattanooga

Richard A. Crooker
Kutztown University

William A. Dando
Indiana State University

Larry E. Davis
College of St. Benedict

Ronald W. Davis
Western Michigan University

Cyrus B. Dawsey
Auburn University

Frank Day
Clemson University

M. Casey Diana
University of Illinois at Urbana-Champaign

Stephen B. Dobrow
Farleigh Dickinson University

Steven L. Driever
University of Missouri, Kansas City

Sherry L. Eaton
San Diego City College

Femi Ferreira
Hutchinson Community College

Helen Finken
Iowa City High School

Eric J. Fournier
Samford University

Anne Galantowicz
El Camino College

Hari P. Garbharran
Middle Tennessee State University

Keith Garebian
Ontario, Canada

Laurie A. B. Garo
University of North Carolina, Charlotte

Jay D. Gatrell
Indiana State University

Carol Ann Gillespie
Grove City College

Nancy M. Gordon
Amherst, Massachusetts

Noreen A. Grice
Boston Museum of Science

Johnpeter Horst Grill
Mississippi State University

Charles F. Gritzner
South Dakota State University

C. James Haug
Mississippi State University

Douglas Heffington
Middle Tennessee State University

Thomas E. Hemmerly
Middle Tennessee State University

Jane F. Hill
Bethesda, Maryland

Carl W. Hoagstrom
Ohio Northern University

Catherine A. Hooey
Pittsburg State University

Robert M. Hordon
Rutgers University

Kelly Howard
La Jolla, California

Paul F. Hudson
University of Texas at Austin

Huia Richard Hutton
University of Hawaii/Kapiolani Community College

Raymond Pierre Hylton
Virginia Union University

Solomon A. Isiorho
Indiana University/Purdue University at Fort Wayne

Ronald A. Janke
Valparaiso University

Albert C. Jensen
Central Florida Community College

Jeffry Jensen
Altadena, California

Bruce E. Johansen
University of Nebraska at Omaha

Kenneth A. Johnson
State University of New York, Oneonta

Walter B. Jung
University of Central Oklahoma

James R. Keese
California Polytechnic State University, San Luis Obispo

Leigh Husband Kimmel
Indianapolis, Indiana

Denise Knotwell
Wayne, Nebraska

James Knotwell
Wayne State College

Grove Koger
Boise Idaho Public Library

Alvin S. Konigsberg
State University of New York at New Paltz

Doris Lechner
Principal, Educational Reference Publishing, LLC

Steven Lehman
John Abbott College

Denyse Lemaire
Rowan University

Dale R. Lightfoot
Oklahoma State University

Jose Javier Lopez
Minnesota State University

James D. Lowry, Jr.
East Central University

Jinshuang Ma
Arnold Arboretum of Harvard University Herbaria

Dana P. McDermott
Chicago, Illinois

Thomas R. MacDonald
University of San Francisco

Robert R. McKay
Clarion University of Pennsylvania

Nancy Farm Männikkö
L'Anse, Michigan

Carl Henry Marcoux
University of California, Riverside

Christopher Marshall
Unity College

Rubén A. Mazariegos-Alfaro
University of Texas/Pan American

Christopher D. Merrett
Western Illinois University

John A. Milbauer
Northeastern State University

Randall L. Milstein
Oregon State University

Judith Mimbs
Loftis Middle School

Karen A. Mulcahy
East Carolina University

B. Keith Murphy
Fort Valley State University

M. Mustoe
Omak, Washington

Bryan Ness
Pacific Union College

Kikombo Ilunga Ngoy
Vassar College

Joseph R. Oppong
University of North Texas

Richard L. Orndorff
University of Nevada, Las Vegas

Bimal K. Paul
Kansas State University

Nis Petersen
New Jersey City University

Mark Anthony Phelps
Ozarks Technical Community College

John R. Phillips
Purdue University, Calumet

Alison Philpotts
Shippensburg University

Julio César Pino
Kent State University

Timothy C. Pitts
Morehead State University

Carolyn V. Prorok
Slippery Rock University

P. S. Ramsey
Highland Michigan

Robert M. Rauber
University of Illinois at Urbana-Champaign

Ronald J. Raven
State University of New York at Buffalo

Neil Reid
University of Toledo

Susan Pommering Reynolds
Southern Oregon University

Nathaniel Richmond
Utica College

Edward A. Riedinger
Ohio State University Libraries

Mika Roinila
West Virginia University

Thomas E. Rotnem
Brenau University

Joyce Sakkal-Gastinel
Marseille, France

Helen Salmon
University of Guelph

Elizabeth D. Schafer
Loachapoka, Alabama

Kathleen Valimont Schreiber
Millersville University of Pennsylvania

Ralph C. Scott
Towson University

Guofan Shao
Purdue University

Wendy Shaw
Southern Illinois University, Edwardsville

R. Baird Shuman
University of Illinois, Champaign-Urbana

Sherman E. Silverman
Prince George's Community College

Roger Smith
Portland, Oregon

Robert J. Stewart
California Maritime Academy

Toby R. Stewart
Alamosa, Colorado

Ray Sumner
Long Beach City College

Paul Charles Sutton
University of Denver

Glenn L. Swygart
Tennessee Temple University

Sue Tarjan
Santa Cruz, California

Robert J. Tata
Florida Atlantic University

John M. Theilmann
Converse College

Virginia Thompson
Towson University

Norman J. W. Thrower
University of California, Los Angeles

Paul B. Trescott
Southern Illinois University

Robert D. Ubriaco, Jr.
Illinois Wesleyan University

Mark M. Van Steeter
Western Oregon University

Johan C. Varekamp
Wesleyan University

Anthony J. Vega
Clarion University

William T. Walker
Chestnut Hill College

William D. Walters, Jr.
Illinois State University

Linda Qingling Wang
University of South Carolina, Aiken

Annita Marie Ward
Salem-Teikyo University

Kristopher D. White
University of Connecticut

P. Gary White
Western Carolina University

Thomas A. Wikle
Oklahoma State University

Rowena Wildin
Pasadena, California

Donald Andrew Wiley
Anne Arundel Community College

Kay R. S. Williams
Shippensburg University

Lisa A. Wroble
Redford Township District Library

Bin Zhou
Southern Illinois University, Edwardsville

REGIONS

OVERVIEW

THE HISTORY OF GEOGRAPHY

The moment that early humans first looked around their world with inquiring minds was the moment that geography was born. The history of geography is the history of human effort to understand the nature of the world. Through the centuries, people have asked of geography three basic questions: What is Earth like? Where are things located? How can one explain these observations?

Geography in the Ancient World

In the Western world, the Greeks and the Romans were among the first to write about and study geography. Eratosthenes, a Greek scholar who lived in the third century BCE, is often called the "father of geography and is credited with first using the word geography (from the Greek words *ge*, which means "earth," and *graphe*, which means "to describe"). The ancient Greeks had contact with many older civilizations and began to gather together information about the known world. Some, such as Hecataeus, described the multitude of places and peoples with which the Greeks had contact and wrote of the adventures of mythical characters in strange and exotic lands. However, the ancient Greek scholars went beyond just describing the world. They used their knowledge of mathematics to measure and locate. The Greek scholars also used their philosophical nature to theorize about Earth's place in the universe.

One Greek scholar who used mathematics in the study of geography was Anaximander, who lived from 610 to 547 BCE. Anaximander is credited with being the first person to draw a map of the world to scale. He also invented a sundial that could be used to calculate time and direction and to distinguish the seasons. Eratosthenes is also famous for his mathematical calculations, in particular of the circumference of Earth, using observations of the Sun. Hipparchus, who lived around 140 BCE, used his mathematical skills to solve geographic problems and was the first person to introduce the idea of a latitude and longitude grid system to locate places.

Such early Greek philosophers as Plato and Aristotle were also concerned with geography. They discussed such issues as whether Earth was flat or spherical and if it was the center of the universe, and debated the nature of Earth as the home of humankind.

Whereas the Greeks were great thinkers and introduced many new ideas into geography, the Roman contribution was to compile and gather available knowledge. Although this did not add much that was new to geography, it meant that the knowledge of the ancient world was available as a base to work from and was passed down across the centuries. Geogra-

---

CURIOSITY: THE ROOT OF GEOGRAPHY

The earliest human beings, as they hunted and gathered food and used primitive tools in order to survive, must have had detailed knowledge of the geography of their part of the world. The environment could be a hostile place, and knowledge of the world meant the difference between life and death. Human curiosity took them one step further. As they lived in an ancient world of ice and fire, human beings looked to the horizon for new worlds, crossing continents and spreading out to all areas of the globe. They learned not only to live as a part of their environment, but also to understand it, predict it, and adapt it to their needs.

phy in the ancient world is often said to have ended with the great work of Ptolemy (Claudius Ptolemaeus), who lived from 90 to 168 CE. Ptolemy is best known for his eight-volume *Guide to Geography*, which included a gazetteer of places located by latitude and longitude, and his world map.

Geography in China

The study of geography also was important in ancient China. Chinese scholars described their resources, climate, transportation routes, and travels, and were mapping their known world at the same time as were the great Western civilizations. The study of geography in China begins in the Warring States period (fifth century BCE). It expands its scope beyond the Chinese homeland with the growth of the Chinese Empire under the Han dynasty. It enters its golden age with the invention of the compass in the eleventh century CE (Song dynasty) and peaks with fifteenth century CE (Ming dynasty) Chinese exploration of the Pacific under admiral Zheng He during the treasure voyages.

Geography in the Middle Ages

With the collapse of the Roman Empire in the fifth century CE, Europe entered into what is commonly known as the Early Middle Ages. During this time, which lasted until the fifteenth century, the geographic knowledge of the ancient world was either lost or challenged as being counter to Christian teachings. For example, the early Greeks had theorized that Earth was a sphere, but this was rejected during the Middle Ages. Scholars of the Middle Ages believed that the world was a flat disk, with the holy city of Jerusalem at its center.

The knowledge and ideas of the ancient world might have been lost if they had not been preserved by Muslim scholars. In the Islamic countries of North Africa and the Middle East, some of the scholarship of the ancient world was sheltered in libraries and universities. This knowledge was extensively added to as Muslims traveled and traded across the known world, gathering their own information.

Among the most famous Muslim geographers were Ibn Battutah, al-Idrisi, and Ibn Khaldun. Ibn Battutah traveled east to India and China in the fourteenth century. Al-Idrisi, at the command of King Roger II of Sicily, wrote *Roger's Book*, which systematically described the world. Information from *Roger's Book* was engraved on a huge planisphere (disk), crafted in silver; this once was considered a wonder of the world, but it is thought to have been destroyed. Ibn Khaldun (1332-1406) is best known for his written world history, but he also was a pioneer in focusing on the relationship of human beings to their environment.

The Age of European Exploration

Beginning in the fifteenth century, the isolation of Europe came to an end, and Europeans turned their attention to exploration. The two major goals of this sudden surge in exploration were to spread the Christian faith and to obtain needed resources. In 1418 Prince Henry the Navigator established a school for navigators and began to gather the tools and knowledge needed for exploration. He was the first of many Europeans to travel beyond the limits of the known world, mapping, describing, and cataloging all that they saw.

The great wave of European exploration brought new interest in geography, and the monumental works of the Greeks and Romans—so carefully preserved by Muslim scholars—were rediscovered and translated into Latin. The maps produced in the Middle Ages were of little use to the explorers who were traveling to, and beyond, the limits of the known world. Christopher Columbus, for example, relied on Ptolemy's work during his voyages to the Americas, but soon newer, more accurate maps were drawn and, for the first time, globes were made. A particularly famous map, which is still used as a base map, is the Mercator projection. On the world map produced by Gerardus Mercator (born Geert de Kremer) in 1569, compass directions appear as straight lines, which was a great benefit on navigational charts.

When the age of European exploration began, even the best world maps crudely depicted only a few limited areas of the world. Explorers quickly began to gather huge quantities of information, making detailed charts of coastlines, discovering new continents, and eventually filling in the maps of those continents with information about both the natural and human features they encountered. This age of exploration is often said to have ended when Roald Amundsen planted the Norwegian flag at the South Pole in 1911. At that time, the world map became complete, and human beings had mapped and explored every part of the globe. However, the beginning of modern geography is usually associated with the work of two nineteenth century German geographers: Alexander von Humboldt and Carl Ritter.

The Beginning of Modern Geography

The writings of Alexander von Humboldt and Carl Ritter mark a leap into modern geography, because these writers took an important step beyond the work of previous scholars. The explorers of the previous centuries had focused on gathering information, describing the world, and filling in the world map with as much detail as possible. Humboldt and Ritter took a more scientific and systematic approach to geography. They began not only to compile descriptive information, but also to ask why: Humboldt spent his lifetime looking for relationships among such things as climate and topography (landscape), while Ritter was intrigued by the multitude of connections and relationships he observed within human geographic patterns. Both Humboldt and Ritter died in 1859, ending a period when information-gathering had been paramount. They brought geography into a new age in which synthesis, analysis, and theory-building became central.

European Geography

After the work of Humboldt and Ritter, geography became an accepted academic discipline in Europe, particularly in Germany, France, and Great Britain. Each of these countries emphasized different aspects of geographic study. German geographers continued the tradition of the scientific view, using observable data to answer geographic questions. They also introduced the concept that geography could take a chorological view, studying all aspects, physical and human, of a region and of the interrelationships involved.

The chorological view came to dominate French geography. Paul Vidal de la Blache (1845-1918) was

---

NATIONAL GEOGRAPHIC SOCIETY AND GEOGRAPHIC RESEARCH

In 1888 the National Geographic Society was founded to support the "increase and diffusion of geographic knowledge" of the world. In its first 110 years, the society funded more than five thousand expeditions and research projects with more than 6,500 grants. By the 1990s it was the largest such foundation in the world, and the results of its funded projects are found on television programs, video discs, video cassettes, and books, as well as in the *National Geographic* magazine, established in 1888. Its productions are cutting-edge resources for information about archaeology, ethnology, biology, and both cultural and physical geography.

---

the most prominent French geographer. He advocated the study of small, distinct areas, and French geographers set about identifying the many regions of France. They described and analyzed the unique physical and human geographic complex that was to be found in each region. An important concept that emerged from French geography was "possibilism." German geographers had introduced the notion of environmental determinism—that human beings were largely shaped and controlled by their environments. Possibilism rejected the concept of environmental determinism, asserting that the relationship between human beings and the environment works in two directions: The environment creates both limits and opportunities for people, but people can react in different ways to a given environment, so they are not controlled by it.

British geographers, influenced by the French approach, conducted regional surveys. British regional studies were unique in their emphasis on planning and geography as an applied science. From this work came the concept of a functional region—an area that works together as a unit based on interaction and interdependence.

American Geography

Prior to World War II, only a small group of people in the United States called themselves geographers. They were mostly influenced by German ideas, but the nature of geography was hotly debated. Two schools of geographers were philosophical adversaries. The Midwestern School, led by Richard Hartshorne, believed that description of unique regions was the central task of geography.

The Western (or Berkeley) School of geography, led by Carl Sauer, agreed that regional study was important, but believed it was crucial to go beyond description. Sauer and his followers included genesis and process as important elements in any study. To understand a region and to know where it is going, they argued, one must look at its past and how it got to its present state.

In the 1930s, environmental determinism was introduced to U.S. geography but ultimately was rejected. Although geography in both Europe and the United States was essentially an all-male discipline, the United States produced the first famous woman geographer, Ellen Churchill Semple (1863-1932).

World War II illustrated the importance of geographic knowledge, and after the war came to an end in 1945, geographers began to come into their own in the United States. From the end of World War II to the early 1960s, U.S. geographers produced many descriptive regional studies.

In the early 1960s, what is often called the quantitative revolution occurred. The development of computers allowed complex mathematical analysis to be performed on all kinds of geographic data, and geographers began to analyze a wide range of problems using statistics. There was great enthusiasm for this new approach to geography at first, but beginning in the 1970s, many people considered a purely mathematical approach to be somewhat sterile and thought it left out a valuable human element.

In the 1980s and 1990s, many new ways to look at geographic issues and problems were developed, including humanism, behaviorism, Marxism, feminism, realism, structuration, phenomenology, and postmodernism, all of which bring human beings back into focus within geographical studies.

Geography in the Twenty-first Century

Geography increasingly uses technology to analyze global space and answer a wide range of questions related to a host of concerns including issues related to the environment, climate change, population, rising sea levels, and pollution. The Geographic Information System (GIS), in particular, provides a powerful way for people trained in geography to understand geographic issues, solve geographic problems, and display geographic information. Geographers continue to adopt a wide variety of philosophies, approaches, and methods in their quest to answer questions concerning all things spatial.

Wendy Shaw

MAPMAKING IN HISTORY

Cartography is the science or art of making maps. Although workers in many fields have a concern with cartography and its history, it is most often associated with geography.

Maps of Preliterate Peoples

The history of cartography predates the written record, and most cultures show evidence of mapping skills. The earliest surviving maps are those carved in stone or painted on the walls of caves, but modern preliterate peoples still use a variety of materials to express themselves cartographically. For example, the Marshall Islanders use palm fronds, fiber from coconut husks (coir), and shells to make sea charts for their inter-island navigation. The Inuit use animal skins and driftwood, sometimes painted, in mapping. There is a growing interest in the cartography of early and preliterate peoples, but some of their maps do not fit readily into a more traditional concept of cartography.

Mapping in Antiquity

Early literate peoples, such as those of Egypt and Mesopotamia, displayed considerable variety in their maps and charts, as shown by the few maps from these civilizations that still exist. The early Egyptians painted maps on wooden coffin bases to assist the departed in finding their way in the afterlife; they also made practical route maps for their mining operations. It is thought that geometry developed from the Egyptians' riverine surveys. The Babylonians made maps of different scales, using clay tablets with cuneiform characters and stylized symbols, to create city plans, regional maps, and "world" maps. They also divided the circle in the sexigesimal system, an idea they may have obtained from India and that is commonly used in cartography to this day.

The Greeks inherited ideas from both the Egyptians and the Mesopotamians and made signal contributions to cartography themselves. No direct evidence of early Greek maps exists, but indirect evidence in texts provides information about their cosmological ideas, culminating in the concept of a perfectly spherical earth. This they attempted to measure and divide mathematically. The idea of climatic zones was proposed and possibly mapped, and the large known landmasses were divided into first two continents, then three.

Perhaps the greatest accomplishment of the early Greeks was the remarkably accurate measurement of the circumference of Earth by Eratosthenes (276-196 BCE). Serious study of map projections began at about this time. The gnomonic, orthographic, and stereographic projections were invented before the Christian era, but their use was confined to astronomy in this period. With the possible single exception of Aristarchus of Samos, the Greeks believed in a geocentric universe. They made globes (now lost) and regional maps on metal; a few map coins from this era have survived.

Later Greeks carried on these traditions and expanded upon them. Claudius Ptolemy invented two projections for his world maps in the second century CE. These were enormously important in the European Renaissance as they were modified in the light of new overseas discoveries. Ptolemy's work is known mainly through later translations and reconstructions, but he compiled maps from Greek and Phoenician travel accounts and proposed sectional maps of different scales in his *Geographia*. Ptolemy's prime meridian (0 degrees longitude) in the Canary Islands was generally accepted for a millennium and a half after his death.

Roman cartography was greatly influenced by later Greeks such as Ptolemy, but the Romans themselves improved upon route mapping and surveying. Much of the Roman Empire was subdivided by instruments into hundredths, of which there is a cartographic record in the form of marble tablets. In Rome, a small-scale map of the world known to the Romans was made on metal by Marcus Vipsanius Agrippa, the son-in-law of Augustus Caesar, and displayed publicly. This map no longer exists, however.

Cartography in Early East Asia

As these developments were taking place in the West, a rich cartographic tradition developed in Asia, particularly China. The earliest survey of China (Yu Kung) is approximately contemporaneous with the oldest reported mapmaking activity of the Greeks. Later, maps, charts, and plans accompanied Chinese texts on various geographical themes. Early rulers of China had a high regard for cartography—the science of princes. A rectangular grid was introduced by Chang Heng, a contemporary of Ptolemy, and the south-pointing needle was used for mapmaking in China from an early date.

These traditions culminated in Chinese cartographic primacy in several areas: the earliest printed maps (about 1155 CE), early printed atlases, and terrestrial globes (now lost). Chinese cartography greatly influenced that in other parts of Asia, particularly Korea and Japan, which fostered innovations of their own. It was only after the introduction of ideas from the West, in the Renaissance and later, that Asian cartographic advances were superseded.

Islamic Cartography

A link between China and the West was provided by the Arabs, particularly after the establishment of Islam. It was probably the Arabs who brought the magnetized needle to the Mediterranean, where it was developed into the magnetic compass.

Some scholars have argued that the Arabs were better astronomers than cartographers, but the Arabs did make several clear advances in mapmaking. Both fields of study were important in Muslim science, and the astrolabe, invented by the Greeks in antiquity but developed by the Arabs, was used in both their astronomical and terrestrial surveys. They made and used many maps, as indicated by the output of their most famous cartographer, al-Idrisi (who lived about 1100–1165). Some of his work still exists, including a zonal world map and detailed charts of the Mediterranean islands.

At about the same time, the magnetic compass was invented in the coastal cities of Italy, which gave rise to advanced navigational charts, including information on ports. These remarkably accurate charts were used for navigating in the Mediterranean Sea. They were superior to the European maps of the Middle Ages, which often were concerned with religious iconography, pilgrimage, and crusade. The scene was now set for the great overseas discoveries of the Europeans, which were initiated in Portugal and Spain in the fifteenth century.

In the next four centuries, most of the coasts of the world were visited and mapped. The early, projectionless navigational charts were no longer adequate, so new projections were invented to map the enlarged world as revealed by the European overseas explorations. The culmination of this activity was the development of the projection, in 1569, of Gerardus Mercator, which bears his name and is of special value in navigation.

Early Modern Mapmaking

Europeans began mapping their own countries with greater accuracy. New surveying instruments were invented for this purpose, and a great land-mapping activity was undertaken to match the worldwide coastal surveys. For about a century, the Low Countries of Belgium, Luxembourg, and the Netherlands dominated the map and chart trades, producing beautiful hand-colored engraved sheet wall maps and atlases.

France and England established new national observatories, and by the middle of the seventeenth century, the Low Countries had been eclipsed by France in surveying and making maps and charts. The French adopted the method of triangulation of Mercator's teacher, Gemma Frisius. Under four generations of the Cassini family, a topographic survey of France more comprehensive than any previous survey was completed. Rigorous coastal surveys were undertaken, as well as the precise measurement of latitude (parallels).

The invention of the marine chronometer by John Harrison made it possible for ships at sea to determine longitude. This led to the production of charts of all the oceans, with England's Greenwich eventually being adopted as the international prime meridian.

Quantitative, thematic mapping was advanced by astronomer Edmond Halley (1656–1742) who produced a map of the trade winds; the first published magnetic variation chart, using isolines; tidal charts; and the earliest map of an eclipse. The Venetian Vincenzo Coronelli made globes of greater beauty and accuracy than any previous ones. In the German lands, the study of map projections was vigorously pursued. Johann H. Lambert and others invented a number of equal-area projections that were still in use in the twentieth century.

Ideas developed in Europe were transmitted to colonial areas, and to countries such as China and Russia, where they were grafted onto existing cartographic traditions and methods. The oceanographic explorations of the British and the French built on the earlier charting of the Pacific Ocean and its islands by native navigators and the Iberians.

Nineteenth Century Cartography

Cartography was greatly diversified and developed in the nineteenth century. Quantitative, thematic mapping was expanded to include the social as well as the physical sciences. Alexander von Humboldt used isolines to show mean air temperature, a method that later was applied to other phenomena. Contour lines gradually replaced less quantitative methods of representing terrain on topographic maps. Such maps were made of many areas, for example India, which previously had been poorly mapped.

Extraterrestrial (especially lunar) mapping, had begun seriously in the preceding two centuries with the invention of the telescope. It was expanded in the nineteenth century. In the same period, regular national censuses provided a large body of data that could be mapped. Ingenious methods were created to express the distribution of population, diseases, social problems, and other data quantitatively, using uniform symbols.

Geological mapping began in the nineteenth century with the work of William Smith in England, but soon was adopted worldwide and systematized, notably in the United States. The same is true of transportation maps, as the steamship and the railway increased mobility for many people. Faster land travel in an east-west direction, as in the United States, led to the official adoption of Greenwich as the international prime meridian at a conference held in Washington, D.C., in 1884. Time zone maps were soon published and became a feature of the many world atlases then being published for use in schools, offices, and homes.

A remarkable development in cartography in the nineteenth century was the surveying of areas newly occupied by Europeans. This occurred in such places as the South American republics, Australia, and Canada, but was most evident in the United States. The U.S. Public Land Survey covered all areas not previously subdivided for settlement. Property maps arising from surveys were widely available, and in many cases, the information was contained in county and township atlases and maps.

Modern Mapping and Imaging

Cartography was revolutionized in the twentieth century by aerial photography, sonic sounding, satellite imaging, and the computer. Before those developments, however, Albrecht Penck proposed an ambitious undertaking—an International Map of the World (IMW). Cartography historically had been a nationalistic enterprise, but Penck suggested a map of the world in multiple sheets produced cooperatively by all nations at the scale of 1:1,000,000 with uniform symbols. This was started in the first half of the twentieth century but was not completed, and was superseded by the World Aeronautical Chart (WAC) project, at the same scale, during and after World War II.

The WAC project owed its existence to flight information made available following the invention of the airplane. Both photography and balloons were developed before the twentieth century, but the new, heavier-than-air craft permitted overlapping aerial photographs to be taken, which greatly facilitated the mapping process. Aerial photography revolutionized land surveys—maps could be made at less cost, in less time, and with greater accuracy than by previous methods. Similarly, marine surveying was revolutionized by the advent of sonic sounding in the second half of the twentieth century. This enabled mapping of the floor of the oceans, essentially unknown before this time.

Satellite imaging, especially continuous surveillance by Landsat since 1972, allows temporal monitoring of Earth. The computer, through Geographical Information Systems (GIS) and other technologies, has greatly simplified and speeded up the mapping process. During the twentieth century, the most widely available cartographic product was the road map for travel by automobile.

Spatial information is typically accessed through apps on computers and mobile devices; traditional maps are becoming less common. The new media also facilitate animated presentations of geographical and extraterrestrial distributions. Cartographers remain responsive to the opportunities provided by new technologies, materials, and ideas.

Norman J. W. Thrower

MAPMAKING AND NEW TECHNOLOGIES

The field of geography is concerned primarily with the study of the curved surface of Earth. Earth is huge, however, with an equatorial radius of 3,963 miles (6,378 km.). How can one examine anything more than the small patch of earth that can be experienced at one time? Geographers do what scientists do all of the time: create models. The most common model of Earth is a globe—a spherical map that is usually about the size of a basketball.

A globe can show physical features such as rivers, oceans, the continents, and even the ocean floor. Political globes show the division of Earth into countries and states. Globes can even present views of the distant past of Earth, when the continents and oceans were very different than they are today. Globes are excellent for learning about the distributions, shapes, sizes, and relationships of features of Earth. However, there are limits to the use of globes.

How can the distribution of people over the entire world be described at one glance? On a globe, the human eye can see only half of Earth at one time. What if a city planner needs to map every street, building, fire hydrant, and streetlight in a town? To fit this much detail on a globe, the globe might have to be bigger than the town being mapped. Globes like these would be impossible to create and to carry around. Instead of having to hire a fleet of flatbed trucks to haul oversized globes, the curved surface of the globe can be transformed to a flat plane.

The method used to change from a curved globe surface to a flat map surface is called a map projection. There are hundreds of projections, from simple to extremely complex and dating from about two thousand years ago to projections being invented today. One of the oldest is the gnomonic projection. Imagine a clear globe with a light inside. Now imagine holding a piece of paper against the surface of the globe. The coastlines and parallels of latitude and meridians of longitude would show through the globe and be visible on the paper. Computers can do the same thing because there are mathematical formulas for nearly all map projections.

Geometric Models for Map Projections

One way to organize map projections is to imagine what kind of geometric shape might be used to create a map. Like the paper (a plane surface) against the globe described above, other useful geometric shapes include a cone and a cylinder. When the rounded surface of any object, including Earth, is flattened there must be some stretching, or tearing. Map projections help to control the amount and kinds of distortion in maps. There are always a few exceptions that cannot be described in this way, but using geometric shapes helps to classify projections into groups and to organize the hundreds of projections.

Another way to describe a map projection is to consider what it might be good for. Some map projections show all of the continents and oceans at their proper sizes relative to one another. Another type of projection can show correct distances between certain points.

Map Projection Properties

When areas are retained in the proper size relationships to one another, the map is called an equal-area map, and the map projection is called an equal-area projection. Equal-area (also called equivalent or homolographic) maps are used to measure areas or view densities such as a population density.

If true angles are retained, the shapes of islands, continents, and oceans look more correct. Maps made in this way are called conformal maps or conformal map projections. They are used for navigation, topographic mapping, or in other cases when it is important to view features with a good representation of shape. It is impossible for a map to be both equal-area and conformal at the same time. One or the other must be selected based on the needs of the map user or mapmaker.

One special property—distance—can only be true on a few parts of a map at one time. To see how far it is between places hundreds or thousands of miles apart, an equidistant projection should be used. There will be several lines along which distance is true. The azimuthal equidistant projection shows true distances from the center of the map outward. Some map projections do not retain any of these properties but are useful for showing compromise views of the world.

Modern Mapmaking

Modern mapmaking is assisted from beginning to end by digital technologies. In the past, the paper map was both the primary means for communicating information about the world and the database used to store information. Today, the database is a digital database stored in computers, and cartographic visualizations have taken the place of the paper map. Visualizations may still take the form of paper maps, but they also can appear as flashes on computer screens, animations on local television news programs, and even on screens within vehicles to help drivers navigate. Communication of information is one of the primary purposes of making maps. Mapping helps people to explore and analyze the world.

Making maps has become much easier and the capability available to many people. Desktop mapping software and Internet mapping sites can make anyone with a computer an instant cartographer. The maps, or cartographic visualizations, might be quite basic but they are easy to make. The procedures that trained cartographers use to make map products vary in the choice of data, software, and hardware, but several basic design steps should always take place.

First, the purpose and audience for whom the map is being made must be clear. Is this to be a general reference map or a thematic map? What image should be created in the mind of the map reader? Who will use the map? Will it be used to teach young children the shapes of the continents and oceans, or to show scientists the results of advanced research? What form will the cartographic visualization take?

SLIDING ROCKS GET DIGITAL TREATMENT

Dr. Paula Messina studied the trails of rocks that slide across the surface of a flat playa in Death Valley, California. The sliding rocks have been studied in the past, but no one had been able to say for certain how or when the rocks moved. It was unclear whether the rocks were caught in ice floes during the winter, were blown by strong winds coming through the nearby mountains, or were moved by some other method.

Messina gave the mystery a totally digital treatment. She mapped the locations of the rocks and the rock trails using the global positioning system (GPS) and entered her rock trail data into a geographic information system (GIS) for analysis. She was able to determine that ice was not the moving agent by studying the pattern of the trails. She also used digital elevation models (DEM) and remotely sensed imagery to model the environment of the playa. She reported her results in the form of maps using GIS' cartographic output capabilities. While she did not solve completely the mystery of the sliding rocks, she was able to disprove that winter ice caused the rocks to slide along together in rafts and that there are wind gusts strong enough to move the biggest rock on the playa.

Will it be a paper map, a graphic file posted to the Internet, or a video?

The answers to these questions will guide the cartographer in the design process. The design process can be broken down into stages. In the first stage of map design, imagination rules. What map type, size and shape, basic layout, and data will be used? The second stage is more practical and consists of making a specific plan. Based on the decisions made in the first stage, the symbols, line weights, colors, and text for the map are chosen. By the end of this stage, there should be a fairly clear plan for the map. During the third stage, details and specifications are finalized to account for the production method to be used. The actual software, hardware, and methods to be used must all be taken into consideration.

What makes a good map? Working in a digital environment, a mapmaker can change and test various designs easily. The map is a good one when it communicates the intended information, is pleasing to look at, and encourages map readers to ask thoughtful questions.

New Technologies

Mapping technology has gone from manual to magnetic, then to mechanical, optical, photochemical, and electronic methods. All of these methods have overlapped one another and each may still be used in some map-making processes. There have been recent advances in magnetic, optical, and most of all, electronic technologies.

All components of mapping systems—data collection, hardware, software, data storage, analysis, and graphical output tools—have been changing rapidly. Collecting location data, like mapping in general, has been more accessible to more people. The development of the Global Positioning System (GPS), an array of satellites orbiting Earth, gives anyone with a GPS receiver access to location information, day or night, anywhere in the world. GPS receivers are also found in planes, passenger cars, and even in the backpacks of hikers.

Satellites also have helped people to collect data about the world from space. Orbiting satellites collect images using visible light, infrared energy, and other parts of the electromagnetic spectrum. Active sensing systems send out radar signals and create images based on the return of the signal. The entire world can be seen easily with weather satellites, and other specialized satellite imagery can be used to count the trees in a yard.

These great resources of data are all stored and maintained as binary, computer-readable information. Developments in laser technology provide large amounts of storage space on media such as optical disks and compact disks. Advances in magnetic technology also provide massive storage capability in the form of tape storage, hard drives, and cloud storage. This is especially important for saving the large databases used for mapping.

Computer hardware and software continue to become more powerful and less expensive. Software continues to be developed to serve the specialized needs that mapping requires. Just as word processing software can format a paper, check spelling and grammar, draw pictures and shapes, import tables and graphics, and perform dozens of other functions, specialized software executes maps. The most common software used for mapping is called Geographic Information System (GIS) software. These systems provide tools for data input and for analysis and modeling of real-world spatial data, and provide cartographic tools for designing and producing maps.

Karen A. Mulcahy

THE CONTINENT

Africa is the second-largest of the world's great landmasses. Together with the large island of Madagascar off its southeast shoreline, Africa covers more than 11.7 million square miles (30.3 million sq. km.), extending from the southern shores of the Mediterranean Sea in the north to the Cape of Good Hope in the south.

Physical Features

Specialists in plate tectonics and continental drift suggest that in the distant geologic past, Africa was part of the enormous land mass of Gondwanaland. According to this theory, Gondwanaland broke into several segments, or plates, one of which, the Indo-Australian plate, moved northward. When it collided with the southeastern edges of the Eurasian plate, the Himalaya Mountains were pushed up by what eventually became India. A second segment of Gondwanaland moved westward to form the continent of Africa.

Tectonic theory argues that Africa's movement has not ended, and that the entire continent is pushing against the southern portions of the Eurasian land mass, subjecting the latter to inevitable earthquake dangers. One of the world's most impressive mountain chains, the Atlas Mountains in northwestern Africa, was thrust upward in an early geologic stage of Africa's continental drift north-

The Cape of Good Hope looking towards the west, from the coastal cliffs above Cape Point, overlooking Dias beach. (Diego Delso)

13

AFRICA

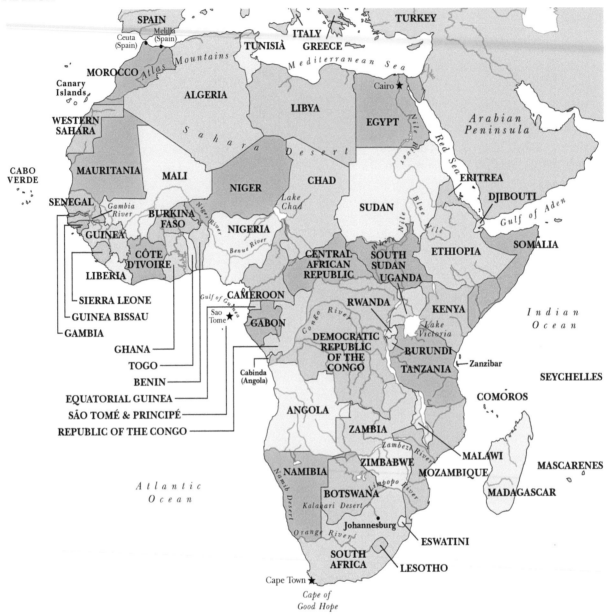

ward. An even more spectacular Eurasian counterpart of tectonic uplift—the European Alps—marks the southwestern tip of the Eurasian plate.

The effect of continental drift on the land mass of Africa formed two immense bodies of water to its west and east—the southern Atlantic and Indian oceans. It also created two nearly landlocked seas to its north—the Mediterranean and Red seas. Other effects of plate tectonics can be traced in land formations within the African continent itself, notably the Great Rift Valley and major zones of former or continuing volcanic activity.

The long oceanic coastlines of Africa, on both the Atlantic and Indian Ocean sides, comprise several geographically distinct subregions bearing specific names. The entrance to the Red Sea is known as the Gulf of Aden, marking the sharp tip called the Horn of Africa, now part of the country of Somalia.

The waters between the continent and the island of Madagascar are called the Mozambique Channel. The channel reaches depths of almost 12,000 feet (3,660 meters) between Mozambique proper and the mid-channel Comoros Island archipelago. Probably the best-known geographical subregion at the southern tip of Africa is the Cape of Good Hope, one of the earliest sites of European colonization following Vasco da Gama's historic voyage around the cape into the Indian Ocean in 1498.

The Atlantic coast of Africa shows the apparent evidence of tectonic theory's claim concerning the continent's former geological connection with the Atlantic coast of South America: The outward bulge of northwest Africa and the curved indentation of the Gulf of Guinea very nearly fit South America's eastern coastline.

Climatic Features and Vegetation Zones

Although the center of Africa straddles the equator, the continent's northern portion, bounded by the Mediterranean Sea, extends to 37° north latitude. Cape Bon, in modern Tunisia, thus shares features of the temperate climate zone and vegetation of Southern Europe.

Africa's southern tip, at the Cape of Good Hope, reaches as far as 35° south latitude, where Cape Agulhas demonstrates all the features of the temperate climate zone of the Southern Hemisphere. This vastness of geographical extent, coupled with varied conditions brought about by altitude or the coastal effects of ocean currents, means that several broad climate zones characterize different regions. Different climatic conditions bring about notable variations in vegetation, both from north to south and in vast pockets in different geographical regions.

Most of Africa is characterized by one of two climatic zones. Most of the northern half, covered by the Sahara Desert, and the southwest tip, extending between 35° and 23° south latitude, have either semiarid steppe or arid desert conditions. Vast expanses of what climatologists call a tropical continental climate extend from west to east, almost all the way across the continent south of the Sahara into the eastern zone of the Horn of Africa. The Horn itself (specifically, modern day Somalia) has a high degree of aridity as a side effect of the north-

HOW AFRICA GOT ITS NAME

It is claimed that in antiquity, the Greeks called the continent Libya and the Romans called it Africa. The term could have come from the Latin *aprica* ("sunny") or the Greek *aphrike* ("without cold"). Africa referred mainly to the northern coast of the continent that represented a southern extension of Europe. It is believed that the Romans called the area south of their settlements Afriga, or Land of the Afrigs, the name of a Berber settlement south of Carthage.

east trade winds. Thus, it is a desertlike pocket that has much in common with the Saharan zone.

The vast tropical continental zone includes most of the interior, running south from the edges of the Sahara to about 25° south latitude (modern Zimbabwe, just north of South Africa). This region enjoys moderate levels of rainfall, enough to sustain typical savanna plant and animal life.

Savanna vegetation consists of vast stretches of grasslands and medium-sized brush and trees. This is the zone of numerous grazing animal herds and the predators or scavengers that live off victims taken from the herds. Very large animals, notably the elephant, hippopotamus, and rhinoceros, also live in this region of Africa.

Equatorial Africa runs inland from the western coast both north and south of the equator proper, from the level of southern Nigeria in the north and from the northern Angolan coast at about 10° south latitude. The typical patterns of equatorial climate and vegetation—heavy amounts of rainfall and tropical rainforest flora—do not extend as a belt all the way across Africa because of the highland geographical features that run from Ethiopia through East Africa (Kenya and Uganda) down into the southern reaches of the continent. The highland zones share the climatic features and flora and fauna of the savanna regions. They also constitute the most important geological area of Africa associated with early or current volcanic activity.

In areas where mountains are a dominant part of the landscape, they often form natural barriers to rain clouds, creating localized subpatterns in Africa's broad north-to-south climatic zones. One example of this phenomenon occurs on the island of Madagascar. The central highland range there contributes to a tropical maritime climate, with luxurious vegetation, on the eastern half of the island, where the majority of the rainfall coming from the Indian Ocean falls.

On the western half of Madagascar, tropical savanna vegetation prevails, the product of lower levels of rainfall beyond the central mountains. The Great Escarpment running along Africa's southwestern coast, coupled with the effects of the Atlantic Ocean's Benguela Current, create an even more extreme pattern. A substantial portion of Southern Africa's inland climate, especially that of the relatively high-elevation Kalahari Desert basin (more than 3,000 feet or about 1,000 meters), is as dry as the interior of the great Sahara in the north.

The Sahara itself is the largest desert surface in the world, covering 3.6 million square miles (9.3 million sq. km.). Within its vast expanse are both fairly elevated mountain outcroppings and the lowest elevation in the interior of the continent—the Qattara Depression in the western desert area of Egypt, which is more than 400 feet (122 meters) below sea level. The name "Sahara" derives from the Arabic word for desert. Although there are some interior areas (approximately 15 percent of the total surface) that correspond to the popular image of desert sand dunes, about 70 percent of the Sahara is covered by bare rock or gravel-strewn surfaces, often referred to as hamada.

The bareness of large stretches of the Sahara is partly the product of severe strong winds that blow from the northeastern sections of Africa (where high-pressure conditions prevail during most of the year) toward the low-pressure zones of the equatorial regions of the continent. These winds do not carry moisture. As they pass into the warmer tem-

Ubari Oasis in the Wadi Al Hayaa District of the Fezzan region in southwestern Libya. (Sfivat)

The "canyon of rats" and the "canyon of monkeys" in Isalo National Park, Madagascar. (Moongateclimber)

perature zones of the desert, the winds themselves warm up, drying out the areas through which they pass. In the somewhat more populated regions on the northern and southern fringes of the Sahara, the seasonal phenomenon of desert wind has been given local names, such as the sirocco, the khamsin, and the harmattan.

Levels of precipitation vary over the vast expanse of the Sahara. To the north and south of the main desert, transitional steppe areas may receive about 10 inches (254 millimeters) of rain annually, which sustains only low shrub growth and some seasonal grasses. The average rainfall for the core area of the desert never exceeds 5 inches (127 millimeters) per year. When rainfall comes, it is often only after long dry periods, so the hardened surface of the desert cannot absorb the rain. Flash flooding is often the result.

Local geological formations can create a situation that brings the water table close to the surface, resulting in oasis conditions in a restricted area. Because subsurface water follows fissures in the rock

formations, green oases may appear in a chainlike sequence over a relatively wide expanse in the middle of otherwise totally arid surroundings.

Two important Atlantic Ocean currents, the Canary Current on the northwest coast and the Benguela Current running from the southwest tip along the Namibian coast northward into the Gulf of Guinea, strongly affect the climate of the regions through which they pass, creating particularly arid conditions along sections of Africa's western coasts. If the enormous arid west-to-east band, including the Sahara, in northern Africa, could be enough to explain the low rainfall of the northwest coast even without the Canary Current, the Benguela Current is a prime contributor to the aridity of the western half of the southern tip of Africa. Whereas Namibia receives less than 16 inches (400 millimeters) of rain per year, the half of southern South Africa facing the Indian Ocean can receive up to three-and-a-half times that amount.

Rainfall levels less than 16 inches (400 millimeters) characterize all of North Africa except the Mediterranean coastal areas of Morocco, Algeria, and Tunisia and the mountainous zones in each of these three Maghreb (northwest Africa) countries. Rainfall in the broad belt of tropical continental, or savanna, territory south of the Sahara, like the eastern African highlands and countries located south of the equatorial lowland forested areas, is more difficult to predict. It ranges from 16 to 55 inches (400 to 1,400 millimeters), with averages somewhere in between.

The danger of drought in savanna zones, however, is a recurrent fact in African climatology. Rainfall levels in the relatively restricted central equatorial zone, and also the West African coastal areas from Côte d'Ivoire (Ivory Coast) through Nigeria, are almost invariably above 55 inches (1,400 millimeters) annually.

Mountainous Zones

Although the Atlas and Aures ranges of northern Africa are extensive mountainous systems, no other area on the continent has major chains that could be compared to Asia's Himalayas, South America's Andes, or North America's Sierras or Rocky Mountains. Geologically speaking, Africa is perhaps the least mountainous continent of the globe.

Deep in the interior of the Saharan zone, in the modern states of Niger, Algeria, and Chad, high desert areas are broken by rather major mountain masses in geographically limited zones. The best known of these are the Tibesti Mountains in Libya and the Hoggar Mountains in southern Algeria. Elevations in the latter reach nearly 10,000 feet (3,050 meters). Another striking zone of African mountains is found along the long geological split in Earth's crust known as the Great Rift Valley.

Major Rivers and Deltas

Three of the world's longest rivers are found in Africa: the Nile, the Congo, and the Niger, located respectively, in the northeast, central equatorial, and western Sudanic regions of the continent. The Nile is the longest river in the world, longer than the

The River Nile near Aswan, Egypt. (Alchemica)

Pont Kennedy (Kennedy Bridge) over the Niger River, in Niamey, Niger. (NigerTZai)

Amazon in South America. Running more than 4,180 miles (6,725 km.) from its sources in Central Africa to the shores of the Mediterranean Sea in Egypt, it drains an area of nearly 1.14 million square miles (3 million sq. km.). Although this total drainage area is significant—measuring only about 190,000 square miles (500,000 sq. km.) less than that of the Mississippi River, but more than twice the drainage area of the Niger River—it scarcely matches the Amazon River, which drains more than 2.7 million square miles (7 million sq. km.). Despite its length, the Nile ranks last among the world's ten major rivers in the volume of water it discharges into the sea.

When the two branches of the Nile fan out several miles north of Cairo, Egypt, they contribute to a triangular mosaic of canals that for centuries have irrigated the fields of the Nile Delta lands. At the terminal points of both branches of the river, the historic ports of Damietta and Rosetta have been visited by traders since ancient times. The smallness of these ports contributed to the growth, from ancient times forward, of the neighboring delta emporium of Alexandria.

This history and familiarity of the Nile Delta can be contrasted with the situation of the Niger River delta. Modern maps of the fanlike form of the Niger delta on the coast of Nigeria show the city of Port Harcourt and Bonny Island near where the easternmost branch of the river enters the Gulf of Guinea. Unlike the two ancient Nile Delta cities, Port Harcourt came into existence only in the middle of the nineteenth century, when British explorers realized that the complex marshy wastes along the Nigerian coast were in fact the terminal point of the Niger itself.

As for the sources of the Niger River, early explorers noted another geographical peculiarity. The Niger rises from waters that originate on the eastern slopes of a watershed near the western coasts of Africa, in modern-day Guinea and Mali. The mountainous land mass of the Fouta Djallon , near the modern borders dividing Senegal and Guinea from Mali, explains this division of the watershed so close to the Atlantic Ocean.

Several relatively short rivers due west from the watershed of the Niger flow directly into the Atlantic Ocean. The best known of these is the Gambia River. The Senegal River is the longest river originating near the Niger's own sources but flowing in an entirely different direction. This river flows northwest from its sources in Mali to form the northern border dividing modern Senegal from Mauritania.

The Niger River initially runs northeast from Mali's capital of Bamako until it bends back to the southeast. Two historic cities, Timbuktu and Gao, are located in the area of the Niger Bend. Traditionally, both took advantage of their position along the northernmost reaches of the Niger to become trade termini for caravans across the Sahara.

The Congo River is Earth's fourth-longest. Although only about 370 miles (600 km.) longer than the Niger River, it discharges almost eight times as much water into the Atlantic Ocean—1,400 billion cubic meters annually. The volume of water carried by the Congo River is second only to the Amazon River, for largely the same reasons: Both drain inland equatorial watersheds where heavy rains occur during most of the year. A considerable number of major tributaries feed water into the Congo, including the Aruwimi, Ubangi, and Sangha rivers north of the Congo's east-to-west course, and the Tshuapa and Kasai rivers to the south. Each of these major tributaries draws water from several lesser tributaries that cover the central equatorial zone.

The Congo River also differs from the Niger and Nile rivers in not having a prominent delta formation where it enters the ocean. The cities of Kinshasa (formerly Léopoldville), the capital of the Democratic Republic of the Congo, and Brazzaville, the capital of the Republic of the Congo, are located somewhat inland from the coast, separated from it by the famous Livingstone Falls.

The principal rivers that enter the Atlantic Ocean south of the Congo River are less well known. The Cuanza River flows out of the mountain mass to the north of the Kalahari basin, watering the lowlands of modern Angola near its capital of Luanda. Much farther south is the westward-flowing Orange River, fed by the Vaal River. Both of these have their sources far away in the drainage system west of the Drakensberg range of South Africa, mountains that nearly reach the shores of the eastern tip of South Africa. The Orange River is the only main river flowing across the southern portion of the arid Kalahari Desert.

Malinke fisher women on the Niger River, Niandankoro, Kankan Region, in eastern Guinea. (Julien Harneis)

Lake Chad as seen from Apollo 7. (NASA)

The narrow lowland plains along Africa's eastern coasts—running from Durban, South Africa (located beneath the high peaks of the Drakensberg range), all the way to the tip of the Horn of Africa in Somalia—are crossed by several rivers that drain into the Indian Ocean. The most important of these is the Zambezi River, whose sources lie deep in the interior. The Zambezi reaches the Indian Ocean on the coast of Mozambique. One of its tributaries, the Luangwa River, flows generally southward away from a point midway between lakes Malawi and Tanganyika, both located in the long Great Rift Valley region. Neither lake receives water from the Luangwa River, despite their closeness to it.

North of Mozambique, several shorter rivers empty into the Indian Ocean. The Tana River flows directly from the slopes of Mount Kenya to the shoreline north of the Kenyan city of Mombasa.

Finally, the lengthy Webi Shebelle River flows directly out of the mountain massifs of Ethiopia and heads toward the Indian Ocean near Somalia's capital city of Mogadishu. Just short of the ocean, it bends southward to follow the coast and dissipates in the arid lowlands without actually flowing into the ocean.

Major Inland Lakes

Several inland lakes in Africa are big enough to be seen on most maps of the entire continent. Located in the arid interior of the Sahara, Lake Chad is the only major body of water in the northern third of Africa. It lies in the center of a major inland basin called the Bodélé Depression, and receives drainage from an immense area reaching as far northwest as the Hoggar Mountains in Algeria and the Adamawa Plateau in Cameroon to the south. Two rivers flow westward toward it from the Maraoné

Open-pit diamond mine (known as the Big Hole or Kimberley Mine) in Kimberley, South Africa. (Irene2005)

Highlands and Ennedi Plateau in Chad, although their waters do not always reach as far as the present lake. Waters from two important rivers to its southeast, the Logone (originating in the Adamawa Plateau) and Chari rivers, are more certain sources of runoff. In earlier geological times, the expanse of Lake Chad was much greater.

In the eastern third of the continent, where the north-south geological effects of the Great Rift Valley are evident, one finds not only Lake Victoria at the level of the equator between the two eastern branches of the Great Rift Valley, but also the long, narrow waters of lakes Tanganyika and Malawi. The complexity of geological structures in the area around Lake Victoria makes this region a center of volcanic activity, and therefore the home of some of Africa's most best-known mountain peaks.

Mineral Wealth

The richest zones of mineral wealth are found south of the equator, especially in the area running from the southeastern corner of the Democratic Republic of the Congo (formerly Zaire) through Zimbabwe and along the eastern zones of South Africa.

South Africa is one of the world's major sources of gold, diamonds, platinum, and palladium. Some of the same areas of South Africa have major coal deposits. Other mineral exports include copper, antimony, chromium, and uranium.

Beyond Southern Africa, mining, on a somewhat smaller scale, is an essential part of the export economy of several other regions. In northwest and West Africa, Mauritania, Côte d'Ivoire (Ivory Coast), Nigeria, and Angola all extract and exploit some minerals. In restricted areas, notably in the Congo region and Gabon in equatorial Africa, sources of uranium are of considerable importance to the world market. Guinea is the world's third-largest producer of bauxite. For copper production, the Democratic Republic of the Congo is fifth in the world, while Zambia is seventh. Zimbabwe is the world's fifth-largest producer of lithium, and Namibia is eighth. Zambia is the world's second-largest source of emeralds. Emeralds are also mined in Egypt (the leading source of emeralds in ancient times) and Madagascar. Namibia and Madagascar are Africa's leading sources of rubies.

After the gold and diamonds of South Africa, the most important resources extracted from the ground beneath Africa are petroleum and its by-products. The biggest producers of petroleum in Africa are Libya, Nigeria, and Egypt.

Agriculture

Certain areas of Africa have moved toward agricultural crop diversification in response to demands in the world market. Nevertheless, the continent can be divided into general regions of predominant staple crops, which provide the basic subsistence food sources in those regions. Production of those crops depends in large measure on broad climatic conditions. Maize (corn) can be grown as a staple crop only in the southern third of Africa, whereas wheat (which requires predictable seasonal rains) is restricted to regions north of the Sahara. The arid Saharan zone, stretching eastward to the Horn of Africa (Ethiopia and Somalia), can only support sorghum and millet. The wettest regions of Africa, mainly the coastlands of the Gulf of Guinea and equatorial Africa itself, depend primarily on root crops such as potatoes, yams, and manioc as staples.

Major Cities

Most of Africa's major cities are of relatively recent origin, and their striking growth in the twenty-first

Rice terraces in Madagascar. (Mariusba)

century is a reflection of several demographic factors. Economic conditions, especially migration of unemployed people from rural areas to presumed job markets in the cities, have created many zones of extensive crowding on the outskirts of relatively undeveloped urban systems. Almost every African capital city shares this characteristic.

A few major cities have longer histories and have developed the recognizable attributes of major urban centers. Among these, the South African cities of Cape Town and Johannesburg owe much of their original development to the heavy emphasis placed on concentrating economic infrastructure during the long period of European colonial rule. Both, however, show signs of central urban modernity, surrounded on all sides by overwhelming problems created by the sizable rural-to-urban migration. Urban migrants came under the largely open condi-

tions in the period following the transition to African majority rule in South Africa in the 1990s.

Africa's largest urban complex is on the opposite end of the continent. Cairo, in Egypt, founded in 969 CE, is now Africa's most populated city, with a 2019 population of 20.4 million. As has occurred in many other less massive African urban centers, Cairo's population has more than doubled in the past twenty years.

Africa's second-largest city, Lagos, Nigeria, has grown at an astronomical rate. In 1950, its population stood at about 325,000. By 2000, it exceeded 7.2 million, and by 2019, it had nearly doubled to 13.9 million. This rapid pace of growth has strained the city's transportation, utilities, housing, and port facilities to the breaking point.

Byron Cannon

DISCUSSION QUESTIONS: THE CONTINENT

Q1. Where does Africa rank size-wise among the continents of the world? What influence does Africa's tectonic activity continue to have on Europe and Asia? What conclusion have geologists reached about the relationship between Africa and the South American landmass?

Q2. Into what general areas do geographers divide Africa? In what essential ways do northernmost Africa and southernmost Africa differ? What is meant by the "Horn of Africa"?

Q3. What is the Great Rift Valley? Where is it located in Africa? How does it manifest itself? What is the Great Rift Valley's geological significance?

THE OFFSHORE ISLANDS

The islands that lie off the coast of Africa fall into five geographical groups: Macaronesia (four archipelagos in the North Atlantic); islands of the Gulf of Guinea; isolated islands of the mid-Atlantic ridge; islands off Africa's east coast; and islands in the Indian Ocean. From tiny, barren Ascension in the Atlantic Ocean to enormous Madagascar in the Indian Ocean, they display almost as much variety as the continent itself. However, they have several factors in common. All were colonized by European nations, and most have suffered from a lack of resources. Many have relied on a plantation-style economy stressing only one or two crops, making them vulnerable to shifts in world markets. Tourism has provided several islands with an economic boost, but tourists can threaten the very cultures that attract them.

Macaronesia Islands

Macaronesia lying off the northwest and west coasts of Africa include Madeira, the Canary Islands to the south, and Cape Verde farther south still. Along with the Azores, which lie in the North Atlantic about 1,000 miles (1,600 km.) west of Portugal,

Pico Papagaio rises behind the town of Santo Antonio on Principe Island, São Tomé and Príncipe. (David Stanley from Nanaimo, Canada)

these islands are known as Macaronesia, a term derived from the Greek words for "large" and "island." All are the peaks of volcanoes, several of which remain active.

Madeira lies about 350 miles (563 km.) from the African coast of Morocco. The Canary Islands are much closer to Africa, and, when the weather is clear, an observer on the island of Fuerteventura can see the coast of southern Morocco. The islands of Cape Verde lie much farther south, about 300 miles (480 km.) off the coast of West Africa.

Lack of rainfall is a problem in most of Macaronesia. Only two of the Canary Islands have rivers that run year-round, and the two islands nearest Africa resemble the arid North African coast opposite them. Settlers in Madeira and the Canaries have built complex systems of canals and tunnels to distribute water evenly, enabling many temperate and subtropical fruits and vegetables to be grown. The islands of Cape Verde are barren and chronically short of water.

Only the Canary Islands seem to have been inhabited when European explorers reached them, although all may have been visited from time to time by early seafarers. Today, Madeira is a part of Portugal and the Canary Islands are part of Spain. Large numbers of tourists visit these islands, and the residents of the Canary Islands enjoy a middle-income economy. Once a Portuguese colony and an important stop for ships, Cape Verde became independent in 1975. As a result of its harsh climate and lack of resources, however, it has never recovered the modest economic status it once enjoyed.

The Gulf of Guinea Islands

Under the bulge of West Africa, a chain of volcanic islands stretches southwestward in an almost straight line into the Gulf of Guinea, a continuation of a range that includes Mount Cameroon on the continent. Nearest Africa is Bioko, an important part of the nation of Equatorial Guinea and the site of its capital, Malabo. Next in line and extending to the equator are the islands that make up the nation of São Tomé and Príncipe. At the end of the chain lies tiny Annobón, also part of Equatorial Guinea.

Africa's Islands Become Free

Africa's history has been immeasurably influenced by the intervention of European countries, many of which occupied African lands—usually forcibly—and enslaved the indigenous peoples. Although independence has not necessarily brought freedom or economic progress to Africa, it has returned the fate of the continent to its own people. Following are the dates on which and the countries from which Africa's islands gained their independence:

1960 Madagascar becomes independent of France.

1963 Zanzibar becomes independent of Great Britain; a year later it joins Tanganyika to form the United Republic of Tanzania.

1968 Equatorial Guinea (including Annobón and Bioko) becomes independent of Spain.

1968 Mauritius becomes independent of Great Britain.

1975 Cape Verde becomes independent of Portugal.

1975 Comoros becomes independent of France.

1975 São Tomé and Príncipe becomes independent of Portugal.

1976 Seychelles becomes independent of Great Britain.

The climate of these islands is tropical wet, and all are renowned for their spectacular beauty. However, the islands can be unhealthy, and in Bioko, malaria and tuberculosis are endemic. Crops such as cocoa and coffee have sustained the islands' economies from time to time, although both countries now import most of their food.

Sparsely populated when the Portuguese arrived in the fifteenth century, the islands in the Gulf of Guinea have been colonies of Spain or Portugal for most of their modern history. Bioko and São Tomé were important slave-trading stations, and even after the abolition of slavery, forced labor remained the norm in São Tomé and Príncipe, as it did in much of Portuguese Africa.

Three Isolated Islands

Rising from the undersea range of mountains known as the Mid-Atlantic Ridge are the three iso-

Anse Lazio is situated in the northwest of Praslin Island, considered by Lonely Planet to be the best beach on Praslin, and one of the best in the archipelago. (Svein-Magne Tunli)

lated islands of Ascension, St. Helena, and Tristan da Cunha. All three islands are volcanic, but only the last has experienced an active eruption in recent times. Ascension and St. Helena lie in the Tropics; Ascension is relatively barren, but St. Helena was heavily forested until imported goats destroyed much of its vegetation.

Tristan da Cunha lies far to the south, 1,600 miles (2,575 km.) from St. Helena. Uninhabited until their discovery in the early sixteenth century by Portuguese explorers, the islands have sometimes benefited from their isolation. Ships routinely stopped at St. Helena for fresh water in the nineteenth century, and the island was chosen by the British as a place of exile for defeated French emperor Napoleon Bonaparte. Ascension has become an important telecommunications center. It and Tristan da Cunha are dependencies of St. Helena; St. Helena is a dependency of the United Kingdom.

Off Africa's East Coast

Madagascar, Africa's largest island, is the fourth-largest island in the world—after Greenland, New Guinea, and Borneo. Nearby are the Comoros, islands lying at the northern end of the Mozambique Channel between Madagascar and the nation of Mozambique. Still farther to the north and hugging the coast of Africa are the small islands of Zanzibar and Pemba, both part of the nation of Tanzania. Lying east of the "horn" of East Africa is the island of Socotra, part of Yemen, a country in the southwestern part of the Arabian Peninsula.

Geologists believe that Madagascar was once connected to Africa and India as part of the prehistoric continent of Gondwanaland. The island apparently became separated during the late Cretaceous Period, some 65 million years ago, allowing many unique plants and animals to evolve. The first human inhabitants of Madagascar arrived more than 2,000 years ago and apparently originated in many places, from Indonesia and Malaya to Africa itself.

Because of its size and geography, Madagascar has a variety of climates. The island's east coast experiences a hot, wet monsoon season from November through April, although its southwestern region remains almost perpetually dry. Comoros experiences the same monsoon on a more uniform basis. Both nations grow rice for domestic consumption, and the burning of forests for its cultivation has led to serious soil erosion on Madagascar. Madagascar and Comoros grow such crops as coffee for export, and are important sources of flavorings and spices such as vanilla and cloves. Comoros is the world's second-largest producer of vanilla and the largest producer of ylang-ylang, a fragrant ingredient in perfume.

Colonized by the French in the nineteenth century, Madagascar became independent in 1960 and Comoros in 1975. Both countries face dire economic problems. Comoros has few natural resources, and it and Madagascar are among the poorest countries in Africa.

Low-lying and once disease-ridden, Zanzibar has few resources of its own, but its position on the coast of East Africa has enabled it to profit as a trading depot for the region. Once the center of the East African slave trade, Zanzibar was persuaded by the British to abolish the trade in 1873. The British subsequently declared a protectorate over the island, which regained its independence in 1963. The following year, Zanzibar joined Tanganyika to become the nation of Tanzania.

In the Indian Ocean, a handful of islands lie scattered east of Madagascar. These include the Mascarene Islands—Mauritius, Rodrigues, and Réunion—just north of the Tropic of Capricorn, and the nation of Seychelles, located just south of the equator. The Mascarenes are volcanic, and a volcano on Réunion remains active. The islands of Seychelles are largely granite, although the group includes a number of low-lying coral atolls, most of them uninhabited.

All the islands in this region have a tropical monsoonal climate. Like Madagascar and Comoros,

Coco Island, sea turtle. (Dara)

they are home to many unique and often fragile species of plants and animals. The *coco de mer*, at 49 pounds (22 kilograms) the heaviest seed in the world, grows on a species of palm in Seychelles. The dodo, a large flightless bird, was found on Mauritius until it was slaughtered to extinction for its meat.

Although they were undoubtedly known to traders and seafarers, the Mascarenes and Seychelles remained largely uninhabited until settled a few centuries ago by Europeans—Mauritius in the late sixteenth century, Réunion in the mid-seventeenth century, and Seychelles in the mid-eighteenth century. Their populations today are descendants not only of their original settlers but of slaves and traders drawn from throughout the Indian Ocean area.

Mauritius and Seychelles received their independence from Great Britain in 1968 and 1976, respectively. Réunion remains a self-governing overseas department of France. Although sugarcane has long been a traditional crop on Mauritius and Réunion, tourism now plays an increasingly important role on all the islands.

Grove Koger

DISCUSSION QUESTIONS: THE OFFSHORE ISLANDS

Q1. What is meant by the term "Macaronesia"? Where is it located? What is the primary geological similarity shared by the Macaronesian islands?

Q2. On what coast of Africa are the Gulf of Guinea islands found? To which countries do the various islands belong? Are any of the islands independent countries? Which three islands lie far out in the Atlantic but are associated with Africa?

Q3. Where does Madagascar rank among the world's largest islands? What impact did Madagascar's long separation from the African continent have on the island's plants and animals? Why does Madagascar experience a variety of climates?

PHYSICAL GEOGRAPHY

OVERVIEW

CLIMATE AND HUMAN SETTLEMENT

"Everyone talks about the weather," goes an old saying, "but nobody does anything about it." If everyone talks about the weather, it is because it is important to them—to how they feel and to how their bodies and minds function. There is plenty they can do about it, from going to a different location to creating an artificial indoor environment.

Climate

The term "climate" refers to average weather conditions over a long period of time and to the variations around that average from day to day or month to month. Temperature, air pressure, humidity, wind conditions, sunshine, and rainfall—all are important elements of climate and differ systematically with location. Temperatures tend to be higher near the equator and are so low in the polar regions that very few people live there. In any given region, temperatures are lower at higher altitudes. Areas close to large bodies of water have more stable temperatures. Rainfall depends on topography: The Pacific Coast of the United States receives a great deal of rain, but the nearby mountains prevent it from moving very far inland. Seasonal variations in temperature are larger in temperate zones.

Throughout human history, climate has affected where and how people live. People in technologically primitive cultures, lacking much protective clothing or housing, needed to live in mild climates, in environments favorable to hunting and gathering. As agricultural cultivation developed, populations located where soil fertility, topography, and climate were favorable to growing crops and raising livestock. Areas in the Middle East and near the Mediterranean Sea flourished before 1000 BCE.

Many equatorial areas were too hot and humid for human and animal health and comfort, and too infested with insect pests and diseases.

Improvements in technology allowed settlement to range more widely north and south. Sturdy houses and stables, internal heating, and warm clothing enabled people to survive and be active in long cold winters. Some peoples developed nomadic patterns, moving with herds of animals to adapt to seasonal variations.

A major challenge in the evolution of settled agriculture was to adapt production to climate and soil conditions. In North America, such crops as cotton, tobacco, rice, and sugarcane have relatively restricted areas of cultivation. Wheat, corn, and soybeans are more widely grown, but usually further north. Winter wheat is an ingenious adaptation to climate. It is sown and germinates in autumn, then matures and is harvested the following spring. Rice, which generally grows in standing water, requires special environmental conditions.

Tropical Problems

Some scholars argue that tropical climates encourage life to flourish but do not promote quality of life. In hot climates, people do not need much caloric intake to maintain body heat. Clothing and housing do not need to protect people from the cold. Where temperatures never fall below freezing, crops can be grown all year round. Large numbers of people can survive even where productivity is not high. However, hot, humid conditions are not favorable to human exertion nor (it is claimed) to mental, spiritual, and artistic creativity. Some tropical areas, such as South India, Bangladesh, Indonesia, and Central Africa, have developed large populations living at relatively low levels of income.

Slavery

Efforts to develop tropical regions played an important part in the rise of the slave trade after 1500 CE. Black Africans were kidnapped and forceably transported to work in hot, humid regions. The West Indian islands became an important location for slave labor, particularly in sugar production. On the North American continent, slave labor was important for producing rice, indigo, and tobacco in colonial times. All these were eclipsed by the enormous growth of cotton production in the early years of U.S. independence. It has been estimated that the forced migration of Africans to the Americas involved about 1,800 Africans per year from 1450 to 1600, 13,400 per year in the seventeenth century, and 55,000 per year from 1701 to 1810. Estimates vary wildly, but at least 12 million Africans were forced to migrate in this process.

European Migration

Migration of European peoples also accelerated after the discovery of the New World. They settled mainly in temperate-zone regions, particularly North America. Although Great Britain gained colonial dominion over India, the Netherlands over present-day Indonesia, and Belgium over a vast part of central Africa, few Europeans went to those places to live. However, many Chinese migrated throughout the Nanyang (South Sea) region, becoming commercial leaders in present-day Malaysia, Thailand, Indonesia, and the Philippines, despite the heat and humidity. British emigrants settled in Australia and New Zealand, Spanish and Italians in Argentina, Dutch (Boers) in South Africa—all temperate regions.

Climate and Economics

Most of the economic progress of the world between 1492 and 2000 occurred in the temperate zones, primarily in Europe and North America. Climatic conditions favored agricultural productivity. Some scholars believe that these areas had climatic conditions that were stimulating to intellectual and tech-

> ### IRELAND'S POTATO FAMINE AND EUROPEAN EMIGRATION
>
> Mass migration from Europe to North America began in the 1840s after a serious blight destroyed a large part of the potato crop in Ireland and other parts of Northern Europe. The weather played a part in the famine; during the autumns of 1845 and 1846 climatic conditions were ideal for spreading the potato blight. The major cause, however, was the blight itself, and the impact was severe on low-income farmers for whom the potato was the major food.
>
> The famine and related political disturbances led to mass emigration from Ireland and from Germany. By 1850 there were nearly a million Irish and more than half a million Germans in the United States. Combined, these two groups made up more than two-thirds of the foreign-born U.S. population of 1850. The settlement patterns of each group were very different. Most Irish were so poor they had to work for wages in cities or in construction of canals and railroads. Many Germans took up farming in areas similar in climate and soil conditions to their homelands, moving to Wisconsin, Minnesota, and the Dakotas.

nological development. They argue that people are invigorated by seasonal variation in temperature, sunshine, rain, and snow. Storms—particularly thunderstorms—can be especially stimulating, as many parents of young children have observed for themselves.

Climate has contributed to the great economic productivity of the United States. This productivity has attracted a flow of immigrants, which averaged about 1 million a year from 1905 to 1914. Immigration approached that level again in the 1990s, as large numbers of Mexicans crossed the southern border of the United States, often coming for jobs as agricultural laborers in the hot conditions of the Southwest—a climate that made such work unattractive to many others.

Unpredictable climate variability was important in the peopling of North America. During the 1870s and 1880s, unusually favorable weather encouraged a large flow of migration into the grain-producing areas just west of the one-hundredth meridian. Then came severe drought and much agrarian distress. Between 1880 and 1890, the combined population of Kansas and Nebraska increased by about a million, an increase of 72 percent. During the 1890s, however, their combined population was virtually constant, indicating that a large out-migration was offsetting the natural increase. Much of the area reverted to pasture, as climate and soil conditions could not sustain the grain production that had attracted so many earlier settlers.

Climate variability can be a serious hazard. Freezing temperatures for more than a few hours during spring can seriously damage fruits and vegetables. A few days of heavy rain can produce serious flooding.

Recreation and Retirement

Whenever people have been able to separate decisions about where to live from decisions about where to work, they have gravitated toward pleasant climatic conditions. Vacationers head for Caribbean islands, Hawaii, the Crimea, the Mediterranean Coast, even the Baltic coast. "The mountains" and "the seashore" are attractive the world over. Paradoxically, some of these areas (the Caribbean, for instance) have monotonous weather year-round and thus have not attracted large inflows of permanent residents. Winter sports have created popular resorts such as Vail and Aspen in Colorado, and numerous older counterparts in New England. Large numbers of Americans have retired to the warm climates in Florida, California, and Arizona. These areas then attract working-age adults who earn a living serving vacationers and retirees. Since these locations are uncomfortably hot in summer, their attractiveness for residence had to await the coming of air conditioning in the latter half of the twentieth century.

Human Impact on Climate

Climate interacts with pollution. Bad-smelling factories and refineries have long relied on the wind to disperse atmospheric pollutants. The city of Los Angeles, California, is uniquely vulnerable to atmospheric pollution because of its topography and wind currents. Government regulations of automobile emissions have had to be much more stringent there than in other areas to keep pollution under control.

Human activities have sometimes altered the climate. Development of a large city substitutes buildings and pavements for grass and trees, raising summer temperatures and changing patterns of water evaporation. Atmospheric pollutants have contributed to acid rain, which damages vegetation and pollutes water resources. Many observers have also blamed human activities for a trend toward global warming. Much of this has been blamed on carbon dioxide generated by combustion, particularly of fossil fuels. A widespread and continuing rise in temperatures is expected to raise water levels in the oceans as polar icecaps melt and change the relative attractiveness of many locations.

Paul B. Trescott

FLOOD CONTROL

Flood control presents one of the most daunting challenges humanity faces. The regions that human communities have generally found most desirable, for both agriculture and industry, have also been the lands at greatest risk of experiencing devastating floods. Early civilization developed along river valleys and in coastal floodplains because those lands contained the most fertile, most easily irrigated soils for agriculture, combined with the convenience of water transportation.

The Nile River in North Africa, the Ganges River on the Indian subcontinent, and the Yangtze River in China all witnessed the emergence of civilizations that relied on those rivers for their growth. People learned quickly that residing in such areas meant living with the regular occurrence of life-threatening floods.

Knowledge that floods would come did not lead immediately to attempts to prevent them. For thousands of years, attempts at flood control were rare. The people living along river valleys and in floodplains often developed elaborate systems of irrigation canals to take advantage of the available water for agriculture and became adept at using rivers for transportation, but they did not try to control the river itself. For millennia, people viewed periodic flooding as inevitable, a force of nature over which they had no control. In Egypt, for example, early people learned how far out over the riverbanks the annual flooding of the Nile River would spread and accommodated their society to the river's seasonal patterns. Villagers built their homes on the edge of the desert, beyond the reach of the flood waters, while the land between the towns and the river became the area where farmers planted crops or grazed livestock.

In other regions of the world, buildings were placed on high foundations or built with two stories on the assumption that the local rivers would regularly overflow their banks. In Southeast Asian countries such as Thailand and Vietnam, it is common to see houses constructed on high wooden posts above the rivers' edge. The inhabitants have learned to allow for the water levels' seasonal changes.

Flood Control Structures

Eventually, societies began to try to control floods rather than merely attempting to survive them. Levees and dikes—earthen embankments constructed to prevent water from flowing into low-lying areas—were built to force river waters to remain within their channels rather than spilling out over a floodplain. Flood channels or canals that fill with water only during times of flooding, diverting water

away from populated areas, are also a common component of flood control systems. Areas that are particularly susceptible to flash floods have constructed numerous flood channels to prevent flooding in the city. For example, for much of the year, Southern California's Los Angeles River is a small stream flowing down the middle of an enormous, 20- to 30-foot-deep (6–9 meters) concrete-lined channel, but winter rains can fill its bed from bank to bank. Flood channels prevent the river from washing out neighborhoods and freeways.

Engineers designed dams with reservoirs to prevent annual rains or snowmelt entering the river upstream from running into populated areas. By the end of the twentieth century, extremely complex flood control systems of dams, dikes, levees, and flood channels were common. Patterns of flooding that had existed for thousands of years ended as civil engineers attempted to dominate natural forces.

The annual inundation of the Egyptian delta by the flood waters of the Nile River ceased in 1968 following construction of the 365-foot-high (111 meters) Aswan High Dam. The reservoir behind the 3,280-foot-long (1,000-meter) dam forms a lake almost ten miles (16 km.) wide and almost 300 miles (480 km.) long. Flood waters are now trapped behind the dam and released gradually over a year's time.

Environmental Concerns

Such high dams are increasingly being questioned as a viable solution for flood control. As human understanding of both hydrology and ecology have improved, the disruptive effects of flood control projects such as high dams, levees, and other engineering projects are being examined more closely.

Hydrologists and other scientists who study the behavior of water in rivers and soils have long known that vegetation and soil types in watersheds can have a profound effect on downstream flooding. The removal of forest cover through logging or clearing for agriculture can lead to severe flooding in the future. Often that flooding will occur many miles downstream from the logging activity. Devastating floods in the South Asian country of Bangladesh, for example, have been blamed in part on clear-cutting of forested hillsides in the Himalaya Mountains in India and Nepal. Monsoon rains that once were absorbed or slowed by forests now run quickly off mountainsides, causing rivers to reach unprecedented flood levels. Concerns about cause-and-effect relationships between logging and flood control in the mountains of the United States were one reason for the creation of the U.S. Forest Service in the nineteenth century.

In populated areas, even seemingly trivial events such as the construction of a shopping center parking lot can affect flood runoff. When thousands of square feet of land are paved, all the water from rain runs into storm drains rather than being absorbed slowly into the soil and then filtered through the watertable. Engineers have learned to include catch basins, either hidden underground or openly visible but disguised as landscaping features such as ponds, when planning a large paving project.

Wetlands and Flooding

Less well known than the influence of watersheds on flooding is the impact of wetlands along rivers. Many river systems are bordered by long stretches of marsh and bog. In the past, flood control agencies often allowed farmers to drain these areas for use in agriculture and then built levees and dikes to hold the river within a narrow channel. Scientists now know that these wetlands actually serve as giant sponges in the flood cycle. Flood waters coming down a river would spread out into wetlands and be held there, much like water is trapped in a sponge.

Draining wetlands not only removes these natural flood control areas but worsens flooding problems by allowing floodwater to precede downstream faster. Even if life-threatening or property-damaging floods do not occur, faster-flowing water significantly changes the ecology of the river system. Waterborne silt and debris will be carried farther. Trying to control floods on the Mississippi River has had the unintended consequence of causing waterborne silt to be carried farther out into the Gulf of Mexico by the river, rather than its being deposited in the delta region. This, in turn, has led to the loss of shore land as ocean wave actions washes soil away, but no new alluvial deposits arrive to replace it.

In any river system, some species of aquatic life will disappear and others replace them as the speed of flow of the water affects water temperature and the amount of dissolved oxygen available for fish. Warm-water fish such as bass will be replaced by cold-water fish such as trout, or vice versa. Biologists estimate that more than twenty species of freshwater mussels have vanished from the Tennessee River since construction of a series of flood control and hydroelectric power generation dams have turned a fast-moving river into a series of slow-moving reservoirs.

Future of Flood Control

By the end of the twentieth century, engineers increasingly recognized the limitations of human interventions in flood control. Following devastating floods in the early 1990s in the Mississippi River drainage, the U.S. Army Corps of Engineers recommended that many towns that had stood right at the river's edge be moved to higher ground. That is, rather than trying to prevent a future flood, the Corps advised citizens to recognize that one would inevitably occur, and that they should remove themselves from its path. In the United States and a number of other countries, land that has been zoned as floodplains can no longer be developed for residential use. While there are many things humanity can do to help prevent floods, such as maintaining well-forested watersheds and preserving wetlands, true flood control is probably impossible. Dams, levees, and dikes can slow the water down, but eventually, the water always wins.

Nancy Farm Männikkö

ATMOSPHERIC POLLUTION

Pollution of the Earth's atmosphere comes from many sources. Some forces are natural, such as volcanoes and lightning-caused forest fires, but most sources of pollution are byproducts of industrial society. Atmospheric pollution cannot be confined by national boundaries; pollution generated in one country often spills over into another country, as is the case for acid deposition, or acid rain, generated in the midwestern states of the United States that affects lakes in Canada.

Major Air Pollutants

Each of eight major forms of air pollution has an impact on the atmosphere. Often two or more forms of pollution have a combined impact that exceeds the impact of the two acting separately. These eight forms are:

1. Suspended particulate matter: This is a mixture of solid particles and aerosols suspended in the air. These particles can have a harmful impact on human respiratory functions.

2. Carbon monoxide (CO): An invisible, colorless gas that is highly poisonous to air-breathing animals.

3. Nitrogen oxides: These include several forms of nitrogen-oxygen compounds that are converted to nitric acid in the atmosphere and are a major source of acid deposition.

4. Sulfur oxides, mainly sulfur dioxide: This sulfur-oxygen compound is converted to sulfuric acid in the atmosphere and is another source of acid deposition.

5. Volatile organic compounds: These include such materials as gasoline and organic cleaning solvents, which evaporate and enter the air in a vapor state. VOCs are a major source of ozone formation in the lower atmosphere.

6. Ozone and other petrochemical oxidants: Ground-level ozone is highly toxic to animals and plants. Ozone in the upper atmosphere, however, helps to shield living creatures from ultraviolet radiation.

7. Lead and other heavy metals: Generated by various industrial processes, lead is harmful to human health even at very low concentrations.

8. Air toxics and radon: Examples include cancer-causing agents, radioactive materials, or asbestos. Radon is a radioactive gas produced by natural processes in the earth.

All eight forms of pollution can have adverse effects on human, animal, and plant life. Some, such as lead, can have a very harmful effect over a small range. Others, such as sulfur and nitrogen oxides, can cross national boundaries as they enter the atmosphere and are carried many miles by prevailing wind currents. For example, the radioactive discharge from the explosion of the Chernobyl nuclear plant in the former Soviet Union in 1986 had harmful impacts in many countries. Atmospheric radiation generated by the explosion rapidly spread over much of the Northern Hemisphere, especially the countries of northern Europe.

Impacts of Atmospheric Pollution

Atmospheric pollution not only has a direct impact on the health of humans, animals, and plants but also affects life in more subtle, often long-term, ways. It also affects the economic well-being of people and nations and complicates political life.

Atmospheric pollution can kill quickly, as was the case with the killer smog, brought about by a temperature inversion, that struck London in 1952 and led to more than 4,000 pollution-related deaths. In the late 1990s, the atmosphere of Mexico City was so polluted from automobile exhausts and industrial pollution that sidewalk stands selling pure oxygen to people with breathing problems became thriving businesses. Many of the heavy metals and organic constituents of air pollution can cause cancer when people are exposed to large doses or for long periods of time. Exposure to radioactivity in the atmosphere can also increase the likelihood of cancer.

In some parts of Germany and Scandinavia in the 1990s, as well as places in southern Canada and the southern Appalachians in the United States, certain types of trees began dying. There are several possible reasons for this die-off of forests, but one potential culprit is acid deposition. As noted above, one byproduct of burning fossil fuels (for example, in coal-fired electric power plants) is the sulfur and nitrous oxides emitted from the smokestacks. Once in the atmosphere, these gases can be carried for many miles and produce sulfuric and nitric acids.

These acids combine with rain and snow to produce acidic precipitation. Acid deposition harms crops and forests and can make a lake so acidic that aquatic life cannot exist in it. Forests stressed by contact with acid deposition can become more susceptible to damage by insects and other pathogens. Ozone generated from automobile emissions also kills many plants and causes human respiratory problems in urban areas.

Air pollution also has an impact on the quality of life. Acid pollutants have damaged many monuments and building facades in urban areas in Europe and the United States. By the late 1990s, the distance that people could see in some regions, such as the Appalachians, was reduced drastically because of air pollution.

The economic impact of air pollution may not be as readily apparent as dying trees or someone with a respiratory ailment, but it is just as real. Crop damage reduces agricultural yield and helps to drive up the cost of food. The costs of repairing buildings or monuments damaged by acid rain are substantial. Increased health-care claims resulting from exposure to air pollution are hard to measure but are a cost to society nevertheless.

It is impossible to predict the potential for harm from rapid global warming arising from greenhouse gases and the destruction of the ozone layer by chlorofluorocarbons (CFCs), but it could be cata-

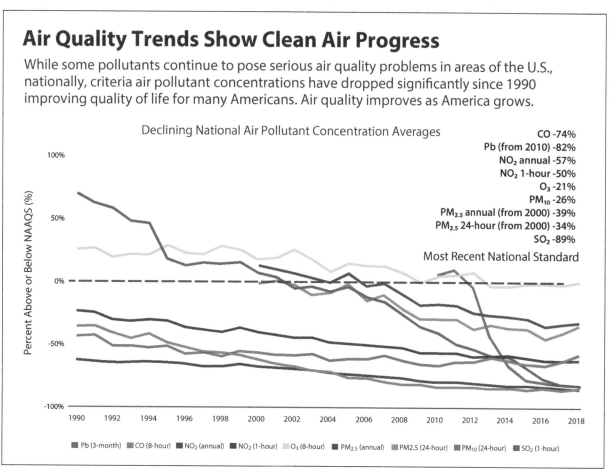

Air Quality Trends Show Clean Air Progress

While some pollutants continue to pose serious air quality problems in areas of the U.S., nationally, criteria air pollutant concentrations have dropped significantly since 1990 improving quality of life for many Americans. Air quality improves as America grows.

Declining National Air Pollutant Concentration Averages

CO -74%
Pb (from 2010) -82%
NO_2 annual -57%
NO_2 1-hour -50%
O_3 -21%
PM_{10} -26%
$PM_{2.5}$ annual (from 2000) -39%
$PM_{2.5}$ 24-hour (from 2000) -34%
SO_2 -89%

Most Recent National Standard

Pb (3-month) CO (8-hour) NO_2 (annual) NO_2 (1-hour) O_3 (8-hour) $PM_{2.5}$ (annual) PM2.5 (24-hour) PM_{10} (24-hour) SO_2 (1-hour)

Source: U.S. Environmental Protection Agency, Our Nation's Air, Status and Trends Through 2018.

strophic. Rapid global warming would cause the sea level to rise because of the melting of the polar ice caps. Low-lying coastal areas would be flooded, or, in the case of Bangladesh, much of the country. Global warming would also change crop patterns for much of the world.

Solutions for Atmospheric Pollution

Although there is still some debate, especially among political leaders, most scientists recognize that air pollution is a problem that affects both the industrialized and less-industrialized world. In their rush to industrialize, many nations begin generating substantial amounts of air pollution; China's extensive use of coal-fired power plants is just one example.

The major industrial nations are the primary contributors to atmospheric pollution. North America, Europe, and East Asia produce 60 percent of the world's air pollution and 60 percent of its food supply. Because of their role in supplying food for many other nations, anything that damages their ability to grow crops hurts the rest of the world. In 2018, about 76 million tons of pollution were emitted into the atmosphere in the United States. These emissions mostly contribute to the formation of ozone and particles, the deposition of acids, and visibility impairment.

Many industrialized nations are making efforts to control air pollution, for example, the Clean Air Act of 1970 in the United States or the international Montreal Accord to curtail CFC production. Progress is slow and the costs of reducing air pollution are often high. Worldwide, bad outdoor air caused an estimated 4.2 million premature deaths in 2016, about 90 percent of them in low- and middle-income countries, according to the World Health Organization. Indoor smoke is an ongoing health threat to the 3 billion people who cook and heat their homes by burning biomass, kerosene, and coal.

In the year 2019 the record of the nations of the world in dealing with air pollution was a mixed one. There were some signs of progress, such as reduced automobile emissions and sulfur and nitrous oxides in industrialized nations, but acid deposition remains a problem in some areas. CFC production has been halted, but the impact of CFCs on the ozone layer will continue for many years. However, more nations are becoming aware of the health and economic impact of air pollution and are working to keep the problem from getting worse.

John M. Theilmann

DISEASE AND CLIMATE

Climate influences the spread and persistence of many diseases, such as tuberculosis and influenza, which thrive in cold climates, and malaria and encephalitis, which are limited by the warmth and humidity that sustains the mosquitoes carrying them. Because the earth is warming as a result of the generation of carbon dioxide and other "greenhouse gases" from the burning of fossil fuels, there is intensified scientific concern that warm-weather diseases will reemerge as a major health threat in the near future.

Scientific Findings

The question of whether the earth is warming as a result of human activity was settled in scientific circles in 1995, when the Second Assessment Report of the Intergovernmental Panel on Climate Change, a worldwide group of about 2,500 experts, was issued. The panel concluded that the earth's temperature had increased between 0.5 to 1.1 degrees Farenheit (0.3 to 0.6 degrees Celsius) since reliable worldwide records first became available in the late nineteenth century. Furthermore, the intensity of warming had increased over time. By the 1990s, the temperature was rising at the most rapid rate in at least 10,000 years.

The Intergovernmental Panel concluded that human activity—the increased generation of carbon dioxide and other "greenhouse gases"—is responsible for the accelerating rise in global temperatures. The amount of carbon dioxide in the atmosphere has risen nearly every year because of increased use of fossil fuels by ever-larger human populations experiencing higher living standards.

In 1998, Paul Epstein of the Harvard School of Public Health described the spread of malaria and dengue fever to higher altitudes in tropical areas of the earth as a result of warmer temperatures. Rising winter temperatures have allowed disease-bearing insects to survive in areas that could not support them previously. According to Epstein, frequent flooding, which is associated with warmer temperatures, also promotes the growth of fungus and provides excellent breeding grounds for large numbers of mosquitoes. Some experts cite the flooding caused by Hurricane Floyd and other storms in North Carolina during 1999 as an example of how global warming promotes conditions ideal for the spread of diseases imported from the Tropics.

Heat, Humidity, and Disease

During the middle 1990s, an explosion of termites, mosquitoes, and cockroaches hit New Orleans, following an unprecedented five years without frost. At the same time, dengue fever spread from Mexico across the border into Texas for the first time since records have been kept. Dengue fever, like malaria, is carried by a mosquito that is limited by temperature and humidity. Colombia was experiencing plagues of mosquitoes and outbreaks of the diseases they carry, including dengue fever and encephalitis, triggered by a record heat wave followed by heavy rains. In 1997 Italy also had an outbreak of malaria. An outbreak of zika in 2015–16, related to a virus spread by mosquitoes, raised concerns re-

garding the safety of athletes and spectators at the 2016 Summer Olympics in Rio de Janeiro and led to travel warnings and recommendations to delay getting pregnant for those living or traveling in areas where the mosquitoes are active.

The global temperature is undeniably rising. According to the National Oceanic and Atmospheric Administration, July 2019, was the hottest month since reliable worldwide records have been kept, or about 150 years. The previous record had been set in July 2017.

The rising incidence of some respiratory diseases may be related to a warmer, more humid environment. The American Lung Association reported that more than 5,600 people died of asthma in the United States during 1995, a 45.3 percent increase in mortality over ten years, and a 75 percent increase since 1980. Roughly a third of those cases occurred in children under the age of eighteen. Asthma is now one of the leading diseases among the young. Since 1980, there has been a 160 percent increase in asthma in children under the age of five.

Heat Waves and Health

A study by the Sierra Club found that air pollution, which will be enhanced by global warming, could be responsible for many human health problems, including respiratory diseases such as asthma, bronchitis, and pneumonia.

According to Joel Schwartz, an epidemiologist at Harvard University, air pollution concentrations in the late 1990s were responsible for 70,000 early deaths per year and more than 100,000 excess hospitalizations for heart and lung disease in the United States. Global warming could cause these numbers to increase 10 to 20 percent in the United States, with significantly greater increases in countries that are more polluted to begin with, according to Schwartz.

Studies indicate that global warming will directly kill hundreds of Americans from exposure to extreme heat during summer months. The U.S. Centers for Disease Control and Prevention have found that between 1979 and 2014, the death rate as a direct result of exposure to heat (underlying cause of death) generally hovered around 0.5 to 1 deaths per million people, with spikes in certain years). Overall, a total of more than 9,000 Americans have died from heat-related causes since 1979, according to death certificates. Heat waves can double or triple the overall death rates in large cities. The death toll in the United States from a heat wave during July 1999 surpassed 200 people. As many as 600 people died in Chicago alone during the 1990s due to heat waves. The elderly and very young have been most at risk.

Respiratory illness is only part of the picture. The Sierra Club study indicated that rising heat and humidity would broaden the range of tropical diseases, resulting in increasing illness and death from diseases such as malaria, cholera, and dengue fever, whose range will spread as mosquitoes and other disease vectors migrate.

The effects of El Niño in the 1990s indicate how sensitive diseases can be to changes in climate. A study conducted by Harvard University showed that warming waters in the Pacific Ocean likely contrib-uted to the severe outbreak of cholera that led to thousands of deaths in Latin American countries. Since 1981, the number of cases of dengue fever has risen significantly in South America and has begun to spread into the United States. According to health experts cited by the Sierra Club study, the outbreak of dengue near Texas shows the risks that a warming climate might pose. Epstein and the Sierra Club study concur that if tropical weather expands, tropical diseases will expand.

In many regions of the world, malaria is already resistant to the least expensive, most widely distributed drugs. According to the World Health Organization (WHO), there were 219 million cases of malaria globally in 2017 and 435,000 malaria deaths, representing a decrease in malaria cases and deaths rates of 18 percent and 28 percent since 2010. Of those deaths, 403,000 (approximately 93 percent) were in the WHO African Region.

Bruce E. Johansen

PHYSIOGRAPHY

The world's second-largest continent, after Asia, Africa covers 11.7 million square miles (30.3 million sq. km.). The Arabian Sea and Indian Ocean surround the continent on the east and south, the Atlantic Ocean on the west, and the Mediterranean Sea on the north. The estimate of its size includes major offshore islands, especially the immense island of Madagascar, whose surface covers 226,658 square miles (587,042 sq. km.). Other significant island complexes are Zanzibar and Pemba, plus the Comoros and Seychelles Islands off Africa's eastern coast, and the Canary and Madeira Islands in the Atlantic off northwest Africa. Individual islands close to the western coast include São Tomé, Príncipe, Annobón, and Bioko in the Gulf of Guinea.

The five traditional regions of Africa are North Africa, East Africa, West Africa, Central Africa, and Southern Africa. It is relatively easy to describe these regions in general terms; to explain the differences among them requires consideration of a number of more complex factors, beginning with an overview of the geological history of the continent.

Elevation Zones

Geologists refer to the northern two-thirds of Africa as "low Africa." Most of the surface of this region is only about 650 feet (200 meters) in elevation. Some low elevations do occur in the southern third (particularly coastal regions), and mountain zones jut out of lower elevations in the northern two-thirds.

"High Africa," with typical elevations of about 300 feet (1,000 meters) or more, begins on the eastern side of the continent in the area of modern Kenya and Uganda and extends through most of the southern third of the continent. Geologists also refer to high Africa as the Southern African

The Atlas Mountains. (Progskynet)

43

PHYSICAL GEOGRAPHY OF AFRICA

superswell, a term that is linked to the tectonic origin of the continent. Tectonics considers the relative buoyancy of the African landmass and other continents as part of the plate system forming the crust of the planet.

North Africa

The vast Sahara Desert divides North Africa from the rest of the continent. This division is further marked in the western half of North Africa by the east-west mountain complex of the Aures and Atlas ranges, running from western Tunisia to the Atlantic coast of Morocco. Mediterranean and Atlantic coastal plains in this region are relatively narrow.

Eastward from Tunisia to the Nile Delta in Egypt, there are some mountain outcroppings, mainly in the Jabal Akhdar of Cyrenaica, but the land generally extends southward to the interior at elevations near sea level. West of the Nile River lies the extensive Qattara Depression, which descends to 440 feet (135 meters) below sea level. These lower regions are not far from either the Mediterranean or the Atlantic; winter rainfall is sufficient to sustain seasonal agriculture throughout most of the western half of coastal North Africa.

The impressive mountain region of the High Atlas in Morocco reaches elevations of nearly 14,000 feet (4,267 meters), surpassed only by eastern African peaks in Ethiopia and Kenya. Southeast of these mountains, and generally southward from points east of Tunisia, lies the desert expanse of the Sahara.

West Africa

On the western half of the continent south of the Sahara, major rivers originating in the Fouta Djallon highlands of Guinea (such as the Niger, the Senegal, and the Gambia) flow toward low coastal zones, saving much of West Africa from the harshness of the desert. The northern part of the vast West Afri-

THE SAHARA

The Sahara is the world's largest desert in the tropical latitudes. Eleven modern African countries, from Mauritania in the west to the Sudan in the east, fall within its boundaries. The Sahara covers more than 3.6 million square miles (9.3 million sq. km.), a good portion of which is totally uninhabited. Although geologists trace the initial stages of its desertification to the early Pliocene era about 5 million years ago, this vast zone has seen different periods ranging from extreme dry conditions to fairly substantial humidity. Pockets of mineral wealth, including iron, copper, zinc, magnesium, chromium, and even gold and silver, occur in restricted subregions of the vast desert.

can region, comprising the modern countries of Senegal, Mali, Burkina Faso, and Niger, is the generally flatter and bleaker half. Zones farther south (the coastal region of Guinea, Sierra Leone, and from Côte d'Ivoire to Nigeria) are more varied both topographically and in terms of vegetation.

Camel caravan going through sand in the Sahara Desert. (Asfour Hamza)

East Africa

To the east, the complicated geology of the rift system creates a major division between the Sahara and the main zone of east Africa from the Horn of Africa in Somalia southward to the Zambezi River. The equator cuts through Africa halfway between the northern mountain region of East Africa and the Zambezi River. Whereas the eastern equatorial zone inland from the Indian Ocean is mountainous —snowcapped Mount Kenya (17,058 feet/5,199 meters) is on the equator—Central Africa at the equator is characterized by the lowlands of the great Congo River basin as well as the most extensive tropical rainforest on the continent.

Southern Africa

Beyond the Zambezi outlet to the Indian Ocean are the first mountains of Southern Africa and Africa's second-largest desert region, the Kalahari. The mountains of Southern Africa along the southeastern coast next to the Indian Ocean are generally not as high as the mountainous zones of northeastern and northwestern Africa. The Drakensberg Range, however, forms an extensive watershed for Southern Africa's major river, the Orange, which flows westward just south of the Kalahari Desert.

Beneath and above these easily observable features of the landscape are major physiographic factors that help explain why each broad region differs from the others. One factor includes geological phenomena; the second is rainfall, which is responsible for both varied vegetation zones and for the major rivers that crisscross Africa's surface.

Tectonic Factors Shaping Africa

Through studies of plate tectonics (the movement of different sections of Earth's crust "floating" over the molten core), scientists have been able to relate the geological origins of Africa to phenomena that created the other continents. Tec-

tonic theory assumes that, in the most distant geological past, a huge single landmass, Gondwanaland, broke into segments in reaction to pressures rising to Earth's surface from its core. The gradual movement of these segments, either in opposite directions (e.g., the presumed separation of North and South America from the other continents) or pressing against one another and lifting (creating tectonic systems like the mountain masses of the Himalayas "pushed" by the Indian subcontinent or the Alps in Europe), can explain how Africa took the form it did.

Geologists have posited at least seven phases of major tectonic activity that shaped Africa. The earliest of these is during the early Permian period (between 290 million and 245 million years ago), preceding the actual breakup of Gondwanaland. It is traceable in some early rock formations in Southern Africa, specifically the Karoo Basin. Another major

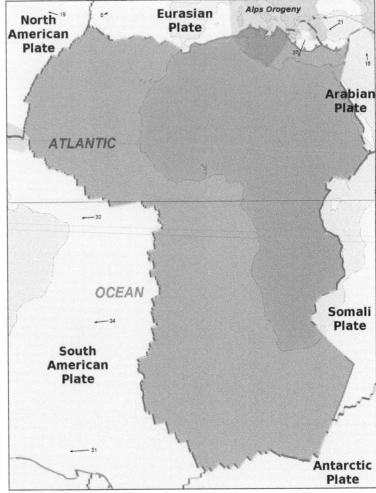

The African Plate. (Alataristarion)

The Great Rift Valley is a name given to the continuous geographic trench, approximately 3,700 miles (6,000 km.) in length, that runs from Lebanon's Beqaa Valley in Asia to Mozambique in South Eastern Africa, most often used to refer to the valley of East Africa.

tectonic period occurred in the late Triassic (225 million to 200 million years ago) and early Jurassic periods, when northwest Africa broke free from North America. Still later, in the mid-Cretaceous period (135 million to 65 million years ago), West Africa separated from South America. The early Cenozoic era (65 million to 2 million years ago) saw the tectonic upheavals associated with the opening of the Red Sea and the East African Great Rift Valley.

East African Rift System

Although technical explanations of African rift systems are still being debated, the observable features of rifting began to attract the attention of geologists as early as the 1830s. Generally speaking, rift systems are regions where extensive faults have caused a combination of upward movement of scarps, or cliffs, and downward movement of basins crushed

between zones of uplifting. The East African system is one of the most easily recognized rift systems in the world. It is clearly connected with the formation of the narrow Red Sea separating Africa from the Arabian Peninsula, and extends southward through some of Africa's most spectacular ecological areas.

The northernmost section of the East African Rift includes the Afar Depression, which is the product of rifting in the Cenozoic era. It appears that these northern segments of the Great African Rift are younger than areas stretching thousands of miles to the south.

The Afar Depression (also called the Afar Triangle) is a V-shaped depressed block that cuts into Ethiopia for 360 miles (580 km.) southward from the Red Sea. To its west and southeast rise major fault scarps leading toward the Ethiopian and Somalian plateaus; to the north, two uplifted blocks, the Danakil Block and the Aisha Block, separate the Afar Depression from the Red Sea and the Gulf of Aden. Its lowest elevation is 509 feet (155 meters) below sea level, the lowest point in Africa. The Ethiopian mountain mass alters the dominant rift pattern until it reappears prominently farther south in modern Kenya.

The rest of the East African Rift System contains a number of subsections that specialists study separately, such as the Gregory Rift, the Albertine Rift, the Lake Tanganyika Rift, and the Lake Malawi Rift zones.

The broad lines of the Gregory Rift in Kenya include phenomenal rift scarps forming the western wall of the Kerio Valley. The highest elevation here is the Elgeyo escarpment, which rises 4,920 feet (1,500 meters) from the valley floor. The rift valley floor itself has high points and low points. At low points of about 1,640 feet (500 meters), basin-like structures gather water in lakes with high saline content, such as Lake Magadi. By contrast, Lake

Naivasha is a freshwater lake at a much higher elevation (6,200 feet/1,890 meters).

The Albertine Rift zone lies due west of Lake Victoria in modern Uganda. It contains three of the best-known rift valley lakes: Albert, Edward, and Kivu, on the western borders of present-day Rwanda. Researchers have traced at least thirty-two basin structures in this area, all the product of extensive subsurface faulting. Perhaps the most spectacular area is northwest of Lake Albert. Here the Bunia scarp, considered a still-active fault system, juts up 4,265 feet (1,300 meters). South of the lake, the giant Rwenzori Block, situated in an area of recent seismic activity, rises to an elevation of 16,795 feet (5,119 meters). By the time the rift system enters the area of Lake Edward on the boundary between the Democratic Republic of the Congo and Uganda, an entire range of recently volcanic mountains, the Virunga Range, dominates the horizon.

Although the next zone, the Tanganyika Rift, shares many of the characteristics of areas to its

Fishing in Lake Victoria. (Npsiegel)

Lake Tanganyika in the Democratic Republic of the Congo. (Abdoulma)

north, including ten separate basins and important scarps to both the west and east, there are no traces of magmatism. Volcanic activity does not reappear until the Rukwa Rift zone (a parallel segment of the Tanganyika Rift) enters the Rungwe volcanic province just north of Lake Malawi. From this point the Lake Malawi Rift zone, which includes the Livingstone Fault scarp on the northeast side of the lake, proceeds another 370 miles (600 km.) southward.

Although signs of faulting beyond the Lake Malawi Rift zone appear in the Southern Mozambique Graben (thus, nearly to the Indian Ocean coast), these belong to a different geological era, and do not form part of the East African rift system.

Lakes

Five of Africa's major natural lakes are located in the East African Rift System: Victoria, Turkana, Albert, Tanganyika, and Malawi. Lake Victoria is the only one of these lakes that contributes to the water runoff system that leads toward the Nile River. A less expansive lake, Tana, lies well to the east of the course of the Nile in Ethiopia. Tana is the most northerly of the rift-related lakes. Lake Chad, by contrast, is the result of long-distance drainage into lower elevations in the central Saharan region, and is relatively shallow from shore to shore. It is also the westernmost major lake on the African continent.

Various regions of Africa contain a number of different types of lakes. Those that have formed in depressions left by volcanic action include Kivu and Chala in Kenya. Soda lakes—those containing high concentrations of sodium bicarbonates that have dissolved out of the acidic basement-complex rocks beneath the lake bottoms—lie to the east and west of Lake Victoria. Another notable example is Lake Magadi in Kenya.

The rift valley system has the deepest lakes in Africa. Lake Tanganyika, the deepest of Africa's lakes in the Eastern Rift area, has a maximum depth of 4,820 feet (1,470 meters), more than twice the maximum depth of Lake Malawi. Lake Victoria is relatively shallow, averaging depths of only 135 feet (41 meters) and a maximum depth of 266 feet (81 meters). This is partly because more than 80 percent of the water entering Lake Victoria comes from rain falling directly on its surface. Because of its vast surface area, much of Victoria's water is lost to evaporation.

Soils of Africa

The nature and quality of surface soils in different regions of Africa vary considerably. Technical experts use four major divisions to classify soils. Two of these make up most of the surface soil of Africa: pedalfers and pedocals.

Pedalfer soils constitute the largest group of soils in Africa. Within this group, ferralsols are probably the most common, appearing mainly in low topographical relief zones and spanning the continent in the central (but not coastal) equatorial and savanna belts. Their principal constituent elements include clay, aluminum oxides, hydroxides, and, as their name suggests, high levels of iron. The process called ferrallitization is typical of tropical regions, where long periods of high temperatures exist in combination with high levels of rainfall, creating extensive leaching. Chemically, ferralsols have a low base content and require large additions of artificial fertilizers before they can sustain more than minor agriculture.

Nitisols, a second soil type in the pedalfer group, are productive tropical soils. Nitisols tend to appear in small areas both west and east from the central ferralsol regions, for example, in the Cameroon highlands and southeastern Nigeria, or in the region west of Lake Victoria. The porous composition and relatively favorable chemical content of nitisols tend to promote growth of deep-rooted plants.

The third most common pedalfer in Africa is the acrisol group. These appear throughout the West African regions. They contain high levels of clays and low content of bases. These soils can maintain agriculture only if the thin surface layer, with its buildup of organic nutrients, is carefully maintained. If the protective natural covering of minor vegetation is removed, acrisol areas are very susceptible to erosion.

Pedocals are the second most common group of soils in Africa. Among the five distinct soils in this category, calcisols and vertisols are most worthy of note. The calcisols dominate the northern third of Africa and are found in the desert zone of Namibia in southwestern Africa. Vertisols are characteristic of the Sahel region south of the Sahara, but also appear in the East African rift zone and in pockets of the eastern edges of southern Africa.

All of these soils are the product of chemical processes that, aided by relative levels of water brought by rainfall, break down constituent minerals on the surface. Thus, rainfall patterns in Africa help to explain a wide variety of organic and nonorganic features of each subregion.

Rainfall and Drainage

Levels of annual precipitation vary greatly from region to region in Africa. Two extremes can be traced, from the lowest levels of theSaharan interior or in the Namibian Desert—near zero—to high levels, near 394 inches (10,000 millimeters) around Mount Cameroon, or in Liberia in West Africa. More typical are annual rainfall levels of 20 to 60 inches (500 to 1,500 millimeters) in the savanna ecosystems immediately south of the Sahara and in large areas of South Africa.

Watershed patterns across Africa lend themselves to an extraordinary variety of tributary systems and, eventually, major rivers. One of the most widely known is the huge Congo watershed basin in the tropical zone, fed by high levels of rainfall. Another lesser-known, but physiographically important, watershed system creates the Niger River, which flows through one of Africa's driest surrounding environments. Each of the continent's other major rivers, including the Nile in the northeast, the Zambezi in the southeast, and the Orange in the southwest, is the product of different sources of drainage that are tied to land formations in the interior.

The total annual water flow from the interior of Africa to the oceans has been estimated to be 892 cubic miles (3,720 cubic km.), with 90 percent of this water flowing westward to the Atlantic, and only 10 percent into the Indian Ocean. Among Africa's major rivers, the Congo carries the largest annual runoff, amounting to 340 cubic miles (1,417 cubic km.), 38 percent of the total for the continent. The Niger and the Ogooué (in the equatorial zone of Gabon) contribute only 7.2 and 4 percent, respectively. From this point, runoff contributions of the next four major rivers drop off considerably, from 25 cubic miles (104 cubic km.) for the Zambezi, through 17.5 cubic miles (73 cubic km.) for the Nile, to 10 cubic miles (42 cubic km.) for West Africa's Volta.

Despite its extraordinary length and the equatorial location of its main watershed, the volume emp-

MAJOR WATERSHEDS OF AFRICA

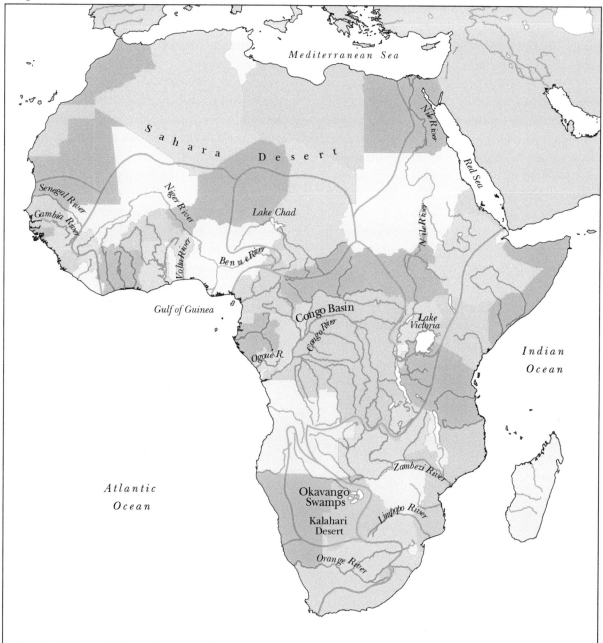

tied by the Nile is reduced by extensive loss to lateral swamps and high levels of evaporation along its course through desert areas.

Large volumes of river water moving from Africa's interior toward the sea (or in the case of Lake Chad, fed by the Chari River, into low elevation basins) can create flood conditions in certain seasons. Although the overall threat of major flooding is less notable in Africa than on other continents, the impact of high waters has consequences not only for the natural environment but also for human populations living along the rivers.

Egypt's very existence as an agricultural society, for example, has depended for millennia on the predictability of Nile flooding. In the case of the Nile and other river systems, the nature of sediments contained in runoff water has both upstream and downstream effects. Specific elements, such as calcium and magnesium, are taken from surface soils and either transferred to downstream ecosystems or deposited in the ocean. High-water seasons on these representative rivers occur in totally different periods of the year.

Mountain Systems

Africa cannot be considered a mountainous continent. Seen in broad profile, its main continental feature resembles a complex of high plateaus and broad river basins. Most mountains that emerged as a result of subsurface tectonic activity formed in the Mesozoic era. Some, especially in Southern Africa, are much older. Several subzones, however, almost all have some mountainous features, although these differ from region to region.

Northwest Africa's Atlas system (including the High Atlas, Middle Atlas, Saharan Atlas, and Rif Mountains) is the product of the folding of base-

ment rock along a rift fracture during the late Cretaceous period. This area and period of mountain formation were common to the Atlas and the European Alps on the other side of the Mediterranean trough.

The mountains of northeast Africa, and specifically those of Ethiopia, are among the most complex systems on the continent, running east-to-west in eastern Ethiopia and north-to-south on both sides of the Great Rift Valley that divides the country almost in two. The highest Ethiopian peaks—Ras Dashen, west of the rift near the Red Sea, and Mount Batu farther south, east of the rift—are more than or near 15,000 feet (4,570 meters).

The mountains of the East African interior owe their origin to volcanic activity. Unlike hot spot formations, however, great peaks like mounts Elgon, Meru, Kenya, Kilimanjaro, and Udzungwa run along the fault system associated with the Rift Val-

The Hoggar Mountains in the region of Asskrem, district of Tamanrasset, southern Algeria. (Agngeoun)

ley. By contrast, the Comoros Islands, halfway between northwestern Madagascar and the coast of East Africa, are considered to be part of the complex of mantle plumes, or hot spots, where volcanic activity remains latent. This apparently was not the case in neighboring Madagascar, where substantial mountain ranges on one side of the island reveal their origins in tectonic uplifting.

Several subzones of the Sahara are characterized by mountainous outcroppings, some very substantial, such as the Hoggar and Tibesti (whose peaks are as high as 10,000 and 11,200 feet (3,050 and 3,415 meters), and the Jebel Marra in the Darfur region of Sudan. The Hoggar and Tibesti mountain zones are presumed to be the product of mantle plumes, where pressures coming from the hot mantle of Earth come relatively close to the surface. In such regions, the likelihood of continuing volcanic activity and the possible formation of mountains is greatest.

The Cameroonian Highlands in West Africa, just above the equator and spilling over into eastern Nigeria, have only a few major peaks (notably Mount Cameroon, at 13,350 feet/4,070 meters). Like the Tibesti and Hoggar regions of the Sahara farther north, the Cameroonian Highlands region is the product of hot-spot volcanic thrusts upward from Earth's mantle. It provides a watershed source for several major rivers, the most important of which is the Benue, the last major tributary of the Niger before it flows into the Gulf of Guinea.

MOUNT KILIMANJARO

Located in northeastern Tanzania, the extinct volcano Mount Kilimanjaro is the highest mountain in Africa. Two peaks, linked by a saddle-like gap, crown its summit. Kibo, at 19,340 feet (5,895 meters), is the highest. Beside it is Mauenzi, which is 17,564 feet (5,354 meters) high. The slopes of Mount Kilimanjaro are covered with a rich volcanic soil that is ideal for cultivating two important crops, coffee and plantains. Despite its nearness to the equator, Kilimanjaro is topped with snow year-round. It takes part of its name from *kilima*, the Swahili word for mountain.

Similarly, ample rainfall in the Fouta Djallon and Nimba Mountains, which are at their highest (a little over 6,000 feet/1,830 meters) in modern Guinea, provides runoff waters that give birth to the Niger itself, as well as the westward flowing Senegal and Gambia rivers.

The southern portion of Africa contains a wide variety of mountains. Older, less dominant complexes associated with what geologists call the Cape Fold Belt and the jutting peaks of the geologically more recent Drakensberg Mountains, running more or less parallel to the Indian Ocean coast, are the result of tectonic uplift and warping at the continent's edges.

Byron D. Cannon

DISCUSSION QUESTIONS: PHYSIOGRAPHY

Q1. In what countries are the Atlas Mountains a prominent geological feature? How do these mountains differ from the terrain elsewhere in Northern Africa? What other notable mountain ranges are there in Africa?

Q2. Which five bodies of waters make up Africa's so-called great lakes? What are their names? Which of them plays a role in the hydrology of the Nile River?

Q3. What are Africa's most common soil types? Which soils are characteristic of certain areas of Africa? Why are certain soil types more subject to erosion than others?

CLIMATOLOGY

Africa's climates derive their characteristics from latitude, the elevation and orientation of the landforms, and ocean currents. No other landmass receives as much total sunshine as Africa, which is the only continent that lies mostly within the tropics. The equator divides the continent almost equally in latitude, creating an arrangement of climatic belts mainly symmetrical on either side of the equator.

Because of the greater east-to-west extent of the continent in the north, large areas in the center of the Sahara are far away from the moderating influence of an ocean. This creates a continentality effect that increases temperature differences between summer and winter and decreases precipitation. The southern part of the continent is much narrower, which allows the maritime influence to extend farther inland.

Africa lacks high mountain ranges that can act as climatic barriers. Both the Himalayas in Asia and the Andes of South America greatly modify the climates of those continents. Africa's great size (about three times that of the United States), coupled with the uniformity of its altitude—ranging generally from 1,000 to 3,000 feet (300 to 915 meters)—creates large climate regions changing gradually from one climate to the next. The few higher mountains have climatic zones according to changes of elevation.

Finally, ocean currents modify temperature and precipitation regimes in Africa. Three ocean currents affect the climates of Africa: The Canary Current in the north and Benguela Current in the south both carry cold water from the temperate zone into the Tropics; the warm Mozambique Current brings warm water from the equatorial region of the Indian Ocean toward the east coast of Africa south of the equator and the east coast of Madagascar. Cold currents chill the winds that blow over them and reduce the amount of moisture that reaches the neighboring shores; warm currents increase rainfall.

Continentality and Maritime Effects

The northern part of Africa is much wider than the area south of the equator. Large regions in the north are more than 900 miles (1,500 km.) away from the oceans, unable to benefit from their moisture and their moderating effect on temperature. Dry land surfaces heat and cool relatively quickly because solar radiation cannot penetrate the solid surface to any meaningful extent. This property is called continentality or the continental effect; it results from the fact that precipitation decreases as distance from the oceans increases, and because the oceans have minimal influence on air temperatures located well inland. Water takes more time than land to heat up and cool down. Coastal areas benefit from the moderating influence of the ocean on air temperature because sea breezes cool down the air temperatures during the summer and warm up the air temperatures during the winter. This is the maritime effect on climate.

Seasons in the Southern Hemisphere are the reverse of those in the Northern Hemisphere. In the south, summer begins about December 22, fall about March 21, winter June 21, and spring approximately September 23. In tropical climates, rainfall usually occurs when the sun is high above the horizon (June 21 to September 22 in the Northern Hemisphere; December 2 to March 20 in the Southern Hemisphere). At the equator, the sun's rays are perpendicular to Earth's surface about March 21 and September 23. These are the times of maximum precipitation. When precipitation is heavy, the ground temperature lowers.

Temperature

In a large area of high temperature, from 10° south latitude to the Tropic of Cancer, average monthly temperature does not reach below 64.4° Fahrenheit (18° Celsius) except on the East African plateaus

MEAN TEMPERATURES AND DISTANCE FROM EQUATOR

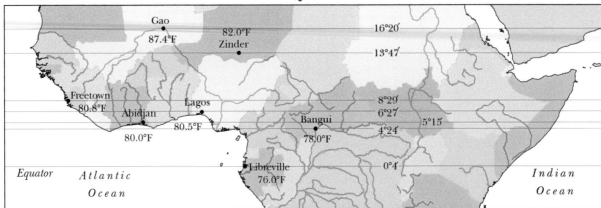

and on the Abyssinian Plateau, where elevations are much higher. In these high-elevation stations, all temperatures are lower than in stations situated in a lowland area at similar latitude. For example, the Ethiopian capital of Addis Ababa is located at 7,620 feet (2,324 meters) above sea level; its yearly average temperature is 61.6° Fahrenheit (16.4° Celsius). Malakal, South Sudan, at 1,272 feet (388 meters) above sea level, has a yearly average temperature of 80.2° Fahrenheit (26.8° Celsius). Beyond this area of high temperature, cooler temperatures are recorded during the winter, the low sun period.

The narrower southern half of the continent is cooler than the north because of the greater influence of the neighboring oceans. As one moves away from the equator seasonal temperature differences increase, but almost nowhere in Africa is the winter a truly cold season. The coldest monthly average recorded at sea level in northernmost Africa is greater than 50° Fahrenheit (10° Celsius); the coldest monthly average in southernmost Africa is never lower than 57.2° Fahrenheit (14° Celsius). In equatorial Africa, where the sky is often overcast, the diurnal (day-to-night) range, although low, can be greater than the annual (seasonal) range; in arid regions, land temperatures drop rapidly at night as a result of radiational cooling, often many degrees within a few hours.

Winds and Air Masses
Africa is divided into two wind belts: the trade winds and the westerlies. The trade winds are created by the bands of high pressure that lie centered on latitudes 30° north and south (subtropical anticyclone or subtropical highs). Because of the rotation of Earth, winds flowing from those high-pressure areas are deflected to their left in the Southern Hemisphere and to their right in the Northern Hemisphere. Therefore, the air moving toward the equator flows from east to west (the easterlies or trade winds), while the air moving toward the poles flows from west to east (the westerlies).

The trade wind zone usually is an area of sunshine and relatively little precipitation, largely because the air has descended from upper levels of the troposphere. As the air descends, the pressure on it increases and its temperature rises. This explains why continental trade winds (those flowing from the north, over the broad expanse of northern Africa) bring meager precipitation, excessive aridity, and extreme evapotranspiration on the continent.

The air masses affected by the continental trade winds are continental tropical (abbreviated cT on weather maps), characterized as dry and warm. Trade winds from the south, blowing from the subtropical anticyclone located over the Indian Ocean, carry a greater amount of humidity. The amount of water vapor they bring to the east coast of the African continent depends on how long the air has moved over water since its descent from high in the troposphere. As the subtropical anticyclone is located in the middle of the Indian Ocean, this component of the trade winds brings heavy precipitation on the east coast of Africa between

CLIMATE REGIONS OF AFRICA

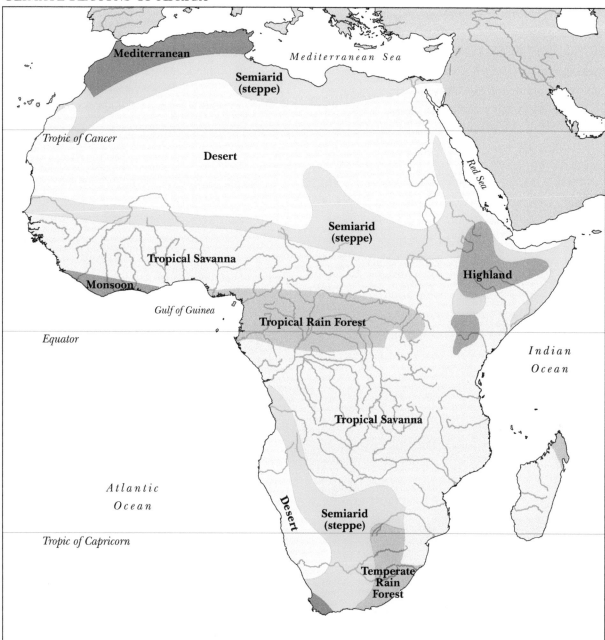

the two Tropics and the east coast of Madagascar. The total precipitation they deliver decreases westward.

Trade winds from subtropical oceanic highs of opposite hemispheres tend to converge along a line or a zone known as the intertropical convergence zone (ITCZ). Because subtropical anticyclones migrate north and south seasonally, this zone is located near the equator in March and September, during the equinoxes.

Ship captains in the trade wind zone during the days of sailing vessels knew these winds well, often detouring far from their course to benefit from those winds' steadiness. The trade winds were the winds that first blew Christopher Columbus and his flotilla to the Caribbean in 1492.

The ITCZ, also called the equatorial low (a low barometric pressure, generating convection and rainfall), accompanies the movement of the high sun from one hemisphere to the other. This migration of the equatorial low is more pronounced in the Northern Hemisphere. It reaches 25° north latitude in July during the Northern Hemisphere summer. The displacement of the equatorial low into the Southern Hemisphere in January is much less important because of its more maritime character. The migration of the intertropical convergence zone across the equator creates a localized monsoon phenomenon. The monsoon is a seasonal wind that blows from the ocean during the summer and affects West Africa particularly. Laden with moisture, these winds generate intense rainfall. In Conakry, the capital city of Guinea, precipitation in July is usually 44.5 inches (1,130 millimeters). This monthly average is greater than the average yearly precipitation in New York City of 42.2 inches (1,072 millimeters).

The westerlies flow out of the subtropical high but toward the cooler temperate zone. Westerlies dominate a zone of about 30° of latitude in width in both hemispheres. They bring moist air from the west of the Mediterranean region during the local winter. The total precipitation they deliver decreases eastward. At times continental polar air masses, labeled cP on weather maps, reach the Mediterranean region and bring bitterly cold dry air. These occasional cold spells damage orange groves, orchards, and flower-producing farms. Maritime polar air masses (cold and moist) are associated with the winter precipitation that occurs over the Mediterranean climate belt.

Precipitation

Generally, precipitation depends on barometric pressure, air masses, and the wind that flows over the continent and moves these air masses. Near the equator, the great number of sunny days, the presence of a low barometric pressure, and active convection—when the air, warmed at the contact of Earth's surface, rises—create the tropical wet climate. In this region, precipitation may reach 400 inches (1,000 millimeters) per year at the foot of the windward side of Mount Cameroon (13,350 feet/4,070 meters). Monrovia, the capital city of Liberia, receives more than 200 inches (500 millimeters) of rain. This represents about five times the yearly precipitation in New York or Philadelphia.

Precipitation in the center of Africa, on the equator, is greater than 80 inches (2,000 millimeters) and occurs all year. That kind of precipitation characterizes the tropical wet climate. There is no dry month in the tropical wet climate because areas with this climate are constantly covered by warm maritime air masses laden with moisture. Rainfall is heaviest in March and in September, when the sun is perpendicular to Earth's surface and solar energy is concentrated.

In general, as one moves away from the equator, the amount of precipitation decreases toward the tropics of Cancer and Capricorn, then increases into the Mediterranean belt. The area between the tropical wet climate and the tropics is bathed in wet, warm maritime tropical air masses (mT on weather maps) during the local summer, and in dry and warm continental tropical air masses (cT) during the local winter.

In addition to a decrease in yearly precipitation, a seasonal pattern of precipitation appears. A rainy season occurs during the local summer (June to September in the Northern Hemisphere, December to March in the Southern Hemisphere). A dry season exists during the local winter. This pattern characterizes the savanna climate, also named the tropical wet-and-dry climate. The yearly rainfall is proportional to the length of the rainy season.

Out of the 5.8 million square miles (15 million sq. km.) of tropical-climate area in Africa, about 60 percent of it receives more than 15 inches (400 millimeters) of rainfall. This is approximately the amount of precipitation that Salt Lake City, Utah, receives yearly. Since farming is possible without irrigation when rainfall exceeds 15 inches yearly, 53 percent of the tropical regions of Africa can be farmed without tapping the groundwater supply.

Across the Tropic of Cancer and the Tropic of Capricorn are subtropical deserts (Sahara in the north and Kalahari in the south) generated by a band of high pressure that prevents rainfall. A belt of drought and moisture called the tropical savanna climate surrounds these deserts. The local summer is rainy, and the local winter is dry. The narrow band of transitional precipitation regime between the savanna and the subtropical desert is called the semiarid steppe or Sahel.

In coastal areas, the summer rainy season of the savanna is sometimes exacerbated, creating the monsoon climate. Farther away from the equator, both to the north and to the south, lies the Mediterranean climate, influenced by westerly winds that bring moisture during the local winter. Only the northernmost and southernmost tips of Africa are beyond the tropical climates' regions. The Mediterranean climate of these two extremities of Africa is characterized by precipitation during the winter and drought during the summer.

Throughout Africa, coasts backed by mountainous highlands receive the highest rainfall. In the few areas where mountains exist, rainfall is higher on their windward side, and temperature decreases as elevation increases. This generates what is called a mountain climate.

Regional Climates

Thus, Africa has seven basic categories of climate: the tropical rainforest (or tropical wet), the tropical savanna (or tropical wet and dry), the monsoon climate, the Sahel climate (semiarid), the subtropical deserts (arid), the Mediterranean climate, and the mountain climate. Each climate has a significant impact on the life that exists there.

Tropical Rainforest

The tropical rainforest (tropical wet) climate, located along the equator, is warm and rainy all year. The central part of Africa, located across the equator, has little variability in temperature; the range in temperatures increases away from the equator. Night is the "winter" of tropical Africa. Diurnal ranges of temperature are greater than the yearly temperature range. Temperatures are warm all year, unless cold currents or elevations bring cooling effects.

The natural vegetation associated with the tropical wet region is the tropical rainforest—a dense, lush forest where gigantic trees can reach heights of 130 to 200 feet (40 to 60 meters), tall ferns can grow to 20 feet (6 meters), and orchids bloom in the upper reaches of the tall trees. As many as forty different species can be found in 2.5 acres (1 hectare). About 1 percent of the light that reaches the highest

trees gets to the ground. Many insects, primates, and reptiles inhabit this climatic region.

Because of deforestation, the tropical rainforest in Africa covers much smaller areas than it . True virgin rainforest, not yet lost to logging, can still be found in the Democratic Republic of the Congo (formerly Zaire). Secondary rainforest, resulting from the regrowth of a forest after clearing the primary or virgin forest, exists along part of the coastal area that surrounds the Gulf of Guinea: Sierra Leone, Liberia, portions of Nigeria, Cameroon, and both Congos. Rivers that flow exclusively in the tropical wet climate are few and short. The longest rivers cross more than one climatic belt and have a complex structure, as does the Congo River or the Ubangi.

Tropical Wet and Dry or Savanna

A transitional belt of drought and moisture, called the tropical savanna, exists between the tropical rainforest climate and the Sahel climate. The rainy season occurs during the local summer, and the dry season during the local winter. Forests do not tolerate the dry season well. In general, the total precipitation of a region impacts the size of trees and the density of natural vegetation. This is why, as the total precipitation decreases away from the equator, the rainforest becomes shorter and less dense, and grassland extends.

The rainforest remains only along permanent rivers, on windward sides of mountains, and in the monsoon climate. It otherwise becomes a new vegetation formation called savanna, comprising grassland with scattered trees. The higher the precipitation, the higher the density of trees in the savanna. This region is home to large herbivores such as the elephant, gazelle, antelope, and giraffe, and carnivores such as the lion, lynx, and cheetah. Rivers originating in this region flow most during the period of high sun and least during the period of low sun.

Monsoon

A variety of tropical climate called the monsoon is found in coastal regions of West Africa (Sierra Leone and Liberia) and Central Africa (portions of Nigeria and Cameroon, Equatorial Guinea, and areas of Gabon and Congo). It has a distinct but short dry

Zebras (Equus quagga bohemi), *Serengeti savanna plains, Tanzania Place: Ngorongoro Crater, Tanzania, Africa. (Gary)*

season during the local winter, when the sun is lower above the horizon. During the summer, winds filled with moisture from the warm oceans over which they have traveled bring heavy precipitation.

Douala, the largest port of Cameroon, receives 150 inches (3,810 millimeters) of rainfall a year, about four times the precipitation of Washington, D.C. Although rainforest grows in these very wet areas, it exhibits fewer plant species per unit area. Indeed, the forest is lower and less dense, and additional light reaches the ground, allowing more light-loving trees to grow.

The Sahel

This region is a narrow zone of semiarid climate where yearly mean rainfall is between 10 and 25 inches (250 and 635 millimeters). The 600-millimeter isohyet (the line connecting locations where the rainfall is 600 millimeters) runs eastward from Dakar. The northern part of Senegal, Mali, Burkina Faso, the southern part of Niger, Chad, and South Sudan, and the western part of Ethiopia are part of this climatic region. The natural vegetation is grassland, also called a steppe. Acacias, umbrella-like trees, grow intermittently.

Rainfall in the Sahel is highly variable from one year to the next. Farmers frequently have attempted to settle and farm areas where rainfall is below the 15 inches (400 millimeters) of rainfall needed for permanent, non-irrigated agriculture or for cattle ranching, but have suffered severe setbacks. Episodes of drought and famine have become much more common in recent years. As recently as 2010, an extreme drought induced a widespread famine in the Sahel.

Subtropical Deserts

North Africa covers about 7.3 million square miles (19 million sq. km.) north of 4° north latitude. The region contains the largest, driest, and hottest desert in the world—the Sahara, where total precipitation is less than 4 inches (100 millimeters) per year.

About 42 percent of this area receives even less than 4 inches of precipitation per year. In the Southern Hemisphere, the Kalahari Desert in Botswana, while not as dry or as hot as the Sahara, exists for the same reason the Sahara does. The subtropical high causes the air to descend from high in the troposphere. As the air descends, it warms up and dries out. Many of the plants in these deserts are ephemerals, germinating and completing their life cycle quickly after a rain. Perennial plants, such as cactus, remain dormant during much of the year. As soon as it rains, they begin to grow and store water in their stems.

The Namib Desert, in Namibia, is a coastal desert generated by the presence of the cold Benguela Current and by its location in the vicinity of the Tropic of Capricorn, where a high pressure is located during the local winter (June to September). North of the Sahara and south of the Namib and Kalahari deserts, there is slightly more precipitation—enough to change these areas into a semiarid steppe, similar to the Sahel in the north.

Mediterranean Climate

A Mediterranean climate prevails in the northwest and southwest of the continent, north of the Sahara and south of the Namib and Kalahari deserts. The Mediterranean climate is the only one in the world characterized by precipitation during the local winter (December 22 to March 20 in the Northern Hemisphere, June 21 to September 22 in the Southern Hemisphere), the season when the sun is low above the horizon. This characteristic is explained by the movement of the ITCZ toward the Southern Hemisphere in December to accompany the high sun period. This migration allows the temperate zone low pressure to influence the Mediterranean region in the north and the southwest corner of Southern Africa. Westerlies take over and deliver maritime polar air masses (mP) to these regions during the local winter.

Average monthly temperatures seldom fall below 50° Fahrenheit (10° Celsius). The lowest temperature in Cape Town, South Africa, located at 38 58'

south latitude, is 53° Fahrenheit (11.7° Celsius) in July. Algiers' coldest month is January, with 51° Fahrenheit (10.6° Celsius). Farmers have adapted to this climate by planting mostly wheat, olive and orange groves, and vineyards. The natural vegetation is the Mediterranean scrub, which consists of widely spaced, hard-leaf evergreen shrub and deciduous trees such as pine and oak. In South Africa, this scrubby vegetation is called *fynbos*; in the Maghreb, it is referred to as maquis.

Mountain Climates

The few high mountains in Africa include: the mountains of Cameroon, the Ethiopian Highlands in the Northern Hemisphere, and the Kilimanjaro and Kenya peaks in East Africa. Windward sides of mountains receive higher precipitation. Temperatures decrease with an increase in elevation; this affects the savanna at the foot of these volcanic peaks. With the increased relative humidity at higher elevation, the savanna turns into open forests, then canopied evergreen forests, and finally to mountain meadows.

Madagascar and Mozambique

The island of Madagascar and the mainland country of Mozambique lie within the belt of the southeast trade winds. Because of the location of the island, hurricanes (called cyclones in this area) frequently strike the populated shores during the Southern Hemisphere in late summer and autumn months. In February 1994, Cyclone Geralda slammed into Madagascar. Geralda—the strongest cyclone to hit the island since 1927—destroyed 95 percent of the harbor of Toamasina.

Denyse Lemaire

Allée des Baobabs near Morondava, Madagascar. (Frank Vassen)

DISCUSSION QUESTIONS: CLIMATOLOGY

Q1. What are the four main variables that influence Africa's climates? What role does Africa's location mostly within the tropics have on the continent's overall climate trend? How does Africa's sunshine quotient compare to those of other continents?

Q2. What are some of the factors that serve to moderate Africa's climate? In what ways do ocean currents influence the climate? What role does the shape of Africa play in continent's climate zones?

Q3. Why are the seasons reversed in Southern Africa compared to Northern Africa? Where does the equator pass through the continent? In what months does Africa receive its maximum precipitation?

HYDROLOGY

Africa has several dominant physical features, including the Great Rift Valley of East Africa and Victoria Falls in Southern Africa, one of the largest waterfalls in the world. The Sahara Desert—the world's largest, stretching from the Atlantic Ocean in the west to the Red Sea in the east—has a great impact on the continent's hydrology.

A third of Africa is desert: The Sahara covers 3.6 million square miles (9.3 million sq. km.), the Kalahari about 360,000 square miles (932,000 sq.

km.), and the Namib 31,000 square miles (81,300 sq. km.). About the size of the United States, the Sahara occupies much of North Africa; the Kalahari and Namib are in Southern Africa.

The highest mountain in Africa, Kilimanjaro in Tanzania, at 19,340 feet (5,895 meters), along with Mount Kenya (17,058 feet/5,199 meters) and the Atlas Mountains (13,665 feet/4,165 meters) in northwest Africa, also influence the continent's hydrology. These mountains either act as obstacles to

The source of the Nile from an underwater spring at the neck of Lake Victoria, Jinja. (Rukn950)

INCREASING DESERTIFICATION OF AFRICA

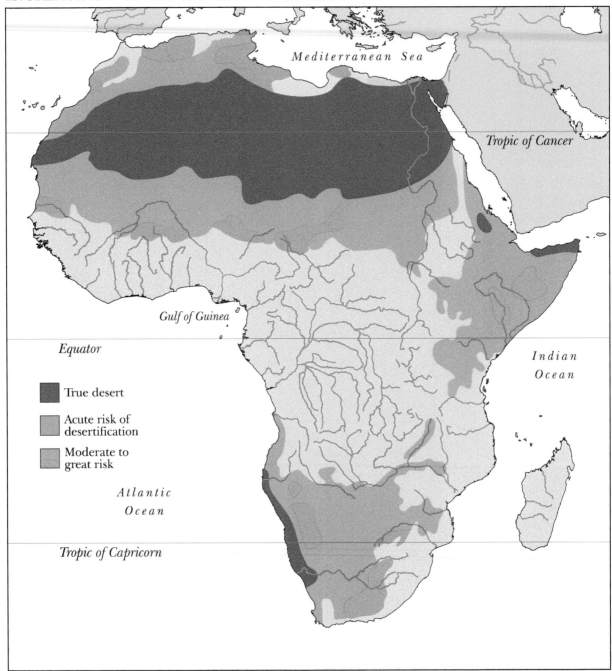

winds carrying moisture from the sea or ocean, causing a rain-shadow effect, or serve as drainage boundaries for rivers or lakes.

With the world's largest desert, the world's longest river, and the lake with the world's largest drainage basin, Africa is hydrologically an enigma. The continent has some of the wettest and driest places in the world. These extremes cause severe drought and flood problems in several parts of Africa.

Despite Africa's size, its people comprised only about 17 percent of the world's population in 2016. However, the continent has one of the fastest-growing populations in the world. A large percentage of its people do not have access to safe drinking water. The availability of water varies from country to country. The devastation brought by droughts and famines in Africa can be truly enormous.

Human Settlement

The largest concentrations of people on the continent live in West Africa, the Great Lakes region of East Africa, the southeast of South Africa, the northwest of North Africa around the Mediterranean Sea, and along the Nile River. The distribution of people in Africa reflects the distribution of surface water, as in other parts of the world. More than 70 percent of Africa's people live within the equatorial and tropical zones, and in the Nile Valley. The population is sparsest in the Sahara and Kalahari deserts.

In 2019, more than 95 percent of Egypt's people lived within a few miles of the Nile River, making it one of the most populous areas in Africa. The Nile River is an example of peoples' dependence on water for survival. From ancient through modern times, the Nile has played a principal role in the life of Egypt.

Sources of Water

Factors that determine the availability of surface water include precipitation, soil type, the rock type beneath the surface, solar radiation, and geologic structure. Rivers, lakes, and rainwater are the major sources of surface water.

The annual precipitation in Africa follows the lines of latitude. Places within 12° north or south of the equator have an annual precipitation of more than 20 inches (500 millimeters). Annual precipitation of more than 79 inches (2,000 millimeters) is typical of Madagascar and the equatorial and tropical regions. Most places above 14° north latitude are either arid or semiarid, with annual precipitation of less than 10 inches (250 millimeters).

Parts of the East African Great Lakes region and southwestern Ethiopia have tropical climates. Rainfall occurs throughout these regions from April to October, and temperatures vary from 60 to 80° Fahrenheit (15.6 to 26.7° Celsius), with relative humidity around 80 percent. The southernmost part of Sudan has a semitropical climate, with 50 inches (1,270 millimeters) of rainfall during the rainy season.

As one moves northward, the rainfall decreases. For example, in the central region of Sudan, rainfall varies from 10 to 21 inches (254 to 533 millimeters) annually. North of Khartoum, the annual rainfall is 5 inches (127 millimeters). This area includes northern Sudan and the Egyptian desert. Precipitation is the source of water in rivers and lakes, as well as of groundwater. More than 90 percent of Africa's surface water is concentrated around the equatorial and tropical regions, except the Nile River, which has a north-south trend.

Rain in a given area affects river flows and water levels in lakes. The lack of adequate precipitation can lead to drought, as is constantly occurring in the Sahel region (the southern fringe of the Sahara). Conversely, too much rainfall can cause severe flooding, as in Mozambique and Madagascar during tropical cyclone season. When this happens, floodwaters can leave hundreds of thousands of people homeless and without adequate safe drinking water.

Rivers

Sometimes called watersheds, drainage basins are areas that supply runoff water to river systems and lakes. For hydrological studies, Africa can be divided into five major drainage basins: the Nile River, Niger River, Congo River, Zambezi River, and Lake Chad. These basins cover about 75 percent of the surface area of the continent.

Most of Africa's rivers rise in the highlands and flow to the seas. The Nile flows northward from the plateaus of east central Africa to the Mediterranean. The Congo and Niger rivers, Africa's second- and third-longest, respectively, drain the waters of west central Africa into the Atlantic Ocean. Most other important rivers in Africa also empty into the Atlantic, except the Limpopo and the Zambezi, which empty into the Indian Ocean.

Often called the Mother of Egypt, the Nile River represents Egypt's past civilizations, its present base, and its future. The Nile gave birth to the ancient civilization that captures the public imagination to the present day.

The Nile is the longest river in the world, with a length of 4,132 miles (6,648 km.); in comparison, the Mississippi is 3,710 miles (5,970 km.) long. The source region of the Nile is in the equatorial lakes, and it empties into the Mediterranean Sea. The Nile's drainage basin consists of approximately 1.2 million square miles (3.1 million sq. km.), and includes parts of Burundi, Egypt, Eritrea, Ethiopia, Kenya, Rwanda, Sudan, South Sudan, Tanzania, Uganda, and the Democratic Republic of the Congo. The annual water discharge of the Nile is about 100,000 cubic feet (2,800 cubic meters) per second at Aswan Dam, with the Ethiopian sub-basin supplying about 86 percent of the total water flow.

The Nile is one of the few exotic rivers in the world. An exotic river is a perennial river that flows through arid or semiarid regions, similar to the Colorado River in the United States. Two sources of water exist for the Nile: Lake Victoria in East Africa is the source water for the White Nile, and Lake Tana in Ethiopia is the source water for the Blue Nile.

The Congo River is the fourth-longest river (2,900 miles/4,670 km.) and has the second-largest drainage basin and discharge volume in the world. Rainfall affects the flows and water levels in rivers. However, where the drainage basin for a river system is large and within the same climatic setting, the weather pattern in an area of the basin may not significantly affect the water level in lakes. This is evident for the Congo River, whose entire drainage basin lies within the equatorial region, where a shortfall in precipitation in one part of the basin has little effect on the flow of the river close to its discharge point.

Victoria Falls during rainy season Africa. (Ludovico Mazzocchi)

VICTORIA FALLS

Discovered by the Scottish missionary-explorer David Livingstone in 1855, Victoria Falls on the Zambezi River marks part of the modern border between modern Zambia and Zimbabwe. These spectacular falls are 5,500 feet (1,676 meters) across, which is twice the width of Niagara Falls. They fall about 355 feet (108 meters), not into a basin, but into a rocky chasm, with sheer cliffs on both sides.

The Congo River system comprises three distinct sections—the upper, middle, and lower Congo. The upper Congo contains confluences, lakes, waterfalls, and rapids. The Congo River has its beginnings in the Democratic Republic of the Congo, where several small rivers unite. Various tributaries of the Congo River include the Lualaba, Luvua, Ubangi, Sangha, and Kasai rivers. The width of the Congo River varies from about 4 miles (6.4 km.) up to 9 miles (14.5 km.) at the mouth of the Mongala River. The Congo discharges its water into the Atlantic Ocean.

The Niger River in West Africa is approximately 2,600 miles (4,183 km.) long with a drainage basin of 0.92 million square miles (2.4 million sq. km.). Its source is the mountainous Fouta Djallon region in Guinea. The river flows northeast toward the Sahara with a middle delta (south of Timbuktu in the country of Mali) where a significant amount of water is lost to evaporation. It then flows southwest before flowing south, emptying into the Atlantic Ocean through the Niger Delta. The Niger River traverses Guinea, Mali, Niger, and Nigeria. The Benue River, a major tributary of the Niger, flows from Mount Cameroon to the west, joining the Niger at Lokoja, Nigeria.

Precipitation ranges from 160 inches (4,100 millimeters) per year close to the river mouth to almost nothing in the desert. Rainfall occurs during the annual monsoon period between July and September, which results in an annual flood that takes several months to make its way down through the river network. Because of climatic variations, the annual river flood occurs at different times in different parts of the basin. In the upper Niger, the high-water discharge occurs in June, and the low-water season is in December.

As the Nile is to Egypt, so the Zambezi River is to Southern Africa. The Zambezi is the eighth-largest river in the world in terms of discharge. It runs through Angola, Namibia, Zambia, Zimbabwe, and Mozambique. Spectacular features associated with this river include Victoria Falls, Lake Kariba, and Mana Pools National Park. Concerns over the quality and quantity of the Zambezi River are growing due to tourism, irrigation, and damming for hydropower.

The Chari River in Central Africa is the major source of water for Lake Chad. The Chari and its major tributary, the Logone River, supply approximately 85 percent of the water input to Lake Chad. The Chari/Logone River system is unusual in that it takes the water approximately six months to get from its source in the Central African Republic to Lake Chad. Because the water arrives during the lake's dry season, Lake Chad's volume is highest during its dry season.

Other rivers in Africa include the Orange, Limpopo, Gambia, Senegal, Volta, Lagone, Ogooué, Sanaga, Cuanza, Jubba, Lugenda, Ruvuma, and Shebelle. Places north of 20° north latitude lack perennial rivers, except for the Nile and small coastal rivers along the Atlantic Ocean. Climate influences the presence of rivers, as most of Africa's rivers are found in the equatorial and tropical regions. Rivers and lakes have a great influence on human settlements, as in the Nile Valley; on agriculture, as in the South Lake Chad Basin; and on industrialization, as in the Zambezi River and the Great Lakes of East Africa.

Lakes

Lakes play a large part in Africa's hydrology. The large surface areas of some of the lakes provide water by evaporation, and they also act as reservoirs that can be used by people, plants, animals, and fish. Large African lakes are found south of the Sahara Desert, most of them in east central Africa (the Great Rift Valley System).

Lake Chad is located south of the Sahara, in a region of Africa ranging from semiarid to arid. It has the largest drainage basin (936,821 square miles/2.4 million sq. km.) of any lake in the world and was once the sixth-largest lake on Earth. The size of the lake has changed over time. In ancient times, it covered about 135,000 square miles (350,000 sq. km.), shrinking to 9,600 square miles (25,000 sq. km.) in the 1960s (about the size of Lake Erie), and to 772 square miles (2,000 sq. km.) in 2000 to 590 square miles (1,530 sq. km.) in 2020. It is shallow (average depth about 5 feet/1.5 m), and the lakebed was dry at least once during the past millennium. The sand dunes that formed in the lakebed then are still visible in the lake. Satellite images have shown the migrating shoreline nature of the lake due to changes in the climate.

Lake Chad is the main source of water in the region, both as surface water and groundwater. It is a closed-basin lake; that is, it has no direct outlet to the sea or other lakes. Unlike most old closed-basin lakes, such as the Great Salt Lake of Utah, its water is fresh, for three main reasons: the Chari/Logone rivers embedded few dissolved solids (salts) into the lake, some dissolved solids precipitate or are absorbed by plants, and about 20 percent of the lake water seeps into the ground, carrying with it the dissolved solids.

Local residents use Lake Chad for transportation, fishing, hunting, salt mining, farming, and drinking. Lake Chad also serves as a political boundary for the four nations that share it (Cameroon, Chad, Niger, and Nigeria). When water levels are low, large portions of the lakebed become available for farming. The practice of farming the lakebed leads to the displacement of people after heavy rainfall. Human activities such as damming and petroleum exploration may pose serious threats to the lake, its ecosystem, and the underlying aquifers.

Lake Victoria is the largest lake in Africa and the second-largest freshwater lake in the world (after North America's Lake Superior). One of the two sources of the Nile River, it is located in Tanzania, Uganda, and Kenya, between the Eastern Rift Valley System and the Indian Ocean. Lake Victoria covers approximately 23,146 square miles (59,947 sq. km.) and is about 223 miles (359 km.) long and 209 miles (337 km.) wide, with more than 4,438 miles (7,141 km.) of coastlines. It is 3,720 feet (1,134 meters) above sea level and reaches a depth of 270 feet (82 meters).

Because of the abundant water supply and fish for food, several million people live within 50 miles (80 km.) of Lake Victoria, making it one of the most densely populated areas in Africa. The cities of Kampala, Entebbe, Kisumu, Jinja, Mwanza, Bukoba, and Shirati are built on or close to its coast. The lake is used for drinking water, transportation, fishing, tourism, and other industries.

Groundwater

Potable water supply is found not only in the rivers and lakes in Africa, but also from groundwater. In some parts of Africa, groundwater is the only source of potable water. The use of groundwater is limited in some areas in Africa, because the groundwater is at a depth of approximately 328 feet (100 meters). Digging for water can be a labor-intensive activity, and without the right drilling equipment, groundwater may not be available. Some groundwater contains a high salt content, making it unfit for use; in urban areas, groundwater can become contaminated by sewage or refuse dumps. Some areas lack a scientific understanding of hydrology and hydrogeology.

Groundwater has become more important as surface water has become polluted with chemicals and disease-causing microbes. Satellite images and various other technologies are employed to locate groundwater resources. When rocks have breaks or cracks in them (referred to as fractures), they pro-

vide pathways for rain and other surface water to infiltrate into groundwater. These fracture zones and drainage patterns can help people find where to drill for groundwater to create a well, called an aquifer (meaning "water bearing").

Three major aquifers exist in the southwest part of the Lake Chad Basin. The first and uppermost aquifer is shallow (less than 49 feet/15 meters). This aquifer generally can be used without drill equipment. The other two aquifers are deep (984 feet/ 300 meters) and require the use of drill equipment. Water from the deep aquifers is hot (122° Fahrenheit/50° Celsius) and flows to the surface without pumping (artesian wells). The waters are left to flow freely and allowed to cool before being used by humans or animals. A large portion of this water is lost through evaporation. When water is extracted from the ground faster than it is replaced or recharged, groundwater depletion occurs. Experts disagree on how well the groundwater in sub-Saharan Africa replenishes itself.

Oases are also important to people in arid regions of Africa. Oases are groundwaters exposed because of wind erosion in the desert. Groundwater is abundant in most of the tropical and equatorial regions of Africa. However, groundwater's contribution to the total water supply is small in some parts of Africa. In South Africa, for example, groundwater accounts for only 13 percent of the water supply. However, two-thirds of the South African population depends on groundwater, compared to 50 percent of the people in the United States. Groundwater resources need to be properly utilized, and care should be taken to protect this valuable resource as well as surface water.

Solomon A. Isiorho

BIOGEOGRAPHY AND NATURAL RESOURCES

OVERVIEW

MINERALS

Mineral resources make up all the nonliving matter found in the earth, its atmosphere, and its waters that are useful to humankind. The great ages of history are classified by the resources that were exploited. First came the Stone Age, when flint was used to make tools and weapons. The Bronze Age followed; it was a time when metals such as copper and tin began to be extracted and used. Finally came the Iron Age, the time of steel and other ferrous alloys that required higher temperatures and more sophisticated metallurgy.

Metals, however, are not the whole story—economic progress also requires fossil fuels such as coal, oil, natural gas, tar sands, or oil shale as energy sources. Beyond metals and fuels, there are a host of mineral resources that make modern life possible: building stone, salt, atmospheric gases (oxygen, nitrogen), fertilizer minerals (phosphates, nitrates, and potash), sulfur, quartz, clay, asbestos, and diamonds are some examples.

Mining and Prospecting

Exploitation of mineral resources begins with the discovery and recognition of the value of the deposits. To be economically viable, the mineral must be salable at a price greater than the cost of its extraction, and great care is taken to determine the probable size of a deposit and the labor involved in isolating it before operations begin. Iron, aluminum, copper, lead, and zinc occur as mineral ores that are mined, then subjected to chemical processes to separate the metal from the other elements (usually oxygen or sulfur) that are bonded to the metal in the ore.

Some deposits of gold or platinum are found in elemental (native) form as nuggets or powder and may be isolated by alluvial mining—using running water to wash away low-density impurities, leaving the dense metal behind. Most metal ores, however, are obtained only after extensive digging and blasting and the use of large-scale earthmoving equipment. Surface mining or strip mining is far simpler and safer than underground mining.

Safety and Environmental Considerations

Underground mines can extend as far as a mile into the earth and are subject to cave-ins, water leakage, and dangerous gases that can explode or suffocate miners. Safety is an overriding issue in deep mines, and there is legislation in many countries designed to regulate mine safety and to enforce practices that reduce hazards to the miners from breathing dust or gases.

In the past, mining often was conducted without regard to the effects on the environment. In economically advanced countries such as the United States, this is now seen as unacceptable. Mines are expected to be filled in, not just abandoned after they are worked out, and care must be taken that rivers and streams are not contaminated with mine wastes.

Iron, Steel, and Coal

Iron ore and coal are essential for the manufacture of steel, the most important structural metal. Both raw materials occur in many geographic regions. Before the mid-nineteenth century, iron was smelted in the eastern United States—New Jersey, New York, and Massachusetts—but then huge hematite deposits were discovered near Duluth, Minnesota, on Lake Superior. The ore traveled by ship to steel mills in northwest Indiana and northeast Illinois, and coal came from Illinois or Ohio. Steel also was made in Pittsburgh and Bethlehem in Pennsylvania, and in Birmingham, Alabama.

After World War II, the U.S. steel industry was slow to modernize its facilities, and after 1970 it had great difficulty producing steel at a price that could compete with imports from countries such as Japan, Korea, and Brazil. In Europe, the German steel industry centered in the Ruhr River valley in cities such as Essen and Düsseldorf. In Russia, iron ore is mined in the Urals, in the Crimea, and at Krivoi Rog in Ukraine. Elsewhere in Europe, the French "minette" ores of Alsace-Lorraine, the Swedish magnetite deposits near Kiruna, and the British hematite deposits in Lancashire are all significant. Hematite is also found in Labrador, Canada, near the Quebec border.

Coal is widely distributed on earth. In the United States, Kentucky, West Virginia, and Pennsylvania are known for their coal mines, but coal is also found in Illinois, Indiana, Ohio, Montana, and other states. Much of the anthracite (hard coal) is taken from underground mines, where networks of tunnels are dug through the coal seam, and the coal is loosened by blasting, use of digging machines, or human labor. A huge deposit of brown coal is mined at the Yallourn open pit mine west of Melbourne, Australia. In Germany, the mines are near Garsdorf in Nord-Rhein/Westfalen, and in the United Kingdom, coal is mined in Wales. South Africa has coal and is a leader in manufacture of liquid fuels from coal. There is coal in Antarctica, but it cannot yet be mined profitably. China and Japan both have coal mines, as does Russia.

Aluminum

Aluminum is the most important structural metal after iron. It is extremely abundant in the earth's crust, but the only readily extractable ore is bauxite, a hydrated oxide usually contaminated with iron and silica. Bauxite was originally found in France but also exists in many other places in Europe, as well as in Australia, India, China, the former Soviet Union, Indonesia, Malaysia, Suriname, and Jamaica.

Much of the bauxite in the United States comes from Arkansas. After purification, the bauxite is combined with the mineral cryolite at high temperature and subjected to electrolysis between carbon electrodes (the Hall-Héroult process), yielding pure aluminum. Because of the enormous electrical energy requirements of the Hall-Héroult method, aluminum can be made economically only where cheap power (preferably hydroelectric) is available. This means that the bauxite often must be shipped long distances—Jamaican bauxite comes to the United States for electrolysis, for example.

Copper, Silver, and Gold

These coinage metals have been known and used since antiquity. Copper came from Cyprus and takes its name from the name of the island. Copper ores include oxides or sulfides (cuprite, bornite, covellite, and others). Not enough native copper occurs to be commercially significant. Mines in Bingham, Utah, and Ely, Nevada, are major sources in the United States. The El Teniente mine in Chile is the world's largest underground copper mine, and major amounts of copper also come from Canada, the former Soviet Union, and the Katanga region mines in Congo-Kinshasa and Zambia.

Silver often occurs native, as well as in combination with other metals, including lead, copper, and gold. Famous silver mines in the United States include those near Virginia City (the Comstock lode) and Tonopah, Nevada, and Coeur d'Alene, Idaho. Silver has been mined in the past in Bolivia (Potosi mines), Peru (Cerro de Pasco mines), Mexico, and Ontario and British Columbia in Canada.

Gold occurs native as gold dust or nuggets, sometimes with silver as a natural alloy called electrum. Other gold minerals include selenides and tellurides. Small amounts of gold are present in sea water, but attempts to isolate gold economically from this source have so far failed. Famous gold rushes occurred in California and Colorado in the United States, Canada's Yukon, and Alaska's Klondike region. Major gold-producing countries include South Africa, Siberia, Ghana (once called the Gold Coast), the Philippines, Australia, and Canada.

THE EXXON VALDEZ OIL SPILL

On March 24, 1989, the tanker Exxon Valdez, with a cargo of 53 million gallons of crude oil, ran aground on Bligh Reef in Prince William Sound, Alaska. Approximately 11 million gallons of oil were released into the water, in the worst environmental disaster of this type recorded to date. Despite immediate and lengthy efforts to contain and clean up the spill, there was extensive damage to wildlife, including aquatic birds, seals, and fish. Lawsuits and calls for new regulatory legislation on tankers continued a decade later. Such regrettable incidents as these are the almost inevitable result of attempting to transport the huge oil supplies demanded in the industrialized world.

Petroleum and Natural Gas

Petroleum has been found on every continent except Antarctica, with 600,000 producing wells in 100 different countries. In the United States, petroleum was originally discovered in Pennsylvania, with more important discoveries being made later in west Texas, Oklahoma, California, and Alaska. New wells are often drilled offshore, for example in the Gulf of Mexico or the North Sea. The United States depends heavily on oil imported from Mexico, South America, Saudi Arabia and the Persian Gulf states, and Canada.

Over the years, the price of oil has varied dramatically, particularly due to the attempts of the Organization of Petroleum Exporting Countries (OPEC) to limit production and drive up prices. In Europe, oil is produced in Azerbaijan near the Caspian Sea, where a pipeline is planned to carry the crude to the Mediterranean port of Ceyhan, in Turkey. In Africa, there are oil wells in Gabon, Libya, and Nigeria; in the Persian Gulf region, oil is found in Kuwait, Qatar, Iran, and Iraq. Much crude oil travels in huge tankers to Europe, Japan, and the United States, but some supplies refineries in Saudi Arabia at Abadan. Tankers must exit the Persian Gulf through the narrow Gulf of Hormuz, which thus assumes great strategic importance.

After oil was discovered on the shores of the Beaufort Sea in northern Alaska (the so-called North Slope) in the 1960s, a pipeline was built across Alaska, ending at the port of Valdez. The pipeline is heated to keep the oil liquid in cold weather and elevated to prevent its melting through the permanently frozen ground (permafrost) that supports it. From Valdez, tankers reach Japan or California.

Drilling activities occasionally result in discovery of natural gas, which is valued as a low-pollution fuel. Vast fields of gas exist in Siberia, and gas is piped to Western Europe through a pipeline. Algerian gas is shipped in the liquid state in ships equipped with refrigeration equipment to maintain the low temperatures needed. Britain and Northern Europe benefit from gas produced in the North Sea, between Norway and Scotland.

Shale oil, a plentiful but difficult-to-exploit fossil fuel, exists in enormous amounts near Rifle, Colorado. A form of oil-bearing rock, the shale must be crushed and heated to recover the oil, a more expensive proposition than drilling conventional oil wells. In spite of ingenious schemes such as burning the shale oil in place, this resource is likely to remain largely unused until conventional petroleum is used up. A similar resource exists in Alberta, Canada, where the Athabasca tar sands are exploited for heavy oils.

John R. Phillips

RENEWABLE RESOURCES

Most renewable resources are living resources, such as plants, animals, and their products. With careful management, human societies can harvest such resources for their own use without imperiling future supplies. However, human history has seen many instances of resource mismanagement that has led to the virtual destruction of valuable resources.

Forests

Forests are large tracts of land supporting growths of trees and perhaps some underbrush or shrubs. Trees constitute probably the earth s most valuable, versatile, and easily grown renewable resource. When they are harvested intelligently, their natural environments continue to replace them. However, if a harvest is beyond the environment's ability to restore the resource that had been present, new and different plants and animals will take over the area. This phenomenon has been demonstrated many times in overused forests and grasslands that reverted to scrubby brushlands. In the worst cases, the abused lands degenerated into barren deserts.

The forest resources of the earth range from the tropical rainforests with their huge trees and broad diversity of species to the dry savannas featuring scattered trees separated by broad grasslands. Cold, subarctic lands support dense growths of spruces and firs, while moderate temperature regimes produce a variety of pines and hardwoods such as oak and ash. The forests of the world cover about 30 percent of the land surface, as compared with the oceans, which cover about 70 percent of the global surface.

Harvested wood, cut in the forest and hauled away to be processed, is termed roundwood. Globally, the cut of roundwood for all uses amounts to about 130.6 billion cubic feet (3.7 billion cubic meters). Slightly more than half of the harvested wood is used for fuel, including charcoal.

Roundwood that is not used for fuel is described as industrial wood and used to produce lumber, veneer for fine furniture, and pulp for paper prod-

ucts. Some industrial wood is chipped to produce such products as subflooring and sheathing board for home and other building construction. Most roundwood harvested in Africa, South America, and Asia is used for fuel. In contrast, roundwood harvested in North America, Europe, and the former Soviet Union generally is produced for industrial use.

It is easy to consider forests only in the sense of the useful wood they produce. However, many forests also yield valuable resources such as rubber, edible nuts, and what the U.S. Forest Service calls special forest products. These include ferns, mosses, and lichens for the florist trade, wild edible mushrooms such as morels and matsutakes for domestic markets and for export, and mistletoe and pine cones for Christmas decorations.

There is growing interest among the industrialized nations of the world in a unique group of forest products for use in the treatment of human disease. Most of them grow in the tropical rainforests. These medicinal plants have long been known and used by shamans (traditional healers). Hundreds of pharmaceutical drugs, first used by shamans, have been derived from plants, many gathered in tropical rainforests. The drugs include quinine, from the bark of the cinchona tree, long used to combat malaria, and the alkaloid drug reserpine. Reserpine, derived from the roots of a group of tropical trees and shrubs, is used to treat high blood pressure (hypertension) and as a mild tranquilizer. It has been estimated that 25 percent of all prescriptions dispensed in the United States contain ingredients derived from tropical rainforest plants. The value of the finished pharmaceuticals is estimated at US$6.25 billion per year.

Scientists screening tropical rainforest plants for additional useful medical compounds have drawn on the knowledge and experience of the shamans. In this way, the scientists seek to reduce the search time and costs involved in screening potentially useful plants. Researchers hope that somewhere in the dense tropical foliage are plant products that could treat, or perhaps cure, diseases such as cancer or AIDS.

Many as-yet undiscovered medicinal plants may be lost forever as a consequence of deforestation of large tracts of equatorial land. The trees are cut down or burned in place and the forest converted to grassland for raising cattle. The tropical soils cannot support grasses without the input of large amounts of fertilizer. The destruction of the forests also causes flooding, leaving standing pools of water and breeding areas for mosquitoes, which can spread disease.

Marine Resources

When renewable marine resources such as fish and shellfish are harvested or used, they continue to reproduce in their environment, as happens in forests and with other living natural resources. However, like overharvested forests, if the marine resource is overfished—that is, harvested beyond its ability to reproduce—new, perhaps undesirable, kinds of marine organisms will occupy the area. This has happened to a number of marine fishes, particularly the Atlantic cod.

When the first Europeans reached the shores of what is now New England in the early seventeenth century, they encountered vast schools of cod in the local ocean waters. The cod were so plentiful they could be caught in baskets lowered into the water from a boat.

At the height of the New England cod fishery, in the 1970s, efficient, motor-driven trawlers were able to catch about 32,000 tons. The catch began to decline that year, mostly as a result of the impact of fifteen different nations fishing on the cod stocks. As a result of overfishing, rough species such as dogfish and skates constitute 70 percent of the fish in the local waters. Experts on fisheries management decided that fishing for cod had to be stopped.

The decline of the cod was attributed to two causes: a worldwide demand for more fish as food and great changes in the technology of fishing. The technique of fishing progressed from a lone fisher with a baited hook and line, to small steam-powered boats towing large nets, to huge diesel-powered trawlers towing monster nets that could cover a football field. Some of the largest trawlers were floating factories. The cod could be skinned, the edible parts cut and quick-frozen for market ashore, and the skin, scales, and bones cooked and ground for animal feed and oil. A lone fisher was lucky to be able to catch 1,000 pounds (455 kilograms) in one day. In contrast, the largest trawlers were capable of catching and processing 200 tons per day.

In the 1990s, the world ocean population of swordfish had declined dramatically. With a worldwide distribution, these large members of the billfish family have been eagerly sought after as a food fish. Because swordfish have a habit of basking at the surface, fishermen learned to sneak up on the swordfish and harpoon them. Fishermen began to catch swordfish with fishing lines 25 to 40 miles (40 to 65 kilometers) long. Baited hooks hung at intervals on the main line successfully caught many swordfish, as well as tuna and large sharks. Whereas the harpoon fisher took only the largest (thus most valuable) swordfish, the longline gear was indiscriminate, catching and killing many swordfish too small for the market, as well as sea turtles and dolphins

As a result of the catching and killing of both sexually mature and immature swordfish, the reproductive capacity of the species was greatly reduced. Harpoons killed mostly the large, mature adults that had spawned several times. Longlines took all sizes of swordfish, including the small ones that had not yet reached sexual maturity and spawned. The decline of the swordfish population was quickly obvious in the reduced landings. But things have changed remarkably, thanks to a 1999 international plan that rebuilt this stock several years ahead of schedule. Today, North Atlantic swordfish is one of the most sustainable seafood choices.

Albert C. Jensen

NONRENEWABLE RESOURCES

Nonrenewable resources are useful raw materials that exist in fixed quantities in nature and cannot be replaced. They differ from renewable resources, such as trees and fish, which can be replaced if managed correctly. Most nonrenewable resources are minerals—inorganic and organic substances that exhibit consistent chemical composition and properties. Minerals are found naturally in the earth's crust or dissolved in seawater. Of roughly 2,000 different minerals, about 100 are sources of raw materials that are needed for human activities. Where useful minerals are found in sufficiently high concentrations—that is, as ores—they can be mined as profitable commercial products.

Economic nonrenewable resources can be divided into four general categories: metallic (hardrock) minerals, which are the source of metals such as iron, gold, and copper; fuel minerals, which include petroleum (oil), natural gas, coal, and uranium; industrial (soft rock) minerals, which provide materials like sulfur, talc, and potassium; and construction materials, such as sand and gravel.

Nonrenewable resources are required as direct or indirect parts of all the products that humans use. For example, metals are necessary in industrial sectors such as construction, transportation equipment, electrical equipment and electronics, and consumer durable goods—long-lasting products such as refrigerators and stoves. Fuel minerals provide energy for transportation, heating, and electrical power. Industrial minerals provide ingredients needed in products ranging from baby powder to fertilizer to the space shuttle. Construction materials are used in roads and buildings.

Location

When minerals have naturally combined together (aggregated) they are called rocks. The three general rock categories are igneous, sedimentary, and metamorphic. Igneous rocks are created by the cooling of molten material (magma). Sedimentary rocks are caused when weathering, erosion, transportation, and compaction or cementation act on existing rocks.

Metamorphic rocks are created when the other two types of rock are changed by heat and pressure. The availability of nonrenewable resources from these rocks varies greatly, because it depends not only on the natural distribution of the rocks but also on people's ability to discover and process them. It is difficult to find rock formations that are covered by the ocean, material left by glaciers, or a rainforest. As a result, nonrenewable resources are distributed unevenly throughout the world.

Some nonrenewable resources, such as construction materials, are found easily around the world and are available almost everywhere. Other nonrenewable resources can only be exploited profitably when the useful minerals have an unusually high concentration compared with their average concentration in the earth's crust. These high concentrations are caused by rare geological events and are difficult to find. For example, an exceptionally rare nonrenewable resource like platinum is produced in only a few limited areas.

No one country or region is self-sufficient in providing all the nonrenewable resources it needs, but some regions have many more nonrenewable resources than others. Minerals can be found in all types of rocks, but some types of rocks are more likely to have economic concentrations than others. Metallic minerals often are associated with shields (blocks) of old igneous (Precambrian) rocks. Important shield areas near the earth's surface are found in Canada, Siberia, Scandinavia, and Eastern Europe. Another important shield was split by the movement of the continents, and pieces of it can be found in Brazil, Africa, and Australia.

Similar rock types are in the mountain formations in Western Europe, Central Asia, the Pacific coast of the Americas, and Southeast Asia. Minerals for construction and industry are found in all three types of rocks and are widely and randomly distributed among the regions of the world.

The fuel minerals—petroleum and natural gas—are unique in that they occur in liquid and gaseous states in the rocks. These resources must be captured and collected within a rock site. Such a site needs source rock to provide the resource, a rock type that allows the resource to collect, and another surrounding rock type that traps the resource. Sedimentary rock basins are particularly good sites for fuel collection. Important fuel-producing regions are the Middle East, the Americas, and Asia.

Impact on Human Settlement

Nonrenewable resources have always provided raw materials for human economic development, from the flint used in early stone tools to the silicon used in the sophisticated chips in personal computers. Whole eras of human history and development have been linked with the nonrenewable resources that were key to the period and its events. For example, early human culture eras were called the Stone, Bronze, and Iron Ages.

Political conflicts and wars have occurred over who owns and controls nonrenewable resources and their trade. One example is the Persian Gulf War of 1991. Many nations, including the United States, fought against Iraq over control of petroleum production and reserves in the Middle East.

Since the actual production sites often are not attractive places for human settlement and the output is transportable, these sites are seldom important population centers. There are some exceptions, such as Johannesburg, South Africa, which grew up almost solely because of the gold found there. However, because it is necessary to protect and work the production sites, towns always spring up near the sites. Examples of such towns can be found near the quarries used to provide the material for the great monuments of ancient Egypt and in the Rocky Mountains of North America near gold and silver mines. These towns existed because of the nonrenewable resources nearby and the needs of the people exploiting them; once the resource was gone, the towns often were abandoned, creating "ghost towns," or had to find new purposes, such as tourism.

More important to human settlement is the control of the trade routes for nonrenewable resources. Such controlling sites often became regions of great wealth and political power as the residents taxed the products that passed through their community and provided the necessary services and protection for the traveling traders. Just one example of this type of development is the great cities of wealth and culture that arose along the trade routes of the Sahara Desert and West Africa like Timbuktu (in present-day Mali) and Kumasi (in present-day Ghana) based on the trade of resources like gold and salt.

Even with modern transportation systems, ownership of nonrenewable resources and control of their trade is still an important factor in generating national wealth and economic development. Modern examples include Saudi Arabia's oil resources, Egypt's control of the Suez Canal, South Africa's gold, Chile's copper, Turkey's control over the Bosporus Strait, Indonesia's metals and oil, and China's rare earth element.

Gary A. Campbell

NATURAL RESOURCES

Africa is one of the largest and least tapped sources of natural resources, including minerals, in the world. Much of Africa's great natural wealth has not been exploited, although some development has occurred on the North African coast and in Southern Africa. Most often, only oil resources have been tapped, and such relatively rare but high-value minerals or metals as diamonds, gold, and chromium have been mined. Few African nations have the economic resources needed to develop their natural and mineral wealth. Such efforts generally have been financed from outside of Africa. Political instability and the lingering presence of former colonial powers has discouraged outside sources from providing funds. The infrastructure—such as good roads and a rail system to transport supplies, raw materials, and finished products—did not exist.

Mineral Resources

Africa's known mineralAfrica is among the world's richest continents for known areas of mineral wealth, which includes coal, petroleum, natural gas, uranium, radium, low-cost thorium, iron ores, chromium, cobalt, copper, lead, zinc, tin, bauxite, titanium, antimony, gold, platinum, tantalum, germanium, lithium, phosphates, and diamonds. The only significantly valuable mineral the continent seems not to have in abundance is silver.

Major deposits of coal exist in four coal basins—in Southern Africa, North Africa, the Democratic Republic of the Congo, and Nigeria. Proven petroleum reserves in North Africa occur in Libya, Algeria, Egypt, and Tunisia. Exploration has been concentrated north of the Hoggar Mountains in southern Algeria. There may also be major Saharan

The pit at the Premier Mine, Cullinan, Gauteng, South Africa. The mine was the source of the 3106 carat Cullinan Diamond, the largest diamond ever found. (Paul Parsons)

RESOURCES OF AFRICA

reserves to the south. The other major oil reserves are in the West African coastal basin—mainly in Nigeria but also in Cameroon, Gabon, and the Democratic Republic of the Congo—and in Angola. Africa's natural gas reserves are concentrated in basins of North Africa and coastal Central Africa.

Southern Africa is one of the world's seven major uranium regions. In South Africa, the joint occurrence of uranium with gold has increased awareness of the presence of uranium and decreased the cost of production. Other countries with significant uranium deposits are Niger, Gabon, the Democratic Republic of the Congo, and Namibia.

Most of Africa's copper is contained in the Central African Copperbelt, which stretches across Zambia and into the Katanga (Shaba) area of the Democratic Republic of the Congo. Other minerals sometimes occur with copper, cobalt being the most common. Africa has about half the world's reserves of bauxite, the chief aluminum ore. Virtually all of this occurs in a major belt stretching some 1,200 miles (1,930 km.) from Guinea to Togo. The largest reserves are in Guinea—the world's third-largest source.

Northern Region

This region comprises Morocco, Algeria, Tunisia, Libya, and Egypt. Morocco, Tunisia, and especially Algeria and Libya have great natural wealth in the form of oil and natural gas. Morocco has natural gas reserves and oil shale resources. Algeria is one of the major producers of oil and gas in Africa, and its full potential in this area has yet to be established. In 2016, the country produced more than 1.7 million barrels of oil per day.

Libya was North Africa's biggest oil producer and one of Europe's biggest North African oil suppliers. Current reserves (2018) are estimated to be 48 billion barrels of crude oil and 13.9 billion cubic meters of natural gas. Production has fallen dramatically in the wake of civil war and unrest. In 2019, Libya averaged more than 1 million barrels of crude oil per day.

Tunisia's oil reserves and production are relatively modest for North Africa. In 2017, its reserves were believed to be 425 million barrels of crude oil and 65 billion cubic meters of natural gas. Crude oil production in 2019 was 48,757 barrels per day. Tunisia is a net importer of crude oil.

Morocco is the world's largest exporter of both phosphate rock and phosphoric acid. Phosphate reserves are estimated to be approximately 50 billion metric tons. Algeria has major deposits of phosphates and iron ore, as well as smaller deposits of coal, lead, zinc, and uranium. Some mercury, copper, and gold are also mined.

Egypt produced approximately 490,000 barrels of oil per day in 2019. It also has modest mineral reserves that include iron ore, petroleum, natural gas, gypsum, phosphate rock, manganese, coal, limestone, salt silica, kaolin, fluorspar, uranium, chromium, ilmenite (an oxide of iron and titanium), tantalum, molybdenum, and some gold.

Sudano-Sahelian Region

This region comprises Mauritania, Western Sahara, Senegal, Mali, Burkina Faso, Niger, Chad, Sudan, and South Sudan.

Western Sahara contains vast reserves of phosphates, which remain largely undeveloped because of political uncertainties. In 2020, Western Sahara, a former Spanish possession, was considered a "non-self-governing territory" by the United Nations. Morocco now controls approximately two-thirds of its area.

Mauritania has become Africa's major supplier of iron ore, producing 13.4 million metric tons in 2013 for 50 percent of the country's exports. Gold and copper are its other important metals. There are also deposits of diamonds, chromium, sulfur, and yttrium.

Senegal has phosphate mining and deposits of iron ore and gold. Burkina Faso is one of Africa's leading gold producers. Nickel and manganese are also mined, as are deposits of copper, iron, phosphate, and tin. Niger is the world's second-largest producer of uranium, producing 3,000 tons per year. Salt and sodium carbonate are mined, and there are deposits of copper, iron ore, phosphate, and tin. Some companies are prospecting for gold.

Chad had estimated oil reserves of 1.5 billion barrels of crude in 2017, and its oil fields yield about 110,000 barrels per day. Minor deposits of sodium carbonate and alluvial diamonds, are mined; marble and uranium resources exist as well. The area around the Tibesti Mountains in northern

---

AFRICAN GOLD AND THE EUROPEAN RENAISSANCE

The portion of western Mali known as the Bambuk has long been a gold-producing region of Africa. In medieval times, traders came to Mali from Morocco and Algeria to trade salt for gold. The gold eventually went to Europe and was used for trade and commerce, enabling Europe to emerge economically from the Dark Ages into the Renaissance. The word "guinea," used in Great Britain for some gold coins, may have originated from the city Djenné, in southeastern Mali.

Chad has deposits of niobium, tantalum, tin, and tungsten. Deposits of iron ore have been found at several locations, and bauxite deposits exist at Koro in the south. Sudan's oil reserves are estimated at 5 billion barrels (2016), and its gas reserves have been estimated at 3 trillion cubic feet (2017). Sudan's deposits of gold, chromite, marble, and mica contribute little to the economy.

For the first decade of the twenty-first century, Sudan's petroleum industry drove the country's economy. But with the secession of South Sudan in 2011, and the consequent loss of nearly 80 percent

of its oil reserves, the Sudanese economy has been reeling. And although the newly independent South Sudan controlled vast oil reserves, it had no oil refineries in its territory. Ten years later, oil accounts for 98 percent of South Sudan's government revenues. South Sudan also has large but undeveloped deposits of gold, copper, iron ore, and zinc chromium.

Gulf of Guinea

This region comprises Guinea-Bissau, Cabo Verde, Gambia, Guinea, Liberia, Sierra Leone, Côte

Searching for tanzanite deep under the ground near the slopes of the highest mountain in Africa. (Africraigs)

Rössing open-pit uranium mine, Namibia. (Ikiwaner)

d'Ivoire (Ivory Coast), Togo, Ghana, Benin, and Nigeria. Guinea-Bissau has some offshore tracts to be explored for possible crude oil reserves, but these and deposits of other mineral resources remain undeveloped.

In 2016, Benin was believed to have about 8 million barrels of proven oil reserves, though no production has occurred since 1998. Benin has unexploited deposits of iron, gold, phosphate, chromium, diamonds, limestone, and titanium. Côte d'Ivoire's oil reserves have been estimated at 100 million barrels, although production peaked in the 1980s and has since tapered down substantially. Côte d'Ivoire also produced gold and diamonds, though most of these deposits are now exhausted. Reserves of iron ore, bauxite, manganese, copper, cobalt, and nickel contribute less than 1 percent to the country's revenue.

Liberia's iron ore, gold, and diamond mining operations are now largely inactive. The country's iron ore reserves are in the billions of tons. It also has underutilized deposits of bauxite, lead, and manganese. Guinea has nearly half of the world's bauxite reserves, iron ore reserves in excess of 3 billion tons, and gold reserves of 1,000 tons of metal. Other substantial mineral wealth includes high-grade iron ore, large diamond and gold deposits, and significant amounts of uranium.. Ghana is Africa's largest gold producer—the world's seventh-largest—and an important source of diamonds as well. The country also has up to 7 billion barrels of proven oil reserves as of 2016, yielding 100,000 barrels per day.

In Sierra Leone, diamonds were first discovered in 1930. Its diamonds continue to attract foreign investment because of their outstanding size and beauty, although today's diamond market is somewhat diminished. Sierra Leone produced 0.5 million carats of diamonds in 2016. The country also has Africa's largest deposits of titanium dioxide ore (rutile). Togo's calcium phosphate mining accounts for half the country's export revenue. Cabo Verde and Gambia have little in the way of natural resources.

Central Region

This region comprises the Republic of the Congo, the Democratic Republic of the Congo, Cameroon, Gabon, Equatorial Guinea, São Tomé and Príncipe, Rwanda, and Burundi. The Democratic Republic of the Congo has large reserves of copper and cobalt—much of the world's reserves of cobalt can be found there. It also has reserves of zinc, lead, gold, tantalum, beryllium, germanium, and lithium, and is the world's fourth-largest producer of diamonds. The Republic of the Congo has estimated current oil reserves of 3 billion barrels (2018) and was producing nearly 310,000 barrels per day in 2019. Large reserves of natural gas are associated with the oil reserves. Small reserves of copper, gold, iron ore, lead, and potash rock (potassium carbonate) are also present.

Cameroon produces 93,000 barrels of crude oil per day. The country also has substantial natural gas reserves as well as deposits of bauxite, rutile, and gold. The Central African Republic has significant deposits of diamonds, gold, copper, iron ore, tin, uranium, and zinc have been found.

Burundi's principal mineral is tin. Deposits of nickel, gold, vanadium, and phosphate have not been developed because of internal strife. Gabon produced about 210,000 barrels of crude oil per day in 2019. Manganese is produced from the Mouana deposits near Moanda and uranium from the Oklo mine. There are also phosphate, niobium, and iron ore.

São Tomé and Príncipe has few or no natural resources. In 2019, Equatorial Guinea produced 227,000 barrels per day of crude oil and 6.8 billion cubic feet of natural gas. Its current oil reserves are estimated at 1.1 billion barrels.

Eastern Region

This region comprises Eritrea, Djibouti, Ethiopia, Somalia, Kenya, Tanzania, and Uganda. With the

Illegal gold mining in Nigeria. (Dame Yinka)

Locomotive No. 303, wagons loaded with iron ore. (Fred P.M. van der Kraaij)

exception of Uganda, East Africa is mineral-poor. Kenya produces sodium carbonate, fluorspar (calcium fluoride), and small quantities of gold, iron ore, garnets, limestone, rutile, and zirconium. Eritrea has high-grade deposits of copper and gold. Ethiopia has reserves of coal, and natural gas reserves estimated to be 24.9 million cubic meters. Gold, manganese, and platinum are mined on a very small scale. Somalia has one of the largest known deposits of gypsum. It also has 5.6 billion cubic meters of proven natural gas reserves. Djibouti has little in the way of natural resources.

Gold is Tanzania's most significant mineral, though there are also deposits of nickel, coal, salt, uranium, lead, and tin. For Uganda, gold exports have soared, and in 2019 were valued at US$1.2 billion. Copper mining is also significant.

Indian Ocean Islands

These islands are Madagascar, Mauritius, Comoros, and the Seychelles. Madagascar may have promising reserves of heavy crude oil and natural gas. Its

chromium deposits have not been mined since 2019. Madagascar was the sixth-leading producer of cobalt in 2019, and is the source of half the world's sapphires. Small-scale gold mining also takes place. Comoros, Mauritius, and the Seychelles have little in the way of natural resources.

Southern Region

This region comprises Angola, Zambia, Zimbabwe, Mozambique, Malawi, Botswana, Namibia, Lesotho, Eswatini (formerly Swaziland), and South Africa. South Africa has vast vanadium and uranium reserves, and accounts for 78 percent of the world's platinum production and 39 percent of the world's palladium, nearly 100 percent of the world's chromium products, 81 percent of the manganese, and about 38 percent of the gold reserves.

South Africa also has substantial reserves of other industrially important metals and minerals, including antimony, asbestos, diamonds, coal, fluorspar, phosphates, iron ore, lead, zinc, uranium, vermiculite, and zirconium. It is the world's largest producer of platinum-group metals, vanadium, and aluminosilicates, and one of the world's top producers of antimony, chromite, diamonds, ferrochrome, ferromanganese, fluorspar, manganese titanium, vermiculite, and zirconium. The country is also a major producer and exporter of bituminous coal, cobalt, nickel, and granite.

In 2019, Zambia was the world's seventh-largest copper-producing nation and the ninth-largest producer of cobalt. It also has deposits of gold, emeralds, limestone, gypsum, feldspar, lead, coal, and tin. Zimbabwe is an important producer of platinum and diamonds, although the country's revenue from minerals is almost entirely lost to corruption.

Angola is the second-leading oil-producing country in Africa (after Nigeria), producing 1.76 million barrels per day in 2019. The country's known recoverable reserves are estimated to total more than 8.2 billion barrels. Angola produces more than half of its oil offshore from its enclave of

Cabinda, with the remainder coming from onshore and offshore sites.

Angola has substantial deposits of diamonds, gold, iron ore, phosphates, manganese, copper, lead, quartz, gypsum, marble, black granite, beryl, zinc, and numerous base and strategic metals such as chromium. Nevertheless, mining production in Angola has been limited to diamonds in the Lunda Norte Province, ornamental stones in the Huíla and Namibe areas, and salt. The country's other mineral resources are largely undeveloped, despite their great potential.

Namibia is the site of the richest alluvial diamond deposits in the world. It has become the world's eighth-leading gem-quality diamond producer, contributing 2 percent of the world's output. In 2018 alone, diamond production exceeded 2.4 million carats, mostly gem quality. Today, Namibia's onshore diamond deposits are becoming depleted but, at the same time, offshore diamond mining has soared. In 2016, Namibia's marine diamond indus-

Wolframite and cassiterite mining in the Democratic Republic of the Congo. (Julien Harneis)

The desert between Assamaka and Arlit with ripples due to the wind showing a uranium mine, open mining, with industrial buildings and a mountain of waste. (Angeline A. van Achterberg)

try produced more than 1.37 million carats of diamonds, with great potential for even larger hauls. Supporting this optimism is the addition of the world's largest custom-built diamond-mining vessel to Namibia's fleet of mining ships, scheduled to take place in the early 2020s.

Namibia contains considerable reserves of uranium, copper, lead, and zinc. Its two largest mines produced 8.4 percent of the world's uranium oxide in 2018. The Haib mineral deposit, situated just north of the South African border near the town of Noordoewer, holds great potential to produce copper, gold, and molybdenum concentrate. Namibia also comprises substantial known reserves of zinc and lead.

Botswana's economy is dominated by diamonds: It is Africa's largest and the world's third-largest producer of diamonds and second-largest producer of gem diamonds. Copper-nickel, soda ash, coal, and gold are also mined. Other known mineral deposits include chromite, feldspar, graphite, gypsum, iron, kaolin, talc, and uranium.

Malawi's coal reserves have yet to be exploited. Mozambique has massive natural gas reserves (more than 2.8 trillion cubic meters in 2018); development of this resource is being fast-tracked by two American companies, with production set to begin in the early 2020s. Some gold, bauxite, graphite, and marble have been mined as well. Eswatini has coal reserves and diamonds, both of which are being mined. Lesotho's mining industry is focused on diamonds; other mineral deposits include uranium, lead, and iron ore.

Water Resources
Although the surface area of Africa constitutes about one-fifth of Earth's entire land surface, the combined annual flow of African rivers is only about 7 percent of the world's river flow reaching the oceans. North Africa's few rivers originate in the

Irrigation of crops, near Gonde, Ethopia. (Niels Van Iperen)

mountains of Algeria and Morocco, and their water is used extensively for irrigation.

From the relatively well-watered areas of the Gulf of Guinea and Central Africa, the Senegal, Niger, Logone-Chari, and Nile rivers flow through the drier inland zones. The Niger River, which originates in the highlands of Guinea, forms a vast floodplain south of Timbuktu in Mali. The Logone and Chari rivers feed Lake Chad. The Nile River, the world's longest, receives more than 60 percent of its water from Ethiopia; however, it originates farther south in the mountains of Burundi.

Several rivers flowing in a roughly southerly direction into the Atlantic Ocean drain the southern part of West Africa. Many flow rapidly over bedrock before entering the coastal plains, draining into the systems of lagoons and creeks along the coast. During the dry season, the upper reaches of these rivers are without water. In Guinea, Sierra Leone, and Liberia, the dry season is short and the rivers flow throughout the year. In the well-watered western part of equatorial Africa, the total average annual flow of the Congo River is enormous: 44 trillion cubic feet (1.25 trillion cubic meters).

Groundwater reserves are important for wells. Large inland depressions in Africa's basement rock sometimes form important groundwater reservoirs, as in Niger, the central Sahara region of Algeria, the Libyan Desert, Chad, the Congo basin, the Karoo area of South Africa, and the Kalahari Desert. The East African plateaus usually contain little or no groundwater, and aquifers (geologic formations containing water) are found only in humid areas where the crystalline rock is weathered or fractured. The chalky shales and dolomitic limestones, which sporadically cover the basement rock of Africa, can contain important aquifers, as in Zambia and South Africa. In the coastal areas of Senegal, Côte d'Ivoire (Ivory Coast), Ghana, Togo, Benin, Nigeria, Cameroon, Gabon, the Republic of Congo, Angola, Mozambique, the East African countries, and Madagascar, aquifers are found in sandstone, limestone, and sand and gravel.

People have long practiced large-scale irrigation, mainly in the northern region, Sudan, South Sudan, South Africa, Mali, Zimbabwe, and Mozambique. Medium-scale irrigation projects have been operated in Madagascar, Senegal, Somalia, and Ethiopia. In Côte d'Ivoire, Burkina Faso, Kenya, Nigeria, Ghana, and Zambia, small-scale and medium-sized projects have been constructed.

Forests

The bulk of North Africa's forestry production generates local firewood, charcoal, and industrial timber. Cork is produced for export from the large areas of cork oak trees that grow near the Mediterranean Sea, especially in the west. Black wattle, introduced from Australia, is widely grown for firewood. Its spread has been greatly encouraged by its harvesting for tannin bark. The eucalyptus is the most widely introduced tree from Australia and grows rapidly under East African conditions.

Grown for firewood and poles, eucalyptus trees occupy the landscape, especially in upland Ethiopia.

By 2020, the timber and forestry industries in Africa have begun to gain a new prominence. In many places new plantation forests have been established, and these have gradually led to the creation of secondary processing and value-addition industries. For those countries lacking mineral resources and with low manufacturing capabilities, the timber industry has become a valuable source of revenue, and figures prominently in strategic development plans. Among the important timber producers in Africa are Liberia, Zimbabwe, the Democratic Republic of the Congo, the Republic of Congo, Gabon, and Cameroon. In Gabon alone, timber exports exceeded US$500 million in 2017. Europe is the primary market for Africa's tropical timber production.

Energy Generation

While substantial growth in energy use has taken place in Africa, by 2020, an estimated 500 million Af-

Logs of Eucalyptus paniculate *cultivated near Johannesburg, South Africa*

ricans still lived without electricity. Steam-powered stations using coal are by far the most common energy-production facilities. Hydroelectricity provides less than 10 percent of the continent's potential. The largest hydroplant is Egypt's Aswan Dam. Africa's single nuclear power station is in South Africa. Electricity consumption in large urban centers has increased considerably. Some countries have extended electrical transmission networks to rural areas or increased the number of isolated, low-powered stations and independent networks. Nevertheless, progress in rural electrification has been extremely slow. Upwards of 60 percent of sub-Saharan Africans have no access to electricity.

Renewable sources of energy may someday help Africa meet its electricity demands, but progress has been slow. The primary reason is cost; solar panels, wind turbines, high-capacity batteries, meters, sockets, cables, and connectors are expensive. With labor costs and overhead factored in, renewable energy remains an expensive proposition for people who are often living on less than $1 per day. Without substantial government subsidies and out-

Tarfaya Wind Farm, Tarfaya, Morocco. (EDxTarfaya)

side financing, for-profit companies would find undertaking such projects nearly impossible. In many areas of economically impoverished Africa, these projects would run at a loss for many years to come.

Despite the financial challenges, renewable energy is beginning to make a dent. Solar energy may hold the greatest potential. Africa enjoys the distinction of being the planet's sunniest continent, and some areas, such as the Sahara Desert, have almost perpetual sunshine. South Africa, Morocco, and Algeria have made the biggest strides in implementing solar technology. In South Africa, state-of-the-art solar facilities provide electricity to hundreds of thousands of households. Morocco is developing its solar-energy infrastructure and, by the mid-2020s, expects it to meet the electricity needs of 40 percent of its people. Elsewhere in Africa, solar power's greatest potential is for limited energy production to facilitate such day-to-day

Workers at Olkaria Geothermal Power Plant. (Lydur Skulason)

needs as small-scale electrification, desalination, water pumping, and water purification.

European investors have shown some interest in building vast solar farms in the North African deserts. Such facilities could provide electricity for local consumption with leftover capacity available for export to Europe. An even more ambitious proposal envisions solar farms covering approximately 0.3 percent of North Africa (an area equivalent to the U.S. state of Maine) that could generate enough energy to satisfy the electricity needs for the entire European Union.

Africa has a large coastline, where wind-power and wave-power resources are abundant, albeit underutilized. South Africa has the most advanced wind-energy infrastructure, with upwards of 52 percent of the country's energy derived via the wind in 2019. Morocco, with large wind farms operating and others under construction, was, in 2013, producing slightly more than 5 percent of its electricity from the wind. The government projects that by

2030, wind energy will supply nearly 10 percent of the country's electricity. And here and there across the continent, stand-alone wind turbines have been built to power individual villages.

Geothermal power holds great potential to provide energy, particularly in East Africa, where the most accessible resources are located. So far, only Kenya has exploited the geothermal potential of the Great Rift Valley. Kenya has several operational geothermal plants, including Africa's oldest (built in 1956) and largest geothermal plants. Kenyan energy officials project that by 2030, geothermal resources will provide 51 percent of the country's energy needs. Ethiopia, Zambia, Eritrea, Djibouti, and Uganda have each developed plans for geothermal facilities. Their projects have invariably stalled due as much to deficient funding as to an absence of the necessary specialized scientific and engineering expertise.

Dana P. McDermott

DISCUSSION QUESTIONS: NATURAL RESOURCES

Q1. How has Africa's mineral wealth contributed to the continent's development? Which areas are richest in precious metals? For which elements is Africa the world's primary source?

Q2. How well is Africa equipped with energy resources? Which countries are important petroleum producers? To what extent has Africa embraced renewable energy sources?

Q3. What are Africa's principal rivers? What makes the Nile unique among the world's rivers? What central roles has the Nile River played in Africa and its peoples? How has irrigation changed the quality of agriculture across the continent?

FLORA

More than any other continent, Africa is tropical. With few exceptions, its vegetation is tropical or subtropical. This is primarily because none of the African continent extends far from the equator, and there are only a few high-elevation regions that support more temperate plants. Listed in order of decreasing land area, the three main biomes of Africa are tropical savanna, subtropical desert, and tropical forest. The vast size and uniqueness of each of these areas have made them familiar to most people. The flora in Southern Africa has been well-studied; the flora of Central and North Africa is less well known.

The subtropical desert biome is the driest of the biomes in Africa and includes some of the driest locations on Earth. The largest desert region is the Sahara in North Africa. It extends from near the west coast of Africa to the Arabian Peninsula and is part of the largest desert system in the world, which extends into south central Asia. A smaller desert region along the western half of Southern Africa includes the Namib Desert, especially near the coast, and the Kalahari Desert, which is primarily inland and east of the Namib Desert.

Where more moisture is available, grasslands predominate, and as rainfall increases, grasslands

Namib desert and ocean. (Robur.q)

gradually become tropical savanna. The difference between a grassland and a savanna is subjective. How many trees does it take for a grassland to be considered a savanna? The grassland/tropical savanna biome forms a broad swath across much of Central Africa and dominates much of East and Southern Africa.

Tropical forests cover a much smaller area of Africa than the other two biomes. They are most abundant in the portions of Central Africa not dominated by the grassland/tropical savanna biome and are not far from the coast of central West Africa. Scattered tropical forest regions also occur along major river systems of West Africa, from the equator almost to Southern Africa.

Subtropical Desert

The subtropical deserts of Africa seem, at first, to be nearly devoid of plants. While this is true for some parts of the Sahara and Namib deserts that are dominated by sand dunes or bare, rocky outcrops, much of the desert biome has a noticeable amount of plant cover. Even in the driest, most inhospitable locations, plants are present. The Sahara is characterized by widely distributed species of plants that are found in similar habitats. The deserts of Southern Africa have a more distinctive flora, with many species endemic to specific local areas.

To survive the harsh desert climate, desert plants use several adaptations: Some, typically called succulents, store water in their leaves and/or stems. Other plants grow near a permanent or semipermanent water source such as a spring, an intermittent stream, or groundwater. A third group is made up of ephemerals: annuals whose seeds germinate when moisture becomes available, quickly mature, set seed, and die. Each of these adaptations is well represented in African plants.

Succulents of the Subtropical Desert

Many succulents are able to retain large amounts of water because they use a specialized type of photosynthesis. Most plants open their stomata (small openings in the leaves) during the day to extract carbon dioxide from the surrounding air, but this leads to high amounts of water loss in a desert envi-

WELWITSCHIA: AFRICA'S STRANGEST PLANT

There are many unusual plants in Africa, but one of the most unusual is the Welwitschia, a resident of the Namib Desert. It has a short, swollen stem only about 4 inches (10 centimeters) high, which terminates in a disc-like structure. Coming off the top of the stem are two strap-like leaves. These two leaves last for the lifetime of the plant and continue to grow very slowly. As they grow, they twist and become shredded, so that an individual plant appears to have many leaves. The reproductive structures rise from the center of the stem, and, instead of flowers, Welwitschia has small cones.

Welwitschia is such a successful survivor of the Namib that it easily lives for hundreds of years. Some specimens have been dated to about 2000 years old. Older plants can reach tremendous sizes, with the top of the stem sometimes reaching 5 feet (1.5 meters) in diameter. Specimens of Welwitschia are extremely difficult to grow in cultivation, requiring special desert conditions and room for the deep taproot.

ronment. Succulents open their stomata at night and, through a special biochemical process, store carbon dioxide until the next day, when it is released inside the plant so photosynthesis can occur without opening the stomata. *Mesembryanthemum*—ice plants and sea figs—is a widespread genus, with species occurring in all of Africa's deserts. It typically has thick, succulent leaves and colorful flowers with numerous parts.

As a further adaptation to prevent water loss, many succulents have no leaves. *Anabasis articulata*, which is widespread in the Sahara Desert, is a leafless succulent with jointed stems. Cacti are only found in North and South America, but a visitor to the Sahara would probably be fooled by certain species in the spurge family that bear a surprising resemblance to cacti. For example, *Euphorbia echinus*, another Saharan plant, has succulent, ridged stems with spines. The most extreme adaptation in succulents is found in the living stones of Southern Africa. Their plant body is reduced to two plump, rounded leaves that are very succulent. They hug the ground, sometimes being partially buried, and

Female cones of a female Welwitschia plant at Welwitschia Plains, Erongo region, Namibia. This plant's cones are about ripe, and some of them are beginning to shed their seeds. (Hans Hillewaert)

have camouflaged coloration so that they blend in with the surrounding rocks and sand, thus avoiding being eaten by grazing animals. Other succulents, such as the quiver tree, attain the size and appearance of trees.

Water-Dependent Plants of the Subtropical Desert
Water-dependent plants are confined to areas near a permanent water source. The most familiar of these plants is the date palm, which is a common sight at desert oases. Date palms have been cultivated for thousands of years for food, but waning interest in date farming has endangered many of these trees. When not properly cared for, they can become water-stressed and often fall prey to a fungus disease called bayoud. Tamarind and acacia are also common where water is available. A variety of different sedges and rushes occur wherever there is

abundant permanent freshwater, the most famous of these being the papyrus or bulrush.

Ephemerals of the Subtropical Desert
Desert ephemerals account for a significant portion of the African desert flora. A majority of the ephemerals are grasses. Ephemerals are entirely dependent on seasonal or sporadic rains. A few days after a significant rain, the desert turns bright green, and after several more days flowers, often in profusion, appear. Some ephemerals germinate with amazing speed, such as the pillow cushion plant, which germinates and produces actively photosynthesizing seed leaves only ten hours after it rains. Reproductive rates for ephemerals, and even for perennial plants, are rapid. Species of morning glory can complete an entire life cycle in three to six weeks.

Tropical Savanna

Tropical savanna ranges from savanna grassland, which is dominated by tall grasses lacking trees or shrubs, to thicket and scrub communities, which are composed primarily of trees and shrubs of a fairly uniform size. The most common type of savanna in Africa is the savanna woodland, which is composed of tall, moisture-loving grasses and tall deciduous or semideciduous trees that are unevenly distributed and generally well spaced. The type of savanna familiar to viewers of African wildlife documentaries is the savanna parkland, which is primarily tall grass with widely spaced trees.

Grasses and Herbs of the African Savanna

Grasses represent the majority of plant cover beneath and between the trees. In some types of savanna, the grass can grow more than 6 feet (1.8 meters) high, giving rise to such names as elephant grass. Although it has been much debated, two factors seem to perpetuate the dominance of grasses: seasonal moisture with a long intervening dry spell, and periodic fires. Given excess moisture and lack of fire, savannas seem inevitably to become forests. Other activities by humans, such as grazing cattle or cutting trees for firewood, also perpetuate, or possibly promote, grass dominance.

A variety of herbs exist in the savanna, but they are easily overlooked, except during flowering periods. Many of them also do best just after a fire, when they are better exposed to the sun and to potential pollinators. Some, like types of hibiscus and coleus, are familiar garden and house plants popular the world over. Vines related to the sweet potato are also common. Many species from the legume, or pea, and sunflower families are present. Wild ginger often displays its showy blossoms after a fire.

Trees and Shrubs of the Savanna

Trees of the African savanna often have relatively wide-spreading branches that all terminate at about the same height, giving the trees a flat-topped appearance. Many are from the legume family, most notably species of *Acacia*, *Brachystegia*, *Julbernardia*,

Sea-fig. (Melissa McMasters)

Tamarind fruits are flattish, beanlike, irregularly curved and bulged pods, cinnamon-brown or grayish-brown externally when mature. The pulp is edible and is said to have medicinal properties. (Dinesh Valke)

and *Isoberlinia*, which, with the exception of *Acacia*, are not well known outside of Africa. There is an especially large number of *Acacia* species ranging from shrubs to trees, many with spines. A few also have a symbiotic relationship with ants that protect them from herbivores. The hashab tree, a type of acacia that grows in more arid regions, is the source of gum arabic.

Although not as prominent, the baobab tree—renowned for its large size and odd appearance—occurs in many savanna regions. It has an extremely thick trunk with smooth, gray bark and can live for up to 2,000 years. Another odd tree is the sausage tree, which gets its name from the large, oblong, sausage-like fruits that dangle from its branches. Many savanna trees also have showy flowers, like the flame tree and the African tulip tree. Both trees are pollinated by sunbirds.

Tropical Forest

The primary characteristics of African tropical forests are their extremely lush growth, high species diversity, and complex structure. The diversity is often so great that a single tree species cannot be identified as dominant in an area. Relatively large trees predominate, growing so close together that their crowns overlap and limit the amount of light that reaches beneath them. A few larger trees, called emergent trees, break out above the thick canopy and may grow to almost twice as high as the average height of the canopy below.

Another layer of smaller trees lives beneath the main canopy. A few smaller shrubs and herbs grow near the ground level, but the majority of the herbs and other perennials are epiphytes, that is, plants that grow on other plants. On almost every available space on the trunks and branches of the canopy trees there are epiphytes that support an entire, unique community. All this dense plant growth is supported by a monsoon climate in which 60 inches (1,500 millimeters) or more of rain often falls annually, most of it in the summer.

The most notable feature of trees in the African tropical forest is their great diversity. Although some species are widespread and relatively common, many are rare and endemic to extremely small areas. Many of the trees belong to families that are found only in the tropics. Larger trees often have extensive buttresses, which are extremely large, flared bases. Buttresses appear to give tropical trees greater stability. A similar adaptation is stilt roots, which arise from buds lower down on certain trees and radiate out and down to the soil.

Among the tallest of the forest trees is the silk cotton tree, one of the sources of kapok (traditionally used to fill life preservers), which can exceed 150 feet (45 meters) in height. Many of the taller trees are valuable as timber. *Ochroma* is the source of the uniquely light balsa wood. Ironwood, a member of the legume family, forms dominant stands in many areas. Ironwood is named for its extremely dense wood, which is resistant to decay and termites, in addition to being difficult to cut. Its wood also has alternating color patterns of light and dark wood, referred to as zebra wood. Because of its strength, it traditionally has been used for heavy duty flooring and in shipbuilding. Some of the more valuable timber trees include iroko and sapele.

Lianas and Epiphytes of the Tropical Forest

Lianas are large, woody vines that cling to trees. Some of these vines hang down near to the ground and were made famous by the Tarzan movies. Many lianas belong to families with well-known temperate vine species, such as the grape, morning glory, and cucumber families. Other related plants are simply climbers and either never become as sturdy as lianas, or remain intimately connected to the trunks of trees. One of these, the strangler fig, is a strong climber that begins life in the canopy.

The fruits are eaten by birds or monkeys, and the seeds are deposited in their feces on branches high in the canopy. The seeds germinate and send a stem

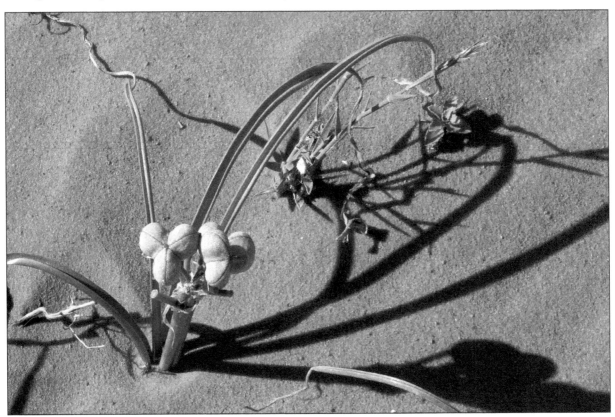

An ephemeral desert plant in Taghit, Bechar, Algeria. (LBM1948)

downward to the ground. Once the stem reaches the ground, it roots; additional stems then develop and grow upward along the trunk of the tree. After many years, a strangler fig can so thoroughly surround a tree that it prevents water and nutrients from flowing up the trunk. Eventually, the host tree dies and rots away, leaving a hollow tube composed essentially of the strangler fig itself. Other climbers include members of the Araceae family, the most familiar being the ornamental philodendron.

The most common epiphytes are bryophytes—lower plants related to mosses—and lichens, which are a symbiotic combination of algae and fungi. The most abundant higher plants are ferns and orchids. As these plants colonize the branches of trees, they gradually trap dust and decaying materials, eventually leading to a thin soil layer that other plants can also use. Accumulations of epiphytes can be so great in some cases that branches break from the weight. Epiphytes are not parasites (although there are some parasitic plants that grow on tree branches) but simply use the host tree for support. They also provide a habitat for a variety of animals, including amphibians, birds, reptiles, and insects. Orchids, of which there are thousands of species, are especially noted for their interactions with insects. Some or-

Pincushion protea. (Jon Sullivan)

chid flowers actually mimic their own pollinators—small species of wasps—and entice males to mate with them. The wasps' attempts at mating result in pollen transfer for the orchid. The epiphyte plant community is probably the least well-understood plant community in the tropical rainforest because of its inaccessibility.

Tropical Forest Floor Plants

Many of the herbs of the forest floor are small and inconspicuous. Grasses are almost entirely absent, and those that do occur there have much broader leaves than usual. Some forest-floor herbs are able to grow in the deep shade beneath the canopy, occasionally being so thoroughly adapted to the low available light that exposure to full sunlight can damage them. Some of the most popular houseplants have come from among these plants because they do not need direct sunlight to survive. Still, the greatest numbers of plants occur beneath breaks in the canopy, where more light is available.

Some of the larger herbs belong to the ginger and arrowroot families. Species of *Costus* possess particularly showy flowers and have often been cultivated in greenhouses. Prayer plants are grown for their beautiful foliage. Members of the pepper family, some of them shrubs and small climbers, also occur here. Leaves of *Piper betle* are wrapped around betel nuts and chewed by many of the indigenous peoples of Africa. The leaves cause excessive production of salivá, and the betel nuts produce a red juice that stains the teeth of habitual chewers. There

---

CENTRAL AFRICAN TROPICAL RAINFORESTS IN PERIL

Although the rainforests of Central Africa are at lower risk of deforestation than are tropical forests in some other parts of the world, they are still in great danger. First, they contain many valuable timber trees that are being exploited actively. Second, these forested areas are in demand for other uses. In some areas, excessive timber harvesting and clearing to make way for oil palm and rubber plantations leaves little untouched forest. Some of the worst damage has occurred in Liberia and the Côte d'Ivoire, where populations are on the rise. Much research is being done to develop sustainable and regenerative ways to harvest trees. In many areas, certain parts of the forest are considered sacred and are jealously guarded from all encroachment by local tribes. Medicinal plants are obtained from these areas, and they also are used as burial grounds that shelter the tribes' ancestral spirits.

is some evidence that betel nut stains on the teeth may help prevent tooth decay.

Other Plant Associations

In addition to the three major biomes, there are several smaller areas in Africa that have some distinctive plant associations. Southern Africa is especially rich in plant life and contains some of the most unusual plant associations on the continent, if not the world.

The Fynbos Biome

The fynbos biome is located in the extreme southwestern and southern parts of Southern Africa. Climatically, the western part is a Mediterranean-type ecosystem with winter rains and summer drought. Moving more to the east, the rains become less seasonal. The literal meaning of the term "fynbos" is "fine-leaved bush," which refers to the dominant vegetation type in this biome. The fynbos is primarily a fire-prone, evergreen shrub land. Many of the shrubs are from the heath or protea families. The grasses are typically evergreen and very wiry. Although the fynbos covers only a relatively small area geographically, it contains about 7,300 species of plants, 80 percent of them endemic.

The Nama Karoo Biome

This biome is also found in Southern Africa and occupies a little more than 22 percent of the region. The nama karoo is a desert-type biome receiving, in its wetter parts, less than 16 inches (410 millimeters) of rain per year, with some areas receiving less than 3 inches (80 millimeters) of rain yearly. As a consequence, the plants display typical desert adaptations. Dwarf shrubs, some of them succulent, dominate much of the landscape. Some areas are dominated by grasses.

Fynbos plants. (Caroline Auzias)

A liana, Secamone alpini, *known as melktou creeper. Indigenous forest in South Africa.*

Many of the plants belong to such familiar plant families as the sunflower, lily, and foxglove. Many of the species from the lily family in this biome are highly prized by gardeners the world over. The mesembryanthemum family is also prominent here, and a number of species have been introduced to other parts of the world. One of these, commonly called the ice plant, was introduced to various parts of California to stabilize sand and soil in freeway medians. Since its introduction, it has escaped extensively and permanently altered some natural sand dune communities, endangering native plants.

The Succulent Karoo Biome
This biome is another desertlike biome, but it receives slightly more rainfall than the nama karoo. As its name implies, most of the species are succulent, with a preponderance of succulent-leaved shrubs from the mesembryanthemum and stonecrop families. There are also succulent monocots, such as al-

oes, and some cactus-like members of the spurge family. This biome also has the highest species richness for any semiarid biome, with more than 5,000 species, 50 percent of them endemic. Also found there are species of living stones discussed under the subtropical desert section.

Other types of low-growing succulents, like living stones, blend in with their surroundings by resembling soil, stones, or animal dung. Some of the shrubs belong to families generally thought of as herbaceous, such as the milkweed, asparagus, and nightshade families. Even the genus *Aloe*, best known to many people by the small succulent, aloe vera, is represented by many shrubs, some of them resembling small trees. Spines are a relatively common feature of many of the succulent herbs and shrubs as well, making them seem superficially like cacti.

The Mediterranean Ecosystem
Only a few places in the world have the combination of climate factors commonly called Mediterranean

101

climate, which is characterized by warm, dry summers and cool, wet winters. Two regions in Africa have this type of climate. The first was discussed in the section on the nama karoo biome. The second area is the part of North Africa that borders the Mediterranean Sea.

Two vegetation associations predominate. Mixed evergreen woodlands composed of holm oak, Aleppo pine, and Aleppo fir thrive in moister areas. At higher elevations, the two conifers are replaced by cedar, although these have disappeared from much of the area after centuries of overexploitation. Olive and mastic trees also exist here, although they are less common than in the past. Shrubs in the Mediterranean ecosystem are primarily evergreen with leathery, drought-tolerant leaves. Some of these shrubs, like holly, also have spines on their leaves. The climate in North Africa has become gradually warmer and drier, endangering many Mediterranean species. Human-caused damage to the vegetation has compounded the problem, with the result that more desertlike species have become established and grassland is all that remains in some areas.

Bryan Ness

Senecio abbreviatus *growing in the Robertson Karoo, South Africa. (S Molteno)*

FAUNA

Known for the enormous diversity and richness of its wildlife, Africa has a greater variety of large ungulates, or hoofed mammals (some ninety species), and freshwater fish (2,000 species) than any other continent. However, probably no group of animals is more identified with Africa than its flesh-eating carnivore mammals, of which there are more than sixty species. In addition to the better-known big, or "roaring," cats, such as lions, leopards, and cheetahs, there are wild dogs, hyenas, servals (long-limbed cats), wildcats, jackals, foxes, weasels, civets, and mongoose.

There are many theories as to why Africa has such an abundance of wildlife, and large wildlife at that. While early North American human societies drove mammoths, giant beavers, and saber-tooth tigers to extinction; early Europeans wiped out lions and rhinos; and Asians domesticated their landscapes, Africans lived in relative accord with creatures that were no less grand or ferocious. Some African folklore places animals on the same footing with people.

Another theory is that the tsetse fly, by spreading sleeping sickness, made much of tropical Africa uninhabitable by humans and protected the wilderness and wildlife from human depredation. Still another possibility may have been the constancy and small size of the African population in comparison with European and Asian numbers—too few people to either overhunt large mammals or exhaust their

Nile crocodile and pelicans at Lake Chamo, Ethiopia. (MauritsV)

HABITATS AND SELECTED VERTEBRATES OF AFRICA

- MEDITERRANEAN
  - Barbary Ape
- Desert Fox
- Dromedary
- Jerboa
- Skink
- Crocodile
- Tortoise
- DESERT
- Lizard
- Gazelle
- Addax
- Hyena
- Baboon
- Civet
- Cheetah
- GRASSLANDS
- Wild Dog
- Kudu
- Lowland Gorilla
- Elephant
- HIGHLANDS
- DESERT
- RAIN FOREST
- Mountain Gorilla
- Chimpanzee
- Nyala
- Leopard
- Hippopotamus
- Lion
- Elephant
- Buffalo
- Bushbuck
- Antelope
- GRASSLANDS
- Bushbaby
- Eland
- Giraffe
- Sidewinder
- Tenrec
- Gemsbok
- Black Rhinoceros
- Hippopotamus
- Chameleon
- Spoonbill
- Lemur
- DESERT
- Zebra
- Wildebeest
- Springbuck
- Flamingo
- Jerboa
- Impala

Mediterranean Sea
Red Sea
Gulf of Guinea
Indian Ocean
Atlantic Ocean

habitats. One more possibility is that the savannas of Africa—grassy plains with scattered tree cover—provide a good habitat for so many ungulates. Humans cannot hunt ungulates easily under these conditions because the humans can be seen and outrun easily. At the same time, a large population of ungulates supports an appreciable population of predators such as lions and scavengers such as hyenas.

Origin of African Fauna

At one time, most African fauna was believed to have originated in the Palearctic realm, that is, Europe, northwest Africa, and much of Asia. There is no doubt that as recently as 15,000 years ago, a milder Saharan climate allowed species such as clariid catfish to reach the river systems of North Africa. Similarly, northern animal life and vegetation seem to have extended far south into the Sahara.

The white rhinoceros evidently coexisted with elklike deer.

The spread of forests during the wetter epochs created separate northern and southern wooded grasslands. This led to the evolution of such closely related northern and southern species of antelope as the kob and puku, the Nile and common lechwe, and the northern and southern forms of white rhinoceros. In earlier periods, the animal life was even more remarkable than in modern times. Fossil deposits have revealed sheep as big as present-day buffalo, huge hippopotamuses, giant baboons, and other types similar to existing species. These "megafauna" probably lived in wetter periods and died out as the climate became drier.

Effects of Human Populations

The fine conditions for Africa's fauna in the mid- to late nineteenth century started to come to an end when European settlers arrived in many parts of Africa. Technologies, in the form of Western medicine and sanitation, sparked a demographic revolution. In places like Rhodesia (now Zimbabwe), the human population exploded twenty-fold during the ninety-year reign of white settlers.

Since 1940 the combined pressures of hunting and habitat destruction have cut wildlife numbers greatly. The antelope known as the Zambian black lechwe, believed to have numbered 1 million in 1900, had been reduced to fewer than 25,000 by 2019. The population of African elephants declined from 2 million in the early 1970s to 415,000 by 2018, largely because of poaching for the ivory trade.

Since the 1960s, poaching has caused a drastic decline in the world's black rhino population to about 5,000 in 2019. The African white rhinoceros stands on the verge of extinction, with only two indi-

Lion (Panthera leo) *lying down in Namibia. (Kavin Pluck)*

*Rough chameleon (*Trioceros rudis*), near Mount Karisimbi, Rwanda. Chameleon was moved from a bush, so is recolouring. (Charles J. Sharp)*

viduals still living in 2019. In West Africa, the continual southward advance of the Sahara Desert has amplified the twin pressures of habitat destruction and human population. The larger fauna that lived there, caught between the desert and the burgeoning human population, are largely gone now.

In Kenya, farmers have long since cleared most of the central part of the country that was once a densely forested region inhabited by wild animals. Some people have invaded national parks for commercial purposes such as logging and cattle grazing, thus forcing wildlife out of their preserved habitats. Some animals have had no alternative but to fight with humans for food and water. Along the Kenyan coast, many people have been attacked and killed by charging hippopotamuses and crocodiles in search of food and space. Similar cases have been common in the highlands, where Kenyans have lost their lives to charging elephants or to leopards and buffalo.

In recent years, human-elephant conflicts in Cameroon have become a major issue. Such con-flicts are more acute in the savanna ecosystem, due to the loss of the elephant's range and habitat following the conversion of natural vegetation to farmland and the logging of large tracts of forests. In 2020, Cameroon still has a herd of elephants estimated to number between 1,000 and 5,000. Poaching for ivory remains a problem there.

Another issue in Cameroon is the 652-mile (1,050-km.) pipeline that traverses tropical rainforests, linking oil fields in landlocked Chad to an export facility in Kribi, Cameroon. The original route of the pipeline was changed to pass through two less-fragile ecosystems, but the new route cuts through tropical areas and provides easier access for poachers to endangered species such as gorillas, chimpanzees, and elephants. Pipeline leaks in 2007 and 2010 particularly impacted underwater habitats.

Mammals

The main group of African herbivores is the antelope, which belong to four subfamilies of the ox family. The first subfamily is further subdivided into the African buffalo and the twist-horned antelope, including the eland (the largest of all antelopes), kudu, nyala, and bushbuck. The second subfamily is the duiker, a small primitive antelope that lives in the thickets, bush, and forests. Other well-known large African herbivores include the zebra, giraffe, hippopotamus, rhinoceros, and African elephant.

Africa's large number of endemic or native mammal species is second only to that of South America. These include several families of the ungulate order *Artiodactyla* (mammals with an even number of toes), such as the giraffe and hippopotamus. Some carnivores—such as civets, their smaller relations, the genets, and hyenas—are chiefly African. The rodent family of jumping hares is endemic, and one order, the aardvark, is exclusively African. Madagascar has a remarkable insect-eating family of mammals called tenrecs—animals with long, pointed snouts. Some tenrecs are spiny and tailless.

Primates

The primates include about forty-five species of Old World monkeys and two of the world's great apes: the chimpanzee and the world's largest ape, the gorilla. The gorilla is present in two subspecies: the lowland gorilla of Central and West Africa and the mountain gorilla of East Africa. The rare mountain gorillas live only in the upland forest on the borders of Uganda, Rwanda, and the Democratic Republic of the Congo. There are two populations: One is in Uganda's Bwindi Impenetrable Forest National Park, where in 2018 approximately 400 gorillas lived; the second is in the Virunga Mountains, where in 2018, an estimated 604 gorillas lived.

Prosimian primates include pottos or African lemurs and galagos, bush babies, or small arboreal lemurs. These and other African lemurs tend to be small and nocturnal. In Madagascar, where there are no true monkeys, the lemurs have occupied all ecological niches, both diurnal and nocturnal, that the monkeys would have taken. Accordingly, the

Vervet monkey, Kruger National Park, South Africa. (flowcomm)

Koppie foam grasshopper (Dictyophorus spumans spumans) *nymph, Walter Sisulu Botanical Gardens, Roodepoort, South Africa. (Charles J. Sharp)*

world's most diverse collection of prosimian lemurs survives in Madagascar.

Reptiles and Amphibians

Most African reptiles have their origins elsewhere—mainly in Asia. These include lizards of the agamid family, skinks, crocodiles, and tortoises. Endemic reptiles include girdle-tailed and plated lizards. Large vipers are common and diverse. Certain species have extremely toxic venom, but they are rarely encountered. One of the most noted is the black mamba. Amphibians also belong mainly to Old World groups. Salamanders and toothed tree frogs are confined to northwest Africa. Abundant and more common frogs and toads include such oddities as the so-called hairy frog of Cameroon, whose hairs are auxiliary respiratory organs.

Birds

Africa is home to about 2,300 bird species, approximately 1,500 of them endemic to the continent. An additional 275 species either reside in northwestern Africa or are winter migrants from Europe. In the past, there may have been as many as 2 billion individual migrants, but their numbers have been reduced considerably by severe droughts and by human land use and predation. Endemic bird species include the ostrich, shoebill, hamerkop, and secretary bird. The many predators of land mammals include eagles, hawks, and owls. Many more, such as storks, waders, and a few species of kingfishers, prey on fish. Even more feed on insects.

Insects

Insects include large butterflies, stick insects, mantises, grasshoppers, safari ants, termites, and dung beetles. Spiders abound throughout the continent, and scorpions and locusts can be plentiful locally. Huge swarms of locusts periodically spread over wide areas, causing enormous destruction to vegetation. Mosquitoes that carry malaria are present wherever there is a body of water. Female blackflies transmit the nematode *Onchocerca volvulus*, a parasitic filarial or threadlike worm. This organism eventually collects in many parts of the body, including the head near the eye. Nematode clusters around the eyes cause vision loss known as "river blindness," which has prevented any significant human habitation in many of Africa's river valleys.

Tsetse flies carry the parasite that causes African sleeping sickness in humans and nagana in livestock. These flies are found in all tropical portions of sub-Saharan Africa. The controversial chemical pesticide, DDT, which is banned throughout North America and Europe, is still being used in some areas of Africa to eliminate mosquitoes and tsetse flies. DDT has been shown to adversely affect birds and fish.

Conservation Efforts

African wildlife has been an important focus of conservation organizations for more than a century. Groups ranging from the World Wildlife Fund for Nature (WWF), the National Geographic Society, Greenpeace Africa, and the African Wildlife Foundation (AWF), among dozens of others, have worked tirelessly to monitor wildlife populations and to preserve their habitats. Research into species and how they interact with people is key to this work. And while there have been some notable bright spots, African wildlife is in serious trouble. Conservationists have documented the precipitous drop in African mammal populations for decades. Many species are either threatened or endangered.

The most highly publicized programs to conserve African wildlife have focused on the elimination of poaching. The World Bank estimates that more than US$1.3 billion were invested globally in tackling illegal wildlife trade between 2010 and 2016. Most of these funds went to protected-area management and law enforcement. These efforts present a huge and continuing challenge. But to many conservationists, the antipoaching efforts mask a much bigger threat to African wildlife: habitat destruction. Every day, wildlife is losing space to agriculture, infrastructure, and urbanization.

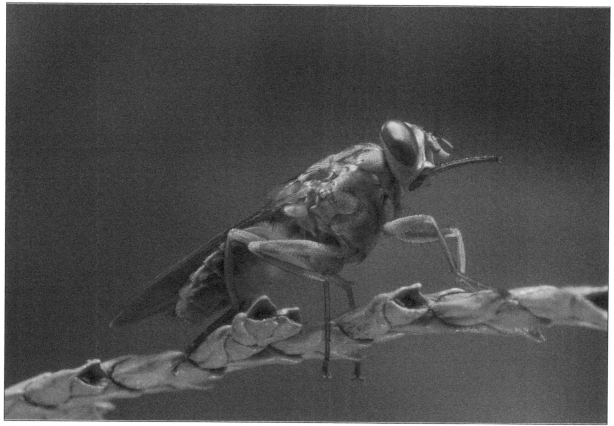

Tsetse fly from Burkina Faso. (International Atomic Energy Agency)

Spotted thick-knee/spotted dikkop (Burhinus capensis) *in burnt grassland at Marievale Nature Reserve, Gauteng, South Africa.*

The African Wildlife Foundation, the leading conservation organization committed exclusively to protecting Africa's wildlife and wildlife habitats, has approached this challenge by training hundreds of people in African communities in the techniques, efforts, and goals of conservation. The AWF's primary aim is land preservation: ensuring that large open landscapes are available for wildlife. A number of large conservation organizations have enacted programs that support existing protected areas, create private land trusts, and work with local community groups.

In recent years, wildlife tourism has emerged as one of Africa's fastest growing industries. In 2018 alone, more than 67 million tourists visited Africa, a 7 percent increase from the year before. And although wildlife tourism has some controversial aspects, it creates jobs, generates income, and ensures large areas of land are maintained for wildlife. Nevertheless, tourism, while often promoted as a pathway to conservation and a driver for local jobs and income, is only possible in some places and at certain times of the year.

The critically endangered black rhinoceros, up to 3.75 meters (12.3 feet) long, is threatened by poaching. (Ikiwaner)

Wildlife conservationists agree that a fundamental problem is that those in the best position to preserve wildlife—residents of local communities —often lack incentives to do so. In many local communities, African wildlife is highly valued, even revered, for its sheer existence. Wildlife is often closely intertwined with the indigenous culture, and its loss can diminish important cultural ties. To turn things around, wildlife management must become a viable option for them.

Governments, tour operators, travel agencies, hotels and lodges, and tourists from overseas reap the primary benefits of wildlife conservation, much more so than do the residents of the local communities. Local people often bear significant costs, and frequently are subject to restricted access to or even physical eviction from conservation areas. They are also most subject to conflict with wildlife, which may eat or trample crops, destroy property, or kill livestock or people. The future of African wildlife hinges to a large extent on how well these challenges are resolved and what policies are enacted that recognize the key role of local people in advancing conservation efforts.

Dana P. McDermott

Endangered Mammals of Africa

Common Name	Scientific Name	Range
Aye-Aye	*Daubentonia madagascariensis*	Madagascar
Cat, black-footed	*Felis nigripes*	Southern Africa
Cheetah	*Acinonyx jubatus*	Africa to India
Chimpanzee	*Pan troglodytes*	Africa
Eland, western giant	*Taurotragus derbianus*	Senegal to Ivory Coast
Gazelle, mountain	*Gazella cuvieri*	Morocco, Algeria, Tunisia
Gorilla	*Gorilla*	Central and western Africa
Hyena, brown	*Parahyaena brunnea*	Southern Africa
Impala, black-faced	*Aepyceros melampus petersi*	Namibia, Angola
Lemurs	*Lemuridae*	Madagascar
Mandrill	*Mandrillus sphinx*	Equatorial West Africa
Rhinoceros, black	*Diceros bicornis*	Sub-Saharan Africa

Source: U.S. Fish and Wildlife Service, U.S. Department of the Interior.

Discussion Questions: Flora/Fauna

Q1. How have geographers classified Africa's biomes? Which biome is the most widespread? How does plant life vary among the biomes?

Q2. Where do rainforests occur in Africa? How do they compare size-wise to Africa's deserts and grasslands? What challenges do Africa's rainforests face?

Q3. How did Africa come to support such a diverse group of large carnivorous mammals? What impact did European explorers and settlers have on Africa's wildlife? What are the main ongoing approaches being taken by wildlife-conservation programs in Africa?

HUMAN GEOGRAPHY

OVERVIEW

THE HUMAN ENVIRONMENT

No person lives in a vacuum. Every human being and community is surrounded by a world of external influences with which it interacts and by which it is affected. In turn, humans influence and change their environments: sometimes intentionally, sometimes not, and sometimes with effects that are harmful to these environments, and, in turn, to humans themselves. Humans have always shaped the world in which they live, but developments over the past few centuries have greatly enhanced this capacity.

Many people feel a sense of alarm about the consequences of widespread adoption of modern technology, including artificial intelligence (AI) and accelerating human population growth in the world. Travel and transportation among the world's regions have been made surer, safer, and faster, and global communication is virtually instantaneous. The human environment is no longer a matter of local physical, biological, or social conditions, or even of merely national or regional concerns—the postmodern world has become a true global community.

Students of human geography divide the human environment into three broad areas: the physical, biological, and social environments. The study of ecology describes and analyzes the interactions of biological forms (mainly plants and animals) and seeks to uncover the optimal means of species cooperation, or symbiosis. Everything that humans do affects life and the physical world around them, and this world provides potentials for and constraints on how humans can live.

As people acquired and shared ever-more knowledge about the world, their abilities to alter and shape it increased. Humans have always had a direct impact on Earth. Even 10,000 years ago, Neolithic people cut down trees, scratched the earth's surface with simple plows, and replaced diverse plant forms with single crops. From this basic agricultural technology grew more complex human communities, and people were freed from the need to hunt and gather. The alteration of the local ecosystems could have deleterious effects, however, as gardens turned eventually to deserts in places like North Africa and what later became Iraq. Those who kept herds of animals grazed them in areas rich in grasses, and animal fertilizer helped keep them rich. If the area was overgrazed, however, destroying important ground cover, the herders moved on, leaving a perfect setup for erosion and even desertification. Today, people have an even greater ability to alter their environments than did Neolithic people, and ecologists and other scientists as well as citizens and politicians are increasingly concerned about the negative effects of modern alterations.

The Physical Environment

The earth's biosphere is made up of the atmosphere—the mass of air surrounding the earth; the hydrosphere—bodies of water; and the lithosphere—the outer portion of the earth's crust. Each of these, alone and working together, affect human life and human communities.

Climate and weather at their most extreme can make human habitation impossible, or at least extremely uncomfortable. Desert and polar climates do not have the liquid water, vegetation, and animal life necessary to sustain human existence. Humans can adapt to a range of climates, however. Mild variations can be addressed simply, with clothing and shelter. Local droughts, tornadoes, hurricanes, heavy winds, lightning, and hail can have devastating effects even in the most comfortable of climates. Excess rain can be drained away to make habitable land, and arid areas can be irrigated. Most people live in temperate zones where weather extremes are rare or dealt with by technological adaptation. Heating and, more recently, air conditioning can create healthy microclimates, whatever the external conditions. Food can be grown and then transported across long distances to supply populations throughout the year.

The hydrosphere affects the atmosphere in countless ways, and provides the water necessary for human and other life. Bodies of water provide plants and animals for food, transportation routes, and aesthetic pleasure to people, and often serve to flush away waste products. People locate near water sources for all of these reasons, but sometimes suffer from sudden shifts in the water level, as in tidal waves (tsunamis) or flooding. Encroachment of salt water into freshwater bodies (salination) is a problem that can have natural or human causes.

The lithosphere provides the solid, generally dry surface on which people usually live. It has been shaped by the atmosphere (especially wind and rain that erode rocks into soil) and the hydrosphere (for example, alluvial deposits and beach erosion). It serves as the base for much plant life and for most agriculture. People have tapped its mineral deposits and reshaped it in many places; it also reshapes itself through, for example, earthquakes and volcanic eruption. Its great variations—including vegetation—draw or repel people, who exploit or enjoy them for reasons as varied as recreation, military defense, or farming.

The Biological Environment

Humans share the earth with over 8 million different species of plants, animals, and microorganisms—of which only about 2 million have been identified and named. As part of the natural food chain, people rely upon other life-forms for nourishment.

Through perhaps the first 99 percent of human history, people harvested the bounty of nature in its native setting, by hunting and gathering. Domestication of plants and animals, beginning about 10,000 years ago, provided humans a more stable and reliable food supply, revolutionizing human communities. Being omnivores, people can use a wide variety of plants and animals for food, and they have come to control or manage most important food sources through herding, agriculture, or mechanized harvesting. Which plants and animals are chosen as food, and thus which are cultivated, bred, or exploited, are matters of human culture, not, at least in the modern world, of necessity.

Huge increases in human population worldwide have, however, put tremendous strains on provision of adequate nourishment. Areas poorly endowed with foodstuffs or that suffer disastrous droughts or blights may benefit from the importation of food in the short run, but cannot sustain high populations fostered by medical advances and cultural considerations.

Human beings themselves are also hosts to myriad organisms, such as fungi, viruses, bacteria, eyelash mites, worms, and lice. While people usually can coexist with these organisms, at times they are destructive and even fatal to the human organism. Public health and medical efforts have eradicated some of humankind's biological enemies, but others remain, or are evolving, and continue to baffle modern science.

The presence of these enemies to health once played a major role in locating human habitations to avoid so-called "bad air" (*mal-aria*) and the breeding grounds of tsetse flies or other pests. The use of pesticides and draining of marshy grounds have alleviated a good deal of human suffering. Human efforts can also control or eliminate biological threats to the plants and animals used for food, clothing, and other purposes.

Social Environments

Human reproduction and the nurturing of young require cooperation among people. Over time, people gathered in groups that were diverse in age if not in other qualities, and the development of towns and cities eventually created an environment in which otherwise unrelated people interacted on intimate and constructive levels. Specialization, or division of labor, created a higher level of material wealth and culture and ensured interpersonal reliance.

The pooling of labor—both voluntary and forced—allowed for the creation of artificial living environments that defied the elements and met human needs for sustenance. Some seemingly basic human drives of exclusivity and territoriality may be responsible for interpersonal friction, violence and, at the extreme, war. Physical differences, such as size, skin, or hair color, and cultural differences, including language, religion, and customs, have often divided humans or communities. Even within close quarters such as cities, people often separate themselves along lines of perceived differences. Human social identity comes from shared characteristics, but which things are seen as shared, and which as differentiating, is arbitrary.

People can affect their social environment for good and ill through trade and war, cooperation and bigotry, altruism and greed. While people still are somewhat at the mercy of the biological and physical environments, technological developments have balanced the human relationship with these. Negative effects of human interaction, however, often offset the positive gains. People can seed clouds for rain, but also pollute the atmosphere around large cities, create acid rain, and perhaps contribute to global warming.

Human actions can direct water to where it is needed, but people also drain freshwater bodies and increase salination, pollute streams, lakes, and oceans, and encourage flooding by modifying riverbeds. People have terraced mountainsides and irrigated them to create gardens in mountains and deserts, but also lose about 24 billion metric tons of soil to erosion and 30 million acres (12 million hectares) of grazing land to desertification each year. These negative effects not only jeopardize other species of terrestrial life, but also humans' ability to live comfortably, or perhaps at all.

Globalization

Humankind's ability to affect its natural environments has increased enormously in the wake of the Industrial Revolution. The harnessing of steam, chemical, electrical, and atomic energy has enabled people to transform life on a global scale. Economically, the Western world still to dominates global markets despite effort of China to capture the crown, and computer and satellite technology have made even remote parts of the globe reliant on Western information and products. Efficient transportation of goods and people over huge distances has eliminated physical barriers to travel and commerce. The power and influence of multinational corporations and national corporations in international markets continues to grow.

Human environmental problems also have a global scope: Extreme weather, changes in ocean temperatures and sea level rise, global warming, and the spread of disease by travelers have become planetary concerns. International agencies seek to deal with such matters, and also social and political concerns once left to nations or colonial powers, such as population growth, the provision of justice, or environmental destruction within a country. Pessimists warn of horrendous trends in population and ecological damage, and further deterioration of human life and its environments. Optimists dismiss negative reports as exaggerated and alarmist, or expect further technological advances to mitigate the negative effects of human action.

Joseph P. Byrne

POPULATION GROWTH AND DISTRIBUTION

The population of the world has been growing steadily for thousands of years and has grown more in some places than in others. On November 2019, the total population of the earth had reached 7.7 billion people. The population of the United States in August 2019 was approximately 329.45 million. India's population in November 2019 was 1.37 billion, making it the world's second most populous country. China's population was about 1.45 billion—about 1 in 5 people on the planet.

How Populations Are Counted

The U.S. Constitution requires that a census, or enumeration, of the population of the United States be conducted every ten years. The U.S. Census Bureau mails out millions of census forms and pays thousands of people (enumerators) to count people that did not fill out their census forms. This task cost about US$5.6 billion in the year 2010, and estimates for the 2020 census have risen to over US$15 billion. Despite this great effort, millions of people are probably not counted in every U.S. census. Moreover, many countries have much less money to spend on censuses and more people to count. Therefore, information about the population of many poor or less-developed countries is even less accurate than that for the population of the United States.

Counting how many people were alive a hundred, a thousand, or hundreds of thousands of years ago is even more difficult. Estimates are made from archaeological findings, which include human skeletons, ruins of ancient buildings, and evidence of ancient agricultural practices. Historical records of births, deaths, taxes paid, and other information are also used. Although it is not possible to estimate the global population 1,000 years ago with great accuracy, it is a fascinating topic, and many people have participated in estimating the total population of the planet through the ages.

History of Human Population Growth

Ancient ancestors of humans, known as hominids, were alive in Africa and Europe around 1 million years ago. It is believed that modern humans (*Homo sapiens sapiens*) coexisted with the Neanderthals (*Homo sapiens neandertalensis*) about 100,000 years ago. By 8000 BCE (10,000 years ago) fully modern humans numbered around 8 million. If the presence of archaic *Homo sapiens* is accepted as the beginning of the human population 1 million years ago, then the first 990,000 years of human existence are characterized by a very low population growth rate (15 persons per million per year).

Around 10,000 years ago, humans began a practice that dramatically changed their growth rate: planting food crops. This shift in human history, called the Agricultural Revolution, paved the way for the development of cities, government, and civilizations. Before the Agricultural Revolution, there were no governments to count people. The earliest censuses were conducted less than 10,000 years ago in the ancient civilizations of Egypt, Babylon, China, Palestine, and Rome. For this reason, historical estimates of the earth's total population are difficult to make. However, there is no argument that human numbers have increased dramatically in the past 10,000 years. The dramatic changes in the growth rates of the human population are typically attributed to three significant epochs of human cultural evolution: the Agricultural, Industrial, and Green Revolutions.

Before the Agricultural Revolution, the size of the human population was probably fewer than 10 million people, who survived primarily by hunting and gathering. After plant and animal species were domesticated, the human population increased its growth rate. By about 5000 BCE, gains in food production caused by the Agricultural Revolution meant that the planet could support about 50 million people. For the next several thousand years, the human population continued to grow at a rate of about 0.03 percent per year. By the first year of the common era, the planet's population numbered about 300 million.

At the end of the Middle Ages, the human population numbered about 400 million. As people lived in densely populated cities, the effects of disease increased. Starting in 1348 and continuing to 1650, the human population was subjected to massive declines caused by the bubonic plague—the Black Death. At its peak in about 1400, the Black Death may have killed 25 percent of Europe's population in just over fifty years. By the end of the last great plague in 1650, the human population numbered 600 million.

The Industrial Revolution began between 1650 and 1750. Since then, the growth of the human population has increased greatly. In just under 300 years, the earth's population went from 0.5 billion to 7.7 billion people, and the annual rate of increase went from 0.1 percent to 1.1 percent. This population growth was not because people were having more babies, but because more babies lived to become adults and the average adult lived a longer life.

The Green Revolution occurred in the 1960s. The development of various vaccines and antibiotics in the twentieth century and the spread of their use to most of the world after World War II caused big drops in the death rate, increasing population growth rates. Feeding this growing population has presented a challenge. This third revolution is called the Green Revolution because of the technology used to increase the amount of food produced by farms. However, the Green Revolution was really a combination of improvements in health care, medicine, and sanitation, in addition to an increase in food production.

Geography of Human Population Growth

The present-day human race traces its lineage to Africa. Humans migrated from Africa to the Middle East, Europe, Asia, and eventually to Australia, North and South America, and the Pacific Islands. It is believed that during the last Ice Age, the world's sea levels were lower because much of the world's water was trapped in ice sheets. This lower sea level created land bridges that facilitated many of the major human migrations across the world.

Patterns of human settlement are not random. People generally avoid living in deserts because they lack water. Few humans are found above the Arctic Circle because of that region's severely cold climate. Environmental factors, such as the availability of water and food and the livability of climate, influence where humans choose to live. How much these factors influence the evolution and development of human societies is a subject of debate.

The domestication of plants and animals that resulted from the Agricultural Revolution did not take place everywhere on the earth. In many parts of the world, humans remained as hunter-gatherers while agriculture developed in other parts of the world. Eventually, the agriculturalists outbred the hunter-gatherers, and few hunter-gatherers remain in the twenty-first century. Early agricultural sites have been found in many places, including Central and South America, Southeast Asia and China, and along the Tigris and Euphrates Rivers in what is now Iraq. The practice of agriculture spread from these areas throughout most of the world.

By the time Christopher Columbus reached the Americas in the late fifteenth century, there were millions of Native Americans living in towns and villages and practicing agriculture. Most of them died from diseases that were brought by European colonists. Colonization, disease, and war are major mechanisms that have changed the composition and distribution of the world's population in the last 300 years.

The last few centuries also produced another change in the geography of the human population. During this period, the concentration of industry in urban areas and the efficiency gains of modern agricultural machinery caused large numbers of people to move from rural areas to cities to find jobs. From 1900 to 2020 the percentage of people living in cities went from 14 percent to just about 55 percent. Demographers estimate that by the year 2025, more than 68 percent of the earth's population will live in cities. Scientists estimate that the human population will continue to increase until the year

2050, at which time it will level out at between 8 and 15 billion.

Earth's Carrying Capacity

Many people are concerned that the earth cannot grow enough food or provide enough other resources to support 15 billion people. There is great debate about the concept of the earth's carrying capacity—the maximum human population that the earth can support indefinitely. Answers to questions about the earth's carrying capacity must account for variations in human behavior. For example, the earth could support more bicycle-riding vegetarians than car-driving carnivores. Questions about carrying capacity and the environmental impacts of the human race on the planet are fundamental to the United Nations' goals of sustainable development. Dealing with these questions will be one of the major challenges of the twenty-first century.

Paul C. Sutton

GLOBAL URBANIZATION

Urbanization is the process of building and living in cities. Although the human impulse to live in groups, sharing a "home base" probably dates back to cave-dweller times or before. The creation of towns and cities with a few hundred to many thousands to millions of inhabitants required several other developments.

Foremost of these was the invention of agriculture. Tilling crops requires a permanent living place near the cultivated land. The first agricultural villages were small. Jarmo, a village site from c. 7000 BCE, located in the Zagros Mountains of present-day Iran, appears to have had only twenty to twenty-five houses. Still, farmers' crops and livestock provided a food surplus that could be stored in the village or traded for other goods. Surplus food also meant surplus time, enabling some people to specialize in producing other useful items, or to engage in less tangible things like religious rituals or recordkeeping.

Given these conditions, it took people with foresight and political talents to lead the process of city formation. Once in cities, however, the inhabitants found many benefits. Walls and guards provided more security than the open country. Cities had regular markets where local craftspeople and traveling merchants displayed a variety of goods. City governments often provided amenities like primitive street lighting and sanitary facilities. The faster pace of life, and the exchange of ideas from diverse people interacting, made city life more interesting and speeded up the processes of social change and invention. Writing, law, and money all evolved in the earliest cities.

Ancient and Medieval Cities

Cities seem to have appeared almost simultaneously, around 3500 BCE, in three separate regions. In the Fertile Crescent, a wide curve of land stretching from the Persian gulf to the northwest Mediterranean Sea, the cities of Ur, Akkad, and Babylon rose, flourished, and succeeded one another. In Egypt, a connected chain of cities grew, soon unified by a ruler using Memphis, just south of the Nile River's delta, as his strategic and ceremonial base. On the Indian subcontinent, Mohenjo-Daro and Harappa oversaw about a hundred smaller towns in the Indus River valley. Similar developments took place about a thousand years later in northern China.

These first city sites were in the valleys of great river systems, where rich alluvial soil boosted large-scale food production. The rivers served as a "water highway" for ships carrying commodities and luxury items to and from the cities. They also furnished water for drinking, irrigation, and waste disposal. Even the rivers' rampages promoted civilization, as making flood control and irrigation systems required practical engineering, an organized workforce, and ongoing political authority to direct them.

Eurasia was still full of peoples who were not urbanized, however, and who lived by herding, pirating, or raiding. Early cities declined or disappeared, in some cases destroyed by invasions from such forces around 1200 BCE. Afterward, the cities of Greece became newly important. Their surrounding land was poor, but their access to the sea was an advantage. Greek cities prospered from fishing and trade. They also developed a new idea, the city-state, run by and for its citizens.

Rome, the Greek cities' successor to power, reached a new level of urbanization. Its rise owed more to historical accident and its citizens' political and military talents than to location, but some geographical features are salient. In some ways, the fertile coastal plain of Latium was an ideal site for a great city, central to both the Italian peninsula and the Mediterranean Sea. There, the Tiber River becomes navigible and crossable.

In other ways, Rome's site was far from ideal. Its lower areas were swampy and mosquito-ridden. The seven hills, with their sacred sites later filled with public buildings and luxury houses, imposed a crazy-quilt pattern on the city's growth. Romans built cities with a simple rectangular plan all over Europe and the Middle East, but their home city grew in a less rational way.

At its peak, Rome had a million residents, a population no other city reached before nineteenth century London. It provided facilities found in modern cities: a piped water supply, a sewage disposal system, a police force, public buildings, entertainment districts, shops, inns, restaurants, and taverns. The streets were crowded and noisy; to control traffic, wheeled wagons could make deliveries only at night. Fire and building collapse were constant risks in the cheaply built apartment structures that housed the city's poorer residents. Still, few wanted to live anywhere but in Rome, their world's preeminent city.

In the Early Middle Ages after the western Roman Empire collapsed, feudalism, based on land holdings, eclipsed urban life. Cities never disappeared, but their populations and services declined drastically. Urban life still flourished for another millenium in the eastern capital of Constantinople. When Islam spread across the Middle East, it caused the growth of new cities, centered around a mosque and a marketplace.

In the twelfth and thirteenth centuries, life revived in Western Europe. As in the Islamic cities, the driving forces were both religious—the building of cathedrals—and commercial—merchants and artisans expanding the reach of their activities. Medieval cities were usually walled, with narrow, twisting streets and a lack of basic sanitary measures, but they drew ambitious people and innovative forces together. Italy's cities revived the concept of the city-state with its outward reach. Venice sent its merchant fleet all over the known world. Farther north, Paris and Bologna hosted the first universities. The feudal system slowly gave way to nation-states ruled by one king.

Modern Cities

Modern cities differ from earlier ones because of changes wrought by technology, but most of today's cities arose before the Industrial Revolution. Until the early nineteenth century, travel within a city was by foot or on horse, which limited street widths and city sizes. The first effect of railroads was to shorten travel time between cities. This helped country residents moving to the cities, and speeded raw materials going into and manufactured goods coming out of the factories that increasingly dotted urban areas. Rail transit soon caused the growth of a suburban ring. Prosperous city workers could live in more spacious homes outside the city and ride rail lines to work every day. This pattern was common in London and New York City.

Factories, the lifeblood of the Industrial Revolution, were built in pockets of existing cities. Smaller cities like Glasgow, Scotland, and Pittsburgh, Pennsylvania, grew as ironworking industries, using nearby or easily transported coal and ore resources, built large foundries there. Neither industrialists nor city authorities worried about where the people working there would live. Workers took whatever housing they could find in tenements or subdivided old mansions.

Beginning in the 1880s, metal-framed construction made taller buildings possible. These skyscrapers towered over stately three- to eight-story structures of an earlier period. Because this technology enabled expensive central-city ground space to house many profitable office suites, up through the 1930s, city cores became quite compacted. Many people believed such skyward growth was the wave of the future and warned that city streets were becoming sunless, dangerous canyons.

Automobiles kept these predictions from fully coming true. As car ownership became widespread, more roads were built or widened to carry the traffic. Urban areas began to decentralize. The car, like rail transit before it, allowed people to flee the urban core for suburban living. Because roads could be built almost anywhere, built-up areas around cities came to resemble large patches filling a circle, rather than the spokes-of-a-wheel pattern introduced by rail lines. Cities born during the automotive age tend to have an indistinct city center, surrounded by large areas of diffuse urban development. The prime example is Los Angeles: It has a small downtown area, but a consolidated metropolitan area of about 34,000 square miles (88,000 sq. km.).

Almost everywhere, urban sprawl has created satellite cities with major manufacturing, office, and shopping nodes. These cause an increasing portion of daily travel within metropolitan areas to be between one edge city and another, rather than to and from downtown. Since these journeys have an almost limitless variety of start points and destinations within the urban region, mass transit is only a partial solution to highway crowding and air pollution problems.

The above trends typify the so-called developed world, especially the United States. Many cities in poor nations have grown even more rapidly but with a different mix of patterns and problems. However, the basic pattern can be detected around the globe, as urban dwellers seek to better their own cir-

URBANIZATION AND DEVELOPING NATIONS

The urban population, or number of people living in cities, in North America accounts for about 75 percent of its total population. In Europe, about 90 percent of the population lives in cities. In developing countries, the urban population is often less than 30 percent. The term "urbanization" refers to the rate of population growth of cities. Urbanization mainly results from people moving to cities from elsewhere. In developing countries, the urbanization rate is very high compared to those of North America or Europe. The high rate of urbanization of these countries makes it difficult for their governments to provide housing, water, sewers, jobs, schools, and other services for their fast-growing urban populations.

cumstances. Today, 55 percent of the world's population lives in urban areas, and that percentage is expected to rise to 68 percent by 2050. Projections show that urbanization combined with the overall growth of the world's population could add another 2.5 billion people to urban areas by 2050, with close to 90 percent of this increase taking place in Asia and Africa, according to a United Nations data set published in May 2018.

Megacities and the Future

In the year 2019 the world had thirty-three megacities, defined as urban areas with a population of 10 million or more. The largest was Tokyo, with an estimated 37.5 million people in 2018, predicted to grow to around 37 million by 2030. Second-largest was Delhi, with more than 28.5 million in 2018 and predicted to grow to around 38.94 million by 2030. Megacities in the United States include New York-Newark with a population of 18.8 million and Los Angeles at 12.5.

Megacities profoundly affect the air, weather, and terrain of their surrounding territory. Smog is a feature of urban life almost everywhere, but is worse where the exhaust from millions of cars mixes with industrial pollution. Some megacities have slowed the problem by regulating combustion technology; none have solved it. Huge expanses of soil pre-

URBAN HEAT ISLANDS

Large cities have distinctly different climates from the rural areas that surround them. The most important climatic characteristic of a city is the urban heat island, a concentration of relatively warmer temperatures, especially at nighttime. Large cities are frequently at least 11 degrees Fahrenheit (6 degrees Celsius) warmer than the surrounding countryside.

The urban heat island results from several factors. Primary among these are human activities, such as heating homes and operating factories and vehicles, that produce and release large quantities of energy to the atmosphere. Most of these activities involve the burning of fossil fuels such as oil, gas, and coal. A second factor is the abundance of heat-absorbing urban materials, such as brick, concrete, and asphalt. A third factor is the surface dryness of a city. Urban surface materials normally absorb little water and therefore quickly dry out after a storm. In contrast, the evaporation of moisture from wet soil and vegetation in rural areas uses a large quantity of solar energy—often more than is converted directly to heat—resulting in cooler air temperatures and higher relative humidities.

empted by buildings and pavements can turn heavy rains into floods almost instantly, and the ambient heat in large cities stays several degrees higher than in comparable rural areas. Recent engineering studies suggest that megacities create instability in the ground beneath, compressing and undermining it.

How will cities evolve? Barring an unforeseen technological or social breakthrough, the current growth and problems will probably continue. The process of megapolis—metropolitan areas blending together along the corridors between them—is well underway in many areas. Predictions that the computer will so change the nature of work as to cause massive population shifts away from cities have not been proven correct. Despite its drawbacks, increasing numbers of people are drawn to urban life, seeking the economic opportunities and wider social world that cities offer.

Emily Alward

PEOPLE

The study of how humans and their prehuman predecessors came to increase in number and eventually spread across the African continent is one of the most dramatic and fascinating stories in history. Archaeologists, paleontologists, and physical anthropologists have discovered, examined, and analyzed. fossil remains, tools, and artifacts. From these studies, researchers have been able to discover what prehumans and humans were like millions of years ago.

Additional scientific fields of study including genetics, botany, zoology, linguistics, and cultural anthropology have provided information on where humans came from, where they went, and when they began to increase in numbers and migrate across thousands of miles of rivers, deserts, grasslands, and forests. Combining all these tools with history and geography has provided the basis for determining how much cultures changed after humans traveled over so vast a space and during so long a period of time.

Over millennia, humans rose from the Great Rift Valley of Africa and followed the Sun across the grasslands of the Sahel (the semiarid belt south of the Sahara) in search of game and dependable sources of water. As they discovered how to make and use stone tools, humans increased in number and began to fill the continent. Then, the Sahara cooled and shared its bounty during a brief wet phase. Humans slowly moved in and learned to domesticate both animals and plants.

San people in Namibia. (Rüdiger Wenzel)

These skills ultimately became the building blocks of an Egyptian culture in North Africa. With the arrival of the Iron Age, the last barrier to human expansion was overcome. People relentlessly expanded across the breadth of sub-Saharan Africa and eventually established a cultural environment that remained much the same until the expansion of Islam and European colonialism.

Origins of Human Population

For many years, people believed that humans came out of a Garden of Eden in the Middle East or the Far East. The English naturalist Charles Darwin, writing in the nineteenth century, suggested that humans must have come from another place. He reasoned that humans were tropical animals and that they must be related to monkeys, since those were the animals most like humans. He subsequently concluded that humans must have originated in Africa, the place that combined the largest tropical landmass with the site of the most varieties of primates.

A number of scientific discoveries in the twentieth century supported Darwin's theory. Robert Broom and Raymond Dart found archaeological evidence of australopithecines in southern Africa. Australopithecines were a type of human-like ancestor called hominids. The australopithecines were an important evolutionary step toward *Homo sapiens*, or modern humans. They were only about 4 feet, 6 inches (137 centimeters) tall, and they stood upright when they walked. More important, their hands were enough like ours to allow them to make simple tools from bone or stone.

Beginning in the 1960s, Mary Douglas Leakey and her husband, Louis B. Leakey, made several additional discoveries in East Africa that helped to place the origin of humans in Africa. They uncovered numerous fossils and artifacts of australopithecines at Olduvai Gorge in the Great Rift Valley. Many of these were from the Lower and Middle Pleistocene geological epochs, between 1.5 million and 2 million years ago.

By examining these remains within the context of where they were found, and what that area was like at that time, the Leakeys determined that small

RACES AND CULTURES IN ANCIENT EGYPT

Although much controversy exists over the racial makeup of ancient Egypt, the important fact is that ancient Egypt thrived because of its diversity. For thousands of years, black Africans from south of the Sahara commingled with the Afro-Asiatic people who lived along the Mediterranean Sea to the north. A similar process took place along the Nile Valley, as the people from the Upper Nile commingled with those living along the Lower Nile. This process was often fraught with conflict, however, which continued for many years.

The growth of Egyptian civilization was driven not only by Asian immigrants from the Near East, as had previously been supposed, but also by a much larger number of immigrants from the west and southwest. As the climate of the Sahara became drier after 2500 BCE, the desert began to slowly reclaim its forests, grasslands, rivers, and lakes, and the people living there began to disperse to its periphery. Many of them moved to the north and west to form a vibrant Berber culture that learned to survive in the harsh dry climate, but most seem to have moved eastward to Egypt. It seems clear that ancient Egyptian culture was not solely a foreign invention imported from the Near East, but an African innovation built upon the experiences and efforts of people of many races and cultures. When the Greek historian Herodotus traveled through Egypt after 450 BCE, he saw clearly that Egypt's cultural roots lay in Africa.

bands of these human-like ancestors had lived along the edges of small ponds, subsisting largely on plants and small animals. After reviewing information on the most recent finds across Africa, the Leakeys decided that three important steps in human evolution had taken place in Africa. First, the original ancestors of all apes had arisen in the Nile region at some point during the Oligocene epoch, between 30 million and 40 million years ago.

Second, the branch of apes from which humans had emerged must have broken away from the others during the late Miocene or early Pliocene epochs, about 12 million years ago. Third, approximately 2 million years ago, toward the end of the Pliocene epoch, "true humans," who had eventually become *Homo sapiens*, had separated from "near humans," or the australopithecines in East Africa. Many scientists came to agree with the Leakeys' conclusion that humans had come out of Africa.

Historical Migration Patterns

The australopithecines began to use simple stone tools about 3 million years ago, during the Stone Age. Over an immense length of time, these tools became heavier and more sophisticated. A great advantage of tools is that they enabled their users to become better hunters. Although they were only able to kill small- or medium-sized animals in the beginning, humans could now kill them more frequently and process them more efficiently. Meat was an important addition to the humans' diet because it allowed them to increase their nutrients and caloric intake. As their diet improved, people began to live longer and have more children. The population slowly increased.

Sometimes families grew rather large. Anthropologists use the term "clan" to describe an extended family of grandparents, children, and grandchildren. When game became too scarce in a particular area, clans typically would divide into smaller groups and move to new areas. This process of growth, division, and migration is similar to what bees do when their hives become too crowded or when food becomes scarce. This process of "hiving"

Mary and Louis Leakey at Olduvai Gorge, Tanzania, Africa. (National Geographic Society)

began to accelerate among the hominids about 1 million years ago as the quantity and quality of stone tools vastly improved.

Many scientists now suggest that the various human species arose in Africa about 200,000 years ago during the Middle Pleistocene epoch. Then, about 50,000 years ago, during the Upper Pleistocene epoch, those human populations began to increase rapidly. This was largely because humans' distant relatives, *Homo neanderthalensis* and *Homo rhodesiensis*, and direct relatives *Homo sapiens*, had learned how to make fire, live in caves, and carry burdens. These skills enabled them to protect and feed larger groups in relative comfort, settle in one area longer, and travel greater distances. Slowly migrating westward from the Great Rift Valley to the Congo basin, along the grasslands north of the Niger and Benue rivers, and then into the Sahel, humans had covered most of sub-Saharan Africa by sometime between 50,000 and 60,000 years ago.

Scientists have learned a great deal about what Africa was like 50,000 years ago. They have determined that many of the outward physical variations that make people from various regions appear different from one another had already occurred. They have also learned that more than half of the humans on Earth during this period were living in sub-Saharan Africa. One of the means researchers have used to determine this is through genetics—the science of heredity and the variation of organisms from generation to generation.

Skull of "Mrs. Ples," Australopithecus africanus, in the Transvaal Museum Pretoria. (José Braga; Didier Descouens)

By studying particular components and attributes of the DNA (deoxyribonucleic acid) of numerous subjects from sub-Saharan Africa, East Asia, and Europe, genetic researchers have determined that populations living in sub-Saharan Africa have a higher degree of in-group diversity than people living in any other part of the world. The high degree of genetic diversity within the African test group led to the conclusion that the human population in sub-Saharan Africa 50,000 years ago had been larger than the human populations of Asia and Europe combined. It was also evidence that the human population there had grown more rapidly than anywhere else. One reason that humans in Africa increased in number was that they had found new and better ways to feed themselves. As their numbers increased, they sought new ways to obtain food and new places to live.

The Pleistocene epoch was beginning to come to a close during the Middle Stone Age, about 35,000 BCE. During this period, *Homo sapiens* began to replace other types of humans across Africa. About 10,000 BCE, as the Middle Stone Age was ending, a climatic change took place in Africa that profoundly affected the future of nearly all humans in Africa.

The Sahara Desert slowly grew cooler during this period, and rain occasionally fell. In time, grasses started to grow, and small rivers began to appear in what had been an inhospitable desert. Humans inevitably followed wild animals into the Sahara. As humans living along the Mediterranean Sea to the north moved down into the Sahara, they began to mingle with those who were moving up from the Sahel in the south.

Eventually, the people who inhabited the Sahara began to domesticate animals and then plants, which heralded the agricultural revolution. This was a significant advantage, because now they could increase the number of calories, they could obtain from the areas they occupied. As their diets improved, their numbers increased once again, and they began to migrate even farther across the continent. Archaeologists have discovered that people were managing herds of wild cattle in Egypt's western desert as early as 7500 BCE, and that Barbary

sheep were being managed in modern-day Libya at about the same time. As they began to domesticate animals, the way people lived began to change as well.

At Nabta Playa in Egypt's Western Desert, huge uncarved stones dating back to about 4500 BCE, have been discovered. These megaliths are interesting for two reasons: First, they were aligned with celestial bodies; and second, ceremonial cattle burials from the same period were found nearby, beneath artificial mounds of rocks called "tumuli." This archaeological find indicates a higher level of social organization and religious ritual among humans than had been found before in Africa. It is unlikely that this is an isolated phenomenon since the domestication of animals had spread across the Sahara by the sixth millennium BCE.

The regular cultivation of edible plants and grains was another important step in human progress. Botanists have discovered that people were experimenting with both sorghum and millet in Nabta Playa by about 7000 BCE. Grinding stones and storage pits for processing wild grains were discovered at Foum el 'Alba in Mali and date back to 4500 BCE. The abundance of wild grains along the small lakes and rivers of the Sahara made it unnecessary to rely heavily on plant cultivation. The Sahara is interesting because it appears that people had domesticated animals in that area thousands of years before they domesticated plants. This is the exact opposite of the pattern of civilization in the Middle East. Nevertheless, people in the Sahara were already exhibiting all of the traits of sedentarism—living in one place year-round—long before they were growing regular crops.

Changing Spatial Distributions

After people were living in clusters of permanent settlements and producing more food than they needed for their own consumption, they had the opportunity to develop specialized skills. Some, for example, became metalworkers or potters, while others began to travel long distances, trading the surplus goods from their own areas with those from other regions. These specialized skills permitted people to enjoy a better lifestyle, because they could exchange the products of their own labor for items they could not manufacture. For example, although one might not be a good potter, one could still possess an excellent figurine by means of exchange.

Between 5500 and 2500 BCE, rain became more frequent in the Sahara. Its wide rivers teemed with fish, and trees covered its hillsides. People were now raising large herds of cattle, as well as millet, sorghum, and other crops. The people of the Sahara were quietly laying the foundation for the Egyptian culture of the Nile Valley.

The Nile Valley only became widely habitable between 8000 and 5000 BCE. During this period, the rich soil from the Upper Nile was slowly deposited over the gravel and silt of the Lower Nile. By about 5200 BCE, the people along the Nile River were finally cultivating regular crops. The seasonal floods of the Nile Valley continually brought more rich soil up from the south and provided the ideal environment for agriculture. As the art of cultivating plants improved, people began to have continually greater surpluses of food available than ever before. Permanent settlements grew larger and closer together than anywhere else in Africa. Eventually, all the different people from north to south along the Nile Valley became unified in the ancient civilization of Egypt.

Natural Barriers to Movement

As early as 2000 BCE, the people of the central Benue River valley in Nigeria were cultivating plants. Unfortunately, the deep equatorial rainforest of the Congo River Basin lay to the south, effectively blocking further movement in that direction. In time, however, the advent of the Iron Age and the technology of ironmaking would overcome that barrier and forever change the people's culture.

At Taruga, Nigeria, iron furnaces have been discovered that date from the eighth century BCE. Furnaces found to the north, at Termit, Niger, may date from as early as 1300 BCE. In fact, this new technol-

Xhosa women, Eastern Cape, South Africa. (South African Tourism)

POPULATION DENSITIES OF AFRICAN COUNTRIES
(BASED ON MID-2020 ESTIMATES)

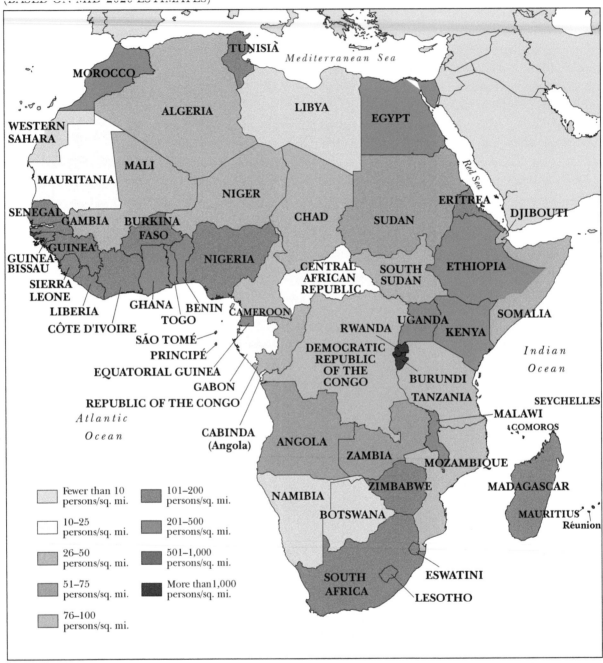

Fewer than 10 persons/sq. mi.	101–200 persons/sq. mi.
10–25 persons/sq. mi.	201–500 persons/sq. mi.
26–50 persons/sq. mi.	501–1,000 persons/sq. mi.
51–75 persons/sq. mi.	More than 1,000 persons/sq. mi.
76–100 persons/sq. mi.	

ogy seems to have emerged independently over a wide span of years and at far-flung sites across sub-Saharan Africa. It appears that iron had come into general use in the Nile Valley, in the Kingdom of Kush, and the city of Axum, by 500 BCE, and the fabrication of iron may have become a major handicraft industry in Meroë by 200 BCE. Archaeologists have also discovered that true steel was being pro-

duced in Tanzania in the middle of the first millennium BCE, and on the other side of the continent at Djenné-Djenno in Mali by 250 BCE. Africa was rapidly changing.

While ironmaking may have come to the people of the central Benue River valley much earlier than generally thought, accepted tradition has it arriving there about 2,000 years ago. With the appearance of iron tools and weapons, the Nok culture arose along the central Benue River in present-day Nigeria. At long last, the great trees could be felled to clear the land for cultivation, timber could be used to shore the walls of an increasing number of mines, and wood could be burned to make charcoal with which to smelt an increasing quantity of iron ore. With iron implements, a small band of people could accomplish more work and feed more people. They also could defeat a larger army with stone or even bronze weapons. Once again, populations increased dramatically, and vast numbers of people began to seek new lands to call home. This time, they would make a new path: to the east through the forests and across the coastal plain to the Indian Ocean, and to the southeast across the Congo, Zambezi, and Limpopo rivers to the Kei River in South Africa.

Tracing Migration Through Linguistics

Linguistics—the study of languages—has allowed scientists to approximate the beginning and the direction of this great migration of people. For many years, the leading expert on African languages was Joseph Greenberg (1915-2001). Greenberg was able to organize all of the languages of Africa into four families: Nilo-Saharan, Niger-Congo, Khoisan, and Afroasiatic. One of the eight branches of the Niger-Congo family is Benue-Niger, which includes Bantu.

The Bantu branch seems to have arisen no more than 2,000 years ago in east central Nigeria, along the Benue and Niger rivers. By comparing the similarities and differences among languages, linguists have been able to trace which languages are related to Bantu. With this knowledge, they determined that the ancestors of the Bantu speakers had begun to move eastward and southeastward through the equatorial rainforest as early as 3000 BCE. To be certain, they collaborated with other scientists.

Much could be learned from evidence of the types of plants and animals the Bantu speakers had domesticated and taken with them on their journey across Africa. It is known that they grew plants that were suited to the warm and wet climate of West Africa—African yams, palm oil, kola nut, bananas, and taro—and that their most numerous domesti-

cated animals were cattle. All their plants were successful as they moved deeper into the forest, but nearly all their cattle soon died because of the tsetse fly. As they pushed into the forests, they began to displace the indigenous people of the Congo rainforest (formerly known as Pygmies). Without crops or domesticated animals, and possessing only stone tools and weapons, the people of the Congo were unable to compete with the Bantu speakers.

As the Bantu-speaking people continued southward, they encountered the Khoisan people . When they met Khoi herders, who had domesticated sheep centuries before, they often bypassed them in search of wetter climates suitable for their crops. It was only years later, after having increased dramatically in number, that they returned to overwhelm the areas they had earlier left behind. When they met the San hunter-gatherers, the results were very much the same as with the other indigenous people: The San were defenseless against the Bantu speakers. Only a small number of the San people now remain in the harsh Namib Desert.

The story was different as the Bantu speakers entered the Western Cape region of South Africa. They seemed to have every advantage: They greatly outnumbered the indigenous Khoi herders, they had far better tools and weapons, and they possessed domesticated plants and animals, but they could not adapt to the Mediterranean-type climate of the Western Cape. Their plants grew best with wet summers, and the rains only fell during the winter in the Western Cape. The Bantu speakers were stopped at the Kei River.

Those Bantu-speaking people who had journeyed east from the Benue and Niger rivers experienced much the same success as those who had traveled to the south. Increasing in number, they emerged from the east edge of the forest sometime after 1000 BCE. Passing the Great Lakes and the Great Rift Valley, they continued east to settle in the wet forests, which were well suited to their crops. In time, they adopted sorghum and millet, which were better suited to the dry plains. Thereafter, they supplanted the remaining Khoi herders. By the end of the first century BCE, they had occupied most of East Africa south of Ethiopia.

Colonialism in Africa

Although the colonial era spanned less than a century, it dramatically changed the spatial structure of the African countries. The modernizing influence of Europe disrupted traditional ways of production and traditional African ways of life. Most of Africa was marginalized from the global economy, and Africa still struggles to achieve social, political, economic, cultural, and spatial development.

Hari P. Garbharran

DISCUSSION QUESTIONS: PEOPLE

Q1. What evidence do paleontologists have to suggest that modern human beings arose in Africa? What role did the Leakeys play in the development of human evolutionary theory? What earlier scientists laid the foundation upon which the Leakeys expanded?

Q2. How did the climate of the Sahara in early historic times differ from what it is now? What factors led the Nile Valley to become an early focus of civilization? What role did the Nile's seasonal floods play in the evolution of a strong agricultural society?

Q3. Who was Joseph Greenberg? Into what four linguistic families did he organize African languages? How did his work inform our understanding of the movements of peoples around the continent thousands of years ago?

POPULATION DISTRIBUTION

Although Africa is Earth's second-largest continent, it is the most sparsely populated of any except Australia. The common division between North Africa and sub-Saharan Africa reflects racial and cultural contrasts. Most North Africans have light skin and European physical characteristics, while most Africans south of the Sahara have dark skin and other features that set them apart from North Africans. North Africa has an almost uniformly Islamic culture, and Islam also dominates the countries that border the Sahara. Farther south, the influence of Islam fades rapidly. In the past, traditional ethnic religions associated with only one ethnic group dominated Africa south of the Sahara. Africa is mainly populated by native Africans, interspersed with Asian and European populations who are minorities in most areas they inhabit.

Although precise data are not available, it is estimated that Africa's population in 1900 was between 115 million and 155 million. About that time, the population declined in the continent, especially in equatorial Africa. Population growth resumed in the 1930s through a combination of factors, including changes in colonial policies and the diffusion of medical and health programs. Between 1950 and 2020, the population grew by more than 760 percent, from 177 million in 1950 to 1.34 billion in 2020. Today, the growth rate stands at approximately 2.5 percent per year.

Modern Distribution of People
Many modern scholars identify the countries of North Africa—Egypt, Libya, Tunisia, Algeria, Morocco, and Western Sahara—with the Arabic-speaking nations of the Middle East. In 2020, North Africa was dominated by an Arab-Islamic culture, and its population totaled about 196 million. The population of sub-Saharan Africa was approximately 1.06 billion, which can be further subdi-vided as follows: West Africa, 381 million; East Africa, 440 million; Central Africa, 179 million; and Southern Africa, 59 million.

The percentage of natural increase for the world's population in 2020 was 1.03 percent. The CIA published statistics listing the countries of the world according to rates of natural increase for the period from 2015 to 2020. The top twelve countries were all in sub-Saharan Africa. Leading the list are Angola, Mali, Malawi, Uganda, Burundi, and Niger. The rates for the top dozen range from 24.4 to 32.7 percent. Mauritius had the lowest rate in Africa at 0.54 percent.

Diffusion of Population
The latest population figures for countries in North Africa differ greatly. For example, Western Sahara has a population of 0.5 million, Libya 5.2 million, Tunisia 10.9 million, Morocco 35.7 million, Algeria 43 million, and Egypt 104 million. Most of these countries cover vast amounts of land, much of it dominated by the Sahara Desert.

Additionally, the Atlas Mountains of Morocco, Algeria, and Tunisia are located in northwestern Africa, between the Mediterranean Sea and the Sahara Desert. In this harsh setting, people generally gravitate toward coastal locations along the Mediterranean Sea or the Atlantic Ocean or settle along rivers. Except in Egypt, more than 90 percent of the populations of North African countries live within 200 miles (320 km.) of the Mediterranean Sea or the Atlantic Ocean.

The ancient land of Egypt, situated at Africa's northeastern corner between the Mediterranean and Red seas, is almost totally dependent on a single river. The Greek historian and geographer Herodotus aptly referred to Egypt as "an acquired country—the gift of the Nile." The Nile River has defined a meandering greenbelt for thousands of

The ever expanding city of Johannesburg is one of Africa's largest and most vibrant urban locations. Johannesburg's warm summer days are coupled with electrifying thunderstorms in the afternoon that help to cool this busy city. Commonly referred to as Joburg, the city is home to 6 million trees which are a treasured site among the growing urbanization and development.

miles across the desert and provides a home for approximately 95 percent of Egypt's people, who live within 12 miles (19 km.) of its banks.

Population densities, generally, are low in sub-Saharan Africa. Less densely populated areas are scattered irregularly throughout the desert and semidesert areas of sub-Saharan Africa and across the grassy and forested areas of tropical bush that encompass most of Africa.

Most of the region's people live in a small number of densely populated areas that together cover only a small part of the region's total area. The most densely populated areas generally are along the coast and in the highlands. These include five core areas south of the Sahara: the coastal belt along the Gulf of Guinea, stretching from southern Nigeria westward to southern Ghana; the savanna grasslands in northern Nigeria; the highland regions surrounding Lake Victoria in Rwanda, Burundi, Tanzania, Kenya, and Uganda; the highlands of Ethiopia; and the eastern coast and parts of the high veld of South Africa.

Fertility Rates

Total fertility rate (TFR) refers to the average number of children women have during their childbearing years, which typically are defined as ages fifteen to forty-nine. Fertility is generally high throughout tropical Africa, with many countries recording TFRs between six and eight. However, national statistics obscure some remarkable variations within countries, where in some areas TFRs are between two and five.

Research has revealed a belt of low rates from southwestern Sudan through the Central African Republic, and into the Democratic Republic of the Congo. Low-fertility areas also include parts of Cameroon and Gabon, isolated locations in the savanna zone of West Africa, parts of Namibia and Botswana inhabited by the San, parts of the East African coast, parts of Ethiopia inhabited by nomadic pastoralists, and the Lake Victoria area. Research shows that in each of these cases there is a close relationship between a low TFR and a particular ethnic group.

Four reasons were noted for relative low rates of fertility in these areas: cultural variations in the length of time a baby is breast-fed; impact of sexually transmitted diseases such as gonorrhea, syphilis, and HIV/AIDS; poor nutrition; and cultural aspects related to marriage. Among African women, the age at first marriage is consistently in the mid-teens, but exceptions to the rule clearly cause reduced fertility. For example, among the Rendille of northern Kenya, cultural practices result in one-third of the women not marrying until their mid-thirties. Among nomadic pastoralists, spousal separation for long periods of time may reduce fertility rates.

Urban and Rural Comparisons

The urban landscapes of Africa south of the Sahara reflect the political, economic, and cultural changes that have affected the region. Towns have become increasingly important as people migrate to them in search of higher wages, improved health and educational facilities, and safer formal and informal job opportunities. Urban areas are growing rapidly and adding extensive tracts of new housing, shantytowns, and industrial and commercial landscapes related to urban functions.

The oldest urban landscapes are found in East and West Africa and are less common from the Congo region southward. In East Africa, the old Arab-Swahili ports that extend from the Red Sea to Mozambique and include Zanzibar, Mombasa, and Malindi are characterized by stone buildings reflecting the pre-European era. Addis Ababa in Ethiopia, Kampala in Uganda, and Kumasi in Ghana are some of the cities that have pre-European buildings, including places of worship and palaces. In West Africa, old urban centers include the trading centers of Kano in Nigeria and Timbuktu in Mali, at the southern end of trade routes across the Sahara. They are still dominated by craft workshops.

European settlement and colonization brought new types of urban centers into Africa. Many were ports to link the new colony to the home country. In the interior, major inland towns developed and were linked to the coast by railroad. Brazzaville, the capital of the Republic of the Congo, and Kinshasa, the capital of the Democratic Republic of the Congo, provide a contrast on opposite banks of the Congo River, reflecting their different French and Belgian colonial heritages.

Other new towns developed as mining settlements, such as Enugu in Nigeria and Johannesburg in South Africa. In the southern region of the Democratic Republic of the Congo and northern Zambia, a series of towns developed around the copper mines. Nairobi in Kenya started as a center where railroad workshops were established along the route from the port of Mombasa to Lake Victoria.

Rural Communities

Most African people live in rural areas. Rural landscapes include cultivated plots, areas where little natural vegetation has been disturbed, mining areas, and human settlements in villages and small towns. In individual countries of sub-Saharan Africa, the estimated percentage of the population living in rural areas ranges from approximately 33 percent (South Africa) to 80 percent or higher (Burundi, Malawi, Niger, and Rwanda). Since the beginning of the twenty-first century, the increasing rate of urbanization is particularly striking in Nigeria: In 2000, 84 percent of the people lived in rural areas. Today, rural dwellers comprise only 48 percent.

Villages and small towns are a growing feature in rural areas of Africa. Most people live in villages, although dispersed homes on individual farms are prevalent in some areas. The typical rural home is a small hut made of sticks and mud, with a thatched roof, no electricity or plumbing, and a dirt floor. Recent trends have seen small homes built of concrete blocks and roofed with corrugated iron. Some villages have evolved into small service centers with shops and market stalls, where food and consumer goods are sold. The governments of Tanzania and Zimbabwe have adopted policies to develop small rural towns as service centers to ease population pressure on the larger cities.

Economics and Population Distribution Patterns

Sub-Saharan Africa has the highest fertility and mortality rates in the world, as well as the highest proportion of young dependents. However, there is both considerable subregional variation and a gradual decline in these trends. Although rapidly growing populations are evident in the region, about 36 percent of the forty-nine sub-Saharan countries have populations under 5 million, and 49 percent have populations under 10 million. Eighteen countries have populations exceeding 25 million; and only two—Nigeria and Ethiopia—have more than 100 million people.

The spatial distributions of population in sub-Saharan Africa coincide with a number of environmental factors—soil, topography, vegetation, and climate; developmental issues—agricultural development, levels of urbanization, and degree of industrialization; and sociopolitical characteristics, such as ethnic disputes, oppressive regimes, and resettlement schemes. The West African coastal strip, for example, is dominated by most of the region's economic, urban, and political centers. Other economic centers—such as the diamond and gold mining centers of South Africa's Witwatersrand region, the Copperbelt of the Democratic Republic of the Congo and Zambia, and the tourist centers of the Lake Victoria borderlands—attract large population clusters.

Dual Economies

Similar to other developing nations, African countries typically have dual economies. Economic activities ranging from mineral extraction and production of tropical crops to energy development are focused on primate cities of each country. These economic activities and the primate cities, except for those in Southern Africa, are externally oriented. The national economies of African countries are under-industrialized and overly dependent on the export of a few primary products, particularly

A group of farmers in Tarfila, Burkina Faso.

minerals and cash crops. Africa's place in the commercial world is mainly that of a producer of foods and raw materials for external sale. Economic benefits from these activities and cities only marginally impact the majority of African populations.

Most Africans are engaged in agriculture, the other segment of the African economy. Agricultural workers range from nomadic herdsmen, to subsistence-level sedentary cultivators who produce crops primarily for their own use, to farmers producing commercial crops such as coffee, cacao, cotton, peanuts, and palm oil for export. Subsistence agriculture is the main occupation in nearly all African countries. Generally, farmers south of the Sahara operate on a subsistence basis and have little cash income. Women do a large share of the farm work and produce 40 to 50 percent of Africa's food, in addition to doing household chores and bearing children.

Almost all African countries are heavily in debt to foreign lenders. For decades, billions of dollars in outside grants and loans have done little to eliminate poverty on the continent. By 2017, virtually every African government was in debt to international lenders such as the International Monetary Fund, (IMF) the World Bank, and private banks. Twenty-four were considered "critically indebted," because they had exceeded the 55 percent debt-to-GDP (gross domestic product) threshold advocated by the IMF. Many countries have had difficulty in meeting just the interest payments on these debts. Some are no longer even servicing their loans. To complicate matters, the amount of economic and humanitarian assistance to the region has slowed considerably just as revenue from most natural resources has dropped and the implications of the COVID-19 pandemic impose even more pressure on cash-strapped countries.

Refugee Issues

In recent decades, many nations in Africa have experienced civil wars and various levels of ethnic strife. One result of such unrest is a massive number of refugees of many different nationalities and eth-

Distribution of humanitarian aid at a refugee camp in the Democratic Republic of the Congo. (Julien Harneis)

nicities. In sub-Saharan Africa alone, the United Nations High Commission for Refugees (UNHCR) estimates that 18 million people qualify as refugees —26 percent of the world's total. Ongoing crises in the Central African Republic, Nigeria, South Sudan, and Burundi, among other countries, are contributing to the soaring number of refugees south of the Sahara.

Since 2016, thousands of refugees have crossed the Mediterranean Sea from Africa in attempts to reach Europe to escape oppression and poverty in their home countries. Most of the African refugees, from countries such as Eritrea, Nigeria, and Somalia, are guided by human smugglers to Libya, where they board boats for the treacherous trip across the Mediterranean. Sometimes, the African refugees are joined by refugees from Syria, who take a circuitous "backdoor" route to Europe via Libya. As of 2019, more than 63,000 refugees had crossed the Mediterranean toward Europe, with the death toll from drowning exceeding 1,000.

In recent years, the terms "climate refugee" or "environmental refugee" have been used to describe people displaced by the effects of climate change. Although climate refugees do not fit the refugee criteria set by the UNHCR, there are nevertheless an estimated 25 million people worldwide who can be so classified. Studies from Senegal and Tanzania, two African countries that are said to be prone to suffering the effects of climate change, suggest that the local people affected by environmental degradation rarely move across international borders. Instead, they adapt to new circumstances by moving short distances for short periods, usually to cities. Further studies are planned to evaluate longer-term effects of climate change on Africa's ongoing trend toward urbanization.

Future Projections

Africa's population is expected to rise 35 percent in the next 15 years, from 1.34 billion today to 1.8 billion in 2035. By that time, Africa will account for approximately half of the world's population growth. Also, by 2035, sub-Saharan Africa will have the youngest population group, though its share of working-age people will increase as well. Just to meet the region's basic food needs in the year 2050, agricultural production would have to increase fivefold. If current trends continue, it is projected that by 2025, half of all Africans will lack basic services such as potable water, food, sanitation facilities, and electricity.

Sub-Saharan countries have begun to change their previous indifferent attitude toward family planning. This change has resulted partly from the three World Population Conferences held in Bucharest, Romania (1974); Mexico City, Mexico (1984); and Cairo, Egypt (1994). The Bucharest conference laid the foundation of the World Population Plan of Action, a framework that provides guidance and serves as the standard reference on population issues.

African leaders have addressed many of the issues covered in the plan of action and pledged to improve the quality of lives of their peoples. One goal was to reduce the regional natural growth rate from 3 percent to 2.5 percent by the year 2000, and to 2 percent by the year 2010. These goals have not been fully met. The growth rate remained slightly above 2.5 percent in 2020. The fertility rate (births per woman) in sub-Saharan Africa was 4.7 in 2018, the highest in the world.

Hari P. Garbharran

Culture Regions

The African continent has remarkable cultural diversity, with many ethnic, linguistic, and religious differences. Africa exceeds all other continents in number of distinct peoples and cultures—a major reason for the numerous ethnic conflicts and repeated political unrest. With only about one-seventh the area of the inhabited world and a population constituting about 17.5 percent of all humankind, Africa contains one-third of the world's languages. More than 1,500 distinct languages are spoken in Africa, and forty of them have 1 million or more speakers. The varieties of indigenous religions are even more numerous. Before the arrival and subsequent domination of Christianity and Islam, each traditional society had developed its own system of faith and ritual intimately related to its distinctive culture.

Dividing this complex cultural mosaic into fifty-four different states (forty-eight south of the Sahara) made it inevitable that some culture groups would be divided by political boundaries. Nation-states—states with only one culture group—are rare in Africa. Multination states—states with multiple culture groups—and multistate nations—culture groups split by political boundaries—are the norm. Thus, identifying culture regions in Africa, broad areas with similar culture characteristics, is no easy task. Nevertheless, eight major traditional cultural regions can be identified in the continent. Below, a brief general description of each region is presented, followed by a spotlight on one or two cultures that characterize the region. The goal is to identify the similarities and differences between culture regions and facilitate comparison.

North Africa (Maghreb)

Extending from the North African coast to the upper reaches of the Niger River, North Africa has strong cultural and ethnic ties with the Arabic Middle East, and in many ways is more a part of that world than of Africa south of the Sahara. North Africa demonstrates the strong relationship between environment and culture. Life in the desert breeds a nomadic existence: grasslands, water scarcities, and sparse populations encourage continuous movement. This affects social organization, such as inheritance laws, and economic and political organization. Perhaps due to the difficult environment, crop production is not as popular as raising animals.

North Africa is the home of the camel, an animal valued for its ability to go long periods without water and to subsist on sparse vegetation, a major asset for transport and swift communication. Other animals are kept in the larger oases or on the fringe of the Nile. Islam is the dominant religion, but indigenous religions persist.

The veil is a common article of clothing for men. Wrapped around the face and head, and leaving only slits for the eyes, it not only protects the nose and mouth against the wind and sand of the desert, but also indicates the status of the wearer. Traveling in caravans for trade across the desert and encountering the dangers inherent therein make skill in warfare, physical prowess, and valor highly coveted values.

The people of North Africa are a mixture of Arab stock, including the Tuaregs and Berbers. The Tuaregs are predominantly Sunni Muslims, monogamous, and a matriarchal society with a strong hierarchy of various classes: nobility, vassals, freed slaves, and slaves. Social status is determined through matrilineal descent, and matrilineal inheritance is the norm. While women retain title to their property after marriage, their husbands must make extensive gifts to their wives' families and assume the costs of the wedding ceremony. The initial residence of the family after marriage is matrilocal—

People of Maghreb, North Africa. From Geschichte des Kostüms *(1905) by Adolf Rosenberg, 1850-1906 and Eduard Heyck, 1862-1941.*

close to the wife's kin. Men begin wearing the veil, which is never removed even in front of family members, at the age of twenty-five. Most people are Muslim, but believe in the continuous presence of various spirits, and divination using the Koran is common. Tuaregs are a multistate nation scattered over Algeria, Libya, Mali, Niger, and Burkina Faso. The bulk of the population lives in Niger and Mali.

There are about 32 million Berbers in North Africa, comprising a multistate nation spread across

Morocco, Algeria, Mauritania, Tunisia, and Libya. Although predominantly Muslim, Berbers are less orthodox and include many elements of pre-Islamic and traditional religious rituals. Traditional Berber occupations are raising sheep and cattle, but increasing numbers raise crops. Other economic activities include flour milling, woodcarving, quarrying of millstones, and production of domestic utensils, agricultural implements, pottery, jewelry, and leather goods. A large amount of Berber art is in the form of jewelry, leather, and finely woven carpets.

The Northeastern Horn

Comprising Ethiopia, Somalia, Djibouti, Eritrea, and Eastern Kenya, the northeastern Horn of Africa is occupied by three groups of people—the Oromo of southwestern Somalia and eastern Kenya, the Somali, and the Afar, who live along the southern slopes of the Red Sea. Cereal agriculture and animal rearing—sheep, horses, and donkeys—predominate, but hunting plays an important subsidiary role. Inheritance of property is patrilineal; marriage is usually monogamous and may be effected by the mock or actual capture of a bride, after which a bride price (gift of cattle) is paid to the wife's family.

Unique to this area are age-grade organizations for men that govern the performance of particular social functions and patterns of behavior. Men of approximately the same age constitute an age grade, and each of the five grades requires eight years to advance. Thus, a son enters each grade exactly forty years after his father did. As members of each grade grow into the next age grade, the functions of the group change, and its members acquire more power and responsibility in their communities. Common membership of age grades helps to reduce the friction between communities and unites them to defend against external threats. Islam is a major unifying factor in this region.

The Upper Nile

This culture region extends from the upper reaches of the Nile River westward to Lake Chad and covers modern Sudan and South Sudan. Primarily pasto-

ral in economy, most people survive on a diet of milk, butter, and fresh blood drawn from cattle (especially the Maasai), supplemented with millet and maize grown for subsistence. While polygamy is permitted, it is rare, because no man can take a second wife until other men his age have been married. Inheritance is patrilineal and domestic residence, patrilocal. Marriage entails the payment of a substantial bride price (forty head of cattle), which can sometimes be reduced by bridal services —labors for the wife's parents or an appropriate relative.

The Maasai are the southernmost Nilotic speakers and are linguistically most directly related to the Turkana and Kalenjin, who live near Lake Turkana in west central Kenya. Maasai are pastoralists and have resisted the urging of the Tanzanian and Kenyan governments to adopt a more sedentary lifestyle. Demanding grazing rights to many of the national parks in both countries, the Maasai routinely ignore international boundaries while moving their great cattle herds across the open savanna with the changing of the seasons.

Cattle are central to Maasai life, providing their food (milk, blood, and meat), their materials (skin for clothes and dung to seal their houses), and their only recognized form of wealth. Rarely killed, cattle are accumulated as a sign of wealth and traded or sold to settle debts. Young men tend the herds and often live in small camps, moving frequently in the constant search for water and good grazing lands. They live in small clusters of huts (kraals or bomas) made of sticks sealed together with cow dung, which provide enclosures for the cattle as well. Maasai also sell their beadwork to the tourists with whom they share their grazing land.

Maasai community politics are embedded in age-grade systems that separate young men and prepubescent girls from the elder men and their wives and children. Marriages are often arranged. All children, whether legitimate or not, are recognized as the property of the woman's husband and his family.

Maasai diviners (*laibon*) are consulted whenever misfortune arises. They also serve as healers, dispensing their herbal remedies to treat physical ailments and ritual treatments to absolve social and moral transgressions. Respected as the best healers in Tanzania, Maasai *laibon* peddle their knowledge and herbs in the urban centers of Tanzania and Kenya. Maasai are best known for their beautiful beadwork, used primarily for ornamentation of the body. Beading patterns are determined by each age set and identify social grades.

The Sudan

Between the tropical forests of West Africa, the Nile Valley, and the Sahara Desert lies the Sudan culture region. Cereal agriculture, sorghum and millet, some hunting, and raising animals—donkeys, sheep, pigs, chickens, guinea fowl, and cattle (the most prestigious)—are the mainstays of the economy. Frequently, cattle are kept mainly as prestige and a form of invested capital; they are slaughtered for food only on rare occasions.

The division of labor follows sex and age structure. Those who are too old to work guard the home and take care of younger children. Men clear and prepare the fields for planting; women cultivate the fields, perform all household chores, and also market the surplus from their farms; children gather firewood and assist parents with chores. Marriage is polygamous and patrilocal, and inheritance is patrilineal. Land is owned by the lineage (family land), belongs to the living, the dead, and the yet unborn, and rights of cultivation are determined by patrilineage. The extended family emphasizes community and mutual support of family members. Islam is the dominant religion—about 96 percent of Senegal's population is Muslim.

The Hausa, numbering about 80 million people, characterize this culture region. The Hausa language, which belongs to the Chad branch of the Afroasiatic language family, is an important lingua franca in West Africa. Hausa culture manifests more specialization and diversification than most of the surrounding peoples.

Subsistence agriculture is the predominant occupation, but tanning, weaving, dyeing, and metalworking are also widespread. Hausas are famous as long-distance itinerant traders and wealthy merchants. Agriculture, scheduled around the May-Oc-

tober rainy season, focuses on millet, maize, Guinea corn, and rice to supply the bulk of the diet. Peanuts, cowpeas, sweet potatoes, cotton, sugarcane, bamboo, tobacco, cassava, and other root crops are grown both for household consumption and as cash crops. Livestock raising (mainly horses, donkeys, goats, sheep, and poultry) is an important economic activity.

Social organization is stratified, based on occupation, wealth, birth, and patron-client ties. Occupational specialties are ranked but tend to be hereditary, and the first son is expected to follow his father's occupation. Patrilineal inheritance and patrilocality are the norm. Polygamy is common, especially among Hausa. In the household division of labor, men are responsible for agriculture, collecting activities, marketing, sewing, laundry, building repairs, and transport. Women cook, clean house, take care of children, pursue their craft specialties, and sometimes engage in trade. Traditional Hausa religion centers on a variety of spirits, both good and bad, and rituals include sacrifices to the spirits and spirit possession.

West Africa's Guinea Coast

This culture region extends along the Guinea Coast from Liberia to southern Nigeria. It supports the cultivation of yams, cassava, plantains, bananas, palm oil, and cocoa, the last two mainly for export. Pigs and chickens are the principal domestic animals; some sheep and donkeys are found in the northern margins. Fishing occurs in the rivers and lakes and along the coastal regions. Periodic (usually weekly) markets provide for the exchange of livestock and cotton products from the north for fish and kola nuts.

Matrilineal inheritance is the norm, which makes the mother's brother, rather than the father, more influential. All members of the matrilineage share responsibility for the debts and legal offenses of its individual members and assume collective responsibility for funeral expenses. Reverence for the dead is particularly important and is demonstrated by periodic sacrifices offered at various shrines to invoke the blessings of the ancestors.

The Asante (also known as Ashanti) combine ancestor worship with the worship of Onyame, the supreme being, and several lesser deities, including the spirits of trees, plants, and animals. Libations of palm wine, liquor, and the blood of sacrificial animals are regular tenets of ancestor veneration and worship.

The Yoruba people of southwestern Nigeria are primarily farmers, growing cocoa and yams as cash crops. Other crops included in their three-year rotational system include cassava, maize, peanuts, cotton, and beans. Their homeland, known as Yorubaland, is characterized by numerous densely populated urban centers with surrounding fields for farming. The Yoruba claim 401 deities known as orisha, and the high god is Olorun. No organized priesthoods or shrines exist in honor of Olorun, but his spirit is invoked to ask for blessings and to confer thanks.

The Yoruba believe that dead ancestors still have influence on Earth, and lineage heads honor all deceased members of the lineage through a yearly sacrifice. Maskers (egungun) appear at funerals and are believed to embody the spirit of the deceased person. The arts of the Yoruba are as numerous as their deities, and many objects are placed on shrines to honor the gods and the ancestors. Beautiful sculpture abounds in wood, brass, and the occasional terra cotta. Varied masking traditions have produced a great diversity of mask forms. Other important arts are pottery, weaving, beadwork, and metalsmithing.

The Congo Basin

This culture region comprises the tropical rainforest area from the Atlantic Ocean to Lake Tanganyika south of Sudan, spanning the area drained by the Congo River and its tributaries. Economic activity ranges from hunting and fishing among the indigenous people to cultivation of manioc, bananas, maize, yams, taro, and sweet potatoes. Principal domestic animals include goats, chickens, and dogs. While men tend livestock and clear the fields for planting, women do all agricultural work.

Metalworking is important here; prominent objects include arrowheads, spear and axe blades, copper rings, bracelets, and bells. Patrilineal inheritance is the norm for inheritance and succession, but because the bride price received by her kinfolk is usually used to obtain a bride for her brother, the sister acquires certain rights over her brother's wife and her children. These rights may include the sister's children claiming the property of her brother, a claim that supersedes the rights of his own children. Religion is based on a supernatural supreme being, and music and dance are major channels of expression.

The Bamileke represent this region quite well. They are part of a larger cultural area known collectively as the Cameroon Grasslands. Within the Bamileke society there are numerous smaller groups who are loosely affiliated and share many similarities while retaining separate identities. Primarily farmers, growing maize, yams, and peanuts as staple crops, they also raise some livestock, including chickens and goats, which play an important role in daily sustenance. Women, who are believed to make the soil more fruitful, are responsible for planting and harvesting the crops. Men usually help with the clearing of the land and practice some hunting.

Authority among the Bamileke, as is the case in most of the western grasslands, is invested in an elected village chief, the Fon, who is supported by a council of elders. Besides his role as dispenser of supreme justice, the Fon is also the de facto owner of all the land that belongs to a given village and holds it in trust on behalf of the people.

While recognizing a supreme god, the Bamileke pay homage to ancestral spirits embodied in the skulls of deceased ancestors. These skulls are carefully protected and preserved. When a family decides to relocate, a dwelling, which first must be purified by a diviner, is built to house the skulls in the new location. Most Bamileke art centers on the Fon and includes statues and carved masks that represent him. They create beadwork as well.

Eastern Africa

This region comprises the peoples of southern Uganda, Rwanda, Burundi, northwestern Tanzania, the Kikuyu of Kenya, the Chagga of the southern slopes of Mount Kilimanjaro, the Swahili on the island of Zanzibar, and the Nyakyusa at the northern end of Lake Malawi. While the northern parts are mostly patrilineal and cattle-based, the southern parts are predominantly matrilineal and almost without cattle. For example, the peoples of Malawi are predominantly matrilineal. Polygamy and divorce are widespread and common. Ancestral worship, the worship of local deities and the supreme god, and witchcraft are all part of the religion. Fertility of the land and the health of the community are all due to the benevolence of the local deities. Because cattle provide a measurement of relative wealth, cattle exchanges between individuals and groups reinforce their interdependence and mutual interests.

The Kikuyu are a Bantu-speaking people of northern Kenya; living in the highlands northeast of Nairobi, they make up the largest culture group in Kenya. Traditionally an agricultural people, the Kikuyu long resided in separate family homesteads raising crops of millet, beans, peas, and sweet potatoes. Some groups also raised animals to supplement their diet, with little or no hunting or fishing. In these family homesteads, the basic social unit consists of a patrilineal group of polygamous men and their wives and children. The Kikuyu have a reputation for being careful with money, a trait that makes them good accountants, traders, and shopkeepers. Kikuyu families own many stalls in the Nairobi tourist market.

Southern Africa

Botswana, Lesotho, Namibia, Zimbabwe, Eswatini, South Africa, and the southern portions of Angola, Mozambique, and Zambia comprise the Southern Africa culture region, whose large groups include the Shona, Swazi, Tswana, and Zulu. A pastoral economy predominates. Arranged marriages are common: Girls are betrothed between two and six

Bayaka people in the Dzanga Sangha Ndoki reserve of the Central African Republic rainforest. (JMGRACIA100)

years of age to adolescent boys and begin living with them from the age of nine, although sexual relations are not initiated until puberty. Instead of a bride price, the groom usually renders bride services for the parents of the bride. Inheritance and descent are bilateral, and patrilocality and matrilocality exist side by side. Some polygamy exists, but not as widely as in other parts of Africa. Polygamous men prefer marrying sisters because they tend to be cooperative and less quarrelsome. Shona and Zulu characterize this culture region.

The Shona of Zimbabwe are best known for their beautifully adorned wooden headrests. Most of their art is either personal or utilitarian. Although they produce no figurative sculpture, they do have a rich tradition of metalworking and woodcarving. Shona are primarily agricultural. Their main crops include maize, millet, sorghum, rice, beans, manioc, peanuts, pumpkins, and sweet potatoes. They raise some cattle, sheep, and chickens, and women usually supplement their income by selling pottery and handwoven baskets. Cows, usually used to pay the bride price, are considered taboo for women, so men must do all of the milking and herding. Men also do some hunting and fishing, and some even work as blacksmiths or carvers by commission.

Men and women both participate in farming. Traditionally, Shona peoples lived in dispersed settlements, usually consisting of one or more elder men and their extended families. Shona peoples believe in two types of spirits—bad spirits that are associated with witchcraft, and good spirits that induce individual talents associated with healing, music, or artistic ability. When illness or problems arise in a Shona family, they consult a traditional healer, *nganga*, who prescribes medicines, charms, and herbs. If these fail to work, the problem is in the realm of the ancestors and appropriate sacrifices and divination are applied.

The Zulu of South Africa are best known for their beadwork and basketry. Rural Zulu raise cattle, and grow corn and vegetables for subsistence purposes. Men are primarily responsible for the cows, which are grazed in the open country, while the women do most, if not all, of the planting and harvesting. Women also own the family house and have considerable economic clout within the family. Zulu religion includes belief in a creator god who does not intervene in day-to-day human affairs. Appeals to the spirit world are made through divination and invocation of the ancestors. Thus, the diviner, who is almost always a woman, plays an important part in the daily lives of the Zulu. Zulus attribute all bad things, including death, to evil sorcery or offended spirits. No misfortune is ever seen as the result of natural causes.

Modernization and African Culture

Traditional African cultures now face many pressures. Values of community support and care, especially for the elderly and children, are crumbling under the weight of an ever-increasing population of poor elders and AIDS orphans. Urban residence and difficult economic crises are slowly weakening extended family and kinship ties and producing nuclear families. Rapid population growth is changing land tenure, and traditional support mechanisms are dying. Nevertheless, in recent years, efforts are being made to rediscover and gain a renewed appreciation of African traditional cultures and spiritualism.

In the twenty-first century, different culture groups are increasingly mixing in cities and even rural areas. Intermarriage among different groups and expanded mobility are slowly erasing differences among regions. Previously, people living in a particular region usually had work and lifestyles in common, but differences now abound. Standards of living and employment options vary not only according to where people live but also due to the opportunities they have had for education and training. Thus, people now identify with many different groups, not only their culture groups or geographic roots. People in large cities share a way of life and culture similar to those of urban people in other parts of the world. No part of the world seems immune to the spread of global culture, spearheaded by modern technologies, particularly television, the Internet, and social media. How Africa's cultures respond to these challenges now will determine the future of Africa's culture regions.

Joseph R. Oppong

EXPLORATION

In the Western world, Africa was often depicted as the "Dark Continent." Until recently, scholars downplayed the importance of indigenous contributions in African history and viewed the continent as a backward land populated by primitive human beings who lived in virtual isolation from one another. Thus, many early studies concluded that African exploration did not begin until the arrival of European explorers in the late fifteenth century. Modern research has dispelled those myths.

Both the northern and eastern shores of Africa had been open to commerce and conquest centuries before Europeans appeared. While Africa may have been virtually unknown and uncharted to the Western world before the coming of Portuguese navigators, a considerable amount of geographical exploration had already taken place. Many African societies had organized expeditions and migrated throughout the continent prior to 1500; consequently, much of the geographic foundation that fa-

BARTHOLOMEW DIAZ ON HIS VOYAGE TO THE CAPE.

An illustration of Bartolomeu Dias's ships, São Cristóvão and São Pantaleão, on the way to the Cape of Good Hope. (Frederick Whymper)

Pyramids of Meröe, Sudan. (Hans Birger Nilsen)

cilitated European exploration was the result of earlier exploration.

Egypt

Archaeological and literary evidence indicates that several Egyptian pharaohs conducted southern expeditions along the Nile River region. Lacking sufficient timber supplies to build seafaring vessels, the Egyptians constructed boats of papyrus reeds and traveled southward to Nubia in search of vital natural resources. Ancient tombs reveal that during Egypt's Old Kingdom period, approximately 2340 BCE, the pharaohs Merenre Nemtyemsaf I and Pepi II Neferkare sent the explorer Harkhuf on four expeditions into the Nile Valley. From his base at Elephantine, Harkhuf journeyed into present-day Sudan and obtained ample supplies of livestock, gold, ivory, ebony, and laborers, and provided the Egyptians with security on their southern frontier. Harkhuf left behind a series of brief descriptions detailing the earliest geographical discoveries of Africa, actions which initiated the exploration of Africa.

After trade routes were established with the Mediterranean world, the Egyptians were able to acquire timber, which enabled them to build ships that could navigate the open seas. When Queen Hatshepsut assumed control of the dynasty during the fifteenth century BCE, she initiated several public works projects and an aggressive trade policy. She sent five large ships to explore the African coastline of the Red Sea and secure relations with the fabled land of Punt along the coast of Somalia. Her tomb, discovered in the late nineteenth century at Deir el-Bahari, across the river from Thebes, reveals that she established lucrative ties with the region and was able to procure much-needed natural resources, including gum, myrrh trees, and animal hides.

Pictorial records list ships, inventories, and slaves, and provide considerable insight into living conditions in Punt. By the time of Hatshepsut's death, the Egyptians had gathered an impressive body of knowledge about Africa's geography that would readily benefit subsequent conquerors. Hatshepsut's activities indicate that the Egyptians did not live in virtual isolation from the rest of the continent and conducted valuable geographical explorations centuries before the first Europeans established contact in Africa.

The Mediterranean World

Egyptian trade with the Mediterranean world also facilitated the opening of Africa's frontiers. As indicated by the archaeological evidence uncovered at the city of Byblos, the pharaohs obtained valuable shipbuilding technology from the Phoenicians and maintained a prosperous link between the North African coast and the eastern Mediterranean Sea. These ties were expanded after the Phoenicians moved westward and founded the city of Carthage in approximately 813 BCE. Located near the present-day city of Tunis, Tunisia, Carthage's success led to other settlements along the Maghreb, or northwestern coastline, eventually spreading into both Morocco and Algeria.

African textiles, foodstuffs, and mercenaries fueled Carthage's economy, and the wealth from gold, tin, and silver deposits was exported out of Africa in exchange for cheap manufactured goods. According to the Greek historian Herodotus, the Egyptians also benefited from Carthage's success, and one pharaoh even financed a Phoenician expedition to circumnavigate the African coast. Although Herodotus claims that these explorers sailed from the Red Sea along the southern Cape and back north through the Strait of Gibraltar, the mission's accomplishments remain questionable. However, by the fifth century BCE, navigators were compiling systematic charts that outlined the continent's vast coastline.

Alexander the Great of Greece invaded and conquered Egypt in 332 BCE. The Hellenistic period lasted for only three centuries, but it had a tremendous impact upon the geographical exploration of Africa. The Greeks centered their power in the new city of Alexandria and forever linked the Nile River and North Africa with Europe. Since this city served as such a vital trade center for the growing regional economy, both merchants and statesmen hoped that the river could be used to exploit the interior of Central Africa. Many Greeks began to search for the river's origins.

In approximately 500 BCE, Hecataeus, the first Greek geographer to visit Egypt, wrote in his text that the Nile River flowed into the River Oceanus and encircled the entire world. Other writers outlined the landscape in the southern Aswan region. The historian Diodorus Siculus traveled to Egypt in 59 BCE, and wrote a book describing the topographical conditions in both Sudan and Ethiopia. Later scholars spent considerable time exploring these ideas, but the theoretical framework that would later inspire adventurers such as David Livingstone and Henry Stanley had been firmly laid down by Greek geographers during the Hellenistic period.

The Romans eventually eliminated Greek power in Africa. Using information that had been gathered from early explorers, they were able to secure control over the entire northern coast by 30 BCE. Local businessmen profited from this relationship and established trading headquarters in Ostia, Rome's outlet to the sea. Financial and military concerns also led to the building of a centralized road system in northern Africa. Surviving records indicate that the Romans constructed an elaborate highway system that linked Carthage with both the southeast and southwest. This network inexorably linked Africans together through a sophisticated internal trade system and exposed various peoples to the goods and services of other Africans. Instead of being isolated from the wider world, North Africa and the Red Sea were already systematically connected to economic and military systems of the wider Mediterranean world.

Islamic Influence

The Muslim invasion of Africa drastically altered the religious, cultural, and territorial landscapes of Africa. Islam quickly spread during the seventh and eighth centuries CE, and became the continent's dominant faith; its influence spread much farther south than that of the Greeks or Romans, into modern-day Nigeria, Uganda, and the Congo region. When the Arabs moved across the Sahara, they encountered considerable resistance, but by 711 CE, all of northwest Africa was under Muslim control. In the process, they captured and destroyed Carthage, built the new Arab city of Tunis and severed the region's ties with the Mediterranean world. These actions sparked an initial clash between Hellenistic and Islamic cultures, but approximately 90 percent of the people ultimately embraced Islam and adopted Arabic languages.

The Muslims also had a historic impact upon Africa's eastern coastline. They established several colonies that evolved into flourishing export cities. Drawn by trade and profits, the Arabs entered into a trade agreement with the Nubian Christian kingdom of Dongola that lasted until the fourteenth century. This agreement secured the Egyptian-Nubian border and granted the Arabs special trade privileges.

Under Islamic influence, Cairo emerged as one of the principal port cities in the world's economy. African natural resources—such as gold and ivory, as well as slaves—were exported in exchange for Arabian textiles and porcelain goods. At the same time, the Arabs opened trading routes between Africa and the Indian Ocean, and increased Africa's integration into the world economic system. Muslim merchants controlled the Indian Ocean until the Portuguese arrived in the fifteenth century, and during this period they initiated contact between India and the African coast. Navigators sailed from the Arabian Sea and launched excursions into Mozambique, Madagascar, and the islands of Pemba, Zanzibar, and Mafia. Trade in gold from the interior gradually opened up additional African frontiers. By the time that the first great wave of European voyagers visited Africa, both the northern and eastern coastal regions had been thoroughly explored.

Western Europe and the Slave Trade

By the end of the sixteenth century, Western European economic and military power dominated the world. The quest for new commercial markets coincided with new scientific, naval, and technological

The Great Mosque of Kairouan, Tunisia, founded in 670, is the oldest mosque in North Africa. (Marek Szarejko)

advancements, which allowed some countries to obtain a commanding influence in Africa. For the first time, Europeans crossed the sub-Saharan frontier and explored the western coastline. To say that these expeditions had an impact on Africa is a gross understatement; some scholars argue that the emergence of the European slave trade destroyed any chance for either African statehood or industrial and commercial success.

While the institution of slavery existed in many African civilizations, the European model quickly surpassed all others in intensity, brutality, and specialization. Approximately 10 million Africans were enslaved and sold in South America, the Caribbean, and North America. Since slaves were valued based on their capabilities, slave traders focused on the most able-bodied members of a community. This forced migration began a destructive process that, from a geographical perspective, had severe sociological, cultural, and political consequences for Africa.

European exploration was driven by a number of factors. Mediterranean merchants had recently regained control of their sea lanes. Profits from Marco Polo's Asian trade motivated others to seek an alternative route to India that would circumnavigate the entire African coast. Europe's Christian Crusades in the Holy Land had generated considerable European interest in Saharan markets, and some Europeans believed that Islamic power could be bypassed and outflanked by exploring southward.

Most importantly, however, European cartographers had made considerable progress in their mapmaking, which generated promising expectations for a new age of explorers. Drawn in 1375 by Abraham Cresques for Charles V of France, the Catalan Atlas identified oases in the desert for launching trade caravans, salt deposits in the Sahara, and gold in Guinea. It provided geographical information on areas such as Mali and medieval Ghana, which had not yet been explored by Europeans, and accurately pinpointed the major cities of Niani, Timbuktu, and Gao.

The first patron of exploration to benefit from this knowledge was Prince Henry the Navigator of Portugal. Hoping to find a practical trade route to

Asia, he initiated a number of voyages into the southwestern Atlantic Ocean. In 1418, an expedition he sent out discovered Madeira Island off the northwestern coast of Africa. Another reached the Azores in 1439. In 1443 Henry launched a series of annual expeditions farther south along the west African coast. His captains obtained helpful information on hydrography, people, culture, and economics.

All of Henry's findings were catalogued by the leading geographers, cartographers, and astronomers of his time and substantially improved each following mission's chances for success. His death heralded a shift by Portugal back toward Morocco, but his ventures transformed knowledge of Africa's geography. Portuguese navigators now had information on the African coast stretching from Gibraltar to Sierra Leone.

Their efforts continued in 1469, when the Portuguese government granted the merchant Fernão Gomes a five-year lease to explore trade possibilities south of Sierra Leone. Gomes agreed to open up 400 miles (640 km.) of coast per year. By 1475 Gomes had navigated an additional 2,000 miles (3,200 km.) of Africa's coastline. He found Bioko (Fernando Po) and São Tomé islands, established contact with African rulers in what later became Ghana and collected gold in such large amounts that the Europeans would later call this area the Gold Coast. This resource, however, prompted a shift in tactics. Lisbon assumed direct control over African trade and established a military presence in Africa.

Portugal continued to expand its awareness of Africa's geography by using foreign agents and launching new naval expeditions. King John II sent Pêro da Covilhã on information-gathering missions to Ethiopia, East Africa, and the Persian Gulf. At the same time, mariner Bartolomeu Dias rounded the Cape of Good Hope in a storm and anchored in Mossel Bay. He went another 170 miles (275 km.) eastward to Algoa Bay before returning to Lisbon.

Both of those missions undoubtedly provided explorer Vasco da Gama with priceless details that allowed him to chart almost the entire African coastline before venturing south for his historic mission to India in 1497. Fulfilling Henry the Navigator's dream for a route around Africa, Vasco da Gama set anchor north of the Cape and visited Mozambique and Mombasa before proceeding on to India in 1498. After he persuaded the Portuguese government to finance additional voyages to East Africa, Portugal—and ultimately the rest of the Western world—possessed sufficient geographical evidence to circumnavigate the entire African continent.

Portugal's decision to build fortresses and practice economic imperialism set a dangerous historical precedent in Africa. Force, rather than diplomacy, would guide affairs. When the Dutch supplanted the Portuguese during the seventeenth century, a thriving slave trade was already in place. Over the next two centuries, this system ignited one of the greatest population movements in history. More than 10 million Africans were uprooted from their homes and enslaved in plantation economies in Brazil, the Caribbean, and later North America. In Africa, this process increased tensions among indigenous peoples, disrupted traditional home life, and stymied African economic development.

The Search for the Nile

With the coastline charted and settled, European explorers undertook several land expeditions in an attempt to unlock the mysteries of Africa's interior. Many geographers speculated about the source of the Nile River and whether a single river or body of water flowed continuously throughout the continent.

From 1768 to 1773, Scottish adventurer James Bruce retraced the source of the Blue Nile. Mungo Park navigated much of the Niger River from 1805 to 1806; English explorer Hugh Clapperton became the first European to cross the Sahara, in the 1820s; and Frenchman René Caillié visited Timbuktu and surveyed much of West Africa. German Heinrich Barth traveled from Tripoli southward to Lake Chad, Gao, and Timbuktu. However, the sources of the Nile River and the Congo Basin were still largely unknown to world geographers.

During the 1850s, two British explorers attempted to locate the origins of the Nile River by traveling west from Zanzibar. Sir Richard Francis

Burton served in Great Britain's Indian army, studied several foreign languages, and was well known as an Islamic scholar for his translation of *The Arabian Nights' Entertainment*.

Lieutenant John Hanning Speke was a well-disciplined soldier with extensive experience surveying the Himalayan Mountains and Tibet. He also possessed considerable communication skills that proved useful when negotiating passage from a hostile African chief into a secluded area. During a joint venture to Somalia in the 1850s, Speke and Burton were ambushed on the coast of Berbera. Both men managed to escape unharmed, but another Indian army officer was killed. Since conditions remained volatile on the Ethiopian coast, they shifted their focus to the Nile's roots.

Three German missionaries had produced a map that sparked considerable interest in Europe. It charted the interior of East Africa, and, based on evidence from merchants who traded in the interior, there were reports of a huge inland lake—Lake Tanganyika. Burton and Speke analyzed this data, presented it to the Royal Geographical Society in London, and obtained funding for a Nile expedition. The men, however, disagreed over which direction they should take. Burton maintained that Tanganyika was the home of the Nile River, while Speke argued that they would find the source at another lake located farther north.

Speke conducted his own investigation and concluded that the river's source must be north of Kazeh in Tanzania. He reached Mwanza on the southern shores of Lake Victoria in 1858, and from 1860 to 1861, he scouted the lake's southern banks until he uncovered the Nile's source. Neither man, however, circumnavigated either inland lake. They eventually parted company and often disagreed over whose contributions unlocked these mysteries, but their expeditions led to two critical geographical discoveries by Europeans in Africa. As a result, it was now possible to chart one of Africa's major waterways from the Mediterranean Sea to the interior of central Africa.

David Livingstone

Scottish missionary David Livingstone also searched for the Nile's origins. Like others, he believed that the Nile emanated from a source deep in the heart of Africa, and while he proved to be unsuccessful, his efforts generated considerable information regarding Africa's interior geography. Born in Scotland, Livingstone attended the University of Glasgow and earned a medical degree. He joined the London Missionary Society and traveled to Southern Africa to work on converting the people to Christianity.

Livingstone's desire for exploration, however, remained his top priority, and during his stay on the continent, he emerged as Europe's leading authority on African geography, customs, and life. He ar-

gued against the continuation of the slave trade, and his successes were largely attributed to his willingness to conduct his affairs in a fair and humane fashion. Upon his death, he was venerated as a true friend of Africa.

Livingstone was the first European to visit certain areas on the continent. Although Central Africa previously had been viewed as an arid desert, Livingstone revealed that the region contained awe-inspiring lakes, rivers, and waterfalls. In 1849 he uncovered Lake Ngami in what later became Botswana. In 1855 he became the first European to witness the beauty of the falls on the Zambezi River, and because of their majestic presence, Livingstone decided to name the waterfalls Victoria, in honor of Britain's queen. He furthered his exploration of

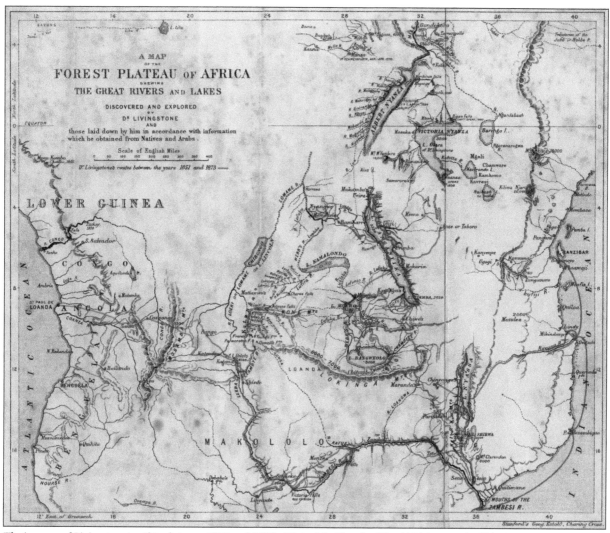

The journeys of Livingstone in Africa between 1851 and 1873. (Map of the Travels of David Livingstone in Africa)

this river and ultimately charted a course across the central Zambezi valley through Angola to Luanda on the Atlantic coast. He organized another mission from 1858 to 1864 and discovered Lake Nyasa in 1859.

During Livingstone's last expedition in the early 1870s, he found the Lualaba River. He incorrectly identified it as part of the Nile River system and failed to appreciate the enormity of his findings. While it would take the work of future explorers to recognize that the Lualaba River was actually the Congo River, Livingstone's struggles unveiled another one of Africa's geographical mysteries. Although he eventually succumbed to malaria and dysentery, and died in Africa in 1873, his achievements sparked an enthusiastic response among Europe's leaders.

Henry Morton Stanley

Born to Welsh parents who placed him in a workhouse, where he remained until age 15, young Henry fled to America and fell into the good graces of an English cotton merchant named Henry Hope Stanley. The Englishman gave the young immigrant his name, helped him finish his education, and provided the guidance that finally resulted in a job as a foreign correspondent for *The New York Herald*. Motivated by fame and fortune, Stanley devised a plan in 1871 to find Livingstone in the interior of Africa.

Due to Livingstone's increasing fame and perilous profession, there was often speculation that he had perished in the jungle. Some of his former workers had started death rumors, and Stanley hoped to scoop the world by either contacting the explorer or verifying his death. He traveled to Livingstone's base camp at Ujiji, on the northeastern coast of Lake Tanganyika, and waited. When Livingstone returned in November 1871, he found Stanley willing to attend to his needs. The reporter produced extra porters and provided the camp with critical supplies, including cotton, cooking pots, medicine, and ammunition.

Since Livingstone's supplies were at a critical shortage, he was grateful to Stanley. He had recently faced a near disaster at Nyangwe on the Lualaba River while traveling under Swahili protection. When he reached the settlement, the Swahili unleashed a violent attack upon the residents. They let loose a barrage upon the marketplace, torched villages, captured slaves, destroyed canoes, and killed hundreds of innocent people. Livingstone felt hopeless to act, especially since he needed supplies from the Swahili in order to return to Ujiji. Livingstone's friends warned him that Stanley was only interested in profiting off his name, but Livingstone was extremely thankful. Over the next four months, both men developed a deep respect for one another.

Although Stanley subsequently published a book, *How I Found Livingstone* (1874), he also became a loyal disciple and served as Livingstone's representative to the press. After Livingstone died in 1873, Stanley arranged to continue the former's explorations of Africa's inland lakes and the Congo Basin, and ultimately helped to find the final pieces of the puzzle surrounding Africa's geography. Many professionals criticized his motives and credibility, but Stanley pursued Livingstone's dream with a vengeance from 1874 to 1877. After leaving from Zanzibar, he traveled north and completed the first successful European circumnavigation of Lake Victoria. In the process, however, he swiftly adopted tactics that differed radically from Livingstone's conciliatory approach toward African people. In April 1875, after being threatened by people at Bumbireh Island, Stanley and his party barely escaped without any casualties. They returned and fired a series of rifle volleys into the tribe, killing fourteen in the skirmish while remaining beyond the range of the tribe's spears and arrows.

In October of the same year, Stanley reached Nyangwe, the Swahili trading post and Livingstone's farthest point on the Lualaba. He persuaded the local leader, Hammid bin Mohammed bin Juma, known as Tippu Tip, from the sound of his guns to guide him on the river and accompany him with 140 soldiers and porters. Tippu Tip was reluctant, but Stanley offered $5,000 and later added $2,600 during the journey. The trip, however, was extremely costly. Stanley's boat had to be disassembled and carried over stretches of jungle;

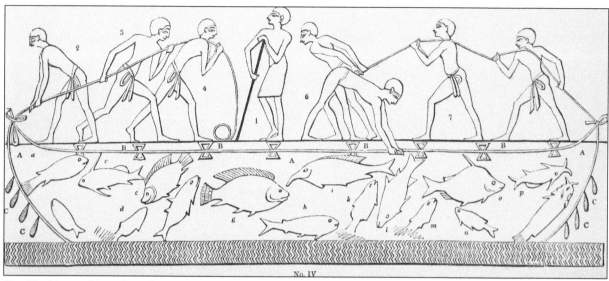

From The history of Herodotus *showing lepidotus and the eel. These are regarded as sacred to the Nile. The people, not priests, ate them both fresh and salted, and fishing with the hook, the bident, and the net, are among the most common representations in the paintings of Thebes and other places.*

food supplies were inconsistent; and typhoid, dysentery, and smallpox decimated the group. On December 19, 1876, Stanley and Tippu Tip survived an attack upon their fortress at Vinya-Njara because of superior firepower, but Stanley lost four men, and thirteen were wounded in the battle. The conflict persuaded Tippu Tip to return home.

Stanley managed to resupply and push farther into the interior. Arriving at a series of menacing waterfalls he would later name the Stanley Falls, several members of his expedition drowned. He lost his boat at the Isangila Falls, yet the surviving members of the party trudged toward the sea. When he arrived at the Atlantic coast in 1877, he had lost 250 women, men, and children to the river, war, and disease. However, he had unlocked the secrets of the fifth-largest river in the world and more or less completed the geographical exploration of Africa. While he clearly demonstrated that the Nile and Congo rivers were not an integrated system, his work indicated that the Congo River surpassed the Nile River in capabilities and could be used to transport goods, services, people, and Christianity into the heart of Africa.

Other Explorers

Other expeditions also contributed to Europe's comprehensive understanding of Africa's geography. British army lieutenant Verney Cameron headed a relief party that failed to arrive before Livingstone's death. Cameron found some of Livingstone's materials and decided to explore the river. Unlike Stanley, he was unable to stay on the river and was forced south, but when he arrived in Benguela off the coast of Angola in 1875, he became the first European to cross Central Africa.

French explorer Pierre Savorgnan de Brazza—after whom Brazzaville was later named—ventured into the uncharted rainforests on the Ogooué River. He was forced to retreat before he could discover where the Ogooué River met the Congo River, but his findings provided more insight into the difficulties surrounding the use of river transportation in central Africa.

These expeditions generated a considerable degree of interest in Africa by the European powers; once they had obtained a clear understanding of the land, they carved Africa up into foreign colonies. Valuable natural resources and profits were channeled away from local merchants and into the hands of European businesses, and military rule destroyed tribal governments. As a result, the nineteenth-century scramble for Africa undermined African political and economic development by ushering in an imperial system, the repercussions of which still resonate today.

Robert D. Ubriaco, Jr.

URBANIZATION

Urbanization involves the movement of people from rural to urban areas. "Urban" generally implies a nonagricultural settlement; with regard to minimum population size, its definition varies from country to country. It is not easy to generalize about Africa since there is so much diversity among its peoples. However, African societies have continually changed at an ever-increasing rate in modern times. Cities have grown, and people have moved back and forth between village and town, resulting in new social groups, occupations, institutions, and forms of communication.

Historical Distribution of Urban Centers

African scholars have been trying to reconstruct much of Africa's history, traditions, customs, and artifacts, which were lost during the colonial period. Efforts in the last three decades of the twentieth century, using folklore, poetry, art objects, archaeological sites, carbon-dating techniques, buildings, and linguistics, have begun reconstructing Africa's past. Evidence suggests that towns and cities have existed for several millennia in Africa. Towns that were major centers of religion, learning, culture, and commercial exchange included Meroë, Napata, Axum,

The Lalibela churches carved by the Zagwe dynasty in the 12th century. (Julien Demade)

MAJOR URBAN CENTERS IN AFRICA

Djenné, Gao, Timbuktu, and Great Zimbabwe. Precolonial towns and cities exhibited rules of social behavior, codes of law, and organized communities. Such centers were characterized by division of labor, class structures, and communication networks.

African scholars generally agree that the earliest known cities in Africa emerged around the central part of the Nile River. For example, Meroë was the capital of the Kush kingdom and existed from the fourteenth to the fourth century BCE. Iron technology, stone masonry, and other crafts and skills diffused from this region to the west and south of Africa. Meroë also had an agricultural economy, which included pastoralism and irrigation agriculture. This early urban civilization also had building and construction technologies, pottery works, textiles, sculptures, and woodwork.

Few precolonial cities survived the colonial era. Among those that endured the ravages of colonialism and still thrive as historic centers, preserving remnants of Africa's rich heritage, are Ibadan and Ife in Nigeria; Mogadishu, Somalia; Kumasi, Ghana; and Addis Ababa, Ethiopia.

Colonial cities invariably sprang up along the coast to facilitate exploitation of resources in the interior. Therefore, most African capitals are coastally located. Modern African cities have a physical structure and design combining features of both indigenous and colonial times, and the character of cities are a function of their historical legacy. Based on their history, seven types of African cities can be distinguished: indigenous, Islamic, colonial, European, apartheid, dual, and hybrid.

Addis Ababa, in Ethiopia, and the Yoruba cities of southwest Nigeria are the best-documented indigenous cities. Yoruba towns of the twelfth and thirteenth centuries were designed around a central palace, encircled by compounds of rectangular courtyards. Such towns were surrounded by high walls and gates that linked main routes to central places.

The impact of urban Islam is still visible in North Africa. Moroccan Islamic cities include Rabat, Marrakesh, and especially Fez,, the world's largest intact medieval city. Islamic cities south of the Sahara include Merca, Somalia; N'Djamena, Chad; Niamey, Niger; and Kano, Zaria, and Sokoto in northern Nigeria. Islamic cities have a permanent central market, mosques, shrines, a citadel, and public baths.

Impact of European Colonization

Colonial cities performed several port functions and developed primarily as administrative outposts and trade centers. Social, spatial, and functional segregation of people and land uses were dominant features of the urban colonial landscape. Such coastal capital cities include Freetown, Sierra Leone; Dakar, Senegal; and Conakry, Guinea.

Cities depicting European culture were transplanted into Africa and developed mainly for European settlement. Africans in those cities were marginalized, took up menial jobs, and did not regard those urban centers as their permanent home. For example, Nairobi, Kenya, catered specifically to Europeans. Other European cities included Harare (founded as Salisbury), Zimbabwe, and Lusaka, Zambia.

Apartheid cities were common in South Africa, and their character reflected the segregation of residential areas by racial type: whites, Coloureds (mixed race), Asians, and Africans. Some commercial and industrial zones also reflected this segregation. Durban, for example, had a white and a nonwhite downtown area. Other apartheid cities included Cape Town, Johannesburg, Pretoria, and Bloemfontein.

Dual cities result from a juxtaposition of two or more characteristics defined above. Sudan's capital Khartoum (colonial)-Omdurman (indigenous) is an example of a dual city. Hybrid cities integrate indigenous and foreign elements. Most modern cities in Africa are hybrids of indigenous, modern, foreign, and colonial development; Accra, Ghana, typifies this type of city.

In-Migration and Push/Pull Factors

On a simple level, migration results when a new location draws ("pulls") a person with the promise of good wages, freedom, land, or peace, while the location where the person lives pushes them away because of low income, repression, overcrowding, or war. However, there are more complex factors at play that determine why and where people move. Such factors include a person's culture, traditions, beliefs, history, and local, national, and international relationships.

Four basic types of migration can be distinguished: primitive, forced, free, and illegal. Primitive migration is associated with preindustrial peoples and includes hunting and gathering groups in Africa that might migrate on a regular basis as the resources of an area are depleted or as game moves. Forced migration refers to situations in which people have little or no alternative but to move. Slavery is an example of such a movement.

In free migration, people make their own decisions and have the option of staying or moving. Eu-

159

ropean overseas expansion to temperate areas such as South Africa involved free migration. Illegal migration results when a country prohibits emigration (moving out) or when people immigrate (enter) into a country without official approval. Generally, illegal migrants consciously violate immigration laws, but illegal migrants can be created as a result of policy changes.

Regional Patterns

In the United States, the minimum population size for a region to be classified urban is 2,500, whereas in Ghana and Nigeria, it is 5,000. Sub-Saharan Africa and some southern and southeastern regions of Asia are the least urbanized regions of the world. Overall, in 2015, 40 percent of Africa's people were urban. North Africa is the most urban (just over 60 percent in 2018), while sub-Saharan Africa's population is only 40 percent, though it is rising rapidly. The degree of urbanization by country ranges from a low of 13 percent in Burundi to a high of 89 percent in Gabon.

The urbanized northwestern fringe of Africa, encompassing Morocco (and the disputed Western Sahara), Algeria, and Tunisia, is known as the Maghreb (place of the west) of the Arab world. The three largest seaport cities there were Algiers, the capital of Algeria, with a population of nearly 3 million (2008); Casablanca, Morocco, with 3.36 million (2015); and Tunis, the capital of Tunisia, with 2.2 million (2014).

North African cities with populations exceeding 1 million by the mid-2010s include Oran, Algeria (1.5 million in 2010) and Fez, Morocco (1.41 million in 2017). Libya's largest city is the seaport capital of Tripoli (about 1.165 million in 2018). More than half of Egypt's population live in towns and cities. Cairo, Egypt's capital, has expanded to the foot of the pyramids, and its population is 9.5 million in 2018. Alexandria, the main Egyptian port, had about 5.2 million people in 2018.

West Africa

Urban population has been increasing in West Africa, and Nigeria has the most impressive urban development in this region. Three of the largest

metropolitan cities in West Africa are clustered in the Yoruba-dominated southwestern portion of Nigeria, in the vicinity of the densely populated belt of commercial cacao production. The largest of these three is Lagos (21 million in 2016). Lagos was Nigeria's capital until 1991, when most governmental functions were moved to a new capital at Abuja in the interior of the country. Northwest of Abuja is Niger's capital city of Niamey, where 1.24 million people lived in 2018.

Immediately west of Nigeria on the Guinea Coast lies Benin's de facto capital, main city, and port, Cotonou (2.4 million). Farther west, in southern Ghana, is that country's capital, Accra (1.67 million). Nearby, Burkina Faso's capital city of

Portuguese explorer Vasco da Gama meeting with the King of Malindi in 1498. The Portuguese Empire ruled Malindi from 1500 to 1630. (Fulviusbsas)

Ouagadougou has about 2.2 million people (2015). Still farther to the west is Bamako (population 1.8 million), the capital of Mali. Dakar, Senegal, on the Atlantic coast, has 1.14 million people. Monrovia (1.01 million), farther down the Atlantic coast, was founded by emancipated American slaves in 1816 and is the capital of Liberia.

The world's largest cacao producer and exporter in the early 2020s was Côte d'Ivoire, whose economic capital and main seaport is Abidjan (3.67 million). Also, on the Guinea Coast is Conakry (1.66 million), the seaport capital of Guinea.

Central Africa

The Central African region has three cities that exceed the 1 million mark in population. The largest of these is Kinshasa (11.8 million), capital of the Democratic Republic of the Congo, which covers an area about one-quarter the size of the United States. Brazzaville, located directly across the Congo River from Kinshasa, is the capital of the Republic of the Congo, with an estimated 2019 population of 2.3 million. Cameroon's inland capital of Yaoundé, with a population of 2.77 million, is connected by rail to the country's seaport of Douala (population 2.76 million).

East Africa

The most productive parts of East Africa are bound together by railways that form a network moving toward the interior from the seaports of Mombasa in Kenya and Dar es Salaam and Tanga in Tanzania. From Mombasa, the main line of the Kenya-Uganda rail link leads inland to Kenya's capital, Nairobi (2019 population, 4.39 million), the most important industrial center in East Africa. Kampala, the capital of Uganda, had a 2019 population of 1.68 million.

Dar es Salaam is the capital, main industrial center, main city, and main port of Tanzania. The city has a population of 4.36 million (2012) and rail connections to Lake Tanganyika, Lake Victoria, and Zambia. Located in the interior of Tanzania is Dodoma, with a population of nearly 411,000 in 2012, the official capital of the country since 1996.

In predominantly rural Sudan, the largest urban district is formed by the capital city of Khartoum (5.49 million). Addis Ababa (3.38 million), the capital of Ethiopia, accounts for the bulk of that country's manufacturing. Mogadishu, the capital of Somalia, had more than 2.45 million people in 2017. Nestled in the central highlands of Madagascar (the fourth-largest island in the world) is Antananarivo (1.6 million), its capital and largest city.

Southern Africa

The most urbanized country in sub-Saharan Africa is South Africa, which has five cities exceeding 1 million in population: Johannesburg, Cape Town, Durban, Pretoria, and Port Elizabeth. Cape Town (3.77 million) is the legislative capital of South Africa; Pretoria (2.47 million) is its administrative capital. The country's leading seaport is Durban (3.72 million), and its largest city is Johannesburg (5.6 million). Port Elizabeth (population, nearly 1 million) is South Africa's tourism-gateway city.

The former Portuguese African possessions of Angola and Mozambique have large coastal cities along the west and east coast of Southern Africa. The seaport, capital, and largest city in Angola is Luanda (2.57 million). Mozambique has the two major port cities of Beira and Maputo (530,604 and 1.08 million, respectively), the latter being the capital.

The middle course of the Zambezi River separates the former British colonies of Zimbabwe and Zambia. The main axis of economic development in Zimbabwe straddles the railway line connecting the largest city and capital, Harare (1.6 million), in the northeast with the second-largest city, Bulawayo, (1.2 million) in the southwest. The Copperbelt region contains the majority of Zambia's manufacturing plants and copper mines and is linked by railway to Zambia's capital and largest city, Lusaka (1.7 million).

Primate Cities

Rapid urbanization in African cities triggered the growth of primate cities. Primate cities are those with populations many times larger than those of any other city in the country. Urban primacy can be

Inside of the Great Enclosure which is part of the Great Zimbabwe ruins. (Jan Derk)

measured in various ways where capital cities dominate the urban landscape, especially in small countries such as Togo, Gambia, and Burundi. Primate cities also are found in larger countries, such as Tanzania and Mozambique, where they previously served as former colonial administrative centers.

The following are some examples of primate cities: Lomé (population, 837,000) is the capital and largest city in Togo, a country about the size of the U.S. state of West Virginia. In Gambia, a country about twice the size of Delaware, Banjul, with a population of 31,300, is the capital. Bujumbura had a 2019 population of just more than 1 million, and is the capital of Burundi, a country close to the size of Maryland. Located on West Africa's finest natural harbor is Freetown (1.05 million), the capital of Sierra Leone, a country about the size of South Carolina.

In most of Africa, European colonists founded a single city in each colony that served as a "head link" to the rest of the colony. A head link is a city that links a country to the rest of the world, as imports and exports flow through it. Roads and railroads from head-link cities provided access to the minerals of Africa or to areas producing tropical crops of cacao (cocoa), peanuts, cotton, or palm oil. Consequently, European-based cities became focal points of each colony in African countries and emerged as primate cities.

The major cities and capitals in African countries are still those that Europeans founded to exploit the continent. Examples include Lagos in Nigeria, Dakar in Senegal, Dar es Salaam in Tanzania, Nairobi in Kenya, and Luanda in Angola.

Besides having a relatively large population size, primate cities generally are characterized by disproportionate amounts of social, cultural, economic, and administrative resources. For example, more than half the jobs in Tanzania's manufacturing sector are concentrated in Dar es Salaam.

Challenges of African Cities

In many African cities, much of the urban population does not live in formal settlements. Many cities

Traffic in Cairo, Egypt. (Erica Chang)

had more than half of their populations living in informal settlements. In some parts of Africa, these settlements are considered illegal, but they seem to have become a permanent feature of the African landscape.

Terminology used to describe informal settlements varies throughout Africa, and includes "squatter communities," "shantytowns," "slums," and other pejorative descriptors. The lack of basic services and infrastructure in these informal communities is compounded by poverty, lack of potable water and sanitation, inadequate health and education services, and other site-specific habitation problems. Nongovernmental organizations and community-based organizations have played an increasing role in providing services to the urban poor. In some of the better-developed areas, such as parts of Southern Africa, there has been an associ-

ated increase in the role of private sector assistance for informal communities.

Generally, African cities are in a crisis situation as a result of the problems associated with rapid and uncontrolled growth. United Nations statistics indicate that urban population growth rates in sub-Saharan Africa in the 2020s were the highest in the world. The average annual growth rate for sub-Saharan cities was 4.1 percent in 2018. From 2015 to 2020, some of the world's fastest annual rates of urbanization occurred in Africa; these include cities in Uganda (5.7 percent), Burundi (5.68 percent), Tanzania (5.22 percent), and Burkina Faso (4.99 percent).

African governments are exceptionally centralized and face many challenges to effectively manage urban areas. Decentralization will allow local communities to participate more effectively in the decision-making process. Local government institutions need to enhance their administrative and technical capacities to coordinate the delivery and operation of services. The financial capability of cities needs to be improved to allow for the provision of adequate urban services. In addition to these city-planning initiatives, regional strategies need to be devised to maximize the benefits of people outside the dominant city, and to slow the influx of migrants to urban areas.

Hari P. Garbharran

POLITICAL GEOGRAPHY

Africa, with a population of more than 1.34 billion in the 2020s, is broadly divided into two regions separated by the Sahara Desert: North, or Arab, Africa, which is culturally part of the Middle East, and sub-Saharan Africa. North Africa extends from the Atlantic shores of northern Mauritania to Egypt's Red Sea coast, with the Mediterranean Sea to the north and the Sahara to the south. A large part of North Africa is also known as the Maghreb ("the West"), meaning that it is the western part of the Arab world. (The Maghreb region includes Mauritania, the disputed territory of Western Sahara, Morocco, Algeria, Libya, and Tunisia; it does not include Egypt and Sudan located in the eastern part of North Africa.) Sub-Saharan Africa is generally divided into four subregions: East Africa, West Africa, Central (or Middle) Africa, and Southern Africa.

A large, mostly desert region of western and north-central Africa, forming a transitional zone between the Sahara and the savannas and encompassing parts of fourteen nations, is known as the Sahel. The easternmost projection of the continent, called the Horn of Africa, is also a distinct geographical and political region; geographically, it contains four countries located within the Horn, and politically, it also includes their immediate neighbors.

As of 2020, Africa had fifty-four sovereign states recognized by the United Nations. Almost all of them are republics except for Morocco, Lesotho, and Eswatini, which are monarchies. Liberia, which had been founded as a colony for freed American slaves and declared its independence in 1847, was the first republic in modern Africa. In the early decades of the post-colonial period, many African republics had been, in effect, one-party states spanning the ideological spectrum from Marxism and African socialism (such as Angola, Algeria, Benin, and Mozambique) to conservatism and state capitalism (such as Gabon, Malawi, and Côte d'Ivoire). These countries have since transitioned to a multiparty system.

Seventeen of the African republics have institutionalized complex systems of local sub-monarchies —traditional hereditary kings and elective tribal chiefs who exercise significant authority over their peoples on the local level. Uganda, for instance, has some thirty such traditional kingdoms, chiefdoms, and principalities. South Africa has more than twenty local traditional rulers, including the powerful kings of the Zulus, who exercise their authority under the Traditional Leadership clause of the national constitution.

Besides sovereign countries, in the 2020s Africa still has ten non-sovereign territories, most of them islands and island groups sprinkled along the continent's coasts. Spain had control over two cities on the north coast of Africa—Ceuta and Melilla— which have the status of Spanish autonomous cities, as well as the Canary Islands (a Spanish autonomous community) and a number of smaller islands. France has control over the Scattered Islands in the Indian Ocean, Mayotte, and Réunion Island as parts of its system of overseas regions, territories, and departments. The island of Saint Helena, Ascension Island, and the archipelago of Tristan da Cunha are British overseas territories. The Madeira archipelago is an autonomous region of Portugal. And Socotra archipelago is a governorate of Yemen.

Early Kingdoms

Many empires rose and fell throughout Africa before European colonization. In North Africa, ancient Egypt under the pharaohs conquered vast areas and spread its civilization along the Mediterranean coast far beyond the Nile River valley. The Egyptian empire eventually was taken over by Arabs, Ottoman Turks, and finally Europeans. In

East Africa, the region of what is now Eritrea and northern Ethiopia achieved political power at an early date. Its imperial roots can be traced back to biblical times when Menelik—the son of King Solomon and the queen of Sheba, according to legend—founded the kingdom of Axum (Aksum). The kingdom, which used the name "Ethiopia" as early as the fourth century CE, benefited from the trade in gold, emeralds, tortoise shells, and especially ivory, which were exported throughout the ancient world, as far as Rome.

The Ghana Empire lasted nearly 1,000 years, at one time covering more than 150,000 square miles (400,000 sq. km.) and controlled the rich trans-Saharan caravan trade. The empire crumbled in the eleventh century, and the Mali Empire became powerful, especially during the reign of Mansa Musa (Musa I). Mali rulers took control of the trans-Saharan caravan trade. However, the empire weakened by 1400 as a result of the overexpansion of territories, which covered parts of ten modern-day countries. It was replaced by the Songhai Empire, which at its peak was one of the largest states in African history. The Kanem-Bornu, Yoruba, and Benin empires soon followed.

Powerful kingdoms in Southern and West-Central Africa established societies rich in culture, music, and traditions, such as the Kongo Kingdom. At its apex in the fifteenth century, it had around 3 million subjects, and it lasted as a largely independent state until the mid-nineteenth century. The Shona and other related peoples of what later became Zimbabwe created several successive states starting with the Kingdom of Mapungubwe in the eleventh century and including the Kingdom of Zimbabwe, the Kingdom of Mutapa, and the Rozvi Empire. During the early nineteenth century, the Zulu Kingdom's strong armies dominated much of southeastern Africa.

It took Europeans hundreds of years to gain effective control of Africa, in contrast to their rapid colonization of the Americas. Portuguese traders, arriving on the coast of West Africa in the mid-fifteenth century, were the first Europeans to discover Africa's potential in ivory, gold, spices, and slaves. The Dutch, British, and French later joined the Por-

tuguese traders. Because of its huge potential for profit, the transatlantic trade soon replaced the trans-Sahara trade. The availability of European and Asian goods led to the deterioration of African handicraft industries. As Africa gradually became a political and economic appendage of Europe, the transatlantic trade, especially the slave trade, affected indigenous African development and altered the political, economic, and cultural composition of the continent.

Slave Trade

The slave trade did not start with the Europeans. Both sub-Saharan African and Arab societies had sold slaves and used slave labor since early antiquity. However, the number of slaves in early times was small compared to the sixteenth- to nineteenth-century transatlantic slave trade. Initially, the Europeans bought a small number of slaves for domestic and farm work. The slave trade increased in size as labor demands grew in the New World, particularly on European sugarcane plantations in the Caribbean. Overall, during the three centuries of the transatlantic slave trade, about 46 percent of all African slaves were brought to the Caribbean islands; 41 percent to Portuguese Brazil; 8 percent to Spanish areas of the New World; and 4 percent to North America.

Most of those slaves came from the western part of Africa, the area between the Niger and Senegal rivers, although large-scale slave trading was introduced in East Africa by Arab traders long before the Europeans brought it to West Africa. Slavery caused a decline in power of the states in the interior savanna and increased the power of the coastal forest states. That part of West Africa became known as the Slave Coast. Early on, some European traders, especially the Portuguese, were directly involved in slave raiding, but for the most part, slaves were sold to European merchants by local tribes and warlords who acquired them through raiding and tribal wars. The demand for more slaves ravaged the interior population and made economic and agricultural development there all but impossible.

The total number of Africans sold to the New World as slaves has been estimated at almost 13 million. About 2 million of them perished while being transported across the Atlantic Ocean in appalling conditions aboard slave ships. Slavery destroyed families, villages, and cultures. It also created enormous misery for the millions of enslaved Africans.

European Colonialism

The slave trade declined at the beginning of the nineteenth century, affecting overall trade among Africa, Europe, and the Americas. Europeans then turned their attention to the exploration of Africa. For the first time, European explorers began to penetrate the interior of the continent, followed by missionaries and merchants. The missionaries wanted to spread Christianity; the merchants wanted to develop new trade links based on the export of palm oil and other raw materials for European industries. Antislavery groups in Europe were able to use this new relationship between Africa and Europe to demand an end to slavery. In 1807 Great Britain outlawed the slave trade; a year later, the Act Prohibiting Importation of Slaves took effect in the United States. (The domestic slave trade within the United States was not affected by this law; in effect, the prohibition of external importation increased the internal slave trade. In addition, some slave smuggling from Africa and the West Indies continued illegally until the end of the 1850s.)

Through military means, Europeans were able to push farther into the interior of Africa and soon gained control over large areas. France gained Tunisia and Algeria, while Great Britain took control of Egypt in the nineteenth century. The Dutch East India Company moved inland from its settlement at Cape Town, in Southern Africa, until that colony and its successors, the settler states, occupied much of modern South Africa. Spain and Portugal joined other Europeans in extending their old coastal holdings.

Europeans gained more political control over Africa as a result of their involvement in trade. The intense competition among European colonialists, who staked claims to profitable parts of Africa, was the beginning of what became known as the "Scramble for Africa." In November 1884, Germany organized a conference to peacefully settle the political partitioning of Africa among Europeans. The 1884-1885 Berlin Conference allowed the French to dominate most of West Africa and the British to control East and Southern Africa. The vast Congo area was given to Belgium's King Leopold II as his private domain. Portugal was given a small dependency in West Africa and two colonies in Southern Africa—Angola and Mozambique. Germany held on to four colonies, the largest

A Bantu servant woman in Mogadishu (1882–1883). (Georges Révoil)

being in the African Great Lakes Region, which was almost three times larger than present-day Germany. Spain and Italy retained small areas.

By 1914 Belgium, Britain, England, France, Germany, Italy, Portugal, and Spain had divided most of Africa among themselves. Ethiopia and Liberia were the only independent African nations remaining. (Liberia, however, lost a large part of its territory to the British and French.) Colonial rule in Africa was established mostly through treaties between the Europeans and the local kings and chiefs, some of whom saw no alternative while others saw colonizers as allies against old enemies and rivals. Ethnic conflicts were exploited by the colonizers, who quashed attempts at organized resistance by using both their own military forces and their African proxies. Ethiopia was the only African nation that successfully resisted military invasion by defeating Italy in the 1887-1889 Italo-Ethiopian War.

Colonial Administrative Procedures

European colonial powers employed different methods to govern their colonies. Most British colonies had a system of indirect rule that was similar to British rule in colonial America before 1776. Indigenous power structures were left in place, and the British ruled through the African chiefs or kings in charge of the areas. Portugal and France used a system of direct rule. This meant that the Africans were ruled by the European administrators who lived in the colonies. Germany ruled through a system of chartered trading and industrial companies, which imposed their own order on the local population but were supported militarily by colonial administrators called Reichskommissars (Imperial Commissioners).

Portugal's direct rule was similar, albeit much harsher, than that of France. The French believed in creating culturally assimilated African elites who

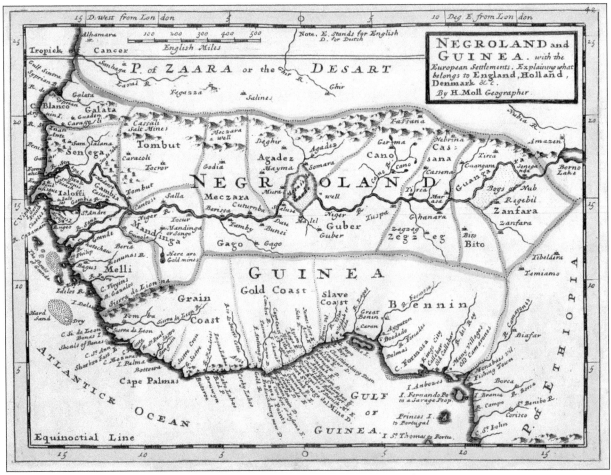

Map of West Africa, ca. 1736, "explaining what belongs to England, Holland, Denmark, etc."

would represent French ideals. Hence, France administered the colonies as overseas territories with representation in the French Parliament. The Portuguese also encouraged the idea of assimilation and treated the colonies as overseas provinces. Belgium followed neither the British policy of indirect rule nor the French policy of elite higher education and assimilation. In the Belgian Congo, even the best-educated Congolese were not allowed to advance past the lowest positions in the administrative hierarchy.

The Europeans, usually in collaboration with church missionaries, created schools to educate the Africans. In most cases, this was to ensure that the Europeans had a suitable labor force to administer the day-to-day activities of the colonies. For effective control, Europeans employed the divide-and-rule method. Colonial governors commonly used rivalries between African tribes to suppress local resistance by turning one tribe or tribal group against the other.

Colonial Impact

Colonial rule did not last long in Africa. Its impact on the continent, however, is felt to this day. Most significantly, colonialism created new political units with boundaries that cut across ethnic homelands in many regions. Colonial rule disrupted African populations and increased ethnic conflicts. It also tied Africa to an economic system based on external needs rather than on local needs. African miners and farmers produced raw materials for European and world markets. The system created Africa's economic dependency on the industrialized nations.

Colonial capitalist practices and the missionaries who came to Africa with the colonists challenged the religious and social traditions that had long been part of African life. As a result, a traditionally strong family and community spirit was eroded and the continuity of cultural heritage was interrupted; new systems of the production and distribution of wealth undercut the order by which Africans had organized their societies and replaced it with an individualistic and exploitative social order. On the other hand, in parts of Africa people gained access, albeit uneven and inadequate, to professional edu-

cation, Western medicine, and public health; future independent nations got a head start in building modern infrastructure. It is still too early, from a historical perspective, to objectively assess the overall economic, social, and cultural impact of colonialism on Africa.

African Nationalism and Pan-Africanism

From the beginning, many Africans resisted colonial rule. In several areas, organized groups demanded independence. However, the lack of broadly accepted African identity and the prevalence of ethnic conflicts prevented Africans from uniting in their resistance to colonialism. After World War II, many Africans who had fought in the war alongside their colonial masters came home to organize independence movements across Africa. In an attempt to present a strong force against colonialism, most of them united under the banner of the political ideology known as African nationalism, which had developed at the end of the nineteenth century and promoted national self-determination for all colonies in Africa and the creation of nation states.

Many of the African leaders of this movement were educated or had lived in the United States, the West Indies, or Europe and were influenced by the ideas of Marcus Garvey, Marxism, and European nationalism; chief among them was Kwame Nkrumah of Ghana (then known as the Gold Coast). Most of them envisioned the future nation states to be established within the existing colonial borders and were convinced that a sense of a single national identity could be created from among diverse and fractured ethnic groups living there. Some others espoused the ideas of pan-Africanism, that is, uniting Africans in supranational federated states, or even one continent-wide federation.

Pan-Africanism and African nationalism became forces of unity in sub-Saharan Africa and played an important part in forcing the decolonization of the continent. In 1963 their leaders, headed by Kwame Nkrumah and Ethiopian Emperor Haile Selassie, founded the Organization of African Unity. The aims of the organization were to eradicate colonialism, defend the newly independent countries, en-

169

courage their political and economic integration, and, when needed, settle conflicts between them.

Pan-Africanism was similar to that of pan-Arabism, which emerged at the same time in the Arab world of North Africa and Western Asia. It led to several attempts by Arab leaders—such as Egypt's Gamal Abdel Nasser and Libya's Muammar Gaddafi—to create a unified Arab state. Nasser's United Arab Republic, however, lasted for only three years and Gaddafi's Federation of Arab Republics, for five.

Independence

Independence came earlier to North African states than to sub-Saharan Africa. In 1922 Egypt became the first African state to become independent (from Great Britain). Libya gained its independence from Italy in 1951. Tunisia and Morocco got their independence from France in 1956. Algeria achieved independence from France in 1962 after an eight-year-long war, in which almost 1 million people on both sides died. Western Sahara became independent from Spain in 1975. Shortly thereafter, Morocco annexed the territory, claiming it as part of Greater Morocco. A national liberation organization, called the Polisario Front, resisted the annexation. In 1976, it proclaimed the establishment of the Sahrawi Arab Democratic Republic (SADR) in a part of the Moroccan-controlled Western Sahara region. (In the 2020s, the SADR has remained a largely unrecognized de facto state.)

In 1957 the Gold Coast became the first sub-Saharan African colony to win its independence, from Great Britain, and renamed itself Ghana. In 1958 Guinea became the first French colony in sub-Saharan Africa to become independent. In 1960 alone, so many African nations became independent that the year became known as the "year of independence." By 1966 a majority of the African colonies were free from colonial rule.

European responses to African liberation struggles varied from peaceful negotiations to armed conflict. In most cases, European colonial powers, exhausted by World War II, were too weak to fight. As a result, they were forced to negotiate with the African leaders. Many sub-Saharan African states,

led by Ghana, Nigeria, Tanganyika (the mainland part of present-day Tanzania), and Uganda, attained independence in this way.

In other parts of Africa, armed struggle was organized. For example, in Kenya, the British were not willing to grant independence because of the large number of British settlers in the colony. Central Kenya's Kikuyu people, whose traditional way of life had been disrupted by British settlement, formed the Mau movement, which fought Britain for independence. The bloody struggle lasted fourteen years until, in December 1963, Kenya gained independence under the leadership of the Kikuyu anticolonial activist Jomo Kenyatta.

Liberation struggles in Southern Africa proved to be difficult due, in part, to the large number of white settlers and the great natural wealth in the region, which the descendants of the colonists were determined to keep for themselves. In Angola and Mozambique, the Portuguese refused to grant independence. Portugal had governed its colonies with an iron fist, allowing colonial authorities to rule everything. There was no way to achieve independence peacefully. In 1956 African nationalists in Angola formed the People's Movement for the Liberation of Angola (MPLA), and in 1962 the Mozambique Front of Liberation (FRELIMO) was formed. The MPLA fought Portugal in Angola for fourteen years before achieving independence in 1975. FRELIMO did the same in Mozambique for ten years, and that colony became independent the same year as did Angola.

In the British crown colony of Southern Rhodesia, white settlers who refused to accept African majority rule demanded independence from Britain in 1961. By 1965, under the leadership of Ian Smith and influenced by events in South Africa, the colony declared itself an independent state under the name Rhodesia. Initially, several African nationalists who joined forces to fight the minority regime of Ian Smith failed because of a lack of political experience. The Zimbabwe African National Union (ZANU) took over the struggle in the early 1970s, and by 1978 controlled large parts of Rhodesia. Consequently, Smith's regime was forced to negotiate with ZANU. Victory came to the African majority in a 1980 one-person, one-vote election won by Robert Mugabe, the leader of ZANU. As long expected, the country's name was changed to Zimbabwe. Mugabe would become one of Africa's longest-ruling leaders, remaining in power until 2017.

South Africa

South Africa has a unique political history within Africa. The Dutch were the first Europeans to form permanent settlements in what is now South Africa, beginning in the mid-seventeenth century. The British arrived in the early nineteenth century. After almost a century of conflicts between the two groups, and as an immediate result of the discovery of gold and diamonds in the interior of South Africa, the British Empire and the Dutch settlers (the latter being known as Boers, or Afrikaners) went to war. The British won the Boer Wars of 1880 and 1899-1902 and were able to colonize all of South Africa. In 1910 South Africa gained independence under the leadership of the white settlers, who made up a small percentage of the country's population.

In 1948 the Afrikaner-dominated National Party (NP) gained control of the government and introduced the policy of apartheid, which was even harsher than Jim Crow laws had been in the United States. Apartheid meant the separation of races, with all privileges given to whites. Opposition to this policy began immediately, first with peaceful protest. The opposition, led by the African National Congress (ANC), intensified in the 1960s and became increasingly violent as the government continued to oppress the black majority. By the 1980s, the ANC, led by Nelson Mandela (who had been serving a sentence of life imprisonment since 1962), gained widespread support; by that time, its armed wing had been conducting open guerrilla warfare against the government. The United States and Western Europe imposed economic sanctions on South Africa aimed at bringing both sides to the negotiating table. In 1990 the NP-led South African government, under F. W. De Klerk, recognized the ANC and released Nelson Mandela after twenty-seven years of imprisonment. In 1994 the first election in which all races could vote was held,

and the ANC won. Mandela became the first black president of South Africa.

Namibia, formerly South West Africa, was Africa's last colony to gain its independence, which it won from South Africa in 1990. It had been a German colony until after World War I, when South Africa, with British support, took it over. In the 1950s, African nationalists in the colony asked the United Nations for help but were largely ignored until 1968, when the United Nations Security Council adopted a resolution officially recognizing the territory under the name Namibia, "in accordance with the desires of its people." A year later, the United Nations declared South Africa's continued occupation of Namibia illegal. The South African government ignored this call, leading to the intensification of an armed struggle by the South West African People's Organization (SWAPO). SWAPO was formed in 1960 but did not become a force to be reckoned with until the 1980s, when it gained support from Angola and Cuba. The militant actions of SWAPO and mounting international pressure persuaded South Africa to give up the colony. In 1990 Sam Nujoma, the leader of SWAPO, was elected as the first president of the country, officially named the Republic of Namibia.

Africa after Independence

African states won the opportunity to govern themselves after independence. These new nations now could control their wealth and use it to benefit their own people and be directly involved in world politics. These gains were limited, however, in part because of the legacy of colonialism. The former rulers did little to prepare the new nations for independence. As a result, many countries lacked the institutional framework for an independent government. The major problem facing these states was their political-geographical structure based on colonial administrative divisions.

Africa's boundaries divide some ethnic groups among two or more nations. For example, most of the Yoruba people live in western Nigeria, but a significant number of them live in neighboring Togo and Benin, which formerly were French colonies. In Central Africa, the lands inhabited by the Kongo people were colonized by Belgium, France, and Portugal. Today they live in three different countries—Angola, the Republic of the Congo, and the Democratic Republic of the Congo.

Throughout Africa, different ethnic groups found themselves in states with peoples of different languages, cultures, and religious backgrounds. These imposed political boundaries have made it difficult for many African countries to generate a common sense of national identity envisioned by the founders of African nationalism. Under the slogans of national unification and stabilization, quite a few post-colonial African leaders turned to authoritarian rule. Such were the cases of Togo's Gnassingbé Eyadéma (who orchestrated the first military coup d'état in post-colonial Africa), Guinea's Ahmed Sékou Touré, Somalia's Siad Barre, and Uganda's brutal despot Idi Amin, among others. In the Central African Republic, the military dictator Jean-Bédel Bokassa even proclaimed himself an emperor.

Ethnic Conflicts and Militant Insurgencies

Once independence was achieved, nation-building became a difficult task. Ethnic allegiances frequently disrupted national unity. Ethnic competition for political positions, government programs, and other benefits were common. Many African leaders who participated in pre-independence politics turned to acquiring political power and amassing wealth. They achieved these mostly by nationalizing huge tracts of communal lands and distributing them among themselves, their families, and their allies, but also by injecting ethnicism into every aspect of African politics. Like many armed liberation movements, most of the political parties in post-colonial sub-Saharan Africa were ethnicity-based.

In Nigeria, for example, the three major ethnic groups—Yoruba, Ibo (Igbo), and Hausa—formed their own political parties, although minorities and others were encouraged to join. In most cases, military regimes were as guilty of ethnicism as their civilian counterparts. Ethnic rivalries soon escalated to renewed armed ethnic conflicts and civil wars in sub-Saharan Africa. In 1967 the Ibo of eastern Nigeria decided to secede and create the Republic of Biafra. The rest of Nigeria united against them, and by 1970, the Nigerian government won the civil war. However, the defeat of the Ibo did not mean that Nigeria had resolved its ethnic as well as religious strife. (The country is roughly half-Muslim and half-Christian.) Since 2002, it has seen sectarian violence by Boko Haram, a jihadist terrorist organization waging an armed rebellion against the Nigerian government and also active in parts of neighboring Chad, Niger, and Cameroon.

Regionalism and religious divisions also brought trouble to the Sudan region. About 75 percent of the people of the region are Muslims, the majority of whom live in the northern and western part of the country of Sudan. The rest are Christians and animists living in the south. The northern Sudanese took over after independence in 1956, with little southern participation. The south, led by the Sudanese People's Liberation Army (SPLA) under the leadership of John Garang, took up arms in the 1970s to demand a separate state. The conflict was fueled by other major wars taking place in the same region, such as the War in Darfur, in western Sudan, between the Sudanese government and that area's groups of Arab descent on one side and non-Arab groups on the other. The ethnic cleansing of non-Arabs rose to the level of genocide. South Sudan finally gained independence in 2011, becoming the most recent African sovereign state recognized by the United Nations. However, in the new country, a civil war between the government and opposition forces, and ethnic violence between South Sudanese rival tribes, broke out before long.

Burundi and Rwanda gained independence from Belgium in 1962 but were ruled by the chiefs of the Tutsi ethnic group, which constitutes a minority in both countries, with the Hutu being the dominant group. After independence, the Tutsi stayed in power in Burundi through the repression of the Hutu. A 1972 Hutu rebellion was met with severe Tutsi repression in which thousands died. In Rwanda, the Hutu majority won control after independence at the expense of the Tutsi. In 1994 Tutsi rebels shot down an airplane carrying the Hutu president, leading to a mass slaughter of the Tutsi, which became known as the Rwandan genocide. The genocide led to a refugee crisis in neighboring Zaire (now the Democratic Republic of the Congo), a cross-border insurgency, and two Congo Wars.

In Somalia, the parliamentary system left by the Italians after independence was a disaster because it emphasized proportional representation—the idea that each social class in the society could have a party with seats in parliament. However, Somalia, unlike Italy, is divided into clans, rather than having a class structure.

Since the years of colonial rule destroyed the old clan divisions, Somalia was difficult to rule. In the early 1990s, the government disintegrated, and rival clans began to run the country. A breakaway region of Somaliland declared its independence from the country in 1991, which has since been seeking international recognition as the Republic of Somaliland. After two decades of state collapse and weak transitional government, the federal government of Somalia was reestablished in 2012, but the country has remained Africa's most violent state due to the activities of the terrorist jihadist group Al-Shabaab as well as the persistent proliferation of local clan militias.

Historically, Liberia had seen competition between the Americo-Liberians and the ethnic majority. The Americo-Liberians are descendants of freed slaves repatriated to Africa from the United States, who make up approximately 5 percent of the population but for a long time controlled much of the wealth. In 1980 a military coup purportedly sought to distribute that wealth among the country's many indigenous ethnic groups. Liberia plunged into a twenty-plus-year period of authoritarian rule, coups, guerilla wars, and civil wars, culminating in the resignation and exile of its twenty-second president, Charles Taylor. (Taylor was later convicted of war crimes by the international Special Court for Sierra Leone for his involvement in that country's bloody civil war over control of diamond fields.) Since then, Liberia has entered an era of stability. Its 2005 presidential elections, the most free and fair in the country's history, brought to power Africa's first female president, Ellen Johnson Sirleaf.

Ethiopia erupted into civil war in the mid-1980s, expelling the communist government in power. The new leadership allowed the former province of Eritrea to secede in 1993. Eritrea became nominally a presidential republic but in effect, a totalitarian, one-party state.

Shortly after Mali achieved independence from France, its northern part became entangled in Tuareg rebellions; they later merged with militant Islamist groups. In 2012, the insurgency culminated in the creation of the unrecognized proto-state of Azawad. France has had thousands of soldiers stationed in Mali since 2013 when it intervened in the conflict to help the Malian government fight the militants. The conflict in northern Mali is part of a larger ongoing Islamist insurgency in the Maghreb and Sahel regions of Africa.

Other ongoing conflicts on the African continent include continued armed unrest on Egypt's Sinai Peninsula, a civil war between rival groups seeking control of post-Gaddafi Libya, Cameroon's long-standing Anglophone separatist movement (the Ambazonia War), and numerous ethnic insurgencies in the Democratic Republic of the Congo.

Tens of thousands of migrants from war-torn and economically underdeveloped parts of Africa have been seeking refuge in Europe in hopes of a better life, embarking on a perilous journey by boat via the Mediterranean Sea. In recent years, this movement, combined with the flight of victims of the Syrian Civil War, has assumed the proportions of a global refugee/migrant crisis.

The political geography of modern Africa has been greatly influenced by the colonial and pre-colonial experience. The unification of ethnic groups with old relationships involving conflict, deep tribal loyalties, diverse languages, and different traditional values has made it difficult to build modern nation states. Nevertheless, territorial disputes between sovereign African nations are relatively minor (compared, for instance, to Asia), and the countries have been increasingly working together to solve common problems. African countries play a leading role in the United Nations peacekeeping efforts and provide nearly half of UN peacekeeping forces deployed worldwide. In recent decades, Af-

A unit of ONLF rebels move from one location to another after being surrounded for a week on a hill top. (Jonathan Alpeyrie)

rica has seen rapid democratic transitions, leading to more responsive, effective, and accountable governments.

In 2002, the Organization of African Unity was replaced by the African Union (AU), a continental organization that includes all African independent states recognized by the United Nations, plus the SADR. The legislative body of the AU is the Pan-African Parliament, which oversees the implementation of the AU's goals of engendering peace, security, good governance, and economic cooperation on the continent. Within the AU, the African Economic Community works to create regional free trade areas and customs unions and, eventually, a continent-wide common market and monetary union somewhat akin to those of the European Union. Regional cooperation among countries has been steadily expanding in all parts of Africa, witnessed by the activities of the thirteen-nation Southern African Development Community, the Intergovernmental Authority on Development involving eight countries of the African Horn region, the Economic Community of West African States, and the Arab Maghreb Union.

Femi Ferreira

ECONOMIC GEOGRAPHY

OVERVIEW

TRADITIONAL AGRICULTURE

Two agricultural practices that are widespread among the world's traditional cultures, slash-and-burn and nomadism, share several common features. Both are ancient forms of agriculture, both involve farmers not remaining in a fixed location, and both can pose serious environmental threats if practiced in a nonsustainable fashion. The most significant difference between the two forms is that slash-and-burn generally is associated with raising field crops, while nomadism as a rule involves herding livestock.

Slash-and-Burn Agriculture

Farmers have practiced slash-and-burn agriculture, which is also referrred to as shifting cultivation or swidden agriculture, in almost every region of the world where the climate makes farming possible. Humans have practiced this method for about 12,000 years, ever since the Neolithic Revolution. Swidden agriculture once dominated agriculture in more temperate regions, such as northern Europe. It was, in fact, common in Finland and northern Russia well into the early decades of the twentieth century. Today, between 200 and 500 million people use slash-and-burn agriculture, roughly 7 percent of the world's population. It is most commonly practiced in areas where open land for farming is not readily available because of dense vegetation. These regions include central Africa, northern South America, and Southeast Asia

Slash-and-burn acquired its name from the practice of farmers who cleared land for planting crops by cutting down the trees or brush on the land and then burning the fallen timber on the site. The farmers literally slash and burn. The ashes of the burnt wood add minerals to the soil, which temporarily improves its fertility. Crops the first year following clearing and burning are generally the best crops the site will provide. Each year after that, the yield diminishes slightly as the fertility of the soil is depleted.

Farmers who practice swidden cultivation do not attempt to improve fertility by adding fertilizers such as animal manures but instead rely on the soil to replenish itself over time. When the yield from one site drops below acceptable levels, the farmers then clear another piece of land, burn the brush and other vegetation, and cultivate that site while leaving their previous field to lie fallow and its natural vegetation to return. This cycle will be repeated over and over, with some sites being allowed to lie fallow indefinitely while others may be revisited and farmed again in five, ten, or twenty years.

Farmers who practice shifting cultivation do not necessarily move their dwelling places as they change the fields they cultivate. In some geographic regions, farmers live in a central village and farm cooperatively, with the fields being alternately allowed to remain fallow, and the fields being farmed making a gradual circuit around the central village. In other cases, the village itself may move as new fields are cultivated. Anthropologists studying indigenous peoples in Amazonia, discovered that village garden sites were on a hundred-year cycle. Villagers farmed cooperatively, with the entire village working together to clear a garden site. That garden would be used for about five years, then a new site was cleared. When the garden moved an inconvenient distance from the village, about once every twenty years, the entire village would move to be closer to the new garden. Over a period of approximately 100 years, a village would make a circle through the forest, eventually ending up close to where it had been located long before any of the present villagers had been born.

In more temperate climates, individual farmers often owned and lived on the land on which they practiced swidden agriculture. Farmers in Finland, for example, would clear a portion of their land, burn the brush and other covering vegetation, grow grains for several years, and then allow that land to remain fallow for from five to twenty years. The individual farmer rotated cultivation around the land in a fashion similar to that practiced by whole villages in other areas, but did so as an individual rather than as part of a communal society.

Although slash-and-burn is frequently denounced as a cause of environmental degradation in tropical areas, the problem with shifting cultivation is not the practice itself but the length of the cycle. If the cycle of shifting cultivation is long enough, forests will grow back, the soil will regain its fertility, and minimal adverse effects will occur. In some regions, a piece of land may require as little as five years to regain its maximum fertility; in others, it may take 100 years. Problems arise when growing populations put pressure on traditional farmers to return to fallow land too soon. Crops are smaller than needed, leading to a vicious cycle in which the next strip of land is also farmed too soon, and each site yields less and less. As a result, more and more land must be cleared.

Nomadism

Nomadic peoples have no permanent homes. They earn their livings by raising herd animals, such as sheep, cattle, or horses, and they spend their lives following their herds from pasture to pasture with the seasons. Most nomadic animals tend to be hardy breeds of goats, sheep, or cattle that can withstand hardship and live on marginal lands. Traditional nomads rely on natural pasturage to support their herds and grow no grains or hay for themselves. If a drought occurs or a traditional pasturing site is unavailable, they can lose most of their herds to starvation.

THE HERITAGE SEED MOVEMENT

Modern hybrid seeds have increased yields and enabled the tremendous productivity of the modern mechanized farm. However, the widespread use of a few hybrid varieties has meant that almost all plants of a given species in a wide area are almost identical genetically. This loss of biodiversity, or, the range of genetic difference in a given species, means that a blight could wipe out an entire season's crop. Historical examples of blight include the nineteenth century Great Potato Famine of Ireland and the 1971 corn blight in the United States.

In response to the concern for biodiversity, there has been a movement in North America to preserve older forms of crops with different genes that would otherwise be lost to the gene pool. Nostalgia also motivates many people to keep alive the varieties of fruits and vegetables that their grandparents raised. Many older recipes do not taste the same with modern varieties of vegetables that have been optimized for commercial considerations such as transportability. Thus, raising heritage varieties also can be a way of continuing to enjoy the foods one's ancestors ate.

In many nomadic societies, the herd animal is almost the entire basis for sustaining the people. The animals are slaughtered for food, clothing is woven from the fibers of their hair, and cheese and yogurt may be made from milk. The animals may also be used for sustenance without being slaughtered. Nomads in Mongolia, for example, occasionally drink horses' blood, removing only a cup or two at a time from the animal. Nomads go where there is sufficient vegetation to feed their animals.

In mountainous regions, nomads often spend the summers high up on mountain meadows, returning to lower altitudes in the autumn when snow begins to fall. In desert regions, they move from oasis to oasis, going to the places where sufficient natural water exists to allow brush and grass to grow, allowing their animals to graze for a few days, weeks, or months, then moving on. In some cases, the pressure to move on comes not from the depletion of food for the animals but from the depletion of a water source, such as a spring or well. At many natural desert oases, a natural water seep or spring provides only enough water to support a nomadic group for a few days at a time.

In addition to true nomads—people who never live in one place permanently—a number of cultures have practiced seminomadic farming: The temperate months of the year, spring through fall, are spent following the herds on a long loop, sometimes hundreds of miles long, through traditional grazing areas; then the winter is spent in a permanent village.

Nomadism has been practiced for millennia, but there is strong pressure from several sources to eliminate it. Pressures generated by industrialized society are increasingly threatening the traditional cultures of nomadic societies, such as the Bedouin of the Arabian Peninsula. Traditional grazing areas are being fenced off or developed for other purposes. Environmentalists are also concerned about the ecological damage caused by nomadism.

Nomads generally measure their wealth by the number of animals they own and so will try to develop their herds to be as large as possible, well beyond the numbers required for simple sustainability. The herd animals eat increasingly large amounts of vegetation, which then has no opportunity to regenerate, and desertification may occur. Nomadism based on herding goats and sheep,

---

DESERTIFICATION

Desertification is the extension of desert conditions into new areas. Typically, this term refers to the expansion of deserts into adjacent nondesert areas, but it can also refer to the creation of a new desert. Land that is susceptible to prolonged drought is always in danger of losing its vegetative ground cover, thereby exposing its soil to wind. The wind carries away the smaller silt particles and leaves behind the larger sand particles, stripping the land of its fertility. This naturally occurring process is assisted in many areas by overgrazing.

In the African Sahel, south of the Sahara, the impact of desertification is acute. Recurring drought has reduced the vegetation available for cattle, but the need for cattle remains high to feed populations that continue to grow. The cattle eat the grass, the soil is exposed, and the area becomes less fertile and less able to support the population. The desert slowly encroaches, and the people must either move or die.

---

for example, has been blamed for the expansion of the Sahara Desert in Africa. For this reason, many environmental policymakers have been attempting to persuade nomads to give up their roaming lifestyle and become sedentary farmers.

Nancy Farm Männikkö

COMMERCIAL AGRICULTURE

Commercial farmers are those who sell substantial portions of their output of crops, livestock, and dairy products for cash. In some regions, commercial agriculture is as old as recorded history, but only in the twentieth century did the majority of farmers come to participate in it. For individual farmers, this has offered the prospect of larger income and the opportunity to buy a wider range of products. For society, commercial agriculture has been associated with specialization and increased productivity.

Commercial agriculture has enabled world food production to increase more rapidly than world population, improving nutrition levels for millions of people.

Steps in Commercial Agriculture

In order for commercial agriculture to exist, products must move from farmer to ultimate consumer, usually through six stages:

1. Processing, packaging, and preserving to protect the products and reduce their bulk to facilitate shipping.

2. Transport to specialized processing facilities and to final consumers.

3. Networks of merchant middlemen who buy products in bulk from farmers and processors and sell them to final consumers.

4. Specialized suppliers of inputs to farmers, such as seed, livestock feed, chemical inputs (fertilizers, insecticides, pesticides, soil conditioners), and equipment.

5. A market for land, so that farmers can buy or lease the land they need.

6. Specialized financial services, especially loans to enable farmers to buy land and other inputs before they receive sales revenues.

Improvements in agricultural science and technology have resulted from extensive research programs by government, business firms, and universities.

International Trade

Products such as grain, olive oil, and wine moved by ship across the Mediterranean Sea in ancient times. Trade in spices, tea, coffee, and cocoa provided powerful stimulus for exploration and colonization around 1500 CE. The coming of steam locomotives and steamships in the nineteenth century greatly aided in the shipment of farm products and spurred the spread of population into potentially productive farmland all over the world. Beginning with Great Britain in the 1840s, countries were willing to relinquish agricultural self-sufficiency to obtain cheap imported food, paid for by exporting manufactured goods.

Most of the leaders in agricultural trade were highly developed countries, which typically had large amounts of both imports and exports. These countries are highly productive both in agriculture and in other commercial activities. Much of their trade is in high-value packaged and processed goods. Although the vast majority of China's labor force works in agriculture, their average productivity is low and the country showed an import surplus in agricultural products. The same was true for Rus-

sia. India, similar to China in size, development, and population, had relatively little agricultural trade. Australia and Argentina are examples of countries with large export surpluses, while Japan and South Korea had large import surpluses. Judged by volume, trade is dominated by grains, sugar, and soybeans. In contrast, meat, tobacco, cotton, and coffee reflect much higher values per unit of weight.

The United States

Blessed with advantageous soil, topography, and climate, the United States has become one of the most productive agricultural countries in the world. Technological advances have enabled the United States to feed its own residents and export substantial quantities with only 3 percent of its labor force engaged directly in farming. In the 2020s there are about 2 million farms cultivating about 1 billion acres. They produced about US$133 billion worth of products. After expenses, this yielded about US$92.5 billion of net farm income—an average of only about US$25,000 per farm. However, most farm families derive substantial income from nonfarm employment.

There is a great deal of agricultural specialization by region. Corn, soybeans, and wheat are grown in many parts of the United States (outside New England). Some other crops have much more limited growing areas. Cotton, rice, and sugarcane require warmer temperatures. Significant production of cotton occurred in seventeen states, rice in six, and sugarcane in four. In 2018, the top 10 agricultural producing states in terms of cash receipts were (in descending order): California, Iowa, Texas, Nebraska, Minnesota, Illinois, Kansas, North Carolina, Wisconsin, and Indiana Typically the top two states in a category account for about 30 percent of sales. Fruits and vegetables are the main exception; the great size, diversity, and mild climate of California gives it a dominant 45 percent.

Socialist Experiments

Under the dictatorship of Joseph Stalin, the communist government of the Soviet Union established a program of compulsory collectivized agriculture in 1929. Private ownership of land, buildings, and other assets was abolished. There were some state farms, "factories in the fields," operated on a large scale with many hired workers. Most, however, were collective farms, theoretically run as cooperative ventures of all residents of a village, but in practice directed by government functionaries. The arrangements had disastrous effects on productivity and kept the rural residents in poverty. Nevertheless, similar arrangements were established in China in 1950 under the rule of Mao Zedong. A restoration of commercial agriculture after Mao's death in 1976 enabled China to achieve greater farm output and farm incomes.

Most Western countries, including the United States, subsidize agriculture and restrict imports of competing farm products. Objectives are to support farm incomes, reduce rural discontent, and slow the downward trend in the number of farmers. Farmers in the European Union will see aid shrink in the 2021–2027 period to 365 billion euros (US$438 billion), down 5 percent from the current Common Agricultural Policy (CAP). Japan's Ministry of Agriculture, Forestry and Fisheries (MAFF) has requested 2.65 trillion yen (roughly US$24 billion) for the Japan Fiscal Year (JFY) 2018 budget, a 15 percent increase over last year. The budget request eliminates the direct payment subsidy for table rice production, but requests significant funding for a new income insurance program, agricultural export promotion, and underwriting goals to expand domestic potato production. In 2019, trade wars with China and punishing tariffs have led to increased subsidies by the U.S. government, totaling US$10 billion in 2018 and US$14.5 billion in 2019.

Problems for Farmers

Farmers in a system of commercial agriculture are vulnerable to changes in market prices as well as the universal problems of fluctuating weather. Congress tried to reduce farm subsidies through the Freedom to Farm Act of 1996, but serious price declines in 1997-1999 led to backtracking. Efforts to increase productivity by genetic alterations, radiation, and feeding synthetic hormones to livestock have drawn critical responses from some consumer groups. Environmentalists have been concerned about soil depletion and water pollution resulting from chemical inputs.

Productivity and World Hunger

Despite advances in agricultural production, the problem of world hunger persists. Even in countries that store surpluses of farm commodities, there are still people who go hungry. In less-developed countries, the prices of imported food from the West are too low for local producers to compete and too high for the poor to buy them.

Paul B. Trescott

MODERN AGRICULTURAL PROBLEMS

Ever since human societies started to grow their own food, there have been problems to solve. Much of the work of nature was disrupted by the work of agriculture as many as 10,000 years ago. Nature took care of the land and made it productive in its own intricate way, through its own web of interdependent systems. Agriculture disrupts those systems with the hope of making the land even more productive, growing even more food to feed even more people. Since the first spade of soil was turned over and the first plants domesticated, farmers have been trying to figure out how to care for the land as well as nature did before.

Many modern problems in agriculture are not really modern at all. Erosion and pollution, for example, have been around as long as agriculture.

However, agriculture has changed drastically within those 10,000 years, especially since the dawn of the Industrial Revolution in the seventeenth century. Erosion and pollution are now bigger problems than before and have been joined by a host of others that are equally critical—not all related to physical deterioration. Modern farmers use many more machines than did farmers of old, and modern machines require advanced sources of energy to unleash their power. The machines do more work than could be accomplished before, so fewer farmers are needed, which causes economic problems.

Cities continue to grow bigger as land—usually the best farmland around—is converted to homes and parking lots for shopping centers. The farmers that remain on the land, needing to grow ever more food, turn to the research and engineering industries to improve their seeds. These industries have responded with recombinant technologies that move genes from one species to another; for example, genes cut from peanuts may be spliced into chickens. This creates another set of cultural problems, which are even more difficult to solve because most are still "potential"—their impact is not yet known.

Erosion

Soil loss from erosion continues to be a huge problem all over the world. As agriculture struggles to feed more millions of people, more land is plowed. The newly plowed lands usually are considered more marginal, meaning they are either too steep, too thin, or too sandy; are subject to too much rain; or suffer some other deficiency. Natural vegetative cover blankets these soils and protects them from whatever erosive agents are active in their regions: water, wind, ice, or gravity. Plant cover also increases the amount of rain that seeps downward into the soil rather than running off into rivers. The more marginal land that is turned over for crops, the faster the erosive agents will act and the more erosion will occur.

Expansion of land under cultivation is not the only factor contributing to erosion. Fragile grasslands in dry areas also are being used more inten-

sively. Grazing more livestock than these pastures can handle decreases the amount of grass in the pasture and exposes more of the soil to wind—the primary erosive agent in dry regions.

Overgrazing can affect pastureland in tropical regions too. Thousands of acres of tropical forest have been cleared to establish cattle-grazing ranges in Latin America. Tropical soils, although thick, are not very fertile. Fertility comes from organic waste in the surface layers of the soil. Tropical soils form under constantly high temperatures and receive much more rain than soils in moderate, midlatitude climates; thus, tropical organic waste materials rot so fast they are not worked into the soil at all. After one or two growing seasons, crops grown in these soils will yield substantially less than before.

Tropical fields require fallow periods of about ten years to restore themselves after they are depleted. That is why tropical cultures using slash-and-burn methods of agriculture move to new fields every other year in a cycle that returns them to the same place about every ten years, or however long it takes those particular lands to regenerate. The heavy forest cover protects these soils from exposure to the massive amounts of rainfall and provides enough organic material for crops—as long as the forest remains in place. When the forest is cleared, however, the resulting grassland cannot provide the adequate protection, and erosion accelerates. Grasslands that are heavily grazed provide even less protection from heavy rains, and erosion accelerates even more.

The use of machines also promotes erosion, and modern agriculture relies on machinery: tractors, harvesters, trucks, balers, ditchers, and so on. In the United States, Canada, Europe, Russia, Brazil, South Africa, and other industrialized areas, machinery use is intense. Machinery use is also on the rise in countries such as India, China, Mexico, and Indonesia, where traditional nonmechanized methods are practiced widely. Farming machines, in gaining traction, loosen the topsoil and inhibit vegetative cover growth, especially when they pull behind them any of the various farm implements designed to rid the soil of weeds, that is, all vegetation except the desired crop. This leaves the soil more exposed to erosive weather, so more soil is carried away in the runoff of water to streams.

Eco-fallow farming has become more popular in the United States and Europe as a solution to reducing erosion. This method of agriculture, which leaves the crop residue in place over the fallow (nongrowing) season, does not root the soil in place, however. Dead plants do not "grab" the soil like live plants that need to extract from it the nutrients they need to live, so erosion continues, even though it is at a slower rate. Eco-fallow methods also require heavier use of chemicals, such as herbicides, to "burn down" weed growth at the start of the growing season, which contributes to accelerated erosion and increases pollution.

Pollution

Pollution, besides being a problem in general, continues to grow as an agricultural problem. With the onset of the Green Revolution, the use of herbicides, insecticides, and pesticides has increased dramatically all over the world. These chemicals are not used up completely in the growth of the crop, so the leftovers (residue) wash into, and contaminate, surface and groundwater supplies. These supplies then must be treated to become useful for other purposes, a job nature used to do on its own. Agricultural chemicals reduce nature's ability to act as a filter by inhibiting the growth of the kinds of plant life that perform that function in aquatic environments. The chemical residues that are not washed into surface supplies contaminate wells.

As chemical use increases, contamination accumulates in the soil and fertility decreases. The microorganisms and animal life in the soil, which had facilitated the breakdown of soil minerals into usable plant products, are no longer nourished because the crop residue on which they feed is depleted, or they are killed by the active ingredients in the chemical. As a result, soil fertility must be restored to maintain yield. Chemical replacement is usually the method of choice, and increased applications of chemical fertilizers intensify the toxicity of this cyclical chemical dependency.

Chemicals, although problematic, are not as difficult to contend with as the increasingly heavy silt load choking the life out of streams and rivers. Accelerated erosion from water runoff carries silt particles into streams, where they remain suspended and inhibit the growth of many beneficial forms of plant and animal life. The silt load in U.S. streams has become so heavy that the Mississippi River Delta is growing faster than it used to. The heavy silt load, combined with the increased load of chemical residues, is seriously taxing the capabilities of the ecosystems around the delta that filter out sediments, absorb nutrients, and stabilize salinity levels for ocean life, creating an expanding dead zone.

This general phenomenon is not limited to the Mississippi Delta—it is widespread. Its impact on people is high, because most of the world's population lives in coastal zones and comes in direct contact with the sea. Additionally, eighty percent of the world's fish catch comes from the coastal waters over continental shelves that are most susceptible to this form of pollution.

Monoculture

Modern agriculture emphasizes crop specialization. Farmers, especially in industrialized regions, often grow a single crop on most of their land, perhaps rotating it with a second crop in successive years: corn one year, for example, then soybeans, then back to corn. Such a strategy allows the farmer to reduce costs, but it also makes the crop, and, thus, the farmer and community, susceptible to widespread crop failure. When the crop is infested by any of an ever-changing number and variety of pests—worms, molds, bacteria, fungi, insects, or other diseases—the whole crop is likely to die quickly, unless an appropriate antidote is immediately applied. Chemical antidotes can do the job but increase pollution. Maintaining species diversity—growing several different crops instead of one or two—allows for crop failures without jeopardizing the entire income for a farm or region that specializes in a particular monoculture, such as tobacco, coffee, or bananas.

Chemicals are not the only modern methods of preventing crop loss. Genetically engineered seeds are one attempt at replacing post-infestation chemical treatments. For example, splicing genes into varieties of rice or potatoes from wholly unrelated species—say, hypothetically, a grasshopper—to prevent common forms of blight is occurring more often. Even if the new genes make the crop more resistant, however, they could trigger unknown side effects that have more serious long-term environmental and economic consequences than the problem they were used to solve. Genetically altered crops are essentially new life-forms being introduced into nature with no observable precedents to watch beforehand for clues as to what might happen.

Urban Sprawl

As more farms become mechanized, the need for farmers is being drastically reduced. There were more farmers in the United States in 1860 than there were in the year 2000. From a peak in 1935 of about 6.8 million farmers farming 1.1 billion acres, the United States at the end of the twentieth century counted fewer than 2.1 million farmers farming 950 million acres. As fewer people care for land, the potential for erosion and pollution to accelerate is likely to increase, causing land quality to decline.

As farmers are displaced and move into towns, the cities take up more space. The resulting urban sprawl converts a tremendous amount of cropland into parking lots, malls, industrial parks, or suburban neighborhoods. If cities were located in marginal areas, then the concern over the loss of farmland to commercial development would be nominal. However, the cities attracting the greatest numbers of people have too often replaced the best cropland. Taking the best cropland out of primary production imposes a severe economic penalty.

James Knotwell and Denise Knotwell

WORLD FOOD SUPPLIES

All living things need food to begin the life process and to live, grow, work, and survive. Almost all foods that humans consume come from plants and animals. Not all of Earth's people eat the same foods, however, nor do they require the same caloric intakes. The types, combinations, and amounts of food consumed by different peoples depend upon historic, socioeconomic, and environmental factors.

The History of Food Consumption

Early in human history, people ate what they could gather or scavenge. Later, people ate what they could plant and harvest and what animals they could domesticate and raise. Modern people eat what they can grow, raise, or purchase. Their diets or food composition are determined by income, local customs, religion or food biases, and advertising. There is a global food market, and many people can select what they want to eat and when they eat it according to the prices they can pay and what is available.

Historically, in places where food was plentiful, accessible, and inexpensive, humans devoted less time to basic survival needs and more time to activities that led to human progress and enjoyment of leisure. Despite a modern global food system, instant telecommunications, the United Nations, and food surpluses at places, however, the problem of providing food for everyone on Earth has not been solved.

According to the United Nations Sustainable Development Goals that were adopted by all Member States in 2015, an estimated 821 million people were undernourished in 2017. In developing countries, 12.9 percent of the population is undernourished. Sub-Saharan Africa has the highest prevalence of hunger; the number of undernourished people increased from 195 million in 2014 to 237 million in 2017. Poor nutrition causes nearly half (45 percent) of deaths in children under five—3.1 million children each year. As of 2018, 22 percent of the global under-5 population were still chronically undernourished in 2018. To meet challenge of Goal 2: Zero Hunger, significant changes both in terms of agriculture and conservation as well as in financing and social equality will be required to nourish the 821 million people who are hungry today and the additional 2 billion people expected to be undernourished by 2050.

World Food Source Regions

Agriculture and related primary food production activities, such as fishing, hunting, and gathering, continue to employ more than one-third of the world's labor force. Agriculture's relative importance in the world economic system has declined with urbanization and industrialization, but it still plays a vital role in human survival and general economic growth. Agriculture in the third millennium must supply food to an increasing world population of nonfood producers. It must also produce food and nonfood crude materials for industry, accumulate capital needed for further economic growth, and allow workers from rural areas to industrial, construction, and expanding intraurban service functions.

Soil types, topography, weather, climate, socioeconomic history, location, population pressures, dietary preferences, stages in modern agricultural development, and governmental policies combine to give a distinctive personality to regional agricultural characteristics. Two of the most productive food-producing regions of the world are North America and Asia. Countries in these regions export large amounts of food to other parts of the world.

Foods from Plants

Most basic staple foods come from a small number of plants and animals. Ranked by tonnage produced, the most important food plants throughout the world are corn (maize), wheats, rice, potatoes, cassava (manioc), barley, soybeans, sorghums and millets, beans, peas and chickpeas, and peanuts (groundnuts).

More than one-third of the world's cultivated land is planted with wheat and rice. Wheat is the dominant food staple in North America, Western and Eastern Europe, northern China, and the Middle East and North Africa. Rice is the dominant food staple in southern and eastern Asia. Corn, used primarily as animal food in developed nations, is a staple food in Latin America and Southeast Africa. Potatoes are a basic food in the highlands of South America and in Central and Eastern Europe. Cassava (manioc) is a tropical starch-producing root crop of special dietary importance in portions of lowland South America, the west coast countries of Africa, and sections of South Asia. Barley is an important component of diets in North African, Middle Eastern, and Eastern European countries. Soybeans are an integral part of the diets of those who live in eastern, southeastern, and southern Asia. Sorghums and millets are staple subsistence foods in the savanna regions of Africa and south Asia, while peanuts are a facet of dietary mixes in tropical Africa, Southeast Asia, and South America.

Food from Animals

Animals have been used as food by humans from the time the earliest people learned to hunt, trap, and fish. However, humans have domesticated only a few varieties of animals. Ranked by tonnage of meat produced, the most commonly eaten animals are cattle, pigs, chickens and turkeys, sheep, goats, water buffalo, camels, rabbits and guinea pigs, yaks, and llamas and alpacas.

Cattle, which produce milk and meat, are important food sources in North America, Western Europe, Eastern Europe, Australia and New Zealand, Argentina, and Uruguay. Pigs are bred and reared for food on a massive scale in southern and eastern Asia, North America, Western Europe, and Eastern Europe. Chickens are the most important domesticated fowl used as a human food source and are a part of the diets of most of the world's people.

Sheep and goats, as a source of meat and milk, are especially important to the diets of those who live in the Middle East and North Africa, Eastern Europe, Western Europe, and Australia and New Zealand.

Water buffalo, camels, rabbits, guinea pigs, yaks, llamas, and alpacas are food sources in regions of the world where there is low consumption of meat for religious, cultural, or socioeconomic reasons. Fish is an inexpensive and wholesome source of food. Seafood is an important component to the diets of those who live in southern and eastern Asia, Western Europe, and North America.

The World's Growing Population

The problem of feeding the world is compounded by the fact that population was increasing at a rate of nearly 82 million persons per year at the end of second decade of the twenty-first century. That rate of increase is roughly equivalent to adding a country the size of Germany to the world every single year.

Also compounding the problem of feeding the world are population redistribution patterns and changing food consumption standards. In the year 2050, the world population is projected to reach approximately 10 billion—4 billion people more than were on the earth in 2000. Most of the increase in world population is expected to occur within the developing nations.

Urbanization

Along with an increase in population in developing nations is massive urbanization. City dwellers are food consumers, not food producers. The exodus of young men and women from rural areas has given rise to a new series of megacities, most of which are in developing countries. By the year 2030, there could be as many as forty-one megacities (cities with populations of 10 million people or more).

When rural dwellers move to cities, they tend to change their dietary composition and food-consumption patterns. Qualitative changes in dietary consumption standards are positive, for the most part, and are a result of copying the diets of what is considered a more prestigious group or positive educational activities of modern nutritional scientists

working in developing countries. During the last four decades of the twentieth century, a tremendous shift took place in overall dietary habits as Western foods became increasingly available and popular throughout the world. While improved nutrition has contributed to a decrease in child mortality, an increase in longevity, and a greater resistance to disease, it is also true that conditions including morbid obesity, Type II diabetes, and hypertension are on the rise.

Strategies for Increasing Food Production

To meet the food demands and the food distribution needs of the world's people in the future, several strategies have been proposed. One such strategy calls for the intensification of agriculture—improving biological, mechanical, and chemical technology and applying proven agricultural innovations to regions of the world where the physical and cultural environments are most suitable for rapid food production increases.

The second step is to expand the areas where food is produced so that areas that are empty or underused will be made productive. Reclaiming areas damaged by human mismanagement, expanding irrigation in carefully selected areas, and introducing extensive agrotechniques to areas not under cultivation could increase the production of inexpensive grains and meats.

Finally, interregional, international, and global commerce should be expanded, in most instances, increasing regional specializations and production of high-quality, high-demand agricultural products for export and importing low-cost basic foods. A disequilibrium of supply and demand for certain commodities will persist, but food producers, regional and national agricultural planners, and those who strive for regional economic integration must take advantage of local conditions and location or create the new products needed by the food-consuming public in a one-world economy.

Perspectives

Humanity is entering a time of volatility in food production and distribution. The world will produce enough food to meet the demands of those who can afford to buy food. In many developing countries, however, food production is unlikely to keep pace with increases in the demand for food by growing populations.

Factors that could lead to larger fluctuations in food availability include weather variations such as those induced by El Niño and climate change, the growing scarcity of water, civil strife and political in-stability, and declining food aid. In developing countries, decision makers need to ensure that policies promote broad-based economic growth—and in particular agricultural growth—so that their countries can produce enough food to feed themselves or enough income to buy the necessary food on the world market.

William A. Dando

AGRICULTURE

Africa's fifty-four countries can be grouped into seven regions on the basis of geographical and climatic homogeneity: Northern Africa, Sudano-Sahelian, Gulf of Guinea, Central Africa, Eastern Africa, Indian Ocean Islands, and Southern Africa. Rainfall—the dominant influence on agricultural output—varies greatly among Africa's countries. Without irrigation, agriculture requires a reliable annual rainfall of more than 30 inches (75 centimeters, 760 millimeters). Only in the Central region does significant annual precipitation allow the practiceof agriculture on a commercial scale. Lack of rainfall is further exacerbated by climate change, which manifests itself in increasing desertification and long periods of drought.

Food output has been growing at a slower rate than the continent's population. In most African countries, more than 50 percent, and often 80 percent, of the population works in agriculture, mainly subsistence agriculture. Yet in 2015, agriculture contributed to a mere 15 percent of the combined gross domestic product (GDP) of Africa, reflecting the low productivity in this sector. Large portions of the continent, such as Mali and the Sudans, have the potential of becoming granaries to the continent and producing considerable food exports, but significant investments in water management, high-yield seeds, mechanization, and fertilizers are required to turn this potential into reality.

Traditional African Agriculture
Traditionally, agriculture in Africa has been labor-intensive subsistence farming in small plots worked byfamily members. New land for farming was obtained by the slash-and-burn method (shifting agriculture). The trees in a forested area would be cut down and burned where they fellThe; t ashes from the burned trees fertilized the soil with needed minerals. Both men and women worked at such farming. Slash-and-burn agriculture is common not only in Africa but also in tropical areas around the world. In these areas, heavy rainfall removes nutrients from the soil. Ultimately, subsequent rainfall washes out the nutrients from the burned trees within two to three years. Then the land is abandoned for a new area.

The type of crops cultivated depends upon the region. In the very dry, yet habitable, parts of Africa —such as the Sudano-Sahelian region that stretches from Senegal and Mali in the west to the Sudans in the east—a key subsistence crop is green (or pearl) millet, a grain. It is ground into a type of flour and can be made into a bread-like substance. Finger millet is native to eastern and southern African re-

Sorghum, a drought-tolerant tropical cereal in Mossurize district of Mozambique. (Ton Rulkens)

SELECTED AGRICULTURAL PRODUCTS OF AFRICA

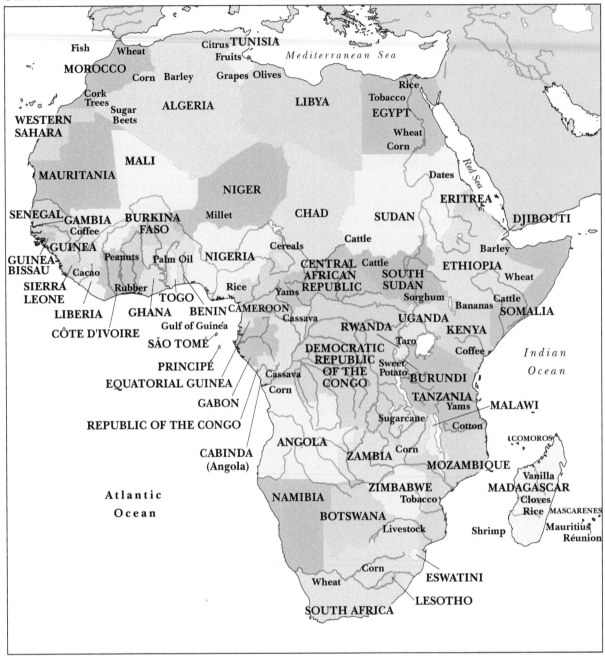

gions, and is still grown there. In moister areas, traditional crops are root and tuber crops, such as yams (which are native to Africa) and cassava, or manioc (which was introduced into Africa from Brazil by Portuguese traders in the sixteenth century). Cassava has an outer surface or skin that is poisonous, but it can be treated to remove the poison. The tuber then can be ground and used to make a bread-like substance.

Sorghum, native to the Sudano-Sahelian region and resistant to drought and high temperatures, was spread throughout the continent by various migrating tribal groups. Along the way, it adapted to a wide range of environments from the highlands of Ethiopia to the semi-arid Sahel. It is widely grown for food, animal fodder, and traditional beer. Special varieties of sorghum are used for supporting yam plants, roof thatching, and making brooms, sleeping mats, baskets, leather dyes (in Morocco), and even strings for traditional musical instruments (in Nigeria).

Processing fruits of the oil palm for cooking oil has been practiced in Africa for 5,000 years. Traditionally, the oil is obtained either by pounding cooked fruits with pestles or by foot-trampling them in troughs. The mash is put in a cloth, and the oil is extracted by twisting and squeezing the cloth.

Animal husbandry, in the form of nomadic herding, is another part of traditional agriculture. Most of the meat produced in sub-Saharan Africa comes from the pastoralist tribes practicing methods that have been passed on from generation to generation. Cattle, goats, and sheep are herded in the Sudano-Sahelian, Eastern, and Southern regions of Africa. This type of pastoral agriculture is highly dependent on the availability of water and exten-sive grassland areas, which have been shrinking because of climate change. Regions that are very moist, such as the Gulf of Guinea, are not suitable for cattle farming because of the tsetse fly, which carries diseases such as sleeping sickness.

Modern Crops

The most widely grown crop is rice, which is cultivated on more than one-third of the water-managed or irrigated cropland in Africa. The African species of rice is thought to have been domesticated in the Niger River basin some 2,000 to 3,000 years ago. The Asian species was introduced into Africa by the Portuguese as early as in the sixteenth century. Since then, it has replaced the African rice everywhere. Cultivated mostly in wetlands and valley bottoms, rice is the most common crop in the hu-

LEADING AGRICULTURAL PRODUCTS OF AFRICAN COUNTRIES WITH MORE THAN 15 PERCENT OF ARABLE LAND

Country	Products	Percent of Arable Land
Burundi	Coffee, cotton, tea, corn, sorghum, sweet potatoes, bananas, manioc, meat, milk, hides	44
Comoros	Vanilla, cloves, perfume essences, copra, coconuts, bananas, cassava, rice, sweet potoatoes	35
Gambia	Peanuts, millet, sorghum, rice, corn, cassava, palm kernels, cattle, sheep, goats	18
Malawi	Tobacco, sugarcane, cotton, tea, coffee, corn, potatoes, groundnuts, cassava, sorghum, pulses, cattle, goats	18
Mauritius	Sugarcane, tea, tobacco, corn, potatoes, bananas, pulses, cattle, goats, fish	49
Morocco	Barley, wheat, citrus, almonds, fish, wine, vegetables, olives, livestock	21
Nigeria	Cocoa, peanuts, palm oil, corn, rice, sorghum, millet, cassava, yams, rubber, bans, rice, kolanut, soybeans, bananas, cattle, sheep, goats, pigs	33
Rwanda	Coffee, tea, pyrethrum (insecticide made from chrysanthemums), white plantains, sweet potatoes, bananas, beans, sorghum, potatoes, livestock	35
Togo	Coffee, cocoa, cotton, yams, cassava, corn, beans, rice, millet, sorghum, meat, fish	38
Tunisia	Olives, tomatoes, dates, oranges, almonds, grain, sugar beets, grapes, poultry, beef, dairy products	19

Source: The World Factbook. Central Intelligence Agency, 2020.

mid areas of the Gulf of Guinea and Eastern Africa. In North Africa, rice is grown mainly in the delta of the Nile River in Egypt. The japonica subspecies is dominant there, and rice is grown only under irrigation.

Wheat cultivation is also irrigated; the grain is grown mostly in Egypt, Morocco, Ethiopia, South Africa, Sudan, Kenya, Tanzania, Nigeria, Zimbabwe, and Zambia. Wheat was the first cereal crop domesticated and grown by ancient Egyptians, and it remains a staple food in modern Egypt. However, in Egypt, and Africa in general, much less wheat is grown than is imported.

Vegetables, including root and tuber crops such as potatoes, sweet potatoes, and beans are present in all regions and in almost every country. Vegetables are grown on about 8 percent of the cultivated areas under water management in Africa. In Algeria, Mauritania, Kenya, Burundi, and Rwanda, they are the most widespread crops under irrigation. Arboriculture (the cultivation of fruit trees and other woody plants), which represents 5 percent of the total irrigated crops, is concentrated in the Northern region and South Africa, mostly for the production of citrus fruits.

Cotton and oilseeds are grown as commercial (cash and export) crops mostly in Egypt, Sudan, and the countries of the Southern region. Africans have grown cotton for domestic needs for thousands of years. In the nineteenth century, cotton was developed into an export crop by European colonialists who needed cotton for Europe's textile industries. Important oilseed crops include sesame, sunflower, peanuts, and cottonseed.

Like many other modern crops, corn (maize) first came to Africa in the sixteenth century with the Europeans; it became one of the key crops grown on the continent during the colonial period. In the early post-colonial period, corn production was significantly expanded: International aid agencies promoted corn because they believed it could solve the food shortage problem the same way that rice had transformed famine-stricken countries of Asia. This did not happen. In Asia, special varieties of rice were used, along with irrigation and fertilizers. In Africa, many farmers could afford neither the

high-yield hybrid corn nor fertilizers, and long periods of drought have resulted in poor harvests. Nevertheless, corn has become a staple food in those many African countries that have remained committed to growing it. In countries such as Malawi, diets based mainly on corn (and containing little animal protein) contribute to malnutrition.

Other commercial crops in Africa are sugarcane, coffee, tea, cocoa, oil and date palm, bananas, tobacco, and cut flowers. Sugarcane is grown in all countries except in the Northern region. Each of the other commercial crops is concentrated in a few countries. Tea, for instance, is grown mostly in Kenya, Malawi, Tanzania, Uganda, Rwanda, and Zimbabwe. Together, they produce more than half a million tons of tea annually, amounting to about 30 percent of world exports.

In 2019, African countries produced 38 percent of their crops (by value) from just 7 percent of their cultivated land, almost all of which is irrigated. This clearly shows a large untapped potential in irrigated agriculture. The largest irrigation projects on the African continent are discussed in the chapter *Engineering Projects*.

North Africa (Northern Region)

This region comprises Morocco, Algeria, Tunisia, Libya, and Egypt. Its agricultural and timber resources are limited by its dry climate. The products from this region are typical of the Mediterranean, steppe, and desert biomes: wheat, barley, olives, grapes, citrus fruits, some vegetables, dates, sheep, and goats.

In 2019, agriculture employed between 20 and 25 percent of the working population in Algeria and Tunisia and fewer in Libya, but about 40 percent in less-urbanized Morocco and as much as 55 percent in Egypt. From about the middle of the twentieth century, North Africa's agricultural production has failed to keep pace with its population growth and remained susceptible to large annual fluctuations. All five countries are net importers of agricultural products, although Morocco is close to self-sufficient. The countries of the Northern region have only small areas of prime arable land and no large reserves suitable for expansion. Cropland occupies about 19 percent of Tunisia, 18 percent of Morocco, 3 percent of Algeria and Egypt, and 1 percent of Libya. (Land-starved and conflict-torn Libya has been facing significant food insecurity.) Some export crops, such as wine grapes, citrus fruits, tobacco, and cotton suffer from strong international competition.

The climate in Egypt causes the cotton fibers to grow extra-long, which allow for high-quality fabrics with exceptional softness and durability. However, the production and export of Egyptian cotton have been in decline since 2004, when the Egyptian government liberalized the cotton sector and left crop rotation to the discretion of producers. As a result, overplanting has started eroding the fertility of the Nile Valley; different varieties of cotton, planted too close to one another, became badly cross-contaminated, and the quality of the cotton declined. Between 2006 and 2016, production of Egyptian cotton fell by 70 percent, and its future remains uncertain. Cotton is an extremely water-intensive crop, and the sustainability of growing it in an arid country such as Egypt has always been questionable. Raising further alarm is a United Nations prediction that within a decade, Egypt could experience water shortages owing to a growing population and climate change.

Morocco produces enough food for domestic consumption, although it must import sugar, coffee, tea, and grains. Forty percent of grains and flour consumed in Morocco are imported from the United States and France. Morocco's exports include cork from its plentiful cork oak forests, cut flowers, tomatoes, zucchini, melons, citrus fruits, and other fruits and vegetables. These high-quality agricultural products are mostly exported to Europe, as is a significant quantity of hashish, an illegal drug made from the resin of the cannabis plant. More than 70 percent of Europe's lucrative hashish market is supplied by Morocco. The country is the world's largest exporter of this drug, which is much more potent than marijuana. Cannabis cultivation areas in Morocco have been expanding since the 1990s.

The Northern region is not a major contributor to the continent's fish catch. Morocco, however, with access to both cool, plankton-rich Atlantic waters and the Mediterranean Sea, has the potential to become one of the largest fish producers in the world. Sardines, bonito, tuna, bream, hake, sea bass, anchovies, and mullet are among the principal commercial species but the country has yet to develop the modern fishing fleets and processing facilities necessary to realize this potential.

Sudano-Sahelian Region

This region comprises Mauritania, the Western Sahara territory (largely controlled by Morocco), Senegal, Gambia, Mali, Burkina Faso, Niger, Chad, Sudan, and South Sudan. This very dry part of Africa supports mostly subsistence farming and nomadic herding. For many people in this region, the primary crops are green millet, sorghum, and corn. Since the twentieth century, devastatingly long droughts have caused famine and starvation. Violent farmer-herder conflicts are widespread. "Sahel" means "shore" in Arabic, denoting this vast land as the "shore" of the Sahara. The Sahel zone has been slowly taken over by the Sahara Desert. Violent conflicts in the Sahel have been attributed to land degradation and food shortages caused by desertification and droughts exacerbated by climate change, and the disgruntlement of livestock herders over development policies that favor crop agriculture at the expense of pastureland.

Parts of Mali, Sudan, and South Sudan have the Niger and Nile rivers flowing through them. These great rivers provide plenty of water for the irrigation of fields. During the very rainy season in Mali—primarily June through September—the Niger River widens into a great, extensive floodplain—ideal conditions for growing rice. The situation in the Sudans is similar: the Blue and White Niles meet at Khartoum to form the Nile River and broad floodplains.

Mali and Burkina Faso are Africa's top producers of cotton. Cotton accounts for 70 percent of Burkina Faso's exports and 10 percent of its GDP. In Mali, cotton contributes 8 percent to the GDP. After gaining independence from France, Mali has continued to use and expand the Office du Niger—a large irrigation project established by the French on the alluvial plains of the Niger River to produce cotton for the textile industry. Facing a struggle with feeding its growing population, Mali has converted large parts of the Office du Niger from cotton to rice.

Since large-scale irrigation projects are expensive for African countries to maintain and expand on their own, the government of Mali has been seeking foreign investment and participation. Several Chinese companies participate in the cultivation of sugarcane within the Office du Niger and the processing of cane into sugar and ethanol. While agricultural development has been making progress in the southern part of Mali, agriculture in the northern part of the country has been challenged by armed conflicts and population loss.

About one-third of the area of Sudan, the largest country on the African continent, is suitable for agriculture. It is one of the top producers of sesame in the world. Sudan's other main crops include peanuts, dates, gum arabic, sorghum, cotton, and sugarcane. Part of the population in Sudan faces food insecurity, but the situation is much worse in South Sudan. Five years of brutal civil war have left more than half of the population facing severe hunger. The current situation is worsened by ongoing farmer-herder conflicts and disease outbreaks in cattle, goats, and sheep—the country's most important agricultural assets. Today, South Sudan is one of the most food-insecure countries in the world.

Mauritania and the territory of Western Sahara have some of the poorest agricultural bases in Africa. Farming in Mauritania is mostly restricted to narrow bands along rivers. Western Sahara has no permanent streams, and the economy of this inhospitable, sparsely populated land is based almost entirely on ocean fishing. In Senegal, Niger, Chad, and Gambia the majority of farmers grow crops only for household subsistence needs, practicing traditional slash-and-burn agriculture Sudano-Sahelian region relies heavily on imports of food.

Gulf of Guinea
This region comprises Guinea-Bissau, Cabo Verde, Guinea, Liberia, Sierra Leone, Côte d'Ivoire, Togo, Ghana, Benin, and Nigeria. With the exception of Nigeria, agriculture there is dominated by the culti-

vation of rice. The percentage of total arable land area ranges from 5 percent in Liberia to almost 50 percent in Togo.

In Ghana, agriculture is dominated by subsistence farming of yams, cassava, and other root crops. However, in the fast-growing industrial centers, consumers prefer rice to these traditional food crops. Ghana is increasingly producing rice, but much of the rice and other foods consumed there are still imported. Ghanaian farmers have to compete with underpriced imported foods because the national currency, the cedi, has been vastly overvalued.

In addition to growing food crops for local consumption, countries such as Ghana and Côte d'Ivoire have been producing significant amounts of high-yield export crops. Cocoa is the main agricultural export of both countries. Côte d'Ivoire, with its fertile land, favorable climate, and abundant water resources is also a leading exporter of coffee, pineapples, and bananas.

Until recently, another source of high export earnings for Côte d'Ivoire was hardwood timber. Timber extraction and the illegal clearing of forest lands by cocoa farmers have caused enormous ecological damage, reducing the country's forest cover from 30 million acres (12 million hectares) in 1960 to 6.9 million acres (2.8 million hectares) in 2017, a loss of more than 75 percent. Today, the logging boom is over, but most Ivorian forests are severely degraded or, at best, in an early stage of secondary growth (natural regeneration).

It is estimated that more than 33 percent of land in Nigeria is under cultivation, and some 70 percent of the country's labor force is employed in agriculture. Increasing rainfall from the semiarid north to the tropically forested south allows for great crop diversity. Practically all crops grown anywhere in Africa are cultivated in Nigeria. However, 80 percent of all these crops are produced by small farmers using inefficient methods of cultivation; large-scale agriculture is underdeveloped.

Among the agricultural cash crops, only cocoa has made a significant contribution to exports. Until the 1990s, Nigeria was among the world's largest exporters of cocoa beans. Its share of the world cocoa market has been substantially reduced because of aging trees, black pod disease, smuggling of beans for the black market, and labor shortage. (Cocoa is among the most labor-intensive crops, along with palm oil and cottonseed. The production of these crops was hit especially hard by the increasing migration of young Africans from rural areas seeking better economic prospects in oil-producing regions or large cities.)

Nigeria was the world's leading exporter of palm oil until it was overtaken by Malaysia in 1971. As Nigeria's production and export of palm products declined, Benin has become the continent's largest palm oil exporter.

By 2018, Benin has also eclipsed Egypt as the premier cotton producer in Africa. More than 70 percent of Benin's export revenues now come from cotton. At the same time, about one-third of Beninese remain extremely poor and rely on subsistence farming for their livelihood. In colonial Guinea, food crops were grown on plantations established by the French. After the country gained independence and these large plantations were nationalized, the government tried running them as collective farms along semi-socialist lines, but they proved unsustainable. Eventually, the land was distributed among smallholders. By the end of the twentieth century, about a half-million small farms were producing twice as much cassava, rice, sweet potatoes, yams, and corn as the state-owned plantations did in the 1970s.

Half of Liberia's landmass is forested, and the country has traditionally exported natural rubber. Since 1926, the largest contiguous rubber tree plantation in the world has been operated in Liberia by the U.S. tire company Firestone.

Togo and Guinea-Bissau are largely self-sufficient in rice and other subsistence crops. Sierra Leone imports a third of its rice, and Cabo Verde imports more than 80 percent of all its food.

Central Region
This region comprises the Central African Republic, Cameroon, the Republic of the Congo, the Democratic Republic of the Congo, Gabon, Equatorial Guinea, Burundi, Rwanda, and São Tomé and Príncipe. Cameroon has 15.3 million acres (6.2 million hectares) of arable land, but only a small part of it is under cultivation, and the country imports 80 percent of its food. At the same time, Cameroon is a leading African exporter of bananas. People living in the highlands of the Central region eat more bananas than anyone else in the world, deriving 35 percent of their daily calories from the fruit.

In the Central region, the percentage of arable land ranges from 0.4 percent for the Republic of the Congo to 43 percent for Burundi and 47 percent for Rwanda. The Democratic Republic of the Congo, Africa's second-largest country, has more available farmland than any other country on the continent and possesses huge agricultural potential. For decades, however, the country has suffered a multitude of crises, from political upheaval and armed conflicts to epidemics of Ebola and cholera. In 2020, only 10 percent of all arable land was cultivated, and about 70 percent of the population lacked access to adequate food. Cassava is harvested from most of the cultivated land, making the country the world's largest consumer of this crop. The neighboring Republic of the Congo was the second-largest. Unfortunately, compared to many other food crops, cassava is a poor dietary source of protein and other essential nutrients, and the rates of malnutrition in both countries are high, especially among children.

Rwanda's government has set a goal of reducing widespread poverty and malnutrition through agriculture. The share of the national budget allocated to agriculture increased from 3 percent in 2006 to 10 percent in 2015. Since 2007, the growth of agricultural production in Rwanda has been averaging 6 percent annually.

The country has largely focused on the production of staple food crops—plantains, cassava, beans, corn, sweet potatoes, bananas, wheat, and rice—rather than crops grown for export. This has helped in reducing dependence on costly imported food and lowering food prices at local markets. Within 10 years, poverty was reduced by 8 percent. In urban areas, more than 85 percent of the population now lives above the poverty line. By providing each of the poorest households with a dairy cow,

Rwanda's One Cow per Poor Family Program has improved the livelihoods of those most in need and alleviated the severity of child malnutrition. The country still has a long way to go in reducing food insecurity, but Rwanda's experience shows how political stability and dedicated agricultural policy can make a sizeable dent in poverty.

Most countries of the Central region, including the Republic of the Congo, the Democratic Republic of the Congo, Gabon, Cameroon, Equatorial Guinea, and São Tomé and Príncipe are rich in oil, and oil extraction has replaced agriculture as the main sector of their economies. They import much of their food with proceeds from oil exports. The agricultural labor force has diminished because of the migration of young people to oil-producing regions.

Resource-poor Burundi is heavily dependent on agriculture, but its subsistence farming cannot sustain this severely overpopulated country that has one of the highest fertility rates in the world. Land is the most valued resource in Burundi and, along with ethnic divisions, it is a major source of ongoing violent conflicts. Unable to grow their own food, large numbers of internally displaced people and Burundian refugees in neighboring countries rely on humanitarian assistance.

Eastern Region

This region comprises Eritrea, Djibouti, Ethiopia, Somalia, Kenya, Uganda, and Tanzania. Agriculture employs about 80 percent of the labor force in Uganda and Ethiopia. Small farms, approximately 2.5 million of them, dominate agriculture in both countries. More than 70 percent of Uganda's land is suitable for agriculture—a high percentage compared to the majority of African countries. Ethiopia has vast areas of fertile land, diverse climates, and generally adequate rainfall.

In four years out of five, the minimum needed rainfall may be expected in most of Uganda and half of Tanzania, but in only 15 percent of Kenya. Somalia receives almost none of the needed minimum. The three countries in the area of Lake Victoria—Kenya, Tanzania, and Uganda—share a large area of strong agriculture. This is especially true in

an arc from western Kenya through Uganda to Bukoba in Tanzania. Food crops here include bananas, sweet potatoes, taro, sesame, and yams. Robusta coffee and cotton are important cash crops. The uplands and mountains also are cultivated intensely. In Kenya, Tanzania, and Uganda, cultivation has spread upward in such highlands. The Irish potato, various species of peas and beans, and wheat and barley are grown there. The lower slopes are suited to Arabica coffee and the higher ones to tea and pyrethrum flowers (used for making a natural insecticide which is in high demand on the global market). About 90 percent of Ugandan and Kenyan tea is also exported.

The three Lake Victoria countries also share the fishery for the Nile perch in the lake, the largest inland fishery in Africa. The Nile perch, which can grow to 6 feet (1.8 meters) and 440 pounds (200 kg.) and has high fat content, was introduced to Lake Victoria in the 1950s by the British colonial authorities. The introduction of this exotic species led to the extinction of native cichlid fishes and many other species. Much of the Nile perch is still exported to Europe, making this fishery an important source of foreign currency for the three countries bordering the lake. The fishery has modern processing facilities and supports several million people. The recent decline in catch rates suggests the effects of overfishing.

Distinct agricultural zones at different elevations are common in Ethiopia. There, the native *ensete* ("false banana") is grown at medium elevations in the forest belt of the south. Its flour is a traditionally important food source. Mediterranean fruits and vines are grown at higher elevations. Highland temperatures make these areas ideal for floriculture (flower farming). Barley, corn, wheat, and the native cereal teff are grown in plowed fields across the country's different regions and ecological zones. Ethiopia's livestock herd is one of the largest in Africa. However, periodic droughts keep decimating the herd.

Ethiopia is Africa's top coffee producer, followed by Uganda, Kenya, and Tanzania. The use of coffee as a beverage began in the ninth century in the part of Ethiopia called Kafa, which gave coffee its name.

By the fifteenth century, it had traveled to North Africa and the Middle East, from which it made its way to Europe. Today, the African continent accounts for only 12 percent of the world's coffee production, but its beans are much prized by coffee connoisseurs worldwide.

Since 2008, Ethiopia has been attracting foreign investors by offering large areas of fertile land for lease. Saudi Arabia, which has very little arable land of its own, grows rice in Ethiopia and imports a significant amount of agricultural products from there. India has invested in the development of sugar production and floriculture; Israel and the European Union are engaged in horticulture and biofuels. High investment flows into the agricultural sector have had a positive impact on the Ethiopian economy. However, in drought years, one in ten Ethiopians still struggles with access to sufficient food. The government has been funding the construction of large dams in order to increase irrigation coverage and rural electrification significantly.

Tanzania has almost four million farms. In addition to coffee, traditional export crops include sesame (of which the country is a leading global producer and exporter), cotton, cashew nuts, tobacco, tea, sisal, and pyrethrum. Tanzania's climatic growing conditions are favorable for the production of a wide range of fruits, vegetables, and flowers. Oilseed crops include both industrial oilseeds (castor seeds) and edible oilseeds (sunflower, peanut, sesame, copra, cottonseed, and soybean). The many spices grown there include peppers, chilies, ginger, onion, coriander, garlic, cinnamon, and vanilla.

Over large areas of eastern Africa, rainfall is inadequate for crop cultivation. This applies to all of Somalia and 70 percent of Kenya,—areas that receive less than 20 inches (500 millimeters) of rain four years out of five. In these areas, the only feasible use of land is pastoralism. In the driest areas of Eritrea along the Red Sea coast, all of Somalia, and northeast Kenya, the principal animal is the Arabian camel. In other areas, cattle are found, along

Ploughing with cattle in southwestern Ethiopia. The nation's agricultural production is overwhelmingly of a subsistence nature. (ILRI/Stevie Mann)

with herds of sheep, goats, and a few donkeys. (The modern donkey, now found nearly worldwide, originated from the African wild ass of Ethiopia, Eritrea, and Somalia.)

Agriculture is not an important factor in the economy of Djibouti, where only 20 percent of the population is rural. Almost all of Djibouti's terrain is flat, barren desert made up of volcanic rock, and rainfall is extremely scant.

Southern Region

This region comprises Angola, Namibia, Zambia, Zimbabwe, Malawi, Mozambique, Botswana, Lesotho, Eswatini, and South Africa. Here, the arable percentage of the total land area ranges from 40 percent in Malawi to just 1 percent in Namibia. With the exception of Mozambique, where cassava predominates, corn is the single major crop in all countries of this region. In Malawi, more than 80 percent of the area harvested is used for corn. Nevertheless, some 10 percent of the country's population faces severe food insecurity year to year. Because of limited mechanization, the corn yield per hectare in Malawi is low, as it is in most other countries of the Southern region except for South Africa. By comparison, South Africa's corn yield averages 3.6 metric tons per hectare (1.5 tons per acre), and in record years, it has been more than 5 tons (more than 2 tons per acre). In Malawi, the average corn yield between 2009 and 2019 was about 1.7 tons per hectare (0.7 ton per acre).

About 13 percent of South Africa's land area can be used for crop production. High-potential arable land comprises only 22.5 percent of the total arable land, or just 3 percent of the territory of South Africa. Four million acres (1.6 million hectares) are under irrigation. The most important factor limiting agricultural production is the availability of water. Rainfall is distributed unevenly across the country, with humid, subtropical conditions occurring in the east, and dry, desert conditions in the west. Almost 50 percent of South Africa's water is used for agricultural purposes. The country also imports significant volumes of water from Lesotho through a system of large dams and tunnels. Varied climatic zones and terrains enable the production of virtu-

ally any type of crop. The largest area of farmland is planted for corn, followed by wheat and, on a lesser scale, oats, sugarcane, and sunflowers.

South Africa is self-sufficient as far as most primary foods are concerned. Millions of tons of corn are produced in South Africa commercially, both on large-scale and small farms. Much of it is used for local human consumption and processed as farm feed, but more than one million tons are exported to neighboring countries, making South Africa the continent's largest exporter of corn. In fact, corn accounts for some three-quarters of the country's total exports. South Africa is well known for the high quality of its fruits, such as apples and citrus, which it also exports. In recent decades, centuries-old South African wines have gained recognition in the world market.

Most of South Africa's land surface (almost 70 percent) is suitable for grazing, and livestock farming is by far the largest agricultural sector in the country, producing wool, mutton, and beef. South Africa's national commercial cattle herd was estimated at 14 million head in 2018.

In the twentieth century, Zimbabwe was among the world's largest tobacco producers. The country was self-sufficient in food production. Its exports of corn and wheat earned it the title "breadbasket of Africa." In the early twenty-first century, the country's longtime autocratic president Robert Mugabe initiated a violent and bloody land reform program, forcibly confiscating large commercial farms owned by white Zimbabweans of European ancestry and splitting them into smaller plots given to subsistence farmers. This led to a sharp decline in the cereal crop output and the complete collapse of Zimbabwe's main agricultural export, tobacco. Zimbabwe's famed beef and dairy industry was spared from the reform because of its strategic importance to the economy, but it also collapsed in the 2000s following repeated outbreaks of foot-and-mouth disease. In recent years, agriculture has been slowly improving. Tobacco production rose after the contract farming system was introduced; the country resumed its beef exports in 2017.

Mozambique is the poorest country in the region. For generations, many of the country's small, hand-cultivated farming households have been headed by women, while men have been seasonally migrating for work out of rural regions. As of 2018, only about 2.5 percent of the country's cultivated land was irrigated. Cassava remains the major staple crop because it is highly drought-tolerant. However, thanks to the Zambezi and Limpopo rivers, the country has the potential to irrigate 7.5 million acres (3 million hectares).

By the 2020s, several countries of the region had developed diversified economies, where the majority of the labor force was employed in non-agricultural activities such as mining, manufacturing, construction, and tourism. These include Angola, Botswana, Eswatini, Namibia, and highly industrialized South Africa. In Lesotho, Malawi, Mozambique, Zambia, and Zimbabwe, agriculture and livestock production employed from 50 to more than 70 percent of the labor force. However, only in Zimbabwe did the agricultural workforce contribute to more than 15 percent of the GDP.

Indian Ocean Islands

This region comprises Madagascar, Mauritius, Comoros, and Seychelles. Since the 1980s, the governments of these countries have sought to diversify the economies beyond their dependence on traditional crop growing. The fisheries sector, including the harvesting of tuna, fishing tourism, aquaculture, and seafood exports, has been the most rapidly growing area of the agricultural economy in the region. In 2020, Mauritius's fisheries and aquaculture sectors represented almost 18 percent of the country's export earnings. This makes up for lost revenues from the ailing vanilla and copra trade. Copra, the dried kernel of the coconut that yields coconut oil, was an important export for all Indian Ocean island countries until the late 1980s, when international markets started switching from high-fat, high-calorie coconut oil to leaner palm oil. Vanilla, of which Madagascar and Comoros are the world's largest producers, has been facing increased competition from synthetic flavorings.

Mauritius has more than 161,000 acres (65,000 hectares) of sugarcane plantations that cover some 85 percent of the cultivated land. The plantations

have consistently produced one of the highest sugarcane and sugar yields in the world. Yet sugar makes up only about 3 to 4 percent of the national GDP. Seychelles has a total land area of only 177 square miles (458.4 sq. km.), of which just 3.4 percent are cultivated. Both countries, however, have some of the highest per capita GDP in Africa, mainly due to tourism, financial services, and commercial fishing. In the fishing sector, earnings are growing annually from processing and licensing fees paid by foreign trawlers to harvest tuna in the countries' territorial waters.

The Comoros's agriculture is heavily weighted toward rice, which is the staple food of the populace. However, most of the Comoros islands have lava-encrusted soil that is unsuitable for agriculture, and much of the rice and other food products consumed by the islanders are imported. This country, too, has been developing fisheries as a main source of export earnings.

Trends

In the twenty-first century, Africa has undergone significant demographic, social, and economic

---

NATURAL HAZARDS OF DEFORESTATION AND DROUGHT

The greatest obstacles to African agriculture are deforestation and drought. Deforestation is the reduction of the amount of land covered by forests. Deforested land usually receives less rain, and the rain it does receive causes more erosion because there are fewer root systems to hold soil in place. Africa was hard hit by droughts in the last three decades of the twentieth century. Droughts in 1970, 1975, 1979, and 1984 devastated much of the Sudano-Sahelian and eastern regions. Famine and death accompanied the drought in many countries, especially Ethiopia. In the drought regions, several hundred thousand people died due to starvation. The eastern region was hit again by drought in the late 1990s. Then, between July 2011 and mid-2012, a severe drought said to be the worst in 60 years caused a severe food crisis across Somalia, Djibouti, Ethiopia and Kenya that threatened the livelihood of 9.5 million people. Droughts have prevented agricultural progress in these regions.

transformations. It has a young and fast-growing urban population and a rising middle class. These trends have been driving changes in food consumption patterns and preferences away from traditional foods such as cassava and corn and toward rice, wheat, and other foods with higher calorie and protein content. However, the rising demand for these foods still outpaces increases in their production, and large parts of the continent still face low food security. It is estimated that up to one-third of the population in sub-Saharan Africa is undernourished. Cereal consumption depends on substantial imports. While many African nations have experienced record industrial growth in recent decades, the overall agricultural production has been growing at a very slow rate: less than 1 percent per year.

The rural populations in many sub-Saharan countries have been unable to move out of poverty, principally because they have not been able to transform their basic economic activity, which is agriculture. Food yields in many countries have changed very little from fifty years ago and represent approximately one-third of the yields achieved in other parts of the world. All this highlights the need for a deep transformation of Africa's agricultural sector. However, this enormous task would require very significant financial and technical inputs as well as strong political will.

Dana P. McDermott

INDUSTRIES

According to *African Economic Outlook*, a publication of the African Development Bank, Africa's overall economic growth has sped up from 2.1 percent in 2016 to 3.4 percent in 2019, with investment accounting for more than half the continent's growth. Much of that investment was in industry and infrastructure. The Outlook highlighted, however, that Africa's growth has been less than uniform or inclusive.

In 2020, the largest economies (by GDP) in North Africa were Egypt, Algeria, and Morocco; in sub-Saharan Africa, the largest economies were Nigeria, South Africa, Ethiopia, Angola, and Kenya. As a regional manufacturing hub, South Africa

In recent years, the World Bank has consistently listed several African countries—including Ethiopia, Ghana, and Côte d'Ivoire—among the fastest-growing economies in the world. Ghana's growth is based on an oil industry that has developed since offshore oil discoveries. In Ethiopia and Côte d'Ivoire, the growth has been spurred by infrastructure projects—large dams in the former and roads, ports, and urban construction in the latter.

Strong economic growth in African countries has not translated into a significant decline in poverty levels, partly due to the inequitable distribution of wealth and rapid population growth. Even in middle-class economies, poverty is widespread, and unemployment levels are high. In some countries, industrial development is also hampered by a lack of resources and trained workers.

Oil Production and Mineral Mining

Oil is the main export of fourteen African countries and the largest export from Africa as a whole; natural gas is the second-largest. Africa's share in the global production of oil and oil products, however, is much smaller than that of the Middle East, Russia, or the United States. As of 2019, only Nigeria has made it into the world's "top ten" list of oil-exporting countries, as number nine. In Africa, Angola is the second-largest oil producer after Nigeria; it is followed by the Republic of the Congo, Equatorial Guinea, and Gabon. Leading natural gas producers are Algeria and Egypt. In countries such as Libya and Gabon, the economy has been totally dependent on the oil sector. In 2018, it represented more than 95 percent of Libya's export earnings and 60 percent of the GDP. The oil sector has accounted for 80 percent of exports of Gabon and 45 percent of its GDP. However, as the country is facing a decline in its oil reserves, the Gabonese government has decided to diversify its economy.

It is the extraction of minerals and metal ores makes Africa's mining industry the largest in the world. Gold and platinum mining is led by South Africa, which is by far the world's largest producer of platinum. (Until 2006, it was also the world's largest gold producer.) As of 2019, Botswana accounted for some 17 percent of the global production of diamonds, and diamond mining contributed one-quarter of the country's GDP. The African continent also accounts for significant shares of the world production of cobalt, manganese, phosphate rock, and uranium. The Democratic Republic of the Congo alone has been producing more than half of the global supply of cobalt.

Both in the oil industry and in the mineral mining sector, the major operations using modern technologies are largely foreign-owned. More than a dozen multinational corporations control oil exploration and production in Nigeria; several large multinationals have been responsible for the current redevelopment of Angola's mining industry. Multinationals often operate through local subsidiaries or joint ventures with local governments, but they restrict local access to their operations and modern technologies.

Two Rivers platinum mine in South Africa. South Africa accounts for 80 percent of global platinum production and a majority of the world's known platinum deposits. (Ryanj93)

Proponents of the involvement of multinational corporations in the exploitation of natural resources in Africa insist that this involvement creates workplaces, alleviates poverty, and stimulates economic growth. Opponents call this involvement a form of neocolonialism aimed at using Africa as a source of raw materials and cheap labor for foreign profit. Critics point out that investment by international corporations only enriches the African elites; these operations also cause ecological damage and have a detrimental social impact. Large-scale mining is unsustainable; in the long run, the countries will be left with depleted resources and underdeveloped enterprises in other, less lucrative sectors of their economies.

Refining Industries
African oil producers export crude oil. Africa's oil refineries are old and need significant upgrading; they satisfy only part of the local demand and, with a few exceptions, are not engaged in export. Efforts of African governments to increase domestic production of oil-based fuels and other finished petroleum products are impeded by fierce competition from well-established non-African oil refiners and fuel traders that have flooded the international market. The growing African demand for gasoline and diesel is largely met by imports. As a result, Africa's top oil producer, Nigeria, has experienced massive fuel shortages when the global prices for crude oil slumped and the country was unable to pay for imported fuels. Under current oil market conditions, international investors are reluctant to invest in building or modernizing refineries in Africa.

The refining of metal ores in Africa is much more advanced than oil refining, and the industry is developing fast. South Africa already has a well-developed refining industry. The country refines most of its gold mine production. Its Rand Refinery, founded in 1920, is the largest integrated single-site precious metals refining and smelting complex in the world. South Africa's refining industry also produces large quantities of ferrochromium, an iron-chromium alloy from which stainless steel is made. chemical industry was founded in the nineteenth century to fill the demand for explosives and chemicals needed for the mining of coal and precious metals. Today, it supplies coal-based fuels, processed petroleum oils, mineral waxes, and various chemicals produced from refined coal tar.

Processing phosphate ore into fertilizers and phosphoric acid for export is a major economic activity in Morocco. South Africa, Egypt, Libya, and Algeria account for 2 percent of the global production of steel and hot-rolled steel products. South Africa and Egypt also produce aluminum, as do Mozambique, Cameroon, Nigeria, and Ghana. Mozambique's two large aluminum smelters are owned by groups of multinational corporations, with the country's government holding a tiny share. The Democratic Republic of the Congo and Zambia lead the continent in the production of refined copper and cobalt, exporting market-grade copper

cathodes and cobalt alloys. Their mines and refining facilities are state owned.

Food Processing

In several African countries, commercial food processing has been a significant component of the economy. Both South Africa and Morocco produce sugar, flour, and various other processed foods for local consumption and export canned fruits and vegetables as well as wine. In Zimbabwe, the dairy industry processes milk from farmers and produces pasteurized milk, butter, mayonnaise, and various milk-based beverages. In Nigeria, large companies produce tens of thousands of bottles of lemonade, fruit juices, beer, wine, and spirits daily. While much

An offshore oil drilling platform off the coast of central Angola. (Paulo César Santos)

of African coffee is exported in the raw, Kenya started exporting shelf-ready coffee in 2016.

Locally dried, salted, and smoked fish is widely consumed in Africa. Lake fish-farming and processing have been growing in Malawi. The Nile perch fisheries in the three countries bordering Lake Victoria—Uganda, Kenya, and Tanzania—have modern refrigeration, processing, and packaging facilities. In Uganda, the trade in lake fish (mostly for export to Europe) contributes 2.2 percent to the GDP. However, biologists warn of overfishing and question the sustainability of the Nile perch industry. Since 2004, Ugandan fish-processing facilities have operated at less than 50 percent of capacity. Competition from cheaper farm-raised tilapia from Vietnam and other Asian countries has contributed to the stagnation of this industry.

Morocco is an exporter of canned ocean fish. Mauritius and Côte d'Ivoire have been exporting processed fish to France and Spain, albeit in small quantities. Senegal has been exporting some processed fish to Asia, principally Japan. In Senegal and several other African nations including Mauritania, Comoros, and Guinea-Bissau, fishing used to be central to the economy. But these countries lack financial resources for the development of modern commercial marine fisheries and seafood-processing capacities. In recent years, they have resorted to obtaining much-needed foreign currency through the granting of fishing licenses to European trawler fleets. Unsustainable overfishing by foreign fleets has played a significant role in large-scale unemployment in West African nations. This, in turn, caused large-scale migration of people from these countries to the oil industries of Nigeria and Ghana, and to Europe.

The food-processing sector is destined to play a huge role in improving both the employment levels and food security in Africa. The tastes of the growing urban and middle class populations have been moving toward processed foods; the demand is currently satisfied mostly by imports.

Consumer Goods Manufacturing

Large-scale manufacturing is led by South Africa and Nigeria in sub-Saharan Africa, and by Egypt and Morocco in North Africa. Egypt is a leading manufacturer of automobiles, chemicals, consumer electronics, home appliances, textiles, and clothing. In Morocco, both the new telecommunications sector and the traditional leather goods and textiles sector have recorded significant economic growth in recent years; automobile and tractor assembly, metal products, and the making of cement and asphalt are also important. Among all twenty-two Arab countries of the Middle East and North Africa, Egypt and Morocco have the largest percentage of GDP that is not derived from oil production.

Algeria manufactures smartphones, tablets, television sets, and air conditioners, mostly for the local market. Algeria's pharmaceutical industry covers about 40 percent of local needs. The country also produces buses, industrial machinery, and military vehicles that are exported to other African countries and the Middle East. Tunisia manufactures and exports car parts.

In sub-Saharan Africa, Nigeria is the largest producer of electronics, pharmaceutical and cosmetic products, textiles, plastics, and building materials such as cement. Nigeria also manufactures cars and buses, some of which are also exported to neighboring countries. In Uganda, diesel trucks are manufactured. Ethiopian garment factories make clothes for some of the leading global clothing brands, but workers in this industry receive some of the lowest wages in the world.

South Africa produces and exports vehicles, machinery (including computers), electrical equipment, plastics, and other manufactured articles. Manufacturing accounts for a significant share of the country's overall economic activity. In recent decades, however, the global competitiveness of South African exports has been in decline, largely due to pressures from lower-cost producers such as China and India.

Tourism and Hospitality

The continent of Africa, with its diverse and pristine landscapes, remarkable wildlife, and numerous cultural and historic sites, offers vast opportunities for tourism. The countries that already have well-developed tourism and hospitality industries include

Egypt, South Africa, Uganda, Morocco, Tunisia, and Algeria. Kenya, Tanzania, Zimbabwe, and Namibia are some of the world's most popular destinations for wildlife tourism. Adventure tourism (such as mountain climbing, river rafting, forest exploration, and camel tourism in the Sahara) has been growing in Tanzania, Côte d'Ivoire, Lesotho, and Chad.

All of the African island nations—Cabo Verde, Madagascar, Comoros, São Tomé and Príncipe, Seychelles, and Mauritius—have either poor or undeveloped natural resources, but they possess pristine beaches and unique flora and fauna. Their economies rely heavily on the tourism and services sector, as do the economies of the African island dependencies of European countries, such as Mayotte, Réunion, the Canary Islands, and the Madeira Islands. In Seychelles, the tourism and services sector has been gradually moving the country toward an upper-middle-income economy.

In recent years, Botswana, Gambia, and Rwanda have been directing significant efforts into advancing tourism, while the traditionally successful tourism industries of Egypt and Tunisia have declined substantially due to outbreaks of terrorism and social turbulence. Instability and insecurity have been undermining the development of the tourism industry in many other countries with a significant potential for tourism, from Libya and Sudan to Mali and Nigeria. Underdeveloped infrastructure—unsafe roads, inadequate water supply and sanitation, poor access to hospitals, and inconsistent electricity—has been the primary challenge to the hospitality sector throughout Africa.

Compared to other world regions, Africa's tourism industry has remained relatively underdeveloped. Of the 1.2 billion people traveling internationally in 2016, only 58 million, or about 5 percent, arrived in Africa. This shows that there is considerable room for growth in Africa's travel and tourism market. Given that the economic and political environment in Africa continues to improve, consumer spending on tourism, hospitality, and recreation in Africa is projected to reach more than US$260 billion by 2030, which would double what it was in 2015.

Energy Sector and Infrastructure

Underdeveloped infrastructure is holding back Africa's economic growth. High transport and energy costs drive up the prices of goods and services, and impede the development of industries. In sub-Saharan Africa, the electrification rate in 2018 was only 45 percent. With its large and powerful rivers, Africa has the highest percentage of untapped technical hydropower potential in the world. As of 2018, only 11 percent of it was utilized. Numerous hydropower dams are either planned or under construction all across the continent. They will significantly increase electricity availability for urban centers and large industries connected to the power grid.

Africa's off-grid DRE (distributed renewable energy) industry is just getting started. This industry provides power for lighting, cooking, heating, and cooling generated by portable wind turbines, solar panels, and biofuels, andis distributed through generators, batteries, or bio-gas cook stoves, that is, independently from the main power transmission grids linked to power plants. In many parts of Africa, the DRE power supply is the most viable solution for electricity shortage and blackouts caused by infrastructure failures. As of 2015, DRE mini-grids were in operation in Mali, Morocco, and Uganda. In the 2020s, the number of jobs in this industry is expected to double in Kenya and increase more than tenfold in Nigeria.

In transportation infrastructure, Africa is far behind not only industrialized Europe and North America but also developing countries such as China and Brazil. For instance, in 2015, China had 0.6 of a mile (1 km.) of railroad for every 15,700 people, and Brazil, for every 7,100 people. Ethiopia had one kilometer of railroad for every 140,000 people.

Infrastructure is the main catalyst for economic growth, but large infrastructure projects are expensive; African nations simply lack the needed financial resources and therefore must borrow funds to pay for them. Since 2005, the government of Angola has used billions of dollars in credit from Portugal, Germany, Spain, Brazil, and China to help rebuild its public infrastructure damaged or undeveloped as a result of the country's twenty-seven-year-long civil war. Overall, Africa's single largest foreign investor remains France, followed by the United Kingdom. (The United States provides funding to African countries predominantly in the form of economic aid rather than in the form of direct investment.)

In recent decades, China has been making large, targeted investments in Africa's key infrastructure areas such as hydropower projects, roads, railways, ports, utilities, and telecommunications. As a result, China's political and diplomatic influence over Africa's fifty-four nations—the largest regional group in the United Nations—has been growing. In many African countries, dams and ports are not only financed with Chinese loans but also built with Chinese contractors and labor. China also secures a significant share of Africa's oil and minerals for Chinese industries. More than 60 percent of Chinese investments have been concentrated in Nigeria and Angola, and are closely linked to their respective oil industries. This is somewhat reminiscent of the strategy of European colonial powers in Africa in the nineteenth century and the first half of the twentieth century; those countries built transportation and irrigation projects in Africa to support cotton production for European textile industries.

The African Development Bank provides concessional loans and grants for infrastructure projects serving the broader public interest, such as sanitation and water supply. The World Bank has approved a regional off-grid DRE electrification project for West Africa and the Sahel. The United Nations Economic Commission for Africa coordinates donor investment in the trans-African highway network. But to sustain Africa's economic development and population growth, its infrastructure requires much more spending. Overall infrastructure funding needs in Africa through 2025 are estimated to reach or exceed US$1.2 trillion. To attract more international investment, most African countries need to lower their cost of doing business and improve the business climate by reducing endemic corruption and strengthening internal security and political stability.

Dana P. McDermott

206

ENGINEERING PROJECTS

When Africa's young, independent nations were embarking on large-scale infrastructure, they were not starting from scratch. Some of these early projects had been initiated or planned in the colonial period. Moreover, the African continent bears a remarkable heritage—a wealth of ancient and medieval infrastructural and town-planning sites and monuments, many of which have been designated by UNESCO as World Heritage Sites. These include the monumental pyramids and temples of ancient Egypt, the ksars (fortified towns) of North Africa, stone buildings of Great Zimbabwe, ancient iron production sites in Burkina Faso, the engineering marvels of medieval Islamic architecture, and coastal ports and trading centers founded long before European colonists set foot in Africa.

Most major engineering achievements in today's Africa relate to water management, for irrigation, consumption, and hydroelectric power; urban planning; mineral and other extractive industries; and the means of moving goods and people. The general situation reflects basic geographical, economic, and historical facts about Africa: fairly low population densities in many areas (although they are increasing rapidly), extreme variability of rainfall in different seasons, and a growing consumption base.

Africa's rivers have the potential to generate more hydroelectric power than all the rest of the world's rivers combined. Indeed, the rivers of the Congo Basin alone could produce one-sixth of the world total, according to some estimates. However, African rivers can also be extremely destructive. With the exception of the Congo system (which is fed by rainy seasons in both hemispheres at different times of the year and thus has a fairly stable rate of flow), major river systems can carry devastating annual floods. Therefore, enormous effort has been applied to controlling these river systems and realizing their power potential.

Dams: The Nile Watershed

The Nile has shaped Egyptian history since ancient times, and even back then, Egyptians had attempted to tame the river by building flood-control dams. The earliest massive dam was constructed in the third millennium BCE at Sadd el-Kafara. It is considered the oldest large dam in the world. (Archaeologists believe that the dam was destroyed by a flood within ten years of construction, before it was finished, because it lacked a spillway.) Another unsuccessful attempt was made in the eleventh century CE. In the late 1890s, shortly after Britain had occupied Egypt and the entire Nile watershed except for Ethiopia, British engineers began to try to regulate the Nile system. By 1900 there were flood-control barrages in the Nile Delta, and in 1902 the first Aswan Dam (known as the Aswan Low Dam) was completed to regulate the rate of annual flooding downstream, and support irrigation and household water supply during the dry season. In later years, that dam had to be raised twice because its initial water storage capacity repeatedly proved insufficient.

When the British occupation began, Egypt was in the throes of rapid population growth and economic expansion. By early in the twentieth century, it was possible to anticipate that demand for water in the Nile watershed could outstrip supply in the short run. British engineers developed plans to turn the entire watershed into a storage and management area. By 1950 this project—actually a group of projects—came to be known as Century Storage. The idea was to store and manage water in such a way that, by the end of the twentieth century, sufficient water would be available in Nile reservoirs to overcome a five-year drought. Dams were completed in Anglo-Egyptian Sudan on both branches of the Nile before World War II, as centerpieces of the Gezira Scheme, one of the largest irrigation projects in the world.

The first large postwar dam project in Africa was the Owen Falls Dam in Uganda, completed in 1954. It was later renamed Nalubaale Hydroelectric Power Station, "Nalubaale" being the Luganda name for Lake Victoria. Because Lake Victoria, Africa's largest lake, functions as a great natural reservoir for the dam, the Nalubaale power station has a huge water storage capacity. The objective was to supply electricity to Uganda and parts of neighboring Kenya and Tanzania. Turbines at the dam had to be upgraded many times to meet demand. A second power station was built nearby in the 1990s farther downstream; the larger Bujagali Power Station was completed between 2007 and 2012. The government of Uganda continues to plan for more dams on the Victoria Nile system.

In the early 1950s, the new republican government in Egypt revived an old British plan for a huge storage dam just upstream from the Aswan Low Dam; it came to be known as the Aswan High Dam project. The undertaking became a notable part of the Cold War economic and ideological contest when Western financiers backed away from the project in protest over Egypt's arms deal with the Soviet Union and were replaced by funding and expertise from the Soviet Union. (Egypt's president Gamal Abdel Nasser also nationalized the Suez Canal, then owned by Britain and France, to use revenues from the canal tolls for additional funding of the dam construction. This led to an armed conflict with Britain, France, and Israel that became known as the Suez Crisis.) Dedicated in 1971, the High Dam at the time was the world's largest rock-fill dam. Standing 365 feet (111 meters) high with a crest length of 12,566 feet (3,830 meters), the dam can impound 4.6 trillion cubic feet (132 billion cubic meters) of water. Its reservoir, named Lake Nasser, is 340 miles (550 km.) in length.

Although the construction of the High Dam, like that of all large dams, had some negative environmental effects, its huge positive economic impact on Egypt is undisputable. Hydroelectricity generated by the dam revolutionized the energy situation in Egypt, and the reservoir allowed hundreds of thousands of acres of land to be opened to cultivation for the first time. While tens of thousands of people had to be moved from the areas that were to be flooded by Lake Nasser, about a half-million families were settled on the newly available arable land around the dam. Egypt's agricultural production increased and diversified; protection from floods and droughts improved, as did the Nile navigation; and a new fishing industry emerged along Lake Nasser. More than twenty ancient monuments and architectural complexes were preserved by relocating them from the reservoir area. Some of them were later granted to countries that helped with the preservation works, including the Temple of Dendur, which is now a major attraction in New York City's Metropolitan Museum of Art.

While still under construction, the Grand Ethiopian Renaissance Dam on the Blue Nile, near Ethiopia's border with Sudan, is already the biggest dam in Africa. Once completed, its reservoir will take between five and fifteen years to fill with water, depending on hydrological conditions and Ethiopia's agreement with downstream Sudan and Egypt over Nile water rights, which is yet to be reached despite a years-long dispute. Ethiopia has already completed another large and controversial dam, Gilgel Gibe III, on the Omo River outside the Nile Basin. Inaugurated in 2016, that dam became one of the tallest in the world; its height is 797 feet (243 meters). It boosted the country's energy output by 85 percent, but it has threatened the already precarious situation of indigenous peoples of the Omo valley, who rely on subsistence agriculture.

Unlike the Aswan Low Dam, which is a buttress (or hollow) dam, and the Aswan High Dam, which is an embankment dam (it is constructed of excavated rock and clay), Ethiopia's dams belong to the RCC (roller-compacted concrete) type, which allows much faster and less expensive construction than traditional methods. The dam-building is financed largely by the government of Ethiopia; the country has set a goal to become a world leader in renewable energy supply.

The Zambezi Watershed

In Southern Africa, British engineers completed the impressive Kariba Dam on the Zambezi River in 1959. At 6.5 trillion cubic feet (185 billion cubic meters), the water storage capacity of its reservoir is the largest in the world. Kariba is a concrete arch structure approximately 425 feet (130 meters) high and carries a two-lane highway linking Zambia and Zimbabwe. It supplies more than half of all the electricity requirements of both countries.

Shortly before Mozambique achieved independence in 1975, its then colonial ruler, Portugal, began the gigantic Cahora Bassa project farther downstream on the Zambezi, a 560-foot (170-meter)-high structure that impounded a reservoir nearly the size of the U.S. state of Rhode Island. Completed in the late 1970s, this dam provides irrigation and flood control for much of central Mozambique. Mozambique itself has been using only about 25 percent of the dam's hydroelectric output; the rest has been supplied to neighboring South Africa per a pre-independence agreement between Portugal and that country.

Civil war broke out in Mozambique in 1979 and lasted for nearly two decades, destroying 80 percent of the power distribution grid. Reconstruction was completed by 1998. The Cahora Bassa project, now co-owned by Mozambique and Portugal, is the world's largest hydroelectric plant, producing energy mainly for export. It uses a high-voltage direct current (HVDC) power transmission system, the first such scheme in Africa, which offers the most efficient means of transmitting large amounts of electricity over long distances. HVDC equipment for Cahora Bassa was built by German and Swiss companies.

The Congo Watershed

The largest hydroelectric power potential in Africa, and perhaps in the world, lies at Inga Falls on the Congo River,, about 25 miles (40 km.) upstream from the Atlantic river port of Matadi, in the Democratic Republic of the Congo. Here the river falls nearly 330 feet (100 meters) with more than 8.7 miles (14 km.) of rapids, at a flow rate of about 1.48 million cubic feet (42,000 cubic meters) per second.

Construction at Inga has been in phases: Inga I was completed in 1972 and provided electricity for a uranium enrichment plant nearby. Inga II, completed in 1982, provided much-needed power for the capital of Kinshasa and the mining industries in

the Katanga region. These two projects brought the total generating capacity at Inga to 1,700 megawatts, an impressive figure but only a tiny fraction of the staggering estimated potential of 43,000 megawatts.

Despite the fact that neither dam has been adequately maintained, and consequently has operated below capacity, plans were developed for building an additional, much larger power station nearby. Called Grand Inga, that project would be the largest hydroelectric power facility in the world, with the potential to produce enough electricity to solve the energy shortage problem for the whole of Central Africa. For a long time, chronic political and military unrest stood in the way, but in recent years bidding for the project has resumed; it involves international consortia with Chinese, German, and South African partners.

West African Dams

The largest storage dam and hydroelectric project in West Africa is at Akosombo in Ghana, on the Volta River. Begun in 1961 and completed in 1965, Akosombo is a rock-fill embankment structure just more than 425 feet (130 meters) high, with a crest length of 2,200 feet (671 meters). The enormous storage reservoir inundated almost 4 percent of the total territory of Ghana. The government of Ghana built Akosombo in partnership with the United Kingdom, the United States, and the World Bank. Additional generating capacity, installed later, enables the dam to supply most of the country's electricity needs and even export some to neighbors. A new fishing industry developed on the sprawling reservoir Lake Volta, the largest artificial reservoir

---

THE EARLIEST ENGINEERS

The most impressive engineering feat in Africa remains the Great Pyramid, built around 2500 BCE. That human-made mountain is constructed of more than 3.2 million blocks laid out with a precision that continues to mystify archaeologists. The scale of the Great Pyramid foreshadows the gargantuan dimensions of more recent undertakings and those likely to be attempted in Africa's future.

---

in the world. However, the land around it is significantly less fertile than the formerly cultivated land that is now under the lake.

On the west coast of Southern Africa, the Caculo Cabaça hydropower plant on Angola's Kwanza River, financed largely by China, was inaugurated in 2017 as part of Angola's ambitious project to double its production of energy, which the country badly needs to support its rapid economic growth.

Irrigation

Seasonal variability of water supply in much of Africa, as well as desert environments in the north and extreme south, have made irrigation and water pipelines as much of a concern on a continental scale as dams and power supply. Many of these schemes have been conceived on a vast scale, and the origins of some hark to colonial times. The French built an irrigation barrage on the Niger River at Sansanding in the 1930s in hopes of revolutionizing irrigated agriculture in what is now Mali. It was part of a monumental public enterprise, called the Office du Niger, created by the French in the colony of West Sudan to produce cotton for the textile industry of France. After Mali gained independence, it continued and expanded the Office du Niger irrigation scheme, which became a Malian government agency. Rice has replaced cotton as the main agricultural crop grown on more than 250,000 acres (100,000 hectares) of lands opened through irrigation.

In the late 1940's1940s, the British tried to turn much of what is now central Tanzania into a scientifically operated peanut complex through irrigation. These and other projects failed miserably, in part because of infertile soils, and partly because the engineers who planned the systems did not understand that successful irrigation schemes are dependent more on micromanagement by thousands of users than they are on simply constructing a dam or main feeder channels. Later projects, which relied more on local planning and were more modest in the anticipated outcome, have done better, but even these are often fairly ambitious engineering projects. For example, the Niono Project in central Mali, administered by the Office du Niger, diverts a large portion of the flow of the Niger through irrigation canals 40 miles (65 km.) long.

Many irrigation projects in Africa have been modeled on the huge Gezira Scheme in Sudan. Conceived by the British in the colonial period, the scheme placed several million hectares into cottoncotton production in the region south of the confluence of the White Nile andthe Blue Nile by damming both rivers close by. As was often the case with major development in colonial times, the Gezira Scheme was designed to develop cash-crop agriculture and compete with cotton production in Egypt. Consequently, the scheme greatly accelerated water consumption on the upper reaches of the Nile watershed, to the dismay of the Egyptians. Overtaken by the government of independent Sudan, the project has been dogged by conflicts over lack of state funds and farming land rights. Still, it remains one of the biggest irrigation projects in the world, covering about half of all irrigated land in Sudan and involving more than 100,000 farmers growing cotton, sorghum, wheat, and vegetables.

Water Supply

As African populations and living standards grow, demand for fresh water is increasing enormously in arid regions. In the Nile watershed, for example, there is demand for every drop available. The Jonglei scheme, begun in the 1980s in deep southern Sudan, was designed to canalize the flow of the Nile through this vast, swampy region, where the Nile loses an estimated 2.4 cubic miles (10 cubic kilometers) of water annually to evaporation. Unrest interrupted the project in the late 1980s as southern Sudan slipped into a secessionist war. Although two-thirds of the Jonglei canal had been excavated, the canal has never been completed, and peoples of newly independent South Sudan have continued to suffer from severe food insecurity.

In South Africa, the Orange-Fish Tunnel, 51 miles (82.5 km.) in length and nearly 18 feet (5.5 meters) in diameter, diverts water from the Hendrik Verwoerd Dam on the Orange River to irrigation projects and cities on the South African coast. A much larger project to draw additional water from the uplands, the Lesotho Highlands Water Project had been largely completed by 2020. Designed to move up to 2.82 million cubic feet (80,000 cubic meters) of water daily into South Africa's leading industrial province, Gauteng, the project has provided the country of Lesotho with a much-needed source of income. It has also significantly contributed to the development of Lesotho's infrastructure, as it involved the building of hundreds of miles of roads. It is one of Africa's largest infrastructure projects, involving five dams and about 124 miles (200 km.) of tunnels.

The largest water transmission project in Africa is the Great Man-Made River in Libya. In the process of intensive exploration in Libya for oil and gas, geologists discovered enormous quantities of water locked underground in Libya's part of the Nubian Sandstone Aquifer System. Originally, the government hoped to develop agriculture in the middle of the Sahara by utilizing this water, as in the case of the Kufra agricultural project in the late 1970s, but the strategy changed to bringing these enormous supplies by pipelines to the populous Mediterranean coast, including the capital of Libya, Tripoli, and other large cities.

A 1,750-mile (2,820-km.)-long system of gravity pipelines and aqueducts includes more than 1,300 deep wells and supplies some 70 percent of Libya's freshwater needs.

Transport

Africa traditionally has had enormous transport problems. River transport is hampered by variable flows, only partly addressed so far by dams. The huge continental coastline has few natural indentations; consequently, harbors must be artificial. In general, these are of moderate size and in places choked with traffic. Many railroads date to colonial times, with a bewildering variety of gauges; modern railroads usually directly serve mine-to-port traffic. Airports are relatively numerous in Africa, but many runways are below international standards. Except in South Africa, the road system, even between major cities, until recently has been only marginal.

To remedy this situation, African nations, with assistance from the African Development Bank, the African Union, and the United Nations Economic Commission for Africa, have started investing in a continental road network consisting of nine trans-African highway corridors. It involves the rehabilitation of existing roadways and the construction of new ones connecting them. The longest of these corridors, the Cairo-Cape Town highway, has been fully paved. The construction of the Yaoundé-Brazzaville highway between the capitals of Cameroon and the Republic of the Congo started in 2016, and in 2020 the first phase of the road became operational. The Trans-Sahelian highway, a 2,800-mile (4,500-km.)-long road that links Dakar in Senegal to Ndjamena in Chad, passes through seven countries, and 80 percent of it was paved by 2020. The Ndjamena-Djibouti highway is also under development. The total length of the nine corridors in the trans-African network is approximately 35,200 miles (56,670 km.).

Urban transportation in the large cities of North Africa has been better developed than in those of sub-Saharan Africa. The first subway in Africa opened in Cairo in 1983, and by 2020, the Cairo Metro had three lines with 65 stations. The capital of Algeria also has a rapid transit system, the Algiers Metro, launched in the 1980s. Tunis, the capital of Tunisia, has a modern light rail system. Morocco opened, in 2011 and 2012, two modern tramway systems in its largest metropolises, Rabat-Salé and Casablanca; both of these systems continue to expand. In West Africa, Côte d'Ivoire in 2017 has embarked on a massive project to improve road infrastructure in its largest urban center, Abidjan.

Among the major transport engineering feats in African history is the Suez Canal, designed by French engineers and excavated in part by Egyptian laborers, before steam- and coal-powered machines were brought in. Archaeologists have shown that several Mediterranean-Red Sea canals, usually running through the Nile Delta, existed in much earlier periods, but the current canal cuts directly through the Isthmus of Suez. It opened in 1869. Approximately 120 miles (193 km.) in length, the canal is a sea-level project that did not require com-

LARGEST STORAGE DAMS IN THE WORLD

Storage Dam	Country	Volume, in Millions of Cubic Meters
Kariba	Zambia/Zimbabwe	185,600
Bratsk	Russia	169,270
Akosombo	Ghana	144,000
Daniel Johnson Dam	Canada	139,800
Guri Dam	Venezuela	135,000
Aswan High Dam	Egypt	132,900
W.A.C. Bennett Dam	Canada	74,000
Krasnoyarsk Dam	Russia	73,000
Zeya Dam	Russia	68,420
Robert-Bourassa Dam	Canada	61,700

Note: Three of the ten largest storage dams in the world are located on the African continent.

plex locks. Constant dredging is required to keep the canal open, and in 2014-2015 the waterway was deepened and widened to allow passage by larger vessels; at the same time, a parallel New Suez Canal was excavated to accommodate two-way traffic.

A highway tunnel has been drilled underneath the canal to connect the Sinai Peninsula with the rest of the country. The world's longest movable steel bridge, 1,100 feet (340 meters) in length, was built for railway traffic over the canal at Al-Fardan in 2001, but it became inoperable after the 2014-2015 canal expansion. In order to resume rail transport and expand automobile traffic, six new railway and road tunnels were commissioned. Several of which were completed in 2020.

The Maputo-Katembe Bridge across an inlet of the Indian Ocean in southern Mozambique is the longest suspension bridge on the African continent. Financed and carried out by the Chinese, the road bridge opened in 2018 and is almost 10,000 feet (3,050 meters) long.

Also built by the Chinese, the Port of Walvis Bay Container Terminal in Namibia was opened in 2019 as part of Namibia's plan to become the leading ocean cargo and passenger gateway in Southern Africa.

Mineral Extraction

Oil-producing land areas of Africa, such as Libya and southern Nigeria, are webbed with the familiar drilling, storage, pipeline, and refinery structures characteristic of the petroleum industry worldwide. Offshore drilling has been rapidly expanding since the 1990s. Some of the largest oil discoveries of the last decades have been made off the west coast of Africa. Among the most recent ones is Senegal's Sangomar deepwater field containing both oil and natural gas. Discovered in 2014, it is expected to start production in 2023. Drilling there had to be done to a total depth of more than 9,000 feet (2,800 meters).

Offshore oilfields were previously discovered in the territorial waters of Nigeria, Angola, Ghana, and Equatorial Guinea. The deepest and most remote of these fields have been exploited using floating production storage and offloading (FPSO) vessels. One of the world's largest FPSOs, Kizomba A, has been operating off the shore of Angola. Built in South Korea in 2004, it is capable of producing and refining 250,000 barrels of crude oil per day and storing a total of 2.2 million barrels. FPSOs are preferred in frontier offshore regions because they are easy to install and do not need a costly pipeline infrastructure to be built between the oilfield and the mainland because the oil is directly offloaded onto tankers.

The mining industries of Southern Africa exploit some of the richest and most strategically important mineral deposits in the world. Mining of gold and diamonds, in particular, presents immense engineering challenges because of the great depth at which they often occur and also, in the case of gold, because of volatile prices that demand ever more production efficiency. At South Africa's Mponeng gold mine, some shafts are more than 6.5 miles (10.5 km.) deep. At such depths, atmospheric pressure is nearly double that on the surface, and the trip to the bottom takes more than an hour. Mponeng holds the record as the world's deepest mine.

Radioactivity and heat from Earth's core raise the temperature to as much as 122°F (50°C). Free oxygen rarely exists so far underground, and must be pumped and pressurized into the shafts, along with slurry ice to keep the temperatures down. Not surprisingly, gold mining in South Africa ranks among the world's most dangerous professions.

Open-pit uranium mining operations at Arlit in Niger and Rössing and Langer Heinrich in Namibia are among the world's largest. Operating since 1976, the Rössing uranium mine is one of the oldest in the world, while Langer Heinrich only started production in 2006. Namibia's significant uranium-producing potential has been constricted by the fluctuating global demand for uranium as well as by the scarcity of water, which is needed in large quantities in the process of uranium mining.

Urban Planning and Design

Modern African architecture has been influenced both by Western styles and local traditions. The Eastgate Centre in Zimbabwe's capital of Harare, designed by the local architect Mick Pearce and opened in 1996, was probably the first large building in the world to use advanced natural ventilation and air conditioning. It consumes just a fraction of energy used for these purposes by similar buildings of its size. (Pierce was inspired by indigenous Zimbabwean masonry and the construction of African termite mounds.)

Konza Technopolis, a flagship project of Kenya's national development plan, is envisioned as a sus-

tainable "smart city" and a world-class technology and education hub, with affordable housing for people living, working, and studying in Konza. A U.S. real estate and economic advisory firm has been leading a team that includes Kenyan, Middle Eastern, American, and other international partners. Dubbed the "Silicon Savannah," the city is to be built by 2030.

The capital of Nigeria, Abuja, is a planned city. It was built mainly in the 1980s as the result of an idea by leaders of independent Nigeria to replace the erstwhile capital Lagos, with its colonial identity, by a brand-new city symbolizing the new nation. In the twenty-first century, Abuja has been the fastest-growing city on the African continent. Egypt, too, has been building a new capital city to replace the overcrowded and traffic-clogged Cairo. The construction started in 2015. The development will be the largest planned city in history, covering up to 270 square miles (700 sq. km.).

The Future

Africa seems to encourage huge projects, some of which have not materialized but may do so in the future. For example, for more than eighty years there has been talk of a railroad across the Sahara, to link West African economies with the north. In 1998 Morocco and Spain signed an agreement to explore a route for a tunnel beneath the Strait of Gibraltar, which might connect with both a trans-Saharan and a trans-North African railroad system. Where the Strait of Gibraltar is narrowest, the water is far too deep to permit a tunnel, and it would have to follow a route to the west some 28 miles (45 km.) in length. Should this project be undertaken, it would be considerably more expensive and ambitious than the tunnel under the English Channel.

The construction of the Bridge of the Horns between the coasts of Djibouti and Yemen across the Bab-el-Mandeb Strait was close to being started when civil unrest in Yemen delayed the development indefinitely. The planned bridge would have been the longest suspension span in the world, and twin eco-cities would be built on either side, both of them running on renewable energy.

Some proposals for future irrigation and power schemes are comparably ambitious. Egypt has long dreamed of constructing a water tunnel or canal from the Mediterranean to the Qattara Depression, a vast, subsea-level region in the northwestern part of the country. The area is extremely hot and dry, and the idea behind the project is that the inflowing seawater from the Mediterranean would be rapidly evaporating and the depression would be constantly filling itself. This indefinitely sustainable inflow would generate hydroelectric power in the process. Still another proposal would turn great areas of the Sahara into a global heat sink for climatic control.

Even if these future dreams do not all come to pass, the progress of civil engineering and infrastructure building on the continent of Africa has markedly accelerated in the twenty-first century.

William T. Walker

TRANSPORTATION

Unlike transportation in the United States, with strategically located connections of interstate road and rail transportation, numerous airports, and long-established urban transit systems, transportation in Africa is problematic. Intercountry networks are poorly developed, connections within countries are similarly limited, and the quality of the system—state of repair, availability, efficiency—varies significantly among countries. Even when reliable transportation infrastructure exists, violent conflicts sometimes make it unusable.

Historical Background

Transportation of goods and people has never been easy in Africa. Harsh weather, difficult terrain, aggressive wildlife, and disease frequently presented major obstacles to transport. For millennia, Africans relied on camels to travel through sandy deserts and on donkeys in other arid and semi-arid climates. Horses had been domesticated and ridden in northern and western parts of Africa long before they became a means of transportation in Europe. In tropical areas, transportation was primarily on foot, and most goods were carried on one's head or back because the tsetse fly prevented the use of animals for hauling. With a markedly seasonal flow, shifting channels, and many cataracts and waterfalls, river transport was a major challenge for dugout canoes and rafts.

Coasts devoid of natural shelter and exposed to heavy surf required expensive breakwaters and artificial harbors. In the rainforest, fallen trees, blocked trails, and washed-out roads and bridges impeded travel. Nevertheless, in ancient times, significant transport networks emerged in many parts of Africa based primarily on paths beaten by human feet. The major modern thoroughfares are based on these paths. In the Nile River valley and the Mediterranean region of North Africa, these paths were upgraded into paved roads by the classical Egyptian and Roman civilizations.

During the colonial era, these old networks were restructured to penetrate the interior from the seaports. Railroads were also established, primarily to serve the commercial and administrative needs of the colonial powers. In the early colonial period, roads were built cheaply and of simple design, reflecting the low traffic densities of the time. Consequently, traffic ran slowly, vehicle life was relatively short, and large sections of the network were closed for varying periods during the rainy season. The seasonal nature of the road network remains an important feature of the system today—the dry season network is several times more extensive than the all-weather network.

Within the first decades after independence, the most intensively used links in the system were upgraded. Gravel or laterite surfaces were replaced, sometimes by two-lane bitumen surfaces. Newly independent countries embarked on large road-building projects. For instance, Niger constructed a highway between the capital Niamey and Lake Chad and named it the Route de l'Unité (the Unity Road); Mauritania built its Route de l'Espoir (Road of Hope) stretching from the capital Nouakchott to the border with Mali. Still, the spatial density of the road network of Africa as a whole has remained lower than that of other world regions, mainly due to the sparse population distribution on the continent. By 2020, only a few motorways (that is, highways designed for high-speed vehicular traffic) existed in sub-Saharan Africa; the Lagos-Ibadan Expressway in Nigeria and the Tema Motorway in Ghana are examples of those. Bridges have replaced ferries, and roads have been realigned to take out sharp bends and steep gradients. Nevertheless, road densities vary significantly among locations, depending on population density, topography, proximity to urban centers, economic activity, and competing forms of transport.

Road Transportation

Roads dominate the transport sector of most African nations, and cover 80 to 90 percent of passenger and freight traffic. Roads in African countries are generally of three classes: primary, secondary, and tertiary or rural roads. Primary and secondary roads are main arterial highways of relatively high standards, connecting major population centers and provincial capitals. Rural roads, which are usually in a poor state of repair, include penetration roads that provide access to potential development areas, provincial roads connecting small districts or communities, and feeder roads that link agricultural areas to market centers directly or by main arterial roads.

Rural travel is filled with difficulties. Roads and seasonal tracks are rarely maintained, and people walk along treacherous paths and footbridges to obtain water and firewood, and to reach markets, schools, and clinics. These same tracks, paths, and footbridges are used to transport export crops and food destined for urban populations. In rural Africa most people walk while carrying their loads, and women frequently carry the bulk of the burden. Other than walking, nonmotorized- vehicles—bicycles, wheelbarrows, and carts, both hand-pulled and animal-drawn—are the primary means of transport. In the 2010s, only a third of rural Africans lived near a drivable road.

Even during the dry season, when roads are generally drivable with little difficulty, motorized transportation is rarely available. Long-distance buses operate only between major cities. In some villages, no car passes for days, while in others, a bush taxi (usually a station wagon or a minivan) arrives once a day. The rainy season compounds the unavailability problem. Road conditions deteriorate so much that in some places, one can get from place to place only by walking along muddy paths and sometimes wading through water. Other problems include broken bridges, flooded roads, washed-out paths, and overgrown trails that provide good hiding places for poisonous snakes.

Urban Transport

Urban transport in much of Africa consists of three main types of transit: government-operated bus systems and different types of taxis—regular taxicabs and ride-hailing services (which are very expensive and used mostly by tourists); privately owned and operated minibuses known by many different local names such as *gbakas* (Côte d'Ivoire), *tro tros* (Ghana), or *matatus* (Kenya); and ubiquitous motorcycle taxis known as *boda bodas* (East Africa), *zemijans* (Benin and Togo), or *kabou-kabous* (Niger). In the 2010s, more than 85 percent of all public transportation in Benin and Chad was served by motorcycle taxis; in Rwanda and Kenya, about 75 percent was served by minibuses. The average for public buses was about 20 percent, but that number was as high as 80 percent in Seychelles and as low as 1 percent in Lesotho.

Government-operated buses are generally overloaded and unreliable, and require users to wait at unsheltered terminals; passenger loading is often disorderly and disorganized, particularly during peak periods. Most trips require multiple transfers, and the buses move slowly, frequently going only 5 to 8 miles (8 to 13 km.) per hour because of the ever-present extreme congestion or deterioration along the few available roads. Street trading along the main routes and failure to enforce regulations against on-street parking contribute to the congestion.

For most poor people in urban areas, motorized transportation is time-consuming, costly, and often unsafe. On average, 50 percent of transport in African cities is by walking or bicycling, but people who take some form of motorized transport can spend a high percentage of household income on mobility. About two-thirds of urban traffic fatalities in sub-Saharan Africa (which are much higher per capita than in developed countries) occur among pedestrians, with half of those fatalities being children. In 2020, local authorities in major Nigerian cities banned the pervasive motorbike taxis (known locally as *okadas*) and tricycle taxis (*kekes*), citing high accident rates caused by poor driver behavior. The ban resulted in long queues at bus stations and shonea spotlight on Nigeria's inadequate public bus system.

To this day, most Africans cannot afford their own vehicles. In 2014, just 3 percent of the population in Uganda and 5 percent in Kenya and Tanzania owned a car (compared to 88 percent in the United States). At the same time, rapidly expanding African cities have experienced explosive growth in numbers of privately owned bicycles and motorized two-wheelers—motorcycles, scooters, and mopeds. In the capital of Burkina Faso, Ouagadougou, these numbers have exceeded 150 motorized two-wheelers and 80 bicycles per 100 households.

The only two African cities with well-developed subway systems are Cairo (Egypt) and Algiers (Algeria). The Cairo Metro opened in 1987 and the Algiers Metro in 2011. Major Algerian cities have light rail or tram systems, as do cities such as Addis Ababa (Ethiopia), Rabat and Casablanca (Morocco), Alexandria (Egypt), Tunis (Tunisia), and Abuja (Nigeria). State-funded light rail transport systems in Lagos (Nigeria) and Abidjan (Côte d'Ivoire) are expected to become operational in the mid-2020s. Intercity commuter rail is developed in major metropolitan areas of Algeria, Morocco, Tanzania, Kenya, and Senegal, and in the four most populous provinces of South Africa.

The Trans-African Highway Network

Progress in road building in Africa since the 1990s has been significant. World Bank and African Development Bank loans, supplemented by national budgets, financed the building and improvement of road networks in many African countries. African governments have been spending, on average, about a quarter of the total financial provision of their national plans on transport development. More than 60 percent of this has been allocated to road construction, improvement, and maintenance. For example, Gabon has invested a substantial portion of its oil revenues in constructing all-weather roads and a railway into the interior. Nevertheless, the road network in Africa in the second decade of the twenty-first century was still inadequate, and only about a quarter of the African road system was up to the standard of all-weather roads.

In 2018, the road length-to-population ratio in sub-Saharan Africa was estimated at 16.7 miles (27 km.) per 10,000 people, compared to 140 miles (225 km.) in the United States. The countries where most roads were in good condition were South Africa, Mauritius, Burkina Faso, Botswana, and the Central African Republic, in that order. The countries with most of their national road networks in poor condition were the Republic of the Congo, Togo, Guinea, and the Democratic Republic of the Congo.

Action on a system of transcontinental highway networks was initiated by the United Nations Economic Commission for Africa (UNECA) in 1971. The plan was to greatly expand the network of strategic continental routes from about 6,210 miles (10,000 km.) to about 37,000 miles (60,000 km.), by both upgrading existing poor roads and building the "missing links." The result would be nine trans-African corridors, some of them following Africa's entire coastline and others strategically criss-crossing the continent. The corridor approach is considered the best way to reduce the time and cost of shipping freight across Africa's huge distances and numerous borders.

The routes of Trans-African Highway (TAH) corridors are as follows: Cairo-Dakar (TAH1); Algiers-Lagos (TAH2, or the Trans-Saharan Highway); Tripoli-Windhoek-Cape Town (TAH3); Cairo-Gaborone-Cape Town (TAH4, the longest of the nine corridors); Dakar-Ndjamena (TAH5, or the Trans-Sahelian Highway); Ndjamena-Djibouti (TAH6); Dakar-Lagos (TAH7); Lagos-Mombasa (TAH8); and Beira-Lobito (TAH9).

In 2020, the individual corridors in this long-term, ongoing project were at different stages of completion. While more than half of the total length has been paved, about a quarter of the Trans-African Highway network still consisted of missing links, mostly in Central Africa.

Railways

In Africa, as in Europe and North America, the major period of railway development extended from the end of the nineteenth century to the end of World War I. This expansion, however, was not co-

Caravan in the desert. Morocco, Sahara. (Sergey Pesterev)

ordinated. Railways were built as common carriers, usually under government ownership, between 1895 and 1914, primarily to stimulate cash production of export crops, facilitate mineral development, and strengthen colonial occupancy.

Most railways were single-line, used light track and simple signaling systems, and avoided embankments and tunnel creation wherever possible. The results were sharp curves, steep gradients, and circuitous routes, and consequently, low running speeds. Moreover, all railroads were narrow-gauge, and these narrow-track gauges varied among countries, as did the braking and coupling systems. Thus, the colonizing powers left a difficult and costly legacy for independent African countries that wanted to link their rail services. As with roads, rail networks have improved considerably since the 1960s. However, the total rail-network size in Africa s as a whole has remained quite small. In 2018, Africa had 51,000 miles (82,000 km.) of railways, compared to nearly 140,000 miles (225,300 km.) in the United States, which is approximately a quarter of

the size of the African continent. Sixteen percent of all African rail track was not operational due to war damage, natural disasters, neglect, or lack of funds.

The early railways were constructed partly to facilitate the administration of interior regions and to bring supplies from ports to central consumption or distribution points and partly—especially in the south—to enable valuable minerals or commodities to reach the coast for export. Instead of forming a network, the resulting rail pattern comprises a series of individual fingers that extend inland from various ports but which seldom connect with other railway lines, limiting their utility. For example, Cameroon's seaport of Douala is connected to the inland capital, Yaoundé, by the Central Railway, which was built during the colonial period. After the country gained independence in 1960, the line was extended to northern Cameroon. As of 2020, the country still had only this one major railroad line and another short one, also originating in Douala. Five of the ten provinces of Cameroon were still devoid of railways, and there were no links to railroads in neighboring countries. However, an extension of the Central Railway to Chad was approved in 2015.

Egypt has the oldest railway network in Africa, but it has had only a few upgrades since its inception. Most of the rail system of Egyptian National Railways is focused on the Nile delta, with lines reaching out from Cairo. Elsewhere in North Africa, the regional railway network among Morocco, Algeria, and Tunisia is somewhat integrated, but all these railroads were developed a century ago, during French colonial rule, and are in a somewhat inferior state. In 2019, the regional Arab Maghreb Union received a grant from the African Development Bank to modernize and expand this network by 2060.

In French-speaking West Africa, a major railway line links the Senegal River port of Saint-Louis with Dakar and extends to Bamako. The railway line linking Abidjan in Côte d'Ivoire through Bobo-Dioulasso to Ouagadougou in Burkina Faso provides landlocked Burkina Faso important rail access to a major seaport. Guinea's railway line, connecting the national capital, Conakry, with Kankan in the interior, has only sporadic service. In neighboring Benin, the railway line links Cotonou with Parakou, and there is a short coastal line linking Cotonou to Porto-Novo, the capital.

Ghana's railway lines are centered in the southern one-third of the country and link Sekondi-Takoradi in the Western Region through mining and forested regions to Kumasi in the Ashanti Region and Accra, the national capital. Another line links Accra with Takoradi in the west, producing what has been locally called the Golden Triangle, reflecting the relative economic development of the region. Nigeria is served by nearly 2,200 miles (3,500 km.) of rail that covers the economically important regions of the country; railways were originally built by the British colonial government. A major project to restore and extend long-neglected Nigerian railways has been underway since 2009. Liberia's three railroads link the ports of Monrovia and Buchanan with the country's iron ore deposits. One of the lines has been closed due to exhaustion of deposits, while another one has been recently re-

LANDMINES

Besides diverting expenditure from maintenance and much-needed new construction, political instability, whether from civil war or ethnic conflicts, disrupts transportation systems. Even after the conflicts are over, roads and railways may remain unused for a long time because of the fear of land mines left over from civil wars. In 2000 most of Angola's roads were known, or suspected, to be mined. Supplies and people had to be flown in to many places, but because some airports were also mined, it was a risky operation. Large tracts of land in Mozambique were also extensively mined. Land mines on many major roads severely disrupted trade and exchange. Mining of roads and insecurity resulting from civil conflicts forced Zimbabwe, Zambia, and other land-locked states to reroute much of their freight through South Africa.

Demining actions carried out by the country have cleared 67,683 miles (108,925 km.) of road, but there are still reported accidents and deaths. In the first quarter of 2019, there were 70 landmine accidents that killed 156 people, including 87 children.

constructed by a multinational steel manufacturing corporation.

In Central Africa, a chain of railways links the Atlantic Ocean with Lake Tanganyika. The Benguela Railway runs almost horizontally along the width of Angola, connecting to the Bulawayo-Port Francqui line at Tenke. Severely damaged during the Angolan Civil War, the railroad was rebuilt and reopened in 2011. The railway line from Kigoma connects the Indian Ocean at Dar es Salaam with branch lines from Tabora to Mwanza on the southern shore of Lake Victoria in Tanzania; the TAZARA Railway, built by Tanzania with Chinese assistance, links Dar es Salaam to Zambia's Central Province. Unlike many other African countries, Tanzanian railways have links with those of neighboring countries; they carry international freight and passengers in transit from Burundi, the Democratic Republic of the Congo, Kenya, Uganda, Zambia, and Rwanda.

The East African Kenya-Uganda line, called the Uganda Railway, was founded in 1895 by the British. It became popular among European and American adventurers traveling to East Africa, as it offered access to wild plateaus of the interior. The railroad was featured in many movies, including the Academy Award-winning *Out of Africa* (1985). The old and decrepit line was closed in 2017. The Mombasa-Nairobi Railway, built between 2014 and 2016 by a Chinese company, now runs parallel to the original Uganda Railway. In 2019, the construction of the new Nairobi-Malaba Railway in Kenya was inaugurated as part of the East African Railway Master Plan.

Sudan has more than 2,920 miles (4,700 km.) of track serving the central and northern parts of the country and giving Khartoum direct access to Port Sudan on the Red Sea and several large towns in the interior. The Addis Ababa-Djibouti Railway, built by the Chinese and opened in 2018, is a modern, standard-gauge railroad that provides landlocked Ethiopia direct access to the port of Doraleh near Djibouti City. (More than 95 percent of all Ethiopia's trade passes through Djibouti.) The Eritrean Railway is a narrow-gauge line linking the Red Sea port of Massawa with Asmara, the capital of Eritrea, and with Keren and Agordat. The line was nonoperational for a quarter-century until it returned to service in 2003 following its rehabilitation.

In Southern Africa, a single system of connected railways links Bulawayo in Zimbabwe with Livingstone to the northwest, Harare to the northeast, and Beira, the Mozambican port on the Indian Ocean. This same system also runs south to Gaborone, capital of Botswana, where it connects with South Africa's system, and also southeast to Maputo, the capital of Mozambique.

South Africa has long-established freight and passenger railways. Large discoveries of gold and diamonds there in the 1860s and 1870s led to the most rapid development of rail network in colonial Africa. Today, only half of the nation's 22,000 miles (36,000 km.) of track is fully utilized. One freight line links Port Elizabeth with Windhoek, the capital of Namibia, and Walvis Bay on the Atlantic Ocean. Another links Cape Town with Pretoria through Bloemfontein and connects to the Bulawayo line.

South Africa's state-owned intercity commuter rail system, Metrorail, has some 1,400 miles (2,253 km.) of rail with more than 470 stations in four provinces and carries up to two million passengers per day. Mainline Passenger Services operates long-distance passenger trains throughout South Africa, carrying some four million passengers per year. Africa's first rapid rail service, the Gautrain, began operation in 2012. Its trains travel at up to 100 miles (160 km.) per hour between Johannesburg, the country's industrial and financial hub, and Pretoria, its administrative capital.

Africa's largest railroad is the Cape to Cairo Railway, initiated at the end of the nineteenth century. Its intended course crosses the continent from Cape Town in South Africa to Port Said in Egypt. However, important sections of the railway have remained incomplete.

In general, the rail network of Africa remains a stark reminder of colonial powers' infrastructure development policies: Most lines are between ports and mines or the largest administrative centers. However, in recent decades there has been a renewed interest in railroads in Africa, prompted by the increasingly detrimental impact of climate and weather on Africa's roadways, serious traffic jams on many roads in the growing population centers, the increase of trade among African states, and the cost-effectiveness of rail transportation across Africa's immense distances.

Air Transport

Well suited to Africa's geographic vastness, air transport has become the primary means of international, and sometimes national, travel in Africa. During the late 1940s and the 1950s, as great advances were made in the extension and improvement of rail and road services, a new transport factor emerged in the introduction of internal and international scheduled air services.

The rapid development of air transport increased the movement of goods and people, and began to open up the hitherto largely closed interior of the continent. Today, Africa has more than 4,000 airports and airfields, but a significant number of them do not meet International Civil Aviation Organization standards and recommended practices. Only a quarter of these airports have paved runways.

In 2015, the top five countries for passenger air travel in Africa were South Africa, Egypt, Morocco, Nigeria, and Algeria. The busiest airports in terms of passenger numbers were O. R. Tambo International Airport (South Africa), Cairo International Airport (Egypt), Cape Town International Airport (South Africa), Aéroport Mohammed V (Morocco), and Addis Ababa Bole International Airport (Ethiopia). Most of them were also major cargo handling hubs, as were Kenya's Jomo Kenyatta International Airport and Nigeria's Murtala Muhammed International Airport. Ethiopian Airlines, South African Airways, EgyptAir, Kenya Airways, and Air Mauritius were among Africa's largest national carriers.

Altogether, in 2015 African airlines carried 79.5 million passengers, representing 2.2 percent of global air passenger transport, and 817,000 tons of air freight, representing 1.6 percent of global air freight. Though the share of Africa in global air transportation was relatively small compared to other regions, it supported 6.8 million jobs and contributed US$72.5 billion to the African economy.

Inland Water Transportation

Historically, throughout the vast interior between the Sahara and the Zambezi River, people or goods have been transported by canoe or boat on the great river systems of the Nile, Senegal, Niger, Congo, Ubangi, and Zambezi rivers, and on the few but large lakes. Large and small dugout canoes are still commonplace in the tropical rainforest and swampy woodland regions where there is plenty of wood for making these boats. These crafts are called *mecoro* in the Okavango Delta and *pirogues* in French-speaking areas. Where conditions were suitable, engine-powered craft later supplemented or displaced paddled or punted canoes. Also notable were the construction of lake ports and the installation of rail ferries across Lake Victoria. Most further development of river transport has been hindered by dams, frequent rapids, and annual seasonal fluctuations in water level, which limit the utility of rivers for transport.

The Congo River is the largest watershed in terms of drainage and discharge in sub-Saharan Africa. The Congo flows across a relatively flat basin that lies more than 1,000 feet (300 meters) above sea level, and meandersextensively through the rainforest. Entry from the Atlantic Ocean is precluded by a series of falls and rapids that make the Congo River only partially navigable. Nevertheless, the Congo River has been the major corridor of travel within the Republic of the Congo and the Democratic Republic of the Congo. Heavily laden barges ferry people and cargo between Kinshasa and Kisangani, and to Brazzaville on the opposite bank of the river. Navigation on the Zambezi is limited because of strong rapids, the basis for the Kariba Dam on the border of Zambia and Zimbabwe, as well as the Cahora Bassa Dam in Mozambique. Transport is more important in the Congo River basin than elsewhere in sub-Saharan Africa. In the Democratic Republic of the Congo, river transport is an important part of the integrated transport system. This system consists of the Matadi to Kinshasa railway, river services from Kinshasa to Kisangani, the rail link from Kisangani to Ukundi, and river services onward from Lualaba to Kindu and then a rail link to Shaba. The other national axis consists of river services from Kinshasa to Ibbo on the Kasai River, a tributary of the Congo, and a rail link onward to Shaba. Transshipments- on both routes are numerous, and transit times are long. This increases costs and makes freight vulnerable to damage and pilfering. The Congo and Oubangoui rivers provide a vital link both for the Republic of the Congo, particularly its northern parts, and for the Central African Republic where Bangui, the capital, is an important river port that also serves neighboring Chad.

On the Niger River, seasonal navigation is possible from the railhead at Kouroussa to Bamako, from Koulikoro to Ansongo, and from Niamey to Yelwa. In the lower reaches, the river is navigable from the Niger Delta to Lokoja all year, and seasonally to Baro and Jebba. On the Benue, the major tributary of the Niger, the open season is June to November to Makurdi; it shortens to six to eight weeks in August and September at Garoua. The Senegal and Gambia rivers, the White Nile from Juba to Khartoum, and the East African lakes, particularly Victoria and Tanganyika, are also used as links in the transport network.

In general, inland navigation is not well coordinated with the other transport systems in the region. For example, while the extensive navigable waterways of the Congo River system are used primarily for inland transport, they lack a direct outlet to the sea. The railways on either bank linking Kinshasa and Brazzaville are narrow-gauge, limited in capacity, and not connected to one another.

Other problems with inland waterways for modern transport include the following: Most rivers are characterized by alternating sections of low gradients and rapids; few rivers are navigable any distance from their mouths, which often have difficult access; unreliable rainfall patterns mean that the rivers may be closed to navigation for part of the year; they often flow through areas of low transport demand; and the location of political boundaries frequently reduces traffic.

Marine Harbors and Ports

Since antiquity, Africa's coastal residents traveled by boat between mainland Africa and its coastal islands. Today, numerous ferry services operate along Afri-

can coasts—in Gabon, Sierra Leone, Guinea-Bissau, and other countries. There are ferries between Limbe (Cameroon) and Calabar (Nigeria), and between Malabo and Bata in Equatorial Guinea. The most important coastal ferry service within Africa is the one in Tanzania, between Dar es Salaam and the island of Zanzibar. Ferry services also carry passengers across the Mediterranean Sea to and from North Africa and Southern Europe; some of the busiest ones are between Tangiers and Melilla in Morocco and Algeciras in Spain and between Tunis in Tunisia and the Italian island of Sicily.

Oceangoing transportation was well developed in ancient Africa, especially along the Mediterranean coast, where ancient Egyptians, Phoenicians, Greeks, and Romans established ports. Excavations at the ancient Egyptian port of *Mersa Gawasis* on the Red Sea unearthed the remains of the oldest seagoing ships ever discovered. The site of the Suez Canal had been on the crossroads of transportation and trade long before the canal itself opened in 1869 and became one of the world's most heavily used shipping routes. Mombasa in Kenya has a centuries-old history as a harbor city. The ports of Alexandria, Tunis, Algiers, and Tripoli grew in impor-

The port of Cape Town remains open 24 hours a day 7 days a week.

tance during the medieval period. Changing technologies of ocean transportation ultimately made many of the historic ports dotting the African coastline obsolete.

Good natural harbors for large ships are scarce in Africa because of the abundance of offshore sandbars and silt-choked river mouths. With a total coastline of more than 18,950 miles (30,500 km.), Africa today has about ninety major ports. Due to the limited availability of good locations for deepwater ports, only a few international ports handle large cargo volumes. In 2016, six of Africa's ports—Damietta/Port Said (Egypt), Port Louis (Mauritius), Tanger-Med (Morocco), Durban, Port Elizabeth, and Cape Town (the latter three in South Africa)—could accommodate post-Panamax-type vessels, a category that includes supertankers and the largest modern container and passenger ships. Durban and Damietta/Port Said were the only ports to have cargo-handling capacities of up to four million to five million twenty-foot equivalent units (TEUs) per year.

Other major African ports in terms of container traffic were Lagos in Nigeria, Alexandria in Egypt, Abidjan in Côte d'Ivoire, Mombasa in Kenya, Port of Richards Bay in South Africa, Dar es Salaam in Tanzania, Beira in Mozambique, Walvis Bay in Namibia, and the Port of Djibouti. In 2016, African ports handled 7.2 percent of worldwide seaborne cargo traffic and about 4 percent of the global container traffic. Passenger ships operated by international cruise lines docked in the ports of South Africa, Egypt, Kenya, Mozambique, Mauritius, Seychelles, Madagascar, Namibia, Senegal, Ghana, and Cabo Verde.

Between 2001 and 2016, total African trade and commerce has increased three-fold. At the same time, various studies indicate that inadequate transport infrastructure has been adding 30 to 40 percent to the costs of goods traded among African countries. According to the United Nations Economic Commission for Africa (UNECA), internal transportation raises the total cost of African exports by one-third, compared to the average of one-tenth for the whole of the developing world. The development of cross-border transportation by land is of particular importance to Africa: Among all continents, it has the largest number of land-locked countries, and land borders account for 84 percent of all African borders.

In the twenty-first century, rapid population growth and industrialization have posed new transportation challenges for the continent of Africa, but these same phenomena can also serve as advantages in the development of transportation. Potentially, the number of jobs in the transportation sector is enormous; also, the fast-paced industries boost the development of transportation infrastructure, and vice versa.

Joseph R. Oppong

COMMUNICATIONS

By the third decade of the twenty-first century, many Africans still did not have access to the Internet or television. Africa's widespread poverty and health-related problems could be better addressed if mass communication systems were more thoroughly developed on the continent. On the other hand, communication systems undoubtedly would be improved if Africa's nations were not so strapped for funds. Still, thanks to cellular phone technology, millions of African citizens have gained access to modern telecommunications for the very first time.

Radio: The African Phenomenon

In the first decades of the twenty-first century, television broadcasting in Africa lagged behind the standards of the industrialized countries of North America, East Asia, and the European Union. Only a few of the largest economies in Africa had television networks of the scope and variety of those found in the U.S. or Europe.

For example, the West African country of Gambia, with a population of 2.17 million in 2020, had three television stations (one state-run, one privately owned, and one online). In contrast, the U.S. state of Idaho, with a population of 1.8 million, had some 32 television stations. Such countries as South Africa, Egypt, Algeria, and Nigeria have developed large and vibrant television markets that broadcasted both in national languages and local dialects. The use of satellite dishes there was widespread, and TV transmissions from neighboring countries and international broadcasters were available. But even in those countries, the radio has remained highly popular.

Radio broadcasting in Africa is much more accessible than television broadcasting, especially in rural areas where access to electricity, the Internet, and television is very limited and often not available at all. For example, in Zambia, access to radio and television in urban areas was about equal (85 percent for radio and 79 percent for TV in 2016), while in rural areas the difference was significant (68 percent for radio and 26 percent for TV). Here, and throughout the many regions of Africa where power supply is unreliable, people depend on battery-operated radio sets.

In the twenty-first century, radio has remained by far the most influential information outlet in Africa, a continent where large segments of the population are still very poor and illiterate. Radio broadcasting is free, radios are cheaper than television sets or computers, and one does not need any level of education to get news and information from the radio, as opposed to print media. The widespread use of cellphones for radio listening has become a uniquely African twenty-first-century phenomenon. In many African countries, these phones come with built-in FM receivers.

Large African countries have numerous local and community radio stations operating alongside national broadcasters. But even in smaller countries, community radio is growing very quickly. With limited or nonexistent access to other forms of mass communication, radio broadcasting in local languages became crucial in building modern communities. However, broadcasting licenses were often granted to those close to the government, which limited the independence of this medium and freedom of speech. In most African countries, from Angola to Zimbabwe, the state controls all broadcast media, both radio and television.

Cellphone: Africa's Communication Lifeline

In Africa, the availability of traditional landline telephone communication was always hampered by outmoded, inadequate cable lines. Only South Africa and parts of some African capitals where the very wealthy live have adequate carrier-equipped open-wire lines, coaxial cables, and telephone exchanges.

At the end of the twentieth century, people in most African nations had little possibility of speaking by telephone to people in other parts of the world, or even to their friends and relatives living in nearby cities and villages. Somalia's 7.5 million people had only 9,000 telephones. In Nigeria, there were only about 405,000 phones in 1995; the population of the country was 108 million. Burkina Faso had a total of about 20,000 phones for its 11.5 million citizens. By comparison, an average American town of 10,000 residents had more than 20,000 phones at the time.

Not surprisingly, as soon as mobile telephony reached a stage when simple cellular phones became affordable, cellphone ownership in Africa exploded. Within twelve years, between 2002 and 2014, the number of cellphones in African countries increased tenfold. In 2002, only 8 percent of people in Ghana owned a mobile phone. That figure reached 83 percent in 2014, and in 2018 there were more than 140 cellphones per 100 Ghanaians. In Burundi, one of the world's least developed countries with less than one landline telephone per 100 inhabitants, the mobile phone penetration reached more than half of the population in 2017. In 2018, the government of the country launched the Burundi Broadband project with the aim of delivering nationwide connectivity by 2025, and newly emerged mobile operators have started to capitalize on the rapidly growing demand.

Most Africans own cheap, low-tech feature phones with a bare minimum of multimedia and Internet capabilities. In 2020, only in Morocco, Algeria, Egypt, South Africa, Nigeria, and Ghana did more than a quarter of the population own a smartphone. However, more and more Africans were using their phones not just for voice calling and text messaging but also for taking pictures and video, listening to the radio, and mobile banking. The mobile banking system M-Pesa was introduced in Kenya in 2007 as an alternative way for the population to store money, send and receive payments, and use microfinancing services. By 2020, it had more than 40 million users in ten countries.

Newspapers

In one-third of African countries, the adult illiteracy rate was about or more than 50 percent in 2018. In Benin, this rate was 73 percent for women and 50 percent for men; in Burkina Faso, it was 57 percent for men and 71 percent for women. Even with such high illiteracy rates, African nations had significant numbers of local and national newspapers, many of which were available both in print and online.

South Africa, where the literacy rate was more than 90 percent for both men and women, had more than fifty daily and weekly newspapers with both print and online editions, as well as more than a dozen online-only newspapers and some 300 regional and community newspapers; the latter were largely free of charge. The South African press had played an important role in that country's nation-building; despite government censorship, many newspapers had openly criticized the apartheid system and covered the anti-apartheid movement. Today, most South African newspapers are in English, but all ten other official languages of the country are represented, including Afrikaans, Zulu, and Xhosa. South Africa's *The Star* and *The Sunday Times*, along with Kenya's *The Standard*, are among the oldest and largest newspapers in sub-Saharan Africa.

In Mozambique, where the illiteracy rate was 55 percent for women and 27 percent for men in 2018, there are several local and national newspapers, but circulation rates were low. This was typical of the many sub-Saharan countries with low literacy and high subscription prices.

In North Africa, the adult literacy rates ranged from 54 percent in Sudan to more than 90 percent in Algeria and Libya, with Egypt and Morocco in the middle. Circulation rates of print media were, in general, higher than in sub-Saharan Africa, as was the readership of online publications. Egypt's venerable *Al-Ahram*, founded in 1875, was the most authoritative daily newspaper not only in Egypt, but perhaps in the whole Arab world.

Throughout Africa, with only a few exceptions, print media are heavily influenced, and in many cases directly controlled, by the government. In 2019, the international organization Reporters

Without Borders, which annually ranks countries of the world according to their level of press freedom, listed Namibia, Ghana, and South Africa as the highest-ranking among all African countries. Eritrea, Sudan, and Djibouti were the lowest-ranking. However, in recent years the number of newspapers, magazines, and online publications with critical views of the government has been growing. The press freedom ratings of Ethiopia and Gambia have dramatically increased within just a few years.

The Internet and Social Media

Few citizens of Africa could access the Internet at the end of the twentieth century, for several reasons. First, African nations did not have the sophisticated telephone connections that existed in other parts of the world. Fiber-optic cable and submarine cable usage were restricted to big coastal cities facing the oceans. The equipment needed for Internet access is expensive, and little of it was manufactured in Africa. Furthermore, the use of computers and the Internet requires well-developed literacy skills. Many Africans were illiterate in their own languages, and few Africans had the knowledge of English that was necessary to use many of the interfaces that support the Internet.

The need for continent-wide planning in order to improve communications and to bring the Internet within the reach of all Africans became clear to African governments by the late 1990s. Nations then began to work together to create a network for their domestic telecommunication needs. As a result of those efforts, the Regional African Satellite Communication Organization (RASCOM) was created. The first Afrocentric RASCOM satellite was built by a Franco-Italian manufacturer and delivered in-orbit in 2007, followed by a second one in 2010. Additionally, by 2020, South Africa had seven of its own satellites in orbit, and Kenya and Egypt had launched their first communications satellites. Satellite ground stations of the VSAT (very-small-aperture terminal) type have become ubiquitous on the African landscape; they offer the most effective way to provide telephone and Internet connections to large rural areas via both African and international satellites. Such efforts

have significantly increased Internet penetration in Africa.

Still, in sub-Saharan Africa, Internet usage has so far lagged behind the rest of the world. In 2019, only about 39 percent of that region's population used the Internet even occasionally, compared to the world average of almost 60 percent. However, in Nigeria and Mali Internet usage was more than 61 percent; in South Africa, 56 percent; and in Kenya, 83 percent. It was even higher in the small but advanced economies of Mauritius and Seychelles, but in many other sub-Saharan countries, it was below 30 percent. That number was even lower among women, older and less-educated people, and persons who did not speak English.

In the North African nations of Tunisia, Morocco, and Libya, the Internet penetration was between 64 and 75 percent. In North Africa in general, the use of the Internet and social media has become widespread among the young and educated urban populations.

Social Effects of Better Communications

Overall, Africa still has a long way to go in the development of communication systems such as television and the Internet. This creates a negative cycle: Africa does not have well-developed communication systems because many of its people are poor and poorly educated, but one reason these Africans are so poor

---

COMMUNICATIONS IN BOTSWANA

In 2019 the 2.25 million citizens of the Southern African nation of Botswana communicated with each other in the country's official language, English, as well as in the local language, Setswana. There are five television stations in Botswana, one of which is state-owned (Botswana TV), along with Now TV, Khuduga HD, Maru TV and EBotswana. There are five local radio stations (RB1, RB2, Duma FM, Gabz FM, and Yarona FM) and thirteen newspapers (*Mmegi*, *Sunday Standard*, *The Telegraph*, *Business Weekly*, *The Botswana Gazette*, *The Voice*, *The Guardian*, *Echo*, *Botswana People's Daily*, *DailyNews*, *Tswana Times*, *Weekend Post*, and *The Monitor*). There were 150 mobile subscriptions registered for every 100 people in 2018, compared to 6 landline subscriptions for every 100 people.

Rural village with mobile phone antenna in The Gambia. Almost every adult Gambian has a mobile phone and uses it quite often.

and poorly educated is because the continent has underdeveloped communications systems.

At the beginning of the twenty-first century, Africa was in the grip of an acquired immunodeficiency syndrome (AIDS) epidemic. That disease became by far the leading cause of death on the continent. All across Africa, particularly in the sub-Saharan region, millions of people had AIDS; yet communication systems were so poor that few of these people knew that their condition made them part of an epidemic. Inadequate communications also prevented people who were stricken with the disease from getting medical help.

Since then, better communication and education have helped mitigate the AIDS epidemic in Africa. This was proven by the efforts of the president of Uganda, Yoweri Museveni. In 1997 the percentage of adults infected with the AIDS virus in that East African country stood at 30 percent. Museveni launched a media and education campaign to help Ugandans understand AIDS. By 2000, the percent-

age of AIDS-infected Ugandan adults had fallen to 9 percent. By 2017, new AIDS infections had fallen by a third across all of East and Southern Africa. It is believed that this drop is due, in part, to a broad education campaign in the media.

In recent decades, social movements have actively sought out and used newly available information and communication technologies. In January 2011, cellphone photos and videos of local protests against Tunisia's corrupt and oppressive regime were spread nationwide. Until then, media control had prevented most Tunisians from knowing about these protests and their violent repression by military forces. The images captured by cellphones helped ignite Tunisia's Jasmine Revolution, which led to the ouster of the country's longtime president. Social media continued to play a vital role in the ensuing anti-government uprisings that engulfed North Africa and much of the rest of the Arab world and became collectively known as the Arab Spring.

African nations have bypassed the landline phone era and jumped directly into the era of the cellphone and the Internet, which have offered new opportunities to hundreds of millions of Africans in areas ranging from education and information to business and finance. Surveys show that just by using online banking services, some 200,000 Kenyan households lifted themselves out of extreme poverty within a few years. Many of those households were headed by widowed, rural women. According to a 2018 poll conducted by the authoritative Pew Research Center, large majorities across Africa believed that the increasing use of the Internet has had a positive influence on education in their country, and more than 60 percent said the same about the economy.

Annita Marie Ward

Gazetteer

Places whose names are printed in SMALL CAPS are subjects of their own entries in this gazetteer.

Aba. Important trade center in eastern NIGERIA (West Africa) Population in 2016 was 2,534,265. It grew in the early twentieth century, when the British established a military base nearby. In 1929 the site of the Aba Riot, a women's revolt against the high taxes imposed by the British administration. Approximately fifty women died in the conflict.

Abidjan. Largest city in CÔTE D'IVOIRE and its capital until 1983. Located on the Atlantic coast of West Africa; population was 3,677,115 in 2020. Côte d'Ivoire's center of commerce and manufacturing; often referred to as the "Paris of Africa" because of its beauty.

Abuja. Capital of NIGERIA since 1991. Total area of 2,824 square miles (7,315 sq. km.) with a 2011 population of 1,235,880. Located 300 miles (480 km.) inland.

Accra. Capital and largest city in GHANA. Located on the Gulf of GUINEA; population in 2010 was 1.594 million. First settled by the Ga ethnic group in 1482, and further developed by the British during colonial rule. Ghana's major international airport is located here.

Adamawa Plateau. An upland area of volcanic origin in West Africa. Extends from central CAMEROON into southeastern NIGERIA and the western part of CENTRAL AFRICAN REPUBLIC. Average elevation is about 3,300 feet (1,000 meters). Savanna vegetation dominates this sparsely populated region. The chief occupation is raising cattle.

Addis Ababa. Capital, economic and manufacturing center, educational center, and largest city of ETHIOPIA. Population was 3,384,569 in 2007. Located on a plateau 8,000 feet (2,440 meters) above sea level, in an area with many streams. Founded in 1887 by the Ethiopian emperor Menelik II; became the national capital two years later. Textiles, food processing, metals, cement, and plywood are its chief industries. Connected by railroad to the port of DJIBOUTI. The name means "new flower" in Amharic.

Adrar Temar. Region running from MOROCCO in NORTH AFRICA to SENEGAL in West Africa. Covering almost 28,000 square miles (72,500 sq. km.) of barren land, oases, and sparse forests in the highland, the region supports nomadic herding. Adrar is a Berber word for "plateau" or "mountain."

Afar Depression. V-shaped depressed block that cuts about 360 miles (600 km.) through ETHIOPIA, southward from the Red Sea.

Agadir. Sixteenth century city on the Atlantic coast in southern part of MOROCCO. Destroyed in 1960 by a devastating earthquake and almost completely rebuilt, it is among Morocco's most modern cities. Has an international seaport and attracts many winter tourists. Population in 2014 was 421,844.

Agulhas Current. Warm, swift ocean current moving south along East Africa's coast. Part of it moves between Africa and MADAGASCAR to form the Mozambique Current. The warm water of the Agulhas Current increases the average temperatures in the eastern part of South Africa.

Ahaggar Mountains. See HOGGAR MOUNTAINS.

Akosombo Dam. Earth-filled dam in West Africa on the VOLTA RIVER in GHANA; similar to the KAINJI DAM in NIGERIA. Is 2,170 feet (660 meters) long and 243 feet (74 meters) above water level. Opened in 1966 at a cost of US$228 million, after almost five years of construction. Supplies hydroelectric power to the ACCRA-Tema region of Ghana; has a capacity of 1,020 megawatts. Also called Volta Dam.

al-Qayrawan. See KAIROUAN.

Albert, Lake. Lake in Central Africa, located in northeast DEMOCRATIC REPUBLIC OF THE CONGO along the border with UGANDA. It is 100 miles (160 km.) long and 22 miles (35 km.) wide, covering about 2,064 square miles (5,300 sq. km.), at an altitude of 2,200 feet (670 meters). Northernmost of the great central African lakes. The SEMLIKI RIVER empties into it from the southwest; the Victoria Nile River in the northeast is its outlet. Discovered in 1864 by Samuel Baker and named after Queen Victoria's consort, Prince Albert. Called Lake Mobutu from 1973 to 1998.

Alexandria. Chief port and second-largest city in EGYPT. Founded in 332 BCE by Alexander the Great, the city's namesake. Population was 5.2 million in 2018. Located on the MEDITERRANEAN SEA, at the western extreme of the NILE delta, in the narrow coastal strip of Egypt that has around 8 inches (200 millimeters) of precipitation per year. It had one of the seven wonders of the ancient world, a massive lighthouse known as the Pharos of Alexandria. From the fourth century BCE to the seventh century BCE, it was the academic center for the Mediterranean world. Its legendary library had nearly a half-million volumes at its pinnacle in the third century BCE, toward the end of which it was destroyed. As Egypt's main port, nearly 80 percent of imports and exports pass through Alexandria. Its diversified manufacturing base includes oil refining, paper, plastics, food processing, and textiles.

Algeria. Largest country in NORTH AFRICA and second largest in Africa. Bounded by the MEDITERRANEAN SEA on the north, MOROCCO on the west, MAURITANIA and MALI on the southwest, NIGER to the east southeast, and LIBYA and TUNISIA on the east. Covers 2,381,741 square miles (6,168,681 sq. km.) population in 2020 was 42,972,878. Its chief cities are its capital, ALGIERS, and ORAN, CONSTANTINE, and Annaba, all in the north. Most of the southern part of the country is sparsely populated desert. OPEC member since 1962, with oil accounting for 20 percent of its GDP and 85 percent of its exports.

Algiers. Largest city and capital of ALGERIA. Situated midway along the coast of the MEDITERRANEAN SEA between MOROCCO and TUNISIA. Has been a thriving community and important port for more than 3,000 years. Its population was 3,915,811 in 2011. The city was controlled by Turkey, then France, until Algeria gained its independence in 1962.

Angola. Republic previously known as Portuguese West Africa, located south of DEMOCRATIC REPUBLIC OF THE CONGO and north of NAMIBIA. Total area of 481,353 square miles (1,246,700 sq. km.) with a 2020 population of 42,972,878. Capital is LUANDA. Most of the country is part of the ANGOLAN PLATEAU. Divided into three major regions: from west to east, the coastal plain, a transition zone, and a vast inland plateau. The low-lying coastal plain varies from about 30 to 90 miles (50 to 150 km.) in width. The transition zone, which consists of a series of terraces or escarpments, is about 90 miles (150 km.) wide in the north, but diminishes to about 19 miles (30 km.) in the center and south. Has a basically tropical climate, although cooler ocean currents make temperatures along the coast fairly temperate. OPEC member since 2007, with oil accounting for half its GDP and nearly 90 percent of its exports. Rich in other natural resources, including diamonds, copper, iron, and uranium. Vegetation ranges from rain forest to dry savanna. After the Portuguese decolonized the area in 1875, a civil war drove out most Portuguese settlers.

Angolan Plateau. Vast plateau covering about two-thirds of ANGOLA. Has an average elevation of 3,000 to 5,000 feet (1,000 to 1,520 meters); highest point is Mount Moco (8,597 feet/2,620 meters). Angola's main rivers, the Caunza and the Cunene, flow from these mountains to the Atlantic Ocean.

Annobón. Tiny volcanic island off Africa in the Gulf of GUINEA, part of EQUATORIAL GUINEA. Located at 1°25′ south latitude, longitude 5°37′east. The island's highest elevation is 2,727 feet (831 meters). It has an area of 7 square miles (17 sq. km.); its population was 5,232 in 2015. Has a high annual rainfall—about 117 inches (2,972 millimeters)—as a result of its proximity to the equator, and is covered with dense equatorial forest. Discovered by Portuguese navigators around the 1470s.

Antananarivo. Capital and largest city of MADAGASCAR, founded about 1625. Located in a central province of the same name, it was the major settlement of the Merina, the island's most organized people. The population in 2015 was 1,610,000.

Ascension Island. Barren island in the ATLANTIC OCEAN. Located about 500 miles (800 km.) south of the equator at 7°57′ south latitude, longitude 14°22′ west. Ascension became a dependency of SAINT HELENA in 1922 and was a refueling point for British planes and ships during the war with Argentina over the Falkland Islands in 1982. Ascension's highest elevation is 2,817 feet (858 meters). It has an area of 34 square miles (88 sq. km.); its population was 806 in 2016.

Asmera. Capital, primary port, and largest city of ERITREA. Had a 2018 population of 896,000. Located near gold and copper mines; local manufacturing includes food processing, textiles,

perfumes, glass, cement, bricks, lumber, and leather.

Aswan. A major city on the Upper NILE RIVER in EGYPT, known for the Aswan High Dam, a major source of hydroelectricity for the country. Construction of the dam created Lake NASSER, the largest body of water in Egypt and SUDAN. Population was 290,327 in 2012.

Aswan High Dam in Egypt.

Asyut. City in central EGYPT, on the NILE RIVER. Located just south of the Al Ibrahimiyah Canal, which furnishes water to several western Nile Valley locales and feeds the Al Fayyum depression, more than 180 miles (290 km.) from Asyut. Dates, sugarcane, and grains are produced in the region. A center for the ancient sect of Coptic Christians and a stronghold for fundamentalist Islam. Population was 389,307 in 2006.

Atlas Mountains. Mountain range in NORTH AFRICA. The RIF and Middle Atlas Mountains run across northern MOROCCO into parts of western ALGERIA and down the center of Morocco. The High Atlas Mountains run through the center of Morocco to the south, where they meet with the Anti-Atlas Mountains that extend toward WESTERN SAHARA. The highest peak is Toubkal (13,665 feet/4,165 meters) in Morocco.

Badagry. Small coastal city in NIGERIA, near the republic of BENIN. Mostly populated by the Yoruba people. A major slave port on the BIGHT OF BENIN between 1711 and 1810. Its beautiful beaches welcome tourists. Population in 2006 was 241,093.

Bamako. Capital, largest city, and financial and industrial center of MALI. The city straddles the NIGER RIVER, which is a major mode of transportation. Population was 1,810,366 in 2009.

Bandundu. Province in southwest DEMOCRATIC REPUBLIC OF THE CONGO. It covers 114,154 square miles (295,658 sq. km.). Population was 8 million in 2010. Drained by the CONGO, KASAI, and Kwango rivers. Commercial center of agricultural products, including palm oil, manioc, and peanuts, mainly for KINSHASA. Population of the city was 137,460 in 2010. Called Banningville until 1966.

Bangui. Capital and main commercial center of the CENTRAL AFRICAN REPUBLIC. Located in the southwest of the country, on the west bank of the UBANGI RIVER near the country's border with the DEMOCRATIC REPUBLIC OF THE CONGO. Population was 734,350 in 2012. Most of the country's industries are located there.

Banjul. Capital and largest, most-developed city of GAMBIA. Founded by the British in 1816 as a port and a base for suppressing the slave trade. Originally called Bathurst, renamed Banjul in 1973. Population was 31,301 in 2013.

Barrage Vert. Long rows of Aleppo pine trees the Algerian government planted in the 1970s along the SAHARAN ATLAS ridge from MOROCCO to the Tunisian border, a distance of 1,500 miles (2,400 km.), to prevent the SAHARA from encroaching on the fertile agricultural areas to its north. This succeeded in containing the desert which, if left alone, might have gradually extended all the way to the MEDITERRANEAN SEA, making ALGERIA and other parts of NORTH AFRICA a wasteland. French phrase for "green barrier." In 2007, a movement titled "The Great Green Wall" revisited and built upon the Barrage Vert initiative. It is expected to be completed in 2030.

Bas-Congo. See KONGO CENTRAL.

Basutoland. Colonial-era name for LESOTHO.

Bechuanaland. Colonial-era name for BOTSWANA.

Beida. Small town on the MEDITERRANEAN SEA about 120 miles (190 km.) east of Benghazi, which once was LIBYA's summer capital. The government has expanded and modernized the city. Population in 2010 was 380,000.

Belgian Congo. See DEMOCRATIC REPUBLIC OF THE CONGO.

Benguela Current. Northward-flowing current along the western coast of Southern AFRICA.

Benin. Small West African nation, formerly a French colony. Total area of 43,484 square miles (112,622 sq. km.) with a population of 12,864,634 in 2020. Capital is PORTO-NOVO, but the port city of COTONOU is the commercial and political capital. Called DAHOMEY until 1975. French is the official language, but Yoruba, Fon, and Adja are also spoken. Considered to be the birthplace of West Indian voodoo and black magic. Administered within the federation of FRENCH WEST AFRICA during the era of colonial rule.

Benin, Bight of. Part of the Gulf of GUINEA in West Africa. Extends approximately 450 miles (720 km.) from the mouth of the VOLTA RIVER to the NIGER RIVER. Fed by the Ogun, Benin, Mono, and Oueme rivers. Its principal ports include ACCRA, GHANA; PORTO-NOVO and COTONOU, BENIN; LOMÉ, TOGO; and LAGOS, NIGERIA. Was known as the Slave Coast throughout the eighteenth century.

Benin Kingdom. Historic kingdom of the Edo-speaking people in southwestern NIGERIA. Not related to the independent republic of BENIN to the west.

Benue River. Chief tributary of NIGER RIVER. It rises in the northern part of CAMEROON and flows west across east-central NIGERIA. It is about 673 miles (1,083 km.) long. The Benue and Niger rivers divide northern Nigeria from southern Nigeria.

Bette Peak. Highest mountain in LIBYA. Located in the TIBESTI MOUNTAINS near the CHAD border, it is 7,500 feet (2,286 meters) high, slightly more than half the height of MOROCCO's Jebel TOUBKAL.

Biafra, Bight of. Part of the Gulf of GUINEA in West Africa. Extends approximately 400 miles (640 km.) from the mouth of the NIGER RIVER in NIGERIA to Cape Lopez in GABON. Fed by the Cross, Niger, and the Sanaga rivers. Its principal ports include MALABO, EQUATORIAL GUINEA; PORT HARCOURT and CALABAR, Nigeria; and DOUALA, CAMEROON.

Bight of Benin. See BENIN, BIGHT OF.

Bight of Biafra. See BIAFRA, BIGHT OF.

Bioko. Large island in the Gulf of GUINEA, part of the nation of EQUATORIAL GUINEA and site of its capital, MALABO. Located about 25 miles (40 km.) off the coast of Africa. The island's highest point, at 9,869 feet (3,008 meters), is Santa Isabel, a volcano that erupted in 1923. The climate is tropical wet. Formerly known as Fernando Po in honor of one of its Portuguese discoverers, the island eventually became part of Spanish Guinea, which gained its independence in 1968. The island was subsequently renamed Bioko in honor of an early king. It covers an area of 779 square miles (2,018 sq. km.) and had a population of 334,463 in 2015.

Bizerte. Northernmost city in TUNISIA, with a population of 142,966 in 2014. Once a French naval base; a port for iron ore and the center of the commercial fishing industry on the coast. When German and Italian forces invaded Tunisia by air from Sicily in World War II, it was the site of their most bloody battles against the Allies. Blockaded by Tunisian troops in 1961, the base was evacuated by the French, who departed from it entirely in 1963.

Bloemfontein. Judicial capital of SOUTH AFRICA and administrative capital of the FREE STATE Province. Most of the region is on the interior plateau, generally level, with an average altitude of 4,480 feet (1,365 meters). Because of low rainfall, there are few rivers and streams. Summers are warm to hot and winters are cold with some frost and snow. Produces maize, wheat, and beef cattle. Founded in 1846 by Afrikaans-speaking farmers who arrived in the 1830s and 1840s and later established a settlement in the area. The name means "fountain of flowers." Harvard and the University of Michigan each opened observatories here in 1927 to take advantage of Bloemfontein's clear atmospheric conditions.

Bo. Administrative center for the southern province of SIERRA LEONE; the capital of the Protectorate of Sierra Leone from 1930 until 1961. An important commercial and educational center for the nation's interior. Population was 174,369 in 2015.

Bongor. Capital of Mayo-Kebbi Prefecture (province) in southwestern CHAD. Located about 149 miles (240 km.) from N'DJAMENA, Chad's capital. Its economy is based on rice, cotton, and fishing. Population was 30,518 in 2010.

Botswana. Republic in Southern Africa. Total area of 224,607 square miles (581,730 sq. km.) with a 2020 population of 2,317,233. Capital is Gaborone, which has a population of 231,592 (2011). The KALAHARI DESERT covers much of the country, and the rest is largely savanna. The economy is largely agricultural, but diamond deposits were discovered in the late twentieth century. Tourism is a growing economic sector. Has one of the world's highest rates of HIV/AIDS infection. Created as the British protectorate of BECHUANALAND in the nineteenth century.

Brazzaville. Capital of the REPUBLIC OF THE CONGO and the country's largest commercial, industrial, and administrative center, and major river port. Located in the southeast of the country on the right bank of the CONGO RIVER, opposite KINSHASA. Terminus of the railroad from POINTE-NOIRE on the Atlantic Ocean, and of navigation on the CONGO and UBANGI RIVER systems. Founded in 1880 by French explorer Pierre Savorgnan de Brazza and named after him. Population was 2.3 million in 2019.

Bujumbura. Capital and largest city of BURUNDI. Located on the northeastern tip of Lake TANGANYIKA, with a 2019 population of 1,092,859. The lake gives Burundi access to the Tanzanian railroad that connects the lake to the INDIAN OCEAN port of DAR ES SALAAM. Fishing, food processing, cotton textiles, handcraftsi, hides, beer, and cement are major local industries.

Bukavu. Capital of South KIVU province in .the eastern part of the DEMOCRATIC REPUBLIC OF THE CONGO. A commercial and industrial center and port. Located on the southwest shore of Lake KIVU at an altitude of 4,768 feet (1,453 meters). Devastated by years of civil war, which has halted tourism; poachers have killed much of the wildlife in the nearby Kahuzi-Biega National Park. Population was 870,954 in 2016. Called Costermansville until 1966.

Burkina Faso. Landlocked country in West Africa; formerly called UPPER VOLTA. Total area of 105,715 square miles (273,800 sq. km.) with a population of 20,835,401 in 2020. OUAGADOUGOU is the capital and largest city. French is the official language, but Mossi and other indigenous languages are spoken. One of the poorest nations in the world, this republic largely depends on economic aid from France.

Burundi. Landlocked country in Africa. With an area of 10,745 square miles (27,830 sq. km.), slightly larger than the U.S. state of Vermont, and a 2020 population of more than 11.8 million, it is one of the most densely populated regions on the continent. Capital is BUJUMBURA. Savanna is the typical natural vegetation. Around 43 percent of the land is arable; about 12 percent is forest, most of which is located in nature preserves. The climate is tropical, but its effects are ameliorated by altitude, as most of the country is a hilly plateau, averaging between about 4,600 and 5,900 feet (1,400 and 1,800 meters) above sea level. Ethnic violence between the empowered minority Tutsis and the Hutus, who make up 85 percent of the population, has been a recurring problem. Burundi's social structure and economy also have been disrupted by its high rate of HIV/AIDS infection.

Cabinda. Province of ANGOLA separated from the rest of the country by the narrow strip of THE DEMOCRATIC REPUBLIC OF THE CONGO along the lower reaches of the CONGO RIVER. The oil-rich coastal enclave has had a simmering independ-

ence movement since the 1960s. Population in 2014 was 716,076.

Cabo Verde. Nation of ten major and five minor islands lying about 320 miles (520 km.) off the coast of West Africa in the North Atlantic Ocean. The capital is Praia. The islands have a subtropical desert climate; the lack of fresh water and other resources has encouraged emigration. Once a Portuguese colony, Cabo Verde became independent in 1975. Cabo Verde comprises 1,557 square miles (4,033 sq. km.), and its highest point, Mount Fogo on the island of Fogo, reaches 9,281 feet (2,829 meters). There were about 583,255 inhabitants in 2020.

Cairo. Capital of EGYPT and the most populous city in Africa. Its population was about 9.5 million in 2018. Long a center for education; Al-Azhar University, founded in 988, is the oldest Islamic university in the world and the oldest continually operated university in the world. The primary publishing center for the Arabic-speaking world; the home of museums, theater, and opera; and the region's financial center. The Egyptian Museum, considered one of the greatest archaeological museums in the world, contains the artifacts from King Tutankhamen's tomb. The pyramids of Giza and the Great Sphinx draw millions of tourists each year. It has been the political capital of Egypt since its founding in 969 by the Fatimids, although the land it sits upon has been settled for 6,000 years. Located just south of the beginning of the NILE Delta, on both banks of the river. Its diversified manufacturing base includes cotton textiles, food processing, automobile assembly, aircraft assembly, chemicals, iron, and steel.

Calabar. Capital of Cross River State in the southeastern part of NIGERIA. Located on an estuary of the Gulf of GUINEA on the left bank of the Calabar River. Population in 2015 was 466,800, the majority of whom belonged to the Efik minority group. A market center for the surrounding area, with trade in palm oil, rubber, and timber.

Caldera de Taburiente, La. The world's largest volcanic crater, approximately 1 mile (1.6 km.)

deep, located on the CANARY ISLAND of La Palma. Site of a Spanish national park since 1954.

Cameroon. Country in west Central Africa, bounded by NIGERIA to the northwest, CHAD to the northeast, the CENTRAL AFRICAN REPUBLIC to the east, the REPUBLIC OF THE CONGO to the southeast, GABON and EQUATORIAL GUINEA to the south, and the Atlantic Ocean to the southwest. Covers 183,568 square miles (475,440 sq. km.), with a population of 27,744,989 in 2020. Capital is YAOUNDÉ. Natural resources include timber, oil, bauxite, iron ore, and rubber. Petroleum is the main export and source of government revenue. Originally created as Kamerun by Germany during the late nineteenth century; after World War I the country was taken away from Germany by the new League of Nations, which divided its administration between Great Britain and France. In 1960 both British and French territories became independent. They united in 1961.

Cameroon Mountain. See Mount Cameroon.

Canal des Pangalanes. Natural channel running about half the length of the east coast of MADAGASCAR. Separated from the INDIAN OCEAN by river silt and sand deposited by ocean currents, it is used for navigation.

Canary Islands. Group of seven major and six minor volcanic islands located in the North ATLANTIC OCEAN off southern MOROCCO, which is 70 miles (113 km.) from the nearest island. The islands nearest the African coast have a desert climate; the rest are subtropical, although water must be conserved for agriculture. When European explorers arrived in the early fifteenth century, they called the indigenous people the Guanches. The Canaries constitute two provinces of the European nation of Spain. Because of their pleasant climate and often luxuriant plant life, much of it introduced, the Canaries are popular with tourists. The islands cover 2,893 square miles (7,492 sq. km.), and their highest point is the volcano PICO DE TEIDE. The population was 2,152,590 in 2019. The islands take their name from the Latin word for "dog," not from the name of a bird.

Cancer, Tropic of. The parallel of latitude that runs 23°30′ north of the equator. It runs through the arid deserts of WESTERN SAHARA, southern ALGERIA, and southern LIBYA, areas that are exceptionally hot in summer, even at the higher elevations.

Cape Agulhas. The true southern tip of the African continent, located in the southwestern part of SOUTH AFRICA. Generally considered to be the dividing line between the INDIAN and the Atlantic oceans.

Cape Bon. Large outcropping of land that juts into the Gulf of Tunis and the MEDITERRANEAN SEA on TUNISIA's northeast coast. The area around it is relatively undeveloped, with none of the large communities that exist west of the Gulf of Tunis.

Cape Coast. Capital of the Central Region in southern GHANA. Population was 169,894 in 2010. First settled by the Portuguese in 1610; was the capital of the British GOLD COAST until the 1870s, when the capital was moved to ACCRA. Its main products include coconuts, cocoa, corn, cassava, and frozen fish.

Cape of Good Hope. Promontory in southwestern SOUTH AFRICA that is generally viewed as the southernmost point in Africa, although CAPE AGULHAS is somewhat farther south. Rises 850 feet (260 meters) above the sea. Its main city is CAPE TOWN, one of South Africa's leading cities and Africa's jumping-off point for travel to Antarctica. The first European to visit the Cape of Good Hope was the Portuguese captain Bartolomeu Dias in 1488.

Cape Peninsula. Long, rocky peninsula on the southwestern side of SOUTH AFRICA that separates the Atlantic Ocean from the INDIAN OCEAN's False Bay. Extending south of CAPE TOWN, the cape is a scenic and popular tourist destination.

Cape Point. Southernmost tip of SOUTH AFRICA's CAPE PENINSULA.

Cape Town. Legislative capital and one of the main cities in SOUTH AFRICA. Cape Town is located in estern Cape province, and had a 2020 population of 4,618,000. Has a Mediterranean climate, with warm, dry summers and cool, wet winters. The main rivers in the area—the Black, the Diep, and the Liesbeek—have been turned into canals. TABLE MOUNTAIN is a leading tourist attraction. TABLE BAY is South Africa's second-largest port and the main departure point for those visiting Antarctica by ship. The harbor offers safe anchorage and fresh supplies to visiting ships. Fishing is an important industry. The cold BENGUELA CURRENT on the west coast is famous for crayfish and lobster. The city has a mixed economy, originally based on agriculture, but now based on industries such as chemicals and food processing The world's largest museum of African American art opened here in 2018.

Capricorn, Tropic of. The parallel of latitude that runs 23°30′south of the equator, passing through the African countries of NAMIBIA, BOTSWANA, SOUTH AFRICA, MOZAMBIQUE, and MADAGASCAR.

Caprivi Strip. Narrow panhandle of NAMIBIA, stretching about 300 miles (480 km.) east from the northeastern corner of the main part of the country to the ZAMBEZI RIVER, where ZAMBIA, ZIMBABWE, and BOTSWANA all nearly touch each other. The strip is a historic anomaly created when Germany pressured Great Britain into allowing it to extend its Southwest Africa territory to the Zambezi.

Carthage. Ancient city that was destroyed by the Romans in 146 BCE. They laced its fertile fields with salt so that nothing could be grown in them. Renamed TUNIS, it later became TUNISIA's capital. A modern, fashionable residential community named Carthage stands on the site of the ancient city 9 miles (14 km.) east of Tunis, dotted with the lavish villas of high-ranking government officials and foreign diplomats.

Casablanca. The largest city in NORTH AFRICA, and the commercial capital of MOROCCO. Located on the Atlantic Ocean about a third of the way south from TANGIER to WESTERN SAHARA, with a population of about 6.86 million (2015). Most international flights into Morocco land there.

Central African Republic. Landlocked country in Central Africa bounded on the south by the DEMOCRATIC REPUBLIC OF THE CONGO, east by SUDAN, and SOUTH SUDAN north by CHAD, west by CAMEROON, and southwest by the REPUBLIC OF THE CONGO. Covers an area of 240,535 square miles (622,984 sq. km.), with a population of 5,990,855 in 2020. Capital is BANGUI. Natural resources include diamonds, uranium, and timber. Coffee, cotton, peanuts, and food crops are grown. Industries include timber, textiles, soap, cigarettes, processed food, and diamond mining. Gained independence from France in 1960.

Ceuta. City on the Tangier Peninsula in the northernmost SPANISH MOROCCO,Protectorate of M across the Straits of Gibraltar from the British crown colony of GIBRALTAR on the Iberian Peninsula. Its Jebel Musa is one of the two so-called PILLARS OF HERCULES. Population was 84,829 in 2019.

Chad. Independent west-central African republic; main borders are with LIBYA on the north, SUDAN on the east, NIGER on the west, and the CENTRAL AFRICAN REPUBLIC on the south. Total area of 495,755 square miles (1,284,000 sq. km.), with a population of 16,877,357 in 2020. Largely a desert country, its official languages are French and Arabic. Its rivers flow north, emptying into Lake CHAD. Administered within FRENCH EQUATORIAL AFRICA during the colonial era, which ended in 1960. Capital is N'DJAMENA.

Chad, Lake. The fourth-largest lake in Africa. Located in west Central Africa in CAMEROON, CHAD, NIGERIA, and NIGER. Its average depth is about 23 feet (7 meters). Has an average area of approximately 590 square miles (1,530 sq. km.).

Chagos Archipelago. Tiny group of atolls in the INDIAN OCEAN. The islands are shared by the United Kingdom and the United States for defense purposes, but are claimed by MAURITIUS. The archipelago covers 23 square miles (60 sq. km.); the sole inhabitants are American and British military personnel.

Chelia, Mount. Highest mountain in the northern half of ALGERIA, at 7,638 feet (2,328 meters). Located in the eastern reaches of the ATLAS MOUNTAINS, in the section known as Kabylia.

Cheliff River. River in the coastal plains of ALGERIA. Although not navigable, it provides water for agricultural irrigation in the fertile coastal areas. Receives runoff from the ATLAS MOUNTAINS.

Chott Djerid. The largest lake in TUNISIA. Located on the edge of the SAHARA, close to the Algerian border and 62.5 miles (about 100 meters) from DJEBEL CHAMBI, Tunisia's highest point. Formed by a 70-mile (113-km.)-deep depression in the ground; often dry. Also known as Shatt al-Jarid.

Comoros. Archipelago of four volcanic islands, located in the northern MOZAMBIQUE CHANNEL. Three of the islands make up the independent nation of Comoros; the fourth island, MAYOTTE, remains a French colony. The islands have a tropical monsoonal climate and frequently experience violent cyclones. After gaining independence from France in 1975, Comoros had a long period of political instability and was invaded by mercenaries several times. Comoros' highest elevation is 7,743 feet (2,360 meters) at Montu Kartala, a volcano that last erupted in 2007. The area of the nation of Comoros is 63 square miles (2,235 sq. km.); its population was 846,281 in 2020.

Conakry. National capital and largest city in GUINEA. Has one of the best natural deepwater harbors on the coast of West Africa. Connected by rail to Kankan, Guinea's second-largest city, which provides access to the NIGER RIVER. Population was 1,660,973 in 2014.

Congo-Brazzaville. See REPUBLIC OF THE CONGO.

Congo Free State. Private colonial empire in Central Africa created by Belgium's King Leopold II during the 1880s. His regime's ruthless exploitation of the country led the Belgian government to take it away from him and transform it into the BELGIAN CONGO in 1908. This colony became independent in 1960 and later became known as the DEMOCRATIC REPUBLIC OF THE CONGO.

Congo River. Main river in Central Africa. Starts from KATANGA province at an elevation of 5,760 feet (1,760 meters) in southeast DEMOCRATIC REPUBLIC OF THE CONGO, near the border with ZAMBIA. Flows northward to KISANGANI, bends westward, and later flows southwest from around BANDAKA to reach the Atlantic Ocean. From where it is joined by the UBANGI RIVER, it forms the border between the Democratic Republic of the Congo and THE REPUBLIC OF THE CONGO. One of the world's longest rivers, about 2,900 miles (4,700 km.), second in Africa only to the NILE RIVER. Its drainage basin of about 1.6 million square miles (4.1 million sq. km.) receives an average of 60 inches (1,525 millimeters) of rain annually. Its rate of flow at its mouth is about 1.45 million cubic feet (41,000 cubic meters) per second, making it second in the world only to the Amazon River. The first European to find its mouth was the Portuguese navigator Diego Cão, in 1482. Navigability is limited because of several cataracts. At one time known as the Zaire River. The upper course is also called the Lualaba River.

Constantine. Ancient city in NORTH AFRICA, now ALGERIA's third-largest city. Located 50 miles (80 km.) south of the MEDITERRANEAN SEA, perched high on chalk cliffs. A deep gorge runs through the center of the dramatic, beautiful city; numerous bridges connect the two parts of the town that are separated by the river. Originally called Cirta; named Constantine by the Roman conquerors after the Roman emperor Constantine. Was the Turkish capital in eastern Algeria. Population was 464,219 in 2018.

Corisco Island. Part of EQUATORIAL GUINEA in West Africa. Located at the mouth of Corisco Bay off the coast of West Africa. Has an area of 5 square miles (13 sq. km.). Initially claimed by Portugal in 1472; taken over by Spain from 1858 to 1968. Agriculture and fishing are the major industries.

Côte d'Ivoire. Former French colony in West Africa. Total area of 124,504 square miles (322,463 sq. km.) with a population of 27,481,086 in 2020. Capital was moved from the coastal city of

239

ABIDJAN to YAMOUSSOUKRO, farther inland, in 1983. After achieving independence in 1960, the country enjoyed rapid economic growth and multiparty democracy until December 1999. For the next fifteen years, the country was torn by military coups and civil war. Democratic elections in 2015 have restored a sense of normalcy. French is the official language. The English name is Ivory Coast. Administered within the federation of FRENCH WEST AFRICA during the era of colonial rule.

Cotonou. Largest city, and the financial and commercial center in BENIN. Located on the Gulf of GUINEA, its business activities attract traders from neighboring countries, such as NIGERIA and TOGO. Population in 2017 was 2,401,067.

Cyrenaica. Region of LIBYA that extends from the MEDITERRANEAN SEA to CHAD. Coast curves away from the SURT DESERT to the Mediterranean Sea. Its 130-mile (210-m.) arch connects the cities of Benghazi and DERNA. Population was 1,613,749 in 2006.

Dahomey. Historic kingdom in West Africa that gave its name to the French colony of Dahomey, which later changed its name to Republic of BENIN.

Dakar. Capital of SENEGAL. Has one of the best harbors on the Atlantic coast of Africa, which was the major supply port for the Allied Powers in Africa during World War II. Called the Gateway to Africa during French colonial rule, when it was the administrative capital of FRENCH WEST AFRICA. Population was 1.4 million in 2013.

Dakhla. North African port city two-thirds of the way down the coast of WESTERN SAHARA toward MAURITANIA. Located just north of the Tropic of CANCER, it is one of the two population centers in this sparsely populated, largely desert region. Population in 2014 was 106,277.

Dar es Salaam. Leading port, financial and educational center, and largest city in TANZANIA. Founded in 1862 as a summer residence for the sultan of ZANZIBAR. During the German colonial period, which began in 1885, it was greatly expanded and became the capital of the German colonies in 1891. Population was 4.36 million in

2012. Its name means "haven of peace." In 1996, the inland city of DODOMA became Tanzania's official capital, although the government bureaucracy has largely remained in Dar es Salaam.

Darnah. See DERNA.

Delagoa Bay. INDIAN OCEAN inlet near the southern tip of MOZAMBIQUE on which the capital city, MAPUTO, stands.

Democratic Republic of the Congo. Central African nation formerly known as Congo-Kinshasa. Straddles the equator between 5° north and 12° south latitude. Total area of 905,354 square miles (2,344,858 sq. km.)—one-quarter the size of the United States—makes it Africa's second-largest nation (after ALGERIA). Has only 25 miles (40 km.) of coastline, on the Atlantic Ocean. Bordered on the west by the REPUBLIC OF THE CONGO and the Angolan enclave of CABINDA; on the north by the CENTRAL AFRICAN REPUBLIC and SOUTH SUDAN; on the east by UGANDA, RWANDA, BURUNDI, and TANZANIA; on the southeast by ZAMBIA; and on the southwest by ANGOLA. Population in 2020 was 101,780,263. French is the official language. Rich in natural and mineral resources, including vast deposits of industrial diamonds, cobalt, and copper. Has the largest forest reserves and the largest hydroelectric potential in Africa. Originally created as the CONGO FREE STATE during the 1880s; became the Belgian Congo in 1908. Secessionist movements have beset the country since independence in 1960. In 1996 the three-decade regime of Mobutu Sese Seko was overthrown by Laurent Kabila with the help of Uganda and Rwanda. Mobutu renamed the country ZAIRE in 1970; Kabila restored its original name in 1997. Civil war continues in the country. Capital is KINSHASA.

Derna. Small coastal town in LIBYA, located east of BEIDA, near the foot of Green Mountain. A quaint town with abundant palm trees and jasmine, whose fragrance perfumes the air. Had a large Italian population from 1915 to 1970, which left Libya because of political pressures.

Also known as Darnah. Population in 2004 was 80,200.

Diego Garcia. Atoll in the CHAGOS ARCHIPELAGO, in the INDIAN OCEAN. Located at 7°20′ south latitude, longitude 72°25′ east, with an area of 11 square miles (28 sq. km.). Site of a United States naval base and space-tracking station.

Dire Dawa. City of eastern ETHIOPIA. Located about midway between ADDIS ABABA and its port, DJIBOUTI. Light manufacturing includes textiles, food processing, and cement production. Population was 440,000 in 2015.

Djanet Oasis. Large oasis in the SAHARA in southeastern ALGERIA, around which a thriving town has developed. Mud and brick houses are built on terraces on the steep hills. Its water supply comes mostly from underground springs. Population in 2008 was 14,655.

Djebel Chambi. Highest mountain in TUNISIA. Located in the west central part of the country, close to the Algerian border, it reaches a height of 5,066 feet (1,544 meters).

Djebel Telertheba. Peak in the HOGGAR MOUNTAINS of southeastern ALGERIA, in NORTH AFRICA. It rises to 8,054 feet (2,455 meters) west of DJANET. Sometimes snowcapped in winter.

Djemila. Ancient Algerian city on the MEDITERRANEAN SEA. The stone arch, temple, and forum of a Roman city stand as testimony to the Roman occupation of the area in ancient times.

Djerba Island. Large island in the MEDITERRANEAN SEA off TUNISIA's east coast, across the Gulf of GABÈS from the city of GABÈS. Legend has it that this island and KERKENNA, 86 miles (138 km.) to the north, are the remains of the lost continent of Atlantis. Djerba Island is also thought by some to be the legendary isle of the Lotus-Eaters in Homer's Odyssey.

Djibouti. Country located on the Gulf of Aden, in northeastern Africa. Total area of 8,958 square miles (23,200 sq. km.), about the size of the U.S. state of New Hampshire, with a population of 921,804 in 2020. Capital city is also called Djibouti, in which most of the country's people live. The land consists primarily of arid plateaus, with mountainous terrain to the north. The country's wettest region receives 15 inches (380 millimeters) of rain per year. Land use centers on grazing, although only one-tenth of the land is suited for this usage. Occasional oases produce some crops for local production. Many of the people are refugees from political unrest in SOMALIA and ETHIOPIA. Most of Djibouti's trade is centered on shipping, as it is the main channel of shipping for its landlocked neighbor, Ethiopia.

A donkey walks through the African savanna in the Republic of Djibouti. (U.S. Air Force photo by Master Sgt. Jeremiah Erickson/Released)

Dodoma. City in the interior of TANZANIA. Located at the intersection of the railway connecting Lake TANGANYIKA to DAR ES SALAAM and the central highway, which links southern Africa to EGYPT. A regional market for the agricultural district surrounding it. An administrative center for the region since its founding by the Germans in 1907. The legislative branch moved there from Dar es Salaam in 1996, although much of the government bureaucracy remains in Dar es Salaam. Population was 410,956 in 2012.

Douala. Main maritime port and a major industrial center in southwest CAMEROON. Located 130 miles (210 km.) west of YAOUNDÉ; connected to the rest of the country by road-and-railroad networks. Capital of German Kamerun (1884-1916) and later of the French portion of CAMEROON (1940-1946). Population was 2,768,400 in 2015.

Drakensberg Mountains. Mountain range on the southeastern coast of Southern Africa, about 700 miles (1,125 km.) long. Among its peaks are Thabana Ntlenyana, the highest in SOUTH AF-

RICA, 11,425 feet (3,482 meters) above sea level. The second-highest is Mont-aux-Sources at 10,823 feet (3,299 meters), where South Africa's scenic Royal Natal National Park is located.

Durban. SOUTH AFRICA's thirdfourth-largest city and its main INDIAN OCEAN seaport. Capital of KwaZULU/NATAL state. Had a 2018 population of 3,720,953. Summers are warm and wet, and winters are mild. The AGULHAS CURRENT warms the coastal waters, and vegetation is luxuriant. One large river, the Mgeni, flows into the sea there. Its harbor is the busiest in South Africa, and it is the center of KwaZulu-Natal's manufacturing industry. Portuguese navigator Vasco da Gama arrived there on December 25, 1487, and named the region Natal after the birth of Christ.

East London. Major port located in the eastern Cape Province of SOUTH AFRICA on the INDIAN OCEAN. Located at the mouth of the Buffalo River. Population was 267,007 in 2011; its primary economic activity is the export of agricultural products.

Eastern Province. See ORIENTALE PROVINCE

Edward, Lake. Lake in the GREAT RIFT VALLEY in east central Africa along the border between UGANDA and the DEMOCRATIC REPUBLIC OF THE CONGO. About 50 miles (80 km.) long, 26 miles (42 km.) wide, and 365 feet (110 meters) deep, covering 830 square miles (2,150 sq. km.). SEMLIKI RIVER is its outlet into Lake ALBERT to the north. Its western shore is within VIRUNGA NATIONAL PARK. Called Lake Idi Amin 1973-1979.

Egypt. Country in northeastern Africa. Total area of 386,662 square miles (9971,001,450 sq. km.), about two-thirds the size of the U.S. state of Alaska. Had a 2020 population of 104,124,440. Capital is CAIRO. Desert covers about 90 percent of the country; the rest includes the Nile Valley, scattered oases, and a narrow strip along the Mediterranean coast that receives 8 inches (200 millimeters) of rain during the winter months. Ninety-nine percent of Egypt's 104 million people live in the Nile Valley, one of the world's most densely populated areas. About 12 percent of the labor force is engaged in agriculture; 34 per-

cent is employed in industry; and 54 percent in services, Crude oil accounts for just under one-quarter of its exports, with plastics second. Tourism is another major industry.

El Aaiún. An international port in NORTH AFRICA and one of two cities in the disputed territory of WESTERN SAHARA. Located just south of the disputed border with MOROCCO. Many of the area's approximately 217,732 (2014) inhabitants live there.

El Bahira. Salt lake in NORTH AFRICA, near TUNIS's harbor. Connected to the MEDITERRANEAN SEA by a 20-foot (6-meter) channel that opened Tunis as a port.

El Borma. Small North African community in the southwest of TUNISIA near the Algerian border. Oil discovered there brought sudden, unexpected prosperity to this desert region.

Équateur Province. Province in northwest DEMOCRATIC REPUBLIC OF THE CONGO. Covers an area of 40,117 square miles (103,902 sq. km.). Capital is BANDAKA. The mean altitude is 1,200 feet (365 meters), and it is covered with dense equatorial forest. Has no railroads and is mostly agricultural.

Equatorial Guinea. Former Spanish colony in west equatorial Africa, astride the equator. Total area of 10,831 square miles (28,051 sq. km.) with a 2020 population of 1,402,985 most of whom are Roman Catholics. Capital is MALABO. The only African nation in which Spanish is the official language. The discovery of oil in 1996 transformed its economy. OPEC member since 2017.

Errachidia. Town in eastern MOROCCO, on the edge of the SAHARA. Its market does a brisk business with those who rove the Sahara, largely Bedouins who travel on camels seeking food, water, and grazing areas for their sheep. Population in 2014 was 92,374.

Erfoud. Desert community in the far eastern part of MOROCCO, almost due east of ESSAOUIRA on the eastern side of the High ATLAS MOUNTAINS. As the last stop before the sand dunes of MERZOUGA begin, it is an important supply center for Bedouins and other nomads who travel the desert looking for food, water, and grazing areas for their sheep and camels. Population in 2004 was 23,637.

Eritrea. Country in Africa, located on the RED SEA. A former province of ETHIOPIA, it gained its independence in 1993 after 30 years of civil war. Total area of 45,406 square miles (117,600 sq. km.), with a 2020 population of 6,081,196, 62 percent of whom live in rural areas. Capital is ASMARA. Plains are found along the coast and in the west; the center and southern regions are plateaus, giving way to mountains in the north. Most of the people are engaged in grazing or subsistence farming. A long-standing border dispute with ETHIOPIA was settled in 2018, and relations between the countries have improved.

Essaouira. Ancient Moroccan port city, located due west of MARRAKESH and north of AGADIR. A popular tourist destination for Moroccans, Europeans, and Americans. Population was 77,966 in 2014.

Eswatini. Small, landlocked state in Southern Africa. Known as Swaziland until 2018. Total area of 6,704 square miles (17,364 sq. km.) with a 2020 population of 1,104,479, most of Bantu origin. Capital is Mbanane. Most of it borders SOUTH AFRICA, with the exception of a border with MOZAMBIQUE in the north. Economy is dominated by subsistence farming, although it has significant iron and asbestos mines, as well as sugar-processing facilities. A British protectorate until 1968. Has one of the highest rates of AIDS infection in the world.

Ethiopia. Landlocked country in Africa. Total area of 426,372 square miles (1,104,300 sq. km.) with a population of 108,113,150 in 2020. Capital is ADDIS ABABA. More than half of the land is on the Ethiopian Plateau, which ranges from 3,280 to 5,510 feet (1,000 to 1,680 meters) in elevation. Bisected by the GREAT RIFT VALLEY. To the north are mountains (Ras Dashen is the highest at 15,157 feet/4,620 meters), and to the west an encroaching desert. Maximum rainfall on the plateau is 71 inches (1,800 millimeters) per year. Roughly 35 percent of the population engages in agriculture, virtually all in subsistence enterprises. Ethiopia is among the world's poorest nations, subject to disastrous droughts and the ex-

panding SAHARA, with little tax base to support the infrastructure necessary to begin intensive industrialization. Coffee comprises more than one-quarter In 2019, Abiy Ahmed, the Ethiopian prime minister, was awarded the Nobel Prize for Peace for ending 20 years of hostility with ERITREA and for enacting domestic reforms.

Shebelle River in Ethiopia. (David Castor)

Fernando Po. See BIOKO.

Frz. Historic inland city in the northern part of MOROCCO, and the most ancient of Morocco's four imperial capitals. In the late eighth century, the indigenous Berber tribesmen there were subdued and converted to the Muslim faith. Fez soon became home to Muslim refugees fleeing from Spain and TUNISIA. Modern Fez is three cities: the ancient city, 1,500 years old; New Fez, which was built on higher ground near the old city when the population was exploding; and the town built by the French in the twentieth century some distance from the older towns. The combined towns had a population of 1.412 million in 2017. Also spelled "Fès."

Fezzan. Desert in southwestern LIBYA. It is inhabited by nomads, who travel to find food, water, and grazing land for their flocks. Although it occupies 33 percent of Libya's landmass, it is home to about 6 percent of the country's population. It contains some large oases in which small communities have flourished, notably SABHA, Murzug, Umm Al-Aranib, and Ghat.

Francophone states. Countries of the MAGHREB that have been most influenced by the French. Most notable are ALGERIA and MOROCCO, in which French is the second language and is used for many governmental affairs.

Free State. Province in central SOUTH AFRICA originally known as the ORANGE FREE STATE, which took its name from a nineteenth century Boer republic. Capital city is BLOEMFONTEIN. Main physical feature is the DRAKENSBERG MOUNTAINS. Primarily an agricultural area with a temperate climate, but there also are gold and diamond mines. The University of the Free State is located in Bloemfontein. See also CONGO FREE STATE.

Freetown. Capital, chief port, and largest city in SIERRA LEONE. Located on the rocky Sierra Leone Peninsula. Population was 1.056 million in 2015. First settled by freed slaves in the late eighteenth century, including black slaves who fought on the side of the British during the American Revolution.

French Equatorial Africa. Created in the 1880s, a colonial administrative region that ceased to exist in 1960, when its component colonies became independent as the CENTRAL AFRICAN REPUBLIC, CHAD, Middle Congo (renamed the REPUBLIC OF THE CONGO) and GABON.

French West Africa. Established in 1895, a colonial administrative region that ceased to exist in 1960, when its component colonies became independent as DAHOMEY (later renamed BENIN), Ivory Coast (later known as CÔTE D'IVOIRE), MALI, MAURITANIA, NIGER, SENEGAL, and UPPER VOLTA (later renamed BURKINA FASO). GUINEA left the federation in 1958.

Funchal. Capital and chief port of a Portuguese district also named Funchal, which consists of the MADEIRA archipelago off the coast of Africa. Founded in 1421, Funchal had a population of 111,541 in 2011.

Gabès. Desert city established around a large North African oasis in TUNISIA. Although on the Mediterranean coast, it is extremely dry, more a desert city than a coastal one. Also known as Qābes. Population was 374,300 in 2014.

Gabès, Gulf of. Gulf formed by an indentation in the MEDITERRANEAN SEA, extending from SFAX to DJERBA ISLAND. Oil strikes there and in EL BORMA drastically changed TUNISIA's economy.

Gabon. Country in west Central Africa. Straddles the equator and is bounded by EQUATORIAL GUINEA and CAMEROON to the north, the REPUBLIC OF THE CONGO to the east and south, and the Atlantic Ocean to the west. Total area of 103,347 square miles (267,667 sq. km.) with a population of 2,230,908 in 2020. Capital is LIBREVILLE. Dense equatorial rainforest covers three-fourths of the country. Natural resources include woods and minerals. One of the world's largest producers of manganese. Rich oil deposits; Gabon joined OPEC in 2016, 21 years after terminating its original membership. Agriculture is a small part of the economy, and the transportation infrastructure is poor. Became independent from France in 1960. The island state of SÃO TOMÉ AND PRÍNCIPE is located offshore.

Gambia. Smallest nation in mainland Africa. Total area of 4,361 square miles (11,295 sq. km.) with a population of 2,173,999 in 2020. Capital is BANJUL. Mandinka, Wolof, and English are spoken. Divided into two by the GAMBIA RIVER and, apart from a small Atlantic Ocean coastline, completely surrounded by SENEGAL. Gambia and Senegal are allied in foreign affairs and defense but their cultural differences prevent unification.

Gambia River. West African river, rising in the Fouta Djallon in GUINEA and flowing west through the nations of GAMBIA and SENEGAL. Enters the Atlantic Ocean near St. Mary's Island,

BANJUL, the Gambian capital. It is about 700 miles (1,125 km.) long.

Garamba National Park. Park in northeast DEMOCRATIC REPUBLIC OF THE CONGO, near the SUDAN border. Established in 1938; covers about 1,900 square miles (4,920 sq. km.). Covered with high-grass savanna with occasional forest galleries. Known for its wildlife, including the white rhinoceros, eland, and giraffe.

Ghana. Country in West Africa, formerly called GOLD COAST by the British. Total area of 92,098 square miles (238,533 sq. km.) with a 2020 population of 29,340,248, nearly half of whom are Ashanti, who speak Akan. Capital and largest city is ACCRA. English is the official language.

Gibraltar. Small British territory on the Iberian Peninsula 14 miles (23 km.) north of the North African coast. In legend, the Rock of Gibraltar is one of the PILLARS OF HERCULES, the other being the Jebel Musa at CEUTA in SPANISH MOROCCO.

Giza. Home to EGYPT's great pyramids, and a suburb of CAIRO. Population was 8.8 million in 2018. The pyramids were built in the twenty-sixth century BCE; the largest, that of Cheops, ranks among the seven wonders of the ancient world. Giza is an educational center, the seat of Cairo University and a number of other institutes, and a manufacturing and industrial center.

Gold Coast. Name of GHANA when it was a British colony.

Great Rift Valley. A continuous fault in earth'sE crust that runs more than 3,000 miles (4,830 km.) through East Africa and Southwest Asia. At

The Great Rift Valley. (Peter Dowley)

its greatest width, it is more than 100 miles (160 km.) wide. Elevations vary from 6,000 feet (1,829 meters) above sea level in KENYA to 1,340 feet (408 meters) below sea level at the Dead Sea (between Israel and Jordan), the lowest point on Earth. Contains a chain of lakes, stretching from Lake NYASA in south-central Africa to the Sea of Galilee in Israel.

Guinea. Republic in West Africa, and the first colony in FRENCH WEST AFRICA to gain independence (1958). Total area of 94,926 square miles (245,857 sq. km.) with a 2020 population of 12,527,440. Capital is CONAKRY. French is the official language. Called French Guinea before independence. West Africa's three major rivers —GAMBIA, NIGER, and SENEGAL—originate in the plateau region. Administered within the federation of FRENCH WEST AFRICA during the era of colonial rule.

Guinea, Gulf of. An arm of the Atlantic Ocean, located in West Africa, between Cape Palmas at the southeastern edge of LIBERIA and Cape Lopez in GABON. Forms two bays: the Bight of BIAFRA and the Bight of BENIN.

Guinea-Bissau. Country located near the western tip of Africa, and the first Portuguese colony in Africa to gain independence (1974). Total area of 13,948 square miles (36,125 sq. km.) with a 2020 population of 1,927,104. Known by the name of its capital city, Bissau, to differentiate it from neighboring GUINEA. Portuguese is the official language. The offshore Bijagóso Archipelago is part of the country. The Fulani and the Balante make up the two largest ethnic groups. Called Portuguese Guinea before independence.

Harare. Capital and largest city of ZIMBABWE, with a population of about 1.6 million people in 2013. Founded as Fort Salisbury (later Salisbury) by a British force occupying the northeast part of the country in 1890. The country's main commercial, financial, and transportation center.

Hoggar Mountains. Mountain range in southeastern ALGERIA. Peaks rise to 9,852 feet (3,003 meters) at Mount TAHAT, the highest point in Algeria, and 8,054 feet (2,455 meters) at DJEBEL TELERTHEBA. Tamanrasset and Tazrouk are located among the range's jagged peaks. The Tropic of Cancer runs through the area, and it is quite hot in summer. The area is dry, and most of the trees that had grown at the higher altitudes have been harvested, leaving the mountains quite bare.

Horn of Africa. Protrusion of northeast Africa extending into the Indian Ocean, at the point where the Arabian Peninsula broke away from Africa. The tip of the Horn is part of Somalia.

Ibadan. Second-largest city in Nigeria (after Lagos). A Yoruba city, it is the capital of Oyo State. Located about 100 miles (160 km.) inland from the Atlantic coast. A commercial city, well served by roads and air routes, and the first African city to have a television station (1959). Population was 3.5 million in 2019.

Indian Ocean. Third-largest body of water on the Earth; forms the coastline of Southern Africa. A tropical ocean, little of which is north of the equator, and it has no inflow of cold water in its northern areas. It has a moderating effect on the southeastern coast of Africa. It is somewhat deeper than the Atlantic Ocean. The Limpopo and Zambezi rivers drain into it.

Inga Falls. Rapids in Central Africa, along the border between Democratic Republic of the Congo and the Republic of the Congo. The site of thirty-two rapids where the Congo River falls 850 feet (260 meters) in about 220 miles (355 km.). The world's largest hydroelectric-dam project is being built there, and will provide electricity to Kinshasa, Republic of the Congo, Katanga, and Zambia, more than 1,000 miles (1,600 km.) away.

Ivory Coast. See Côte d'Ivoire.

Johannesburg. Largest city in South Africa, commercial and financial capital of Africa, and the hub of South Africa's industrial and mining undertakings; capital of Gauteng Province. Its 2019 population was 5.6 million. Located 5,750 feet (1,753 meters) above sea level, it has warm summers and cold to mild winters. In 1886 the discovery of gold triggered one of the biggest gold rushes in the history of the world. Also called Igoli, a Zulu name meaning "City of Gold."

Jos. Capital of the Plateau State in the middle belt of Nigeria. Located on the Delimi River near the center of Jos Plateau, about 4,300 feet (1,300 meters) above sea level. Its average monthly temperature of 69° to 77° Fahrenheit (21° to 25° Celsius), cooler than any other city in Nigeria, attracts tourists. The center of the nation's tin mining industry. Population was 925,000 in 2016.

Jos Plateau. Tropical highland in northern Nigeria. It has an elevation of about 4,300 feet (1,300 meters) and covers an area of almost 3,000 square miles (7,800 sq. km.). Has cooler temperatures and more rainfall annually than the surrounding lowlands.

Kabompo River. River in Southern Africa that flows into the Zambezi from the east. Drains the watershed between the Zambezi and Congo River systems. The Lunga River, which flows into it, is more important for internal navigation than the upper Kabompo.

Kainji Dam. One of the largest dams in the world, located in West Africa. Constructed on the Niger River in western Nigeria at a cost of US$209 million, it opened in 1968 after four years of construction. It extends about 6 miles (10 km.) and has twelve hydroelectric turbines capable of generating 960,000 kilowatts. Supplies electricity to most cities in Nigeria and some neighboring states.

Kainji Lake. Reservoir on the Niger River in western Nigeria (West Africa) created by the Kainji Dam. About 84 miles (135 km.) long and 20 miles (30 km.) wide. Several villages were moved for the creation of the reservoir. The lake supports fishing and irrigation.

Kairouan. City in North Africa, on the Mediterranean Sea halfway down Tunisia's eastern coast. Founded around 670 ce. Its great mosque, completed in the ninth century bce, is a major attraction. Tourists are also drawn there for the hand-woven rugs. Also called al-Qayrawan.

Kalahari Desert. One of the world's major deserts, occupying parts of South Africa, Botswana, and NAMIBIA. Covers 360,000 square miles (932,000 sq. km.). Its appearance is one of sandy, red soil except during the rainy season, when large mud flats appear.

Kalemie. Port on the west shore of Lake TANGANYIKA, south of the outlet of the LUKUGA RIVER, in Central Africa. Located in KATANGA province in southeast DEMOCRATIC REPUBLIC OF THE CONGO. A commercial center and terminus of a railroad from LUBUMBASHI and KINDU. Commercial activities include fishing, textiles, and cement. Population was 146,974 in 2012. Called Albertville until 1966.

Kampala. Largest city, educational center, and political capital of UGANDA. Population was 1,680,600 in 2019. Located near Lake VICTORIA, the largest body of water in this landlocked country, and connected by rail to its primary port, Port Bell. Lake Victoria gives Kampala access to the INDIAN OCEAN port of DAR ES SALAAM, TANZANIA, via rail. Principal manufactured products are textiles, processed food, cement, and cigarettes. Built on the site of a fort founded by the British in 1890.

Kananga. Capital of KASAI OCCIDENTAL province in south central DEMOCRATIC REPUBLIC OF THE CONGO. Main commercial and communication center in the region. Population was 1,271,704 in 2015. Known as Luluabourg until 1972. A mutiny of African troops in the CONGO FREE STATE took place there in 1895.

Kano. Commercial center and capital of Kano State in northern NIGERIA. Located about 500 miles (805 km.) northeast of LAGOS. Population was 3,931,300 in 2016, the majority of whom were Hausa Muslims. It has the largest mosque in Nigeria.

Kariba, Lake. Reservoir forming part of the border between ZAMBIA and ZIMBABWE, in Southern Africa. Created by the Kariba Dam, which was built in the 1950s, it collects the waters of the ZAMBEZI RIVER. The lake is 170 miles (280 km.) long and fills Kariba Gorge; the concrete dam is 2,024 feet (617 meters) long and 419 feet (128 meters) high.

Karisimbi Mountain. Extinct volcano in east DEMOCRATIC REPUBLIC OF THE CONGO. Highest peak (14,787 feet/4,507 meters) of the VIRUNGA range. Located near the border with RWANDA, just northwest of MIKENO Mountain, about 20 miles (32 km.) northeast of Goma. The mountain's top is occasionally covered with snow.

Kasaï-Occidental. Province in south central DEMOCRATIC REPUBLIC OF THE CONGO. Covers 59,746 square miles (154,742 sq. km.) with a population of 5.3 million in 2010. Capital is KANANGA. The KASAI RIVER and its tributaries drain most of the province. The diamond mines of Tshikapa supply the international market with gems.

Kasaï-Oriental. Province in south central DEMOCRATIC REPUBLIC OF THE CONGO. Covers 3,685 square miles (9,545 sq. km.) with a population of 3.1 million in 2015. Capital is MBUJI-MAYI. The diamond mines of Mbuji-Mayi supply the international market with the industrial diamonds.

Kasai River. One of the principal left tributaries of the CONGO RIVER, running mainly in the DEMOCRATIC REPUBLIC OF THE CONGO. Rises on the central plateau of ANGOLA and forms part of the border between Angola and the Democratic Republic of the Congo in the south. Length is 1,338 miles (2,153 km.), of which about 250 miles (400 km.) is in Angola. Its lower course is called the Kwa. Navigable to about 490 miles (790 km.) upstream. Diamonds are washed in the southern Congolese section.

Katanga. Province in southeast DEMOCRATIC REPUBLIC OF THE CONGO. Covers 191,845 square miles (496,877 sq. km.). Population was 5.6 million in 2010. Capital is LUBUMBASHI. Katanga Plateau covers most of the region, with altitude varying between 3,000 and 6,000 feet (900 and 1,800 meters). The most industrialized region of the country, it was developed as a complex of mining and industrial towns as well as transportation and communications networks. Produces copper, tin, cobalt, manganese, zinc, and uranium. In the 1960s, it attempted to secede from the rest of the country.

Kenya. Country straddling the equator in East Africa. Covers 224,081 square miles (580,367 sq.

km.), slightly smaller than the U.S. state of Texas. In 2020, its population was 53,527,936, most of whom live in the south. Its capital city, NAIROBI, is also the financial capital of East Africa. Kenya has a narrow coastal plain that gives way to a series of plateaus and mountains. The highest region is in the center of the country, culminating in the extinct volcano Mount Kenya, Africa's second-highest peak at 17,058 feet (5,199 meters). The highlands are bisected by the Eastern RIFT VALLEY. The northern two-thirds of the country is either desert or steppe. Kenya is the most industrialized nation of East Africa. Its industry is dominated by food processing (especially tea and coffee) and textiles. Refined petroleum and cement are exported to other East African nations. Much of the labor force is employed in service industries, including tourism. However, most people are still engaged in subsistence agriculture.

Kerkenna Island. Large island off the coast of TUNISIA, across from SFAX on the east coast of the mainland. Legend has it that this and DJERBA ISLAND are the remains of the lost continent of Atlantis, which some people think was inundated by a huge tidal wave in ancient times.

Khartoum. Capital and transportation, economic, and educational center of SUDAN. Located at the heart of the Gezira, an enormous region of irrigation agriculture where the White and Blue Niles combine to form the NILE RIVER. Population was about 5.49 million in 2014. Industries include printing, food processing, textiles, and glass manufacturing. The area has had significant settlement for more than 6,000 years. Built in 1821 as an outpost for EGYPT's new southern empire, Khartoum became a British colony soon afterward.

Kigali. Largest city and political capital of RWANDA. Located on a highway that connects it to BURUNDI and UGANDA, its population was 859,332 in 2012. Textiles, processed foods, chemicals, and refined tin are produced there.

Kilimanjaro, Mount. Highest mountain in Africa, located in northeastern TANZANIA. A dormant volcano, with twin peaks situated nearly 6.8 miles (11 km.) apart, the higher at 19,330 feet (5,892 meters) above sea level, the lower at 17,564 feet (5,354 meters). Despite its tropical location, it is high enough to have a permanent snowcap. Its lower regions, owing to ancient lava flows, are quite fertile, producing coffee and plantains. Global warming is partially to blame for the mountain's retreating ice cap.

Kimberley. City in the northern Cape Province of SOUTH AFRICA. Diamonds were discovered there in 1871, and it is one of the largest sources of diamonds in the world. Population was 225,160 in 2011; its economy, while still based on diamond mining, includes iron mining and diamond finishing.

Kindu. River port and capital of MANIEMA province in east central DEMOCRATIC REPUBLIC OF THE CONGO. Located on both banks of the CONGO RIVER, about 390 miles (630 km.) south of KISANGANI. The Congo River is navigable from Kindu to about 100 miles (160 km.) south of Kisangani. Kindu is also a terminus of railroads from LUBUMBASHI and KALEMIE and an important transshipment point. Population was 172,321 in 2012.

Kinshasa. Largest city in Central Africa and the capital and commercial and industrial center of the DEMOCRATIC REPUBLIC OF THE CONGO. Located in the southwest of the country, directly across the CONGO RIVER from BRAZZAVILLE, the capital of the REPUBLIC OF THE CONGO. The terminus of navigation on the Congo River from KISANGANI, and the railroad terminus from MATADI. Population was 11.85 million in 2017. Founded in 1881 by Henry Morton Stanley. Called LÉOPOLDVILLE before 1966.

Kisangani. Capital of ORIENTALE PROVINCE in northeast DEMOCRATIC REPUBLIC OF THE CONGO, and an important commercial and distribution center. Straddles the CONGO RIVER. A major river port, the terminus of the railroad from the south, and the head of steam navigation to KINSHASA. Population was 1.6 million in 2015. Henry Morton Stanley first reached the area in 1877, and established the post five years later. City was called Stanleyville until 1966.

Kivu, Lake. Lake in east DEMOCRATIC REPUBLIC OF THE CONGO, along the border with RWANDA. Highest lake (altitude 4,790 feet/1,460 meters) of the GREAT RIFT VALLEY. It is 60 miles (96 km.) long, 30 miles (50 km.) wide, and 1,558 feet (475 meters) deep, covering about 1,040 square miles (2,700 sq. km.). Contains a large volume of dissolved methane gases. Main ports are BUKAVU and Goma.

Kivu, South. Province in east DEMOCRATIC REPUBLIC OF THE CONGO. It is bordered on the east by Lake KIVU and RWANDA. With altitudes varying from 3,000 to more than 10,000 feet (900 to 3,000 meters), the climate is mild and temperate, and vegetables can be grown. Rich in mineral resources, including gold and tin. Covers 25,120 square miles (65,070 sq. km.) with a population of 5.7 million in 2015.

Kolwezi. Industrial town in KATANGA province in southeast DEMOCRATIC REPUBLIC OF THE CONGO. Population was 572,942 in 2015. Located on the Lubumbashi-Lobito railroad about 80 miles (130 km.) from LIKASI and 145 miles (230 km.) from LUBUMBASHI. Major producer of copper and cobalt.

Kongo Central. Province in southwest DEMOCRATIC REPUBLIC OF THE CONGO. Covers 20,819 square miles (53,920 sq. km.). Population was 5.575 million in 2015. Bordered on the south by ANGOLA, on the west by the Atlantic Ocean and the Angolan enclave of CABINDA, on the north by the REPUBLIC OF THE CONGO.

Kumasi. Capital of the Ashanti region of GHANA. Located in central Ghana, in a dense forest belt. The commercial center for the rich cocoa-producing area. In 1961 Ghana's University of Science and Technology opened there. Population was 3.2 million in 2019.

Kwando River. Largest tributary of the ZAMBEZI RIVER in Southern Africa. Flows in a generally straight course to the southeast. For much of its course, it flows through vast swamps dotted with islands. Also known as the Chobe or the Linyanti River.

KwaZulu/Natal. South African province on the western side of the country on the INDIAN OCEAN. Capital is DURBAN; second major city is PIETERMARITZBURG, the site of the University of Natal. Climate is subtropical; the main agricultural product is sugarcane. The African population is mostly Zulu, and the dominant language is English. Population in 2016 was 11.1 million.

Lagos. Largest city and former national capital of NIGERIA. Located at the southwestern end of Nigeria's Atlantic coastline, it is the country's industrial and commercial center along the busiest seaport. The seat of the Lagos State government. Population was 21 million in 2016.

Las Palmas de Gran Canaria. Capital of Las Palmas province in the CANARY ISLANDS, located on the island of Gran Canaria. Its nearby port, Puerto de la Luz, is the most important in the Canary Islands. Founded in 1478, the city had a population of 378,517 in 2018.

Léopoldville. See KINSHASA.

Lesotho. An independent kingdom completely surrounded by SOUTH AFRICA. Total area of 11,720 square miles (30,355 sq. km.) with a 2020 population of 1,969,334. Capital and only large city is Maseru. The DRAKENSBERG mountain range dominates it and is the source of two important rivers, the Orange and the Tugela. Economy is predominantly agricultural. A large portion of its population is migratory, as many workers travel to South Africa to find employment. Created as the British protectorate of BASUTOLAND during the 1860s.

Liberia. Africa's oldest modern republic. Total area of 43,000 square miles (111,369 sq. km.) with a population of 5,073,296 in 2020. Capital is MONROVIA. Founded by the American Colonization Society, which purchased land in 1822 to create an African homeland for freed American slaves. Has the world's largest merchant navy because of its low taxes and registration of many foreign-owned ships. The Ebola virus claimed 3,500 lives between 2014 and 2015. A long period of political instability plagued the country until 2018.

Libreville. Capital and commercial and educational center of GABON. Located on Gabon's northwest shore and an important seaport on the Gulf of Guinea. A highly industrialized town, with sawmills, plywood and cloth-printing factories, brewing, and shipbuilding. Offshore oil was discovered north of the city in 1970s. Population was 703,904 in 2013. The capital of French Equatorial Africa from 1888 to 1904. Name means "free town."

Libya. The fourth-largest country in Africa. Located in North Africa, it covers 679,362 square miles (1,759,540 sq. km.), about two-and-a-half times the size of the U.S. state of Texas, with a 2020 population of 6,890,535. Bordered on the north by the Mediterranean Sea, on the west by Algeria and Tunisia, on the east by Egypt and Sudan, and on the south by Niger and Chad. Ninety percent desert, with no rivers. Its population has grown rapidly as a result of its low death rate, high birth rate, and prosperity brought by the oil boom that began in the 1960s. OPEC member since 1962. The twenty-first century was marked by political and social unrest in Libya.

Likasi. Industrial town in Katanga province in southeast Democratic Republic of the Congo. Located on the Lubumbashi-Lobito railroad about 65 miles (105 km.) northwest of Lubumbashi. The country's leading center of copper and cobalt refining; also has chemical factories manufacturing sulfuric acid, hydrogen chloride, sodium chlorate, and glycerin. Founded in 1917. Population was 447,449 in 2012.

Lilongwe. Capital of Malawi in Southern Africa. Had a 2020 population of 1,122,000. Founded in 1947; became a city in 1966, when the population was less than 20,000. Population increased as a result of an extensive government construction program. Located near the border of Mozambique and Zambia in the heart of Malawi's agricultural region on the Lilongwe River and is a market center for tobacco and other crops.

Limpopo River. River originating in northern South Africa that flows through, or forms borders of, South Africa, Botswana, Zimbabwe, and Mozambique. Enters the Indian Ocean at Delagoa Bay. It is about 1,000 miles (1,600 km.) long.

Lomami River. Left tributary of Congo River in central Democratic Republic of the Congo. Rises in the southeast of the country in Katanga province, about 15 miles (24 km.) west of Kamina, and flows north about 900 miles (1,450 km.) to join the Congo River about 70 miles (110 km.) west of Kisangani. Navigable upstream for about 250 miles (400 km.).

Lomé. Largest city and major port of Togo, on West Africa's Guinea Coast. Population was 837,437 in 2010.

Lorenço Marques. See Maputo.

Lualaba River. See Congo River.

Luanda. Capital, largest city, and chief port of Angola in Southern Africa. Had a population of about 2.57 million in 2019. Located on the west coast of Africa, along the Atlantic Ocean, it is one of the only cities along a stretch of more than 870 miles (1,400 km.) of Angolan coastline. Average temperatures range between 69.8° Fahrenheit (21° Celsius) in January and 60.8° Fahrenheit (16° Celsius) in June. Hasoil refineries, light industrial facilities such as foundries, sawmills, textile mills, cement, and food processing plants. Fishing is also important. Founded in 1576, it was the main center of Portuguese settlement in Angola.

Luapula River. River in southeast Democratic Republic of the Congo along the border with Zambia, in Central Africa. Originating from the south end of Lake Bangweulu, it flows south then north for 350 miles (560 km.), emptying into Lake Mweru. Considered the upper course of the Luvua River. Navigable upstream from Lake Mweru for more than 60 miles (96 km.).

Lubumbashi. Capital of Katanga province in southeast Democratic Republic of the Congo; the country's second-largest city and main commercial and industrial center. Located on a railroad near the border with Zambia, at an altitude of 5,008 feet (1,526 meters). Population was 1,794,118 in 2015. Established by Belgium as a copper-mining settlement in 1910; called Elisabethville until 1966.

Lufira River. Right tributary of the CONGO RIVER in southeast DEMOCRATIC REPUBLIC OF THE CONGO. Originates in the KATANGA highlands near the border of ZAMBIA, about 70 miles (110 km.) northwest of LUBUMBASHI; runs 300 miles (480 km.) northward to join the Lualaba River at Lake Kisale next to Lake UPEMBA. One of its many rapids is Mwadingusha Falls—the site of one of the largest hydroelectric plants in the country.

Lukuga River. Right tributary of the CONGO RIVER in southeast DEMOCRATIC REPUBLIC OF THE CONGO, and the only outlet for Lake TANGANYIKA. Originates from the west shore of Lake Tanganyika at KALEMIE and runs about 200 miles (320 km.) to empty its water in the Lualaba River.

Lusaka. Capital, chief administrative center, largest city, and the major financial, transportation, and manufacturing hub of ZAMBIA in Southern Africa. Had a population of 1,747,152 in 2010. Located on a high plateau in the south central region, at an altitude of 4,265 feet (1,300 meters). Established by European settlers in about 1905 as a small trading post.

Luvua River. Right tributary of the CONGO RIVER in southeast DEMOCRATIC REPUBLIC OF THE CONGO. Originates from the north end of Lake MWERU and flows 215 miles (345 km.) northwest to join the Lualaba River. Navigable for shallow-draught boats for about 100 miles (160 km.) upstream. A hydroelectric plant was built at Piana-Mwanga, about 120 miles (195 km.) upstream.

Madagascar. Fourth-largest island in the world, an INDIAN OCEAN island nation formerly known as the Malagasy Republic. Located about 250 miles (400 km.) across the MOZAMBIQUE CHANNEL from the coast of southeast Africa. MOZAMBIQUE is its closest mainland neighbor. The Tropic of CAPRICORN cuts across its southern half. Plains lie along the east and west coasts, the latter of which has several natural harbors. A range of highlands closely parallels the east coast, while an unusual channel, the CANAL DES PANGALANES, runs along the same coast. In the north, the Tsaratanana massif rises to Mount Maromokotro, at 9,436 feet (2,876 meters) the island's highest point. The island is drained by several rivers. The home of several highly sophisticated kingdoms, Madagascar was colonized by the French but became independent again in 1960. Madagascar stretches 994 miles (1,600 km.) from north to south, covers 226,658 square miles (587,041 sq. km.), and had a population of 26,955,737 in 2020. The island's economy is basically agricultural, and it exports large amounts of extremely high-quality rice and timber. A favorite place for naturalists, as its isolation has resulted in the evolution of exotic species that are found nowhere else. Although geographically tied to the African continent, its culture is closer to those of France and Southeast Asia. Capital is ANTANANARIVO.

Madeira. Archipelago located about 400 miles (645 km.) from the African coast of MOROCCO. The group's largest island, also called Madeira, rises to an elevation of 6,106 feet (1,861 meters). The northern part of Madeira Island receives most of the rainfall, which is distributed throughout the island by a system of tunnels and canals. The archipelago was uninhabited when it was discovered in the early fifteenth century. The group constitutes the Portuguese province of FUNCHAL, and the province's capital is also known as Funchal. Madeira is noted for its wine production, especially the fortified wine known as Madeira, and is a favorite among tourists for its genial climate. It covers 305 square miles (790 sq. km.) and had an estimated population of 289,000 in 2016.

Maghreb. Area of NORTH AFRICA between the MEDITERRANEAN SEA on the north and the SAHARA on the south. It includes MOROCCO, ALGERIA, TUNISIA, and parts of LIBYA. Most of its inhabitants have a common history, but have not been organized into a cohesive social or political entity.

Maiko National Park. Protected area in eastern DEMOCRATIC REPUBLIC OF THE CONGO, halfway between BUKAVU and KISANGANI. Covers about 4,202 square miles (10,885 sq. km.). The main vegetative cover is the dense equatorial forest. Animal species include gorillas, elephants, leopards, and okapi. The park is not open to tourism.

Majardah. River system in NORTH AFRICA that begins in ALGERIA and drains into the Gulf of TUNIS. The most important river system in Algeria and TUNISIA and the only river in these countries that has water year-round.

Malabo. Capital, largest city, and economic center of EQUATORIAL GUINEA. Located on the northern coast of BIOKO Island (formerly Fernando Po Island) near the equator. Founded in 1827 by the British as Port Clarence. Population was 297,000 in 2018.

Malagasy Republic. See MADAGASCAR.

Malawi. Republic in southeastern Africa. Total area of 45,747 square miles (118,484 sq. km.) with a 2020 population of 21,196,629. Capital is LILONGWE. Extends about 520 miles (840 km.) north to south and varies in width from about 50 to 100 miles (80 to 160 km.). Much of its surface is covered by Lake Malawi, the third-largest lake in Africa. Several plateaus lie to the east and west of Lake Malawi. Has a rainy season from November to April. Almost all of its wealth comes from agriculture. Formerly the British protectorate of Nyasaland.

Mali. Republic in West Africa. Total area of 478,841 square miles (1,240,192 sq. km.) with a population of 19,553,397 in 2020. Capital is BAMAKO. French is the official language. Contains the fabled city of Timbuktu, once known as the center of Islamic learning and commerce. Administered within the federation of FRENCH WEST AFRICA during the era of colonial rule.

Maniema. Province in east central DEMOCRATIC REPUBLIC OF THE CONGO. Covers 51,166 square miles (132,519 sq. km.), with a population of 2 million in 2010. Provincial capital is KINDU. Drained by the CONGO RIVER, called the Lualaba River there. Was the main base for the slave trade in the nineteenth century.

Maputo. Capital and largest city of MOZAMBIQUE in Southern Africa. Had a population of 1.08 million in 2017. Located on DELAGOA BAY, an inlet of the INDIAN OCEAN. Has an excellent harbor; a tourist industry has emerged in part because of the fine sand beaches. Lined with sand dunes and swamps. Climate is mainly tropical. Produces refined petroleum, building materials, clothing, and foods. A cyclone in 2019 devastated Mozambique. Maputo received its current name after Mozambique achieved independence in 1975; known as Lorenço Marques in colonial times.

Maradi. A major commercial and transportation center in the southern part of NIGER. Highways connect it to NIAMEY, Niger's capital, and the city of KANO in NIGERIA. Its chief products are cotton and groundnuts (peanuts). Population was 267,249 in 2012.

Marrakesh. Second-largest city in MOROCCO. Located inland toward the center of Morocco, west of the High ATLAS MOUNTAINS, with a 2014 population of 928,850. It began as a town of tents in an oasis during the eleventh century and has long been a trading center. It is considered the unofficial capital of the southern part of the country.

Mascarene Islands. Small group of islands in the INDIAN OCEAN east of MADAGASCAR that includes MAURITIUS, RODRIGUEZ, and RÉUNION. Named after Pedro Mascarenhas, a Portuguese visitor of the early sixteenth century.

Matadi. Capital of KONGO CENTRAL province in southwestern DEMOCRATIC REPUBLIC OF THE CONGO; the country's major port and a major commercial center. Located about 100 miles (160 km.) from the Atlantic port of Banana and is the farthest point reached by ocean ships. Also the head of the railroad to KINSHASA. Population was 306,053 in 2007. Founded in 1879 by Henry Morton Stanley.

Mauritania. Islamic republic in West Africa, heavily dependent on aid from France. Total area of 397,955 square miles (1,030,700 sq. km.) with a 2020 population of 4,005,475, most of whom are Moors and many of whom are nomads. Capital is Nouakchott. About 80 percent of the country is located in the SAHARA. Islam is the official religion and Arabic is the official language. Administered within the federation of FRENCH WEST AFRICA during the era of colonial rule. Located directly south and east of the disputed WESTERN SAHARA territory (formerly Spanish Sahara). To its east is the southern part of ALGERIA.

Mauritius. Nation of volcanic islands in the INDIAN OCEAN east of MADAGASCAR, part of the MASCARENE group. The nation's capital is PORT LOUIS. The nation of Mauritius includes Mauritius Island, RODRIGUEZ, the Agaléga Islands, and the Cargados Carajos Shoals. Although the climate is tropical monsoonal, irrigation is necessary for organized agriculture. Mauritius was the home of the dodo, a large flightless bird slaughtered to extinction for its meat. It was a British colony until its independence in 1968. The islands' total area is 788 square miles (2,040 sq. km.), and the population was 1,379,365 in 2020.

Mayotte. African island in the COMOROS chain. Because of the island's long association with France and its stronger economic base, its citizens chose to remain attached to France when the rest of the chain declared independence in 1975. Mayotte and several nearby islets cover an area of 144 square miles (374 sq. km.). In 2017 the population was 256,518.

Mbandaka. Capital of ÉQUATEUR PROVINCE in northwest DEMOCRATIC REPUBLIC OF THE CONGO. Located on the left bank of the CONGO RIVER, about 370 miles (595 km.) north northeast of KINSHASA, the nation's capital. A commercial and river communications center. Founded by Henry Morton Stanley in 1883; called Coquilhatville until 1966. Population in 2012 was 345,663.

Mbuji-Mayi. Industrial town and capital of KASAI ORIENTAL province in south central DEMOCRATIC REPUBLIC OF THE CONGO. Population was 3.368 million in 2015. It became a mining town after the discovery of diamonds in the area in 1909, and the region produced up to 10 percent by weight of the world's industrial diamonds until 1990.

Mediterranean Sea. Large body of water that separates NORTH AFRICA from EUROPE. It takes its name from Latin words meaning "in the middle of land"—a reference to its nearly landlocked nature. Covers about 970,000 square miles (3 million sq. km.) and extends 2,300 miles (3,700 km.) from west to east and about 1,000 miles (1,600 km.) from north to south at its widest. Its greatest depth is 16,706 feet (5,092 meters). Major African ports on the Mediterranean include ALGIERS, ALEXANDRIA, BIZERTE, ORAN, PORT SAID, SFAX, TRIPOLI, and TUNIS.

Meknès. City in MOROCCO, located 50 miles (80 km.) directly inland from RABAT, with a 2014 population of 632,079. Renowned for the seventeenth century Bab el-Mansour, one of the most elaborate and richly decorated of Morocco's city gates. Because of its excellent architecture, Meknèshas been called the Versailles of Morocco. In its prime, Meknès had fifty palaces and 16 miles (26 km.) of protective walls with twenty gates.

Melilla. A Spanish community on a cape extending from MOROCCO into the MEDITERRANEAN SEA close to the Algerian border. One of two Spanish possessions in NORTH AFRICA (the other being CEUTA), which are surrounded by Morocco and have been considered by Spain for autonomous status. Population in 2017 was 86,120.

Merzouga. Desert outpost east of ERFOUD, in NORTH AFRICA, where the sand dunes of the SAHARA begin. There, sand drifts over the few roads that exist, much as snow drifts over roads in blizzard conditions in America's midwestern states.

Mikeno. Extinct volcano in east DEMOCRATIC REPUBLIC OF THE CONGO. Second-highest peak (about 14,600 feet/4,450 meters) of the VIRUNGA range. Located near the RWANDA border, just northwest of KARISIMBI MOUNTAIN, in VIRUNGA NATIONAL PARK.

Mobutu, Lake. See Lake ALBERT.

Mogadishu. Capital and largest city of SOMALIA, Africa. Founded around 900 by Arab traders. Until the outbreak of civil war and the ensuing anarchy that paralyzed the country in the 1990s, it was the chief economic center for the nation. The city continues to be wracked by extreme violence. Population was 2,590,000 in 2017.

Mombasa. Second-largest city and the primary port of Kenya. Population was 1,208,333 in 2019. The most significant port of East Africa, with rail connections to Lake Victoria and Dar es Salaam, and roads to most other Indian Ocean ports of the region. Primary ocean port for landlocked Uganda and Rwanda. Founded in the eighth century by Moslem traders.

Monrovia. Capital, chief port, and most developed city in Liberia. Founded in 1822 and named after U.S. president James Monroe. Located at the entrance of the Mesurado River on the Atlantic coast. The city's population reached over 1 million people in 2018; however, it was devastated by civil war during the 1990s and again from 2003 until 2006. Epicenter of the world's largest Ebola outbreak in 2014.

Morocco. A kingdom in North Africa; part of the Maghreb. Total area of 172,414 square miles (446,550 sq. km.). Had a 2020 population of about 35.7 million. Capital is Rabat; largest city is Casablanca. Bordered on the north by the Mediterranean Sea and the Atlantic Ocean, on the west by the Atlantic Ocean, on the east by Algeria, and on the south by Western Sahara. The closest North African country to Europe, being just 14 miles (22 km.) from Gibraltar. Rabat and Casablanca are on the Atlantic Ocean; other important communities—Fez, Marrakesh, Meknes, and Quarzazate—are inland. The Atlas Mountains run down the center of Morocco, and the Rif Mountains run east-west in the north.

Moroni. Capital of the African island nation of Comoros, on Njazidja (also known as Grande Comore) Island. It has been an important regional trading center for centuries. Its population was 54,000 in 2011.

Mount Cameroon. Active volcano and the highest mountain in West Africa; elevation of 13,350 feet (4,070 meters). Located in southwestern Cameroon near the Gulf of Guinea. The rich soils of the mountain slope make it ideal for agriculture. Abundant rainfall makes this area the wettest in Africa.

Mozambique. Southern African republic, directly to the northeast of South Africa. Total area of 308,642 square miles (799,380 sq. km.), with a 2020 population of 30,098,197. Capital is Maputo. Most of the country is coastal lowland with plateaus inland that rise to nearly 8,000 feet (2,436 meters), which is the height of Mount Binga near the western border. Its main river is the Zambezi River, which is dammed behind the Cahora Bassa Dam. The Ruvuma, Save, and Limpopo rivers are also important. Became independent from Portugal in 1975.

Mozambique Channel. Channel of the Indian Ocean stretching between Madagascar and the coast of southeast Africa, near Comoros. It is 186 miles (300 km.) wide at its narrowest point. Several live coelacanths, primitive fish thought to have been extinct for 70 million years, have been caught in this channel.

Mwanza. A major port city on Lake Victoria in Tanzania. Population was 2.7 million in 2012. Products are shipped there from neighboring landlocked Uganda and Burundi for rail transport to Dar es Salaam on the Indian Ocean. Major manufacturing industries are cotton textiles, meat packing, and fishing.

Mweru, Lake. Lake in southeast Democratic Republic of the Congo, along the border with Zambia. It is about 70 miles (110 km.) long, 30 miles (50 km.) wide, and 30 to 50 feet (9 to 15 meters) deep, covering about 700 square miles (100 sq. km.). Receives water from the Luapula River from the south. Its outlet is the Luvua River, which drains into the Congo River. Navigable by small steamboats and barges. Its main ports in the Democratic Republic of the Congo are Kilwa and Pweto.

Naftah. See Nefta.

Nairobi. Capital and largest city of Kenya, and the dominant commercial and financial city of East Africa. Population was more than 4.3 million in 2019. Founded in 1899 by the British and quickly became the colonial administrative center. Located 5,495 feet (1,675 meters) above sea level, it has a relatively cool climate, ample water reserves, and a well-developed infrastructure. Tourism, manufacturing, textile and food processing, and several export industries fuel its economy. Its downtown area is the most modern of any East African city. Economic disparities are evident in the economic segregation of traditional neighborhoods and the existence of shantytowns.

Nakuru. A major transportation center in the GREAT RIFT VALLEY, in the highland region of west central KENYA, near Lake Nakuru. A center for food processing and textiles. Population was 570,674 in 2019.

Namibia. Republic on the Atlantic side of Southern Africa. Total area of 318,261 square miles (824,292 sq. km.) with a 2020 population of 2,630,073. Capital is WINDHOEK. Shares a long border with SOUTH AFRICA, which has long tried to dominate it. Climate is arid; the country includes the forbidding KALAHARI DESERT. Rich in mineral resources; diamonds and other gems are its main exports. Formerly West Africa.

Nasser, Lake. Reservoir in Africa, 350 miles (560 km.) long, created by the building of the Aswan High Dam. Many significant sites disappeared under the water; however, the temples of Abu Simbel—constructed by Rameses II in the thirteenth century BCE—were saved. This was a magnificent technological achievement, as the massive structures were disassembled, moved, and reassembled on higher ground. Named for the leader of EGYPT when the project was undertaken, Gamal Abdel Nasser.

N'Djamena. Capital and largest city of CHAD, with a population of about 951,418 in 2009. Located on the Chari River, at the country's border with northeastern NIGERIA, near Lake Chad.

Nefta. One of the largest oases in TUNISIA. Located in the southwestern part of the country on the shores of SHATT AL-JARID, Tunisia's largest lake, which is often dry because of lack of rain and desert conditions. Nefta is 25 miles (38 km.) from another large oasis, TOZEUR, and is just

slightly farther than that from the Algerian border. Also called Naftah.

Niamey. Capital and largest city of NIGER. Located in the southwestern part of the country on the banks of the NIGER RIVER. Construction of the Kennedy Bridge in 1970 led to the expansion of the city. Population was 1.24 million in 2018.

Niger. Country in West Africa, named after the famous river that flows through part of the country. Total area of 489,191 square miles (1,267,000 sq. km.) with a 2020 population of 22,772,361, most of whom are Muslims. Capital is NIAMEY. Official language is French. Administered within the federation of FRENCH WEST AFRICA during the era of colonial rule.

Niger River. Third-longest river in Africa—2,600 miles (4,185 km.). Rising in the GUINEA highlands near the SIERRA LEONE border, it flows first northeast, then east, and finally southeast across NIGERIA into the Gulf of Guinea. A critical source of water for MALI and NIGER. BAMAKO, capital of MALI, and NIAMEY, capital of NIGER, are located on its banks.

Nigeria. Most populous country in Africa; located on Atlantic coast in West Africa. Total area of 356,669 square miles (923,768 sq. km.) with a population of 214,028,302 in 2020. Capital is ABUJA. English is the official language, but there are 11 major indigenous languages in 240 dialects. The three major ethnic groups are the Yoruba in the west, the Ibo in the east, and the Hausa-Fulani in the north. A major exporter of petroleum and a member of the Organization of Petroleum Exporting Countries (OPEC). since 1971 After achieving independence in 1960, the country waslargely ruled by the military until late 1999, when a civilian government was elected Largest economy in Africa in the third decade of the twenty-first century.

Nile River. The longest river in the world, the Nile has two main branches. The White Nile, its major branch, begins in Lake VICTORIA, travels through UGANDA, SUDAN, SOUTH SUDAN, and EGYPT to the MEDITERRANEAN SEA, a distance of about 4,180 miles (6,725 km.). Near KHARTOUM it is joined by the Blue Nile, which rises in Ethiopia's Lake Tana. Much of the Nile's course takes it through the eastern SAHARA, bringing life to an otherwise inhospitable land. Prosperity in Egypt and northern Sudan is possible only because of this river, which has made both countries major agricultural producers. From the city of Edfu in Upper Egypt ("upper" refers to southern Egypt, up the Nile River) to the beginning of the delta, the river valley averages 14 miles (23 km.) in width. The delta itself is 150 miles (250 km.) wide at the sea. Until the erection of twentieth century dams, the Nile flooded this region annually. Flood cycles reinvigorated the soils with silt, brought fish beyond its banks, and provided enough excess water to be stored for use later in the agricultural year. The damming of the Nile interrupted the flow of water. The Nile was also the prime transportation route for traders and people who lived along its banks. The cataracts, or rock waterfalls, that appear in the river helped insulate early Egyptians from invasion from the south. In 2015, Egypt, Ethiopia, and Sudan signed an agreement stating how they would share water from the river.

North Africa. The area lying south of the MEDITERRANEAN SEA, east of the Atlantic Ocean, north of the SAHARA, and west of EGYPT and SUDAN. It comprises four independent nations, MOROCCO, ALGERIA, TUNISIA, and LIBYA. Its countries are a mixture of Islamic and Western cultures, although the majority religion in the area is Islamic.

Northern Rhodesia. Colonial-era name of ZAMBIA.

Nyamuragira. Active volcano of the VIRUNGA MOUNTAINS in east DEMOCRATIC REPUBLIC OF THE CONGO. Located about 20 miles (32 km.) north of Lake KIVU in the southern part of VIRUNGA NATIONAL PARK. More than 10,000 feet (3,000 meters) high. Most recent eruptions were in 2006 and 2010.

Nyasa, Lake. One of the chain of lakes in the GREAT RIFT VALLEY. Bordered by MOZAMBIQUE, ZAMBIA, and TANZANIA, it covers 8,683 square miles (22,490 sq. km.).

Nyasaland. See MALAWI.

Nyiragongo. Active volcano of the VIRUNGA MOUNTAINS in east DEMOCRATIC REPUBLIC OF THE CONGO. Located near the RWANDA border, northeast of Lake KIVU, about 12 miles (20 km.) north of the city of Goma. A 2002 eruption devastated the city. About 11,400 feet (3,475 meters) high.

Okavango. Southern African river that flows through NAMIBIA and BOTSWANA. About 1,000 miles (1,600 km.) long. It originates in upper ANGOLA and eventually merges with the Cuito River. In the lowlands, it forms the Okavango Delta, which is basically a swamp and home to a number of unique plants. Named after the Okavango people, who inhabit Namibia.

Olduvai Gorge. Region of the GREAT RIFT VALLEY in northern TANZANIA, which has been one of the two most productive sites for uncovering early hominid life on Earth. (The other is Lake TURKANA, also in Africa.) It is about 50 miles (80 km.) long, nearly 300 feet (91 meters) deep.

Omdurman. Largest city in SUDAN. Population was 2,395,159 in 2008. Situated on the NILE RIVER, just north of the confluence of the Blue Nile and White Nile rivers. Located within the Gezira, the major irrigated agricultural region of SUDAN, it shares its metropolitan area with the capital, KHARTOUM.

Oran. Second-largest city in ALGERIA, located 225 miles (360 km.) west of ALGIERS, on high cliff plateaus that plunge to the MEDITERRANEAN SEA. Has more European influence than other Algerian cities. More Christian than Muslim, it once had more cathedrals than mosques. Population in 2010 was 1.5 million.

Orange Free State. Former name of SOUTH AFRICA's FREE STATE province.

Orientale Province. Province in north and northeast DEMOCRATIC REPUBLIC OF THE CONGO. Covers an area of 204,164 square miles (529,397 sq. km.). Bounded on the north by the CENTRAL AFRICAN REPUBLIC, the northeast by SOUTH SUDAN, and the east by UGANDA and Lake ALBERT. Primarily equatorial rainforest; drained by the Uele-Kibali, Aruwimi-Ituri, and CONGO rivers. Produces palm oil, coffee, cocoa, rubber, quinine, and timber. KISANGANI is the provincial capital. Population was 8,197,975 in 2010.

Ouagadougou. Capital and largest city in BURKINA FASO. Founded in the fifteenth century as the capital of the Mossi Kingdom. Known for its prestigious national museum, major craft centers, and textiles. Population was 2.2 million in 2015.

Ouarzazate.. North African desert outpost. Located inland from AGADIR and ESSAOUIRA in MOROCCO, between the high ATLAS MOUNTAINS and the SAHARA.

Pemba. Population in 2014 was 71,067. East African offshore island located a few miles north of ZANZIBAR, from which it was historically ruled. World's leading source of cloves. Population in 2012 was 406,808.

Philippeville. See SKIKDA.

Pico de Teide. Volcano on Tenerife in Spain's CANARY ISLANDS that erupted in 1909. At 12,198 feet (3,718 meters) it is the highest point in Spain and in the ATLANTIC OCEAN. By contrast, Alaska's Denali, the highest peak in North America, rises to 20,3201 feet (6,190 meters).

Pietermaritzburg. City in KWAZULU/NATAL in SOUTH AFRICA. Located on the Msunduzi River near DURBAN. Had a 2011 population of 475,238 and is the site of the University of Natal. Founded in 1838 by Boers fleeing British domination in the Cape Colony.

Pillars of Hercules. Name given to GIBRALTAR on the Iberian Peninsula and Jebel Musa in CEUTA, the northernmost reach of MOROCCO. According to classical mythology, Hercules formed them when he tore a mountain apart to get to Cádiz on Spain's southwest coast.

Pointe-Noire. Main maritime port in southwest REPUBLIC OF THE CONGO. Located about 250 miles (400 km.) west of BRAZZAVILLE. A commercial and industrial center and the terminus of the railroad from Brazzaville. It has a refinery, which processes offshore petroleum. Population was 715,334 in 2007. Capital of FRENCH EQUATORIAL AFRICA 1950-1958.

Port Elizabeth. Major South African port on the INDIAN OCEAN. Located on the southeastern coast of the country, equally close to CAPE TOWN in the south and DURBAN. Also a major rail, airline, and manufacturing center. Has an extensive tourist industry and is close to the Addo Elephant National Park. Named after Elizabeth Donkin, the wife of Rufane Donkin, the governor of the Cape Colony in the 1820s. Population in 2020 was 967,677.

Port Harcourt. Capital of Rivers State in southeast NIGERIA. Located on the Bonny River in the Niger delta. A major industrial center, with exports including petroleum and palm products. Population was 1,865,000 in 2020. In 1915 the British established a port in the city that served eastern and parts of northern Nigeria.

Port Louis. Capital and largest city of MAURITIUS, in the INDIAN OCEAN. Founded in 1735, it has long been an important regional trading center. It had a population of 147,066 in 2018.

Port Said. Resort and fueling station for ships, situated at the juncture of the SUEZ CANAL and the MEDITERRANEAN SEA. Population was 603,787 in 2010. Founded in 1859 as the canal was being constructed. Also a manufacturing center, producing chemicals, processed food, cotton, and cigarettes.

Port Sudan. The only RED SEA port in SUDAN. Population was 489,725 in 2007. It exports cotton, sheep, cattle, and gum arabic.

Porto-Novo. Capital and second-largest city of BENIN. Its busy port is located on a lagoon that connects to the Gulf of GUINEA. Population was 264,320 in 2013.

Portuguese East Africa. See MOZAMBIQUE.

Portuguese Guinea. See GUINEA-BISSAU.

Praia. Capital and largest city of CABO VERDE, built on cliffs overlooking a small natural harbor with its historical museums and other landmarks. The population was 159,050 in 2017.

Pretoria. Administrative capital, transportation and manufacturing center, and fourth-largest city in SOUTH AFRICA; capital of TRANSVAAL Province. Located northeast of JOHANNESBURG. South Africa's two largest universities, the University of Pretoria and the University of South Africa, are located there. Founded in 1855 by

Marthinus Pretorius, the first president of South Africa, who named it for his father, Andries Pretorius, a Boer military leader. Population in 2018 was 2,472,612.

Príncipe. Small African island, part of the nation of São Tomé and Príncipe. Located in the Gulf of Guinea at 1°37′ north latitude, 7°25′ east longitude. It has an area of 45 square miles (117 sq. km.) and a population of 8,420 in 2018.

Qābes. See Gabès.

Quarzazate. North African desert outpost. Located inland from Agadir and Essaouira, between the high Atlas Mountains and the Sahara. Many desert motion pictures have been filmed there. Although somewhat isolated, it is a trading center for Bedouins who wander the Sahara.

Rabat. Capital of Morocco and administrative center of government when the French controlled Morocco. Located on the Atlantic Ocean between Casablanca to the south and Tangier to the north. Many of its 2.1 million inhabitants (2014) are employed in government positions.

Ras Lanuf. Small city on the southwest shore of the Gulf of Sidra, in North Africa, with an oil refinery capable of producing 220,000 barrels of oil a day. Also called Ra's al-Unuf.

Red Sea. Body of water separating northeastern Africa from the Arabian Peninsula. An extension of the Great Rift Valley, formed when the plate on which the Arabian Peninsula is situated tore loose from Africa. It is 157 miles (253 km.) long, and up to 220 miles (354 km.) wide. An important conduit for oil tankers moving from the Arabian Peninsula and Persian Gulf destined for Europe via the Suez Canal, which connects the Red Sea to the Mediterranean Sea.

Réunion. Island in the Indian Ocean east of Madagascar, part of the Mascarene group. Its capital is Saint-Denis. Its highest elevation is 10,069 feet (3,069 meters). Piton de la Fournaise, at a slightly lower elevation, is an active volcano. Réunion is an overseas department of France. Its area is 972 square miles (2,517 sq. km.), and its population was 859,959 in 2020.

Republic of the Congo. Country in west Central Africa; also known as Congo-Brazzaville. Total area of 132,047 square miles (342,000 sq. km.) with a population of 5,293,070 in 2020. Bounded on the west by the Atlantic Ocean, the south by Angola, on the southeast and east by the Democratic Republic of the Congo, on the northeast by the Central African Republic, on the northwest by Cameroon, and on the west and southwest by Gabon. Brazzaville is the capital and most-developed city. Had the first communist government in Africa, but rejected communism in the early 1990s. French is the official language. Administered within French Equatorial Africa during the colonial era, which ended in 1960. Has substantial oil and natural gas reserves; joined OPEC in 2018.

Rhodesia. See Zimbabwe.

Rhumel River. North African river that runs through Constantine, in Algeria's northeastern area near Tunisia, 50 miles (80 km.) inland from the Mediterranean Sea. It has created deep gorges below the chalky cliffs that it has carved out through the centuries.

Rif Mountains. North African mountains running along the coast of the Mediterranean Sea north of the Atlas Mountain chain. They rise steeply and dramatically from the sea in extreme northwestern Morocco, reaching heights of 8,000 feet (2,400 meters).

Rift Valley. See Great Rift Valley.

Rodrigues. Rodriguez Island surrounded by a coral reef in the Indian Ocean, part of the nation of Mauritius. Its highest elevation is about 1,300 feet (400 meters), and it has an area of 40 square miles (104 sq. km.). Population in 2014 was 41,669.

Ruzizi River. River in Central Africa along the borders of the Democratic Republic of the Congo, Rwanda, and Burundi. The outlet of Lake Kivu into Lake Tanganyika. About 100 miles (160 km.) long, with many rapids in its upper course. A hydroelectric power plant near Lake Kivu supplies electricity to Bukavu and the surrounding region.

Rwanda. Landlocked country in Africa. Total area of 10,169 square miles (26,338 sq. km.), with a population of 12,712,431 in 2020. Capital is

KIGALI. The western VIRUNGA MOUNTAINS—with their highest peak, the extinct volcano KARISIMBI, at 14,826 feet (4,519 meters)—slope into a hilly plateau in the center of the country, in which most of the people live. More than 80 percent of the population is rural, virtually all engaging in subsistence agriculture. Like its neighbor BURUNDI, Rwanda has had ethnic strife between the minority Tutsis and the more numerous Hutus.

Rwenzori Mountains. Nonvolcanic mountain range in Central Africa. Located between lakes EDWARD and ALBERT, along the border between UGANDA and the DEMOCRATIC REPUBLIC OF THE CONGO. About 30 miles (50 km.) wide, it extends north-south for 80 miles (130 km.). Its highest point is 16,763 feet (5,109 meters). Snow cover can be found from 14,000 feet (4,270 meters). The entire base has dense equatorial forests with several vegetation zones succeeding each other in altitude. The western side of the range is in Rwenzori National Park.Sabha. Largest oasis in LIBYA, with a permanent population in 2012 of nearly 100,000. Located in the SAHARA, it grew rapidly in the 1960s, when large deposits of oil were discovered in the desert. Irrigation canals were built to bring in water to sustain the workers needed for economic growth. New schools, hospitals, roads, and an airport were also built.

Sahara. Largest desert in the world. Covers 3.6 million square miles (9.3 million sq. km.)—almost the same area covered by the entire United States—and it continues to grow. The desert extends about 3,000 miles (4,830 km.) from the RED SEA to the Atlantic Ocean, and about 745 miles (1,200 km.) from the coast of the MEDITERRANEAN SEA south to the SAHEL, a dry steppe region across West and Central Africa. The desert has three distinct divisions: the Libyan Desert (also known to Egyptians as the Western Desert) extends from the west bank of the NILE RIVER into eastern LIBYA; the Arabian Desert (also known to the Egyptians as the Eastern Desert) runs from the east bank of the Nile River to the Red Sea, roughly to the border of SUDAN; the Nubian Desert lies to the south of the region, en-compassing most of northern Sudan. The effect of high-pressure systems keeps out almost all rain; most of the region receives less than 5 inches (127 millimeters) of rain per year. Aside from the Nile, only occasional oases provide water in the region. Wadis (seasonal river beds) are found near coastal areas where rainfall occurs. The Sahara isalso among the world's hottest places.

Saharan Atlas Mountains. One of two mountain ranges in NORTH AFRICA that divide ALGERIA into three physical regions. Runs from Colomb-Bechar in the west to Annaba in the east.

Sahel. Region in West Africa that is a transition zone between the dry SAHARA in the north and the wet tropical areas in the south. Annual rainfall ranges between 4 and 8 inches (102 and 203 millimeters). Most of the rain falls between June and September. Peanuts and millet are common crops in this region; there is also nomadic herding.

Sahil. A stretch of the coastline of TUNISIA that makes an S-curve to the south, running from the Gulf of TUNIS past CAPE BON. There are two busy harbors in this area. Sahil comes from the same Arabic word for "coast" that gave East Africa's Swahili language its name.

Saint-Denis. Port and capital of the African island of RÉUNION. Located at the mouth of the Saint-Denis River. The population in 2016 was 147,920.

Saint Helena. Island in the South ATLANTIC OCEAN, about 1,200 miles (1,930 km.) from Africa. The capital is Jamestown. Located at 15°58′ south latitude, longitude 5°43′ west. Discovered by Portuguese explorers in 1502, the island is a British dependency. Saint Helena was the final residence of French Emperor Napoleon Bonaparte, who was exiled there 1815-1821 after his defeat by the British at Waterloo. The island's area is 47 square miles (122 sq. km.), and its highest elevation is 2,685 feet (818 meters). In 2016 its population was 4,534. Saint-Louis. Seaport and commercial center for the Senegal River Valley. Located in northwestern SENEGAL, on Saint-Louis Island at the mouth of the SENEgal River. Established by the French in 1659 as a trading post, it is one of the oldest European settlements in West Africa. Population was 176,000 in 2005.

Salonga National Park. Protected area in central DEMOCRATIC REPUBLIC OF THE CONGO. Located halfway between KINSHASA and KISANGANI. Its two sections form the largest reserve in the country, covering 8.9 million acres (3.6 million hectares) Created in 1970. Dense equatorial forest is the main vegetative cover. Among the park's animal and bird species are parrots, elephants, antelope, monkeys, and a unique species of pygmy chimpanzee.

Santa Cruz de Tenerife. Capital of the province of the same name in the CANARY ISLANDS, located on the island of Tenerife. The population was 204,856 in 2018.

São Tomé and Príncipe. The smallest nation in or near Africa, consisting of two small volcanic islands lying just north of the equator in the Gulf of GUINEA. The capital is the city of São Tomé, with a population of 71,868 (2015). A former Portuguese colony, São Tomé and Príncipe became an independent country in 1975. Its main exports have been cocoa, coconut, and coffee, but it has been encouraging tourism to make up for failing production. It covers 372 square miles (964 sq. km.), an area less than one-third the size of the U.S. state of Rhode Island. Its population in 2020 was 211,122.

Semliki River. River in northeast DEMOCRATIC REPUBLIC OF THE CONGO. Starts from Lake EDWARD and runs 130 miles (210 km.) north to Lake ALBERT. Part of its course is entirely in the DEMOCRATIC REPUBLIC OF THE CONGO; the rest forms the border with UGANDA. Most of its course lies in VIRUNGA NATIONAL PARK.

Senegal. West African country. French colony until 1960. It covers an area of 75,955 square miles (196,722 sq. km.) with a population of 15,736,368 in 2020. DAKAR is the capital. French is the official language. It achieved independence in 1960; in 1982, it began fighting a separatist movement in the southern part of the country. Administered within the federation of

FRENCH WEST AFRICA during the era of colonial rule.

Senegal River. West African river forming the boundary between SENEGAL and MAURITANIA. It has two main sources, the Bafing and the Bakoy rivers. It is 1,015 miles (1,635 km.) long and empties into the Atlantic Ocean near SAINT-LOUIS in SENEGAL.

Seychelles. Nation consisting of more than 100 islands and islets in the INDIAN OCEAN northeast of MADAGASCAR. The capital is VICTORIA. The major islands are granitic, rising from a large undersea plateau and reaching no more than 2,968 feet (905 meters) above sea level. The rest of the islands are low-lying coral atolls. The climate is tropical monsoonal. The country covers only 176 square miles (455 sq. km.), making it one of the world's smallest countries. However, because the islands are spread across such a broad area (about 746 miles/1,200 km. between the farthest points), Seychelles claims fishing and mineral rights over 521,235 square miles (1.35 million sq. km.) of the Indian Ocean. The population was 95,981 in 2020. Seychelles was a British colony until 1976. Tourism has become its major source of income.

Sfax. Second-largest city of TUNISIA, located on the MEDITERRANEAN SEA, halfway along Tunisia's coastline, with a population of 330,440 in 2014. Known for the excellent olive oil it produces and exports around the world.

Shatt al-Jarid. See CHOTT DJERID.

Sidra, Gulf of. Indentation in the MEDITERRANEAN SEA, at whose eastern end is Benghazi, LIBYA. Libya claims the entire 150,000 square miles (388,500 sq. km.) of its territory. However, most other countries have recognized only a small fraction of this territory as belonging to Libya. Strained international relations between the United States and Libya worsened when a U.S. naval fleet held maneuvers there and clashed with Libya's armed forces in 1981.

Sierra Leone. West African country, founded in 1787 by a British antislavery group as the first colony for freed slaves in Africa. Total area of 27,699 square miles (1,740 sq. km.), with a population of 6,624,933 in 2020. Capital and largest city is FREETOWN. English is the official language. A decade-long civil war ended in 2002, and the country has had peaceful elections since then.

Sinai Peninsula. Triangular landmass connecting Africa with Asia. It is arid, as the Mediterranean coastal plain gradually rises to a central plateau, with jagged mountains in its extreme south. Sparsely populated but rich in mineral resources, including oil, natural gas, and metallic ores. Its western border houses the SUEZ CANAL. Controlled by EGYPT since antiquity.

Sirte, Gulf of. See SIDRA, GULF OF.

Skikda. Major Mediterranean port city in northeastern ALGERIA. Serves CONSTANTINE, 50 miles (80 km.) to the northwest. Formerly Philippeville. Population in 2008 was 182,903.

Slave Coast. See BIGHT OF BENIN.

Somalia. Country in Africa. Total area of 246,201 square miles (637,657 sq. km.), slightly smaller than the U.S. state of Texas, with a 2020 population of 11,757,124, primarily rural. Capital is MOGADISHU. Almost exclusively desert, with some irrigation agriculture possible in the south along the Jubba and Shebelle rivers Despite more than 1,865 miles (3,000 km.) of coastline, it has few natural harbors. Encompasses the geographic feature known as the HORN OF AFRICA. Most of the population engages in nomadic or seminomadic animal husbandry. The people are socially organized by clans, which establish status and the type of obligations their members have to one another and to outsiders. Warfare among clans competing for political and economic power in the 1990s led to anarchy Continued violence and clan conflict have impoverished the country.

South Africa. A republic in Southern Africa. Total area of 470,693 square miles (1,219,090 sq. km.). The 2020 population of 56,463,617 was 89.7 percent black, 7.8 percent white, the rest mixed race. Administrative capital is PRETORIA; legislative capital is CAPE TOWN; judicial capital is BLOEMFONTEIN. Most of the country is covered by a large plateau, with the DRAKENSBERG MOUNTAINS on the southeast. Part of the KALAHARI DESERT is in the northern area on the border with NAMIBIA. The Orange, Vaal, and LIMPOPO rivers flow through South Africa. Its temperate climate is mostly regulated by trade winds from the INDIAN OCEAN. Has vast mineral resources; the gold mines in WITWATERSRAND are the richest in the world. Main cities are JoHANNESBURG, CAPE TOWN, and DURBAN. There are eleven official languages.

South African Mountains. The main mountain ranges in the interior of SOUTH AFRICA are the DRAKENSBERG, Stormberg, Nuweveld Reeks, Roggeveld, Berge, and Bokkeveld Berg. Several peaks are more than 10,000 feet (3,050 meters) high. To their east is the High Veld, with elevations of 2,000 to 6,000 feet (600 to 1,800 meters). In the southwest part of South Africa lies the Little Karoo; farther north is the Great Karoo, which ultimately merges with the KALAHARI DESERT.

Southern Rhodesia. Colonial-era name of ZIMBABWE.

South Sudan. Republic in the eastern SAHEL region of Africa. Total area of 248,777 square miles (644,329 sq. km.) with a 2020 population of 10,561,244. The capital city is Juba. Gained its independence from SUDAN in 2011 following years of civil war. Juba is the capital. Despite rich mineral reserves, South Sudan is primarily agricultural and remains one of the world's poorest nations.

Spanish Morocco. Two tiny North African communities—CEUTA and MELILLA—surrounded by MOROCCO—which Spain retained when it gave up its claim to its Rio d'Oro colony south of Morocco in the 1970s. Ceuta is located on the MEDITERRANEAN SEA on the northern tip of Africa, 14 miles (20 km.) from GIBRALTAR. Melilla, on a spit of land that juts into the Mediterranean Sea, is close to the Algerian border. Spain has considered granting them autonomy.

Spanish Sahara. See WESTERN SAHARA.

Sudan. Third-largest country in Africa. Total area of 718,723 square miles (1,886,068 sq. km.) with a 2020 population of 45,561,556. Capital is

KHARTOUM. The northern third of the country is located in the section of the SAHARA known as the Nubian Desert, pierced by the White Nile and Blue Nile rivers, which join in the middle of the nation to form the NILE RIVER. Farther south is a semiarid plateau and mountainous region, which gives way to the massive swamp region along the White Nile known as the As Sudd. Sudan lost more than a quarter of its territory with the secession of SOUTH SUDAN in 2011. More than two-thirds of the people engage in cultivation or animal husbandry. Cotton lint and cotton seed are the major exports. Eighty percent of the world's gum arabic is produced here.

Suez Canal. Connection between the MEDITERRANEAN and RED seas, located in northeastern EGYPT. It is 105 miles (169 km.) long, at least 197 feet (60 meters) wide, and able to accommodate ships with 62 feet (19 meters) of drag. An earlier canal was excavated in this area in the thirteen century BCE and maintained irregularly until the eighth century CE. The modern canal was completed in 1869 by a French company. An expansion in 2014-2015 doubled the canal's capacity.

Surt Desert. North African desert that runs across the Mediterranean coast of northern LIBYA, from TRIPOLITANIA to the Tunisian border.

Swaziland. See ESWATINI.

Table Bay. Important harbor on the western side of SOUTH AFRICA. CAPE TOWN is located on its southern shore. Table Mountain, one of South Africa's main peaks, is nearby.

Table Mountain. Mountain overlooking CAPE TOWN, South Africa, (3,549 feet/1,082 meters) that dominates the view from the city center and harbor and provides an excellent view of the city. A cable railway carries visitors to the top.

Tahat, Mount. Highest mountain in ALGERIA, rising to 9,852 feet (3,003 meters). Located in the HOGGAR MOUNTAINS, almost precisely on the TROPIC OF CANCER. Frequently snowcapped for extended periods in winter.

Tanga. Port city in northeastern TANZANIA. The terminus of a railroad to the interior, it gives access to plantation crops such as rubber, tea, and cof-

fee. It also has coastal highways connecting it with DAR ES SALAAM and the Kenyan cities of NAIROBI and MOMBASA. Developed as a port by the Germans in the 1890s. Population was 273,332 in 2012.

Tanganyika. Name of the mainland portion of TANZANIA when the country was under British administration.

Tanganyika, Lake. Lake straddling the DEMOCRATIC REPUBLIC OF THE CONGO's borders with BURUNDI and TANZANIA in the GREAT RIFT VALLEY. Second-largest lake in eastern Africa, and the longest freshwater lake (410 miles/660 km.) and the second-deepest (4,820 feet/1,470 meters) in the world. Its width varies from 10 to 45 miles (16 to 70 km.); it covers 12,700 square miles (32,890 sq. km.). Has a large number of hippopotami and crocodiles. Its ports include BUJUMBURA, BURUNDI; KALEMIE, DEMOCRATIC REPUBLIC OF THE CONGO; and Kigoma, TANZANIA.

Tangier. Port city on the Atlantic Ocean in northern MOROCCO. Population in 2014 was 947,952. Morocco's closest large city to Europe, located less than 40 miles (65 km.) from the Iberian Peninsula.

Tanta. A major city and railroad center in the NILE delta of Africa, situated on the main highway between CAIRO and ALEXANDRIA. Tobacco, food processing, and textiles are the main industries. Population was 528,672 in 2018.

Tanzania. Country in East Africa. Total area of 365,757 square miles (947,300 sq. km.) with a 2020 population of 58,552,845. Capital is DAR ES SALAAM. Dominated by a plateau averaging

Usambara Mountains, Tanzania. (Joachim Huber)

about 3,940 feet (1,200 meters) in height, there is also a coastal plain, and mountain ranges in the northeast and southwest. Site of MOUNT KILIMANJARO, the highest point in Africa. The GREAT RIFT VALLEY runs along Tanzania's western border, demarcated by the lakes formed in this fault. Most of Tanzania's people live on the volcanic soils near Kilimanjaro and the fertile land around Lake NYASA, nearly 70 percent engaging in agriculture, fishing, or forestry. The islands of ZANZIBAR joined with TANGANYIKA in 1964 to form the United Republic of Tanzania.

Taroudant. Ancient walled town in NORTH AFRICA, located west of QUARZAZATE. It has been a desert outpost and trading center for more than 1,000 years. Many Berbers live here and engage in trade with those who wander the SAHARA. Population in 2014 was 80,149.

Tassili-n-Ajjer. Site in the southeastern SAHARA in ALGERIA, where prehistoric artifacts have been found, including the jawbone of a human who lived half a million years ago. Some 4,000 sandstone engravings that are more than 5,000 years old have been collected there, the largest collection of its kind. Primitive rock paintings also have been found in rock formations near there.

Tawzar. See TOZEUR.

Tell Atlas. One of two North African mountain ranges that divide ALGERIA into three physical regions. Located near the MEDITERRANEAN SEA, stretching from west of ORAN to ALGIERS. The fertile coastal area running from Algeria to TUNISIA is known as the Tell. The Tell Atlas to its west draws moisture from the air that provides the area with an abundant supply of water for irrigation.

Tétouan. Tourist resort in northern MOROCCO that faces onto the MEDITERRANEAN SEA. Its population of 380,787 (2014) swells considerably during tourist season and holidays.

Tibesti Mountains. Volcanic-formed mountains in northern CHAD, extending into southern LIBYA and northeastern NIGER. Emi Koussi, the highest peak, has an elevation of 11,204 feet (3,415 meters).

Tobruk. Mediterranean port in northeastern LIBYA. Intense fighting occurred there between British and German forces during World War II. Population in 2011 was 120,000.

Togo. Tiny West African nation that is divided physically and ethnically. Total area of 21,925 square miles (56,785 sq. km.), with a population of 8,608,444 in 2020. Capital is LOMÉ. French is the official language. The Muslims in the north are separated from the southern Togolese by a mountain range. It has a large deposit of phosphates. The country was a colonial creation formed by Germany as Togoland during the 1880s. After World War I, the administration of Togoland was partitioned between Great Britain and France. Most of the original colony was reassembled as the independent nation of Togo in 1960.

Toubkal, Mount. Highest mountain peak in MOROCCO, located in the High ATLAS MOUNTAINS in south central Morocco, it soars to 13,668 feet (4,166 meters)—an altitude about 400 feet (120 meters) lower than that of Pikes Peak in Colorado.

Tozeur. North African oasis in southwestern TUNISIA near the Algerian border. Located on the shores of CHOTT DJERID, Tunisia's largest lake, which is often virtually dry because of limited rainfall and desert conditions. Also called Tawzar. Population in 2014 was 37,370.

Transvaal. Historic region of SOUTH AFRICA between ZIMBABWE in the north and the FREE STATE—from which it is separated by the Vaal River—in the south. The highland region is mostly grassland and hilly savanna. The DRAKENSBERG MOUNTAINS are a main feature of the Transvaal. Kruger National Park is in the eastern portion, near the border with MOZAMBIQUE. Afrikaners (Boers) who settled the region in the mid-nineteenth century created an independent republic. In 1886 gold was found in the WITWATERSRAND area around JOHANNESBURG, and a massive gold rush ensued. Conflict between the Afrikaners and new residents sparked the South African (Boer) War, which the British won. After a period of colonial rule, the

Transvaal joined the Cape Colony, the ORANGE FREE STATE, and Natal as a province in the Union of South Africa. During the early 1990s the province was split into two provinces: Northern Transvaal and Eastern Transvaal, with their capitals at Pietersburg and Nelspruit.

Tripoli. Largest city and capital of LIBYA, located on the MEDITERRANEAN SEA in the northwestern part of the country. It had about 1.165 million inhabitants in 2018, about one-fifth of the country's total population, and grew rapidly during Libya's oil boom of the 1960s.

Tripolitania. Productive agricultural region in which TRIPOLI, the capital of LIBYA, is located. Covering 16 percent of Libya's landmass, its Mediterranean coast extends 186 miles (300 km.) from the SURT DESERT to TUNISIA in the west.

Tristan da Cunha. Archipelago in the South ATLANTIC OCEAN. Its only inhabited island has the same name. Located at 37°15′ south latitude, longitude 12°30′ west. Since 1938, Tristan da Cunha has been a dependency of SAINT HELENA. Eruption of the island's volcano (elevation 6,760 feet/2,060 meters) in 1961 forced the island's 300 inhabitants to flee, but most returned to the island in 1963. Tristan da Cunha covers an area of 40 square miles (104 sq. km.). Population of 293 in 2016.

Tropics. See CANCER, TROPIC OF and CAPRICORN, TROPIC OF.

Tunis. Largest city and capital of TUNISIA. Its population was 2.2 million in 2014. Built on low, chalk-white hills that extend toward a salt lake, EL BAHIRA, close to the harbor. In 1893 the French built a 20-foot (6-meter)-long channel across the Lake of Tunis to the MEDITERRANEAN SEA, opening Tunis as a port.

Tunis, Gulf of. Large, deep-water gulf in northeastern TUNISIA, located 65 miles (104 km.) from Sicily. Connected to TUNIS, the capital of Tunisia, by a 20-foot (6-meter)-wide channel that has turned Tunis into a port.

Tunisia. Smallest country in NORTH AFRICA. Total area of 63,170 square miles (163,610 sq. km.) with a 2020 population of 11,721,177, more than half living in its six largest cities. Capital is TUNIS. Bordered on the north and east by the MEDITERRANEAN SEA, on the west by ALGERIA, on the east by LIBYA, and on the south by Libya and Algeria. Halfway between GIBRALTAR and SUEZ, and 85 miles (137 km.) south of Sicily, Tunisia was an important stop for ancient traders. Its highest point is Mount Chambi at 5,066 miles (1,544 meters).

Tunisian Dorsale. Branch of the ATLAS MOUNTAINS that extends from ALGERIA into TUNISIA. Numerous ranges of the Atlas Mountains continue from Algeria to Tunisia.

Turkana, Lake. One of the chain of lakes in the GREAT RIFT VALLEY, Africa. Having no outlet, it is brackish and has a high rate of evaporation. Nevertheless, it supports several species of fish, as well as hippopotami and crocodiles. It is at the center of one of the richest areas of hominid fossils. Covers 3,500 square miles (9,100 sq. km.).

Ubangi River. Right tributary of the CONGO RIVER in the north and west of the DEMOCRATIC REPUBLIC OF THE CONGO, along the border with the CENTRAL AFRICAN REPUBLIC and the REPUBLIC OF THE CONGO. About 1,400 miles (2,255 km.) long. It results from the union of the Uele and Bomu rivers and joins the Congo River about 60 miles (100 km.) southwest of the town of BANDAKA.

Ubangi-Shari. Colonial-era name of what became the CENTRAL AFRICAN REPUBLIC.

Uganda. Landlocked nation in Africa. Total area of 93,065 square miles (241,038 sq. km.) with a population of 43,252,966 in 2020. Capital is KAMPALA. Land includes swampy lowlands in the south, a fertile plateau covering most of the country, and a steppe and desert area in the north. The GREAT RIFT VALLEY is found in the western section of the country. The RWENZORI Range forms the border with the DEMOCRATIC REPUBLIC OF THE CONGO, and has more than a dozen peaks over 13,000 feet (4,000 meters) in altitude. The highest, Mount Stanley, is Africa's third-highest peak, at 16,765 feet (5,110 meters). Seven peaks have permanent snowcaps, despite being located near the equator. Seventy-six percent of the labor force is engaged in agriculture; sixty-two percent are literate. Main exports are coffee, fish, and gold. There is access by rail to the Kenyan port of MOMBASA, and by boat and rail to the Tanzanian port of DAR ES SALAAM, by way of the southern Lake VICTORIA port of MWANZA. AIDS has been an enormous drain on the economy of Uganda, but infection rates have dropped to under 6 percent of the population.

United Republic of Tanzania. See TANZANIA.

Upemba, Lake. Lake in southeast DEMOCRATIC REPUBLIC OF THE CONGO. An expansion of the CONGO RIVER, it is about 16 miles (25 km.) long and 18 miles (29 km.) wide, covering 190 square miles (500 sq. km.). Much of the lake is overgrown with papyrus. Its eastern side is in Upemba National Park.

Upemba National Park. Protected area in KATANGA province in southeast DEMOCRATIC REPUBLIC OF THE CONGO. Established in 1939 and covers 748 square miles (1,937 sq. km.). Extends eastward from Lake UPEMBA on the CONGO RIVER. Vegetation includes savanna and papyrus swamps. Fauna include zebras, peccaries, buffalo, antelope, crocodiles, and hippopotami.

Upper Volta. Former name of the West African nation of BURKINA FASO.

Victoria. Capital and only city of the nation of SEYCHELLES, located on the island of Mahé. Its population was 26,450 in 2010.

Victoria, Lake. Largest lake in Africa and the world's second-largest freshwater lake, after Lake Superior in North America. Located in KENYA, TANZANIA, and UGANDA. Ringed by productive agricultural districts, it also supports a sizable fishing industry. Around this lake is one of the densest population belts in Africa. It is one of the sources of the NILE RIVER. A dam constructed in 1954 where the river leaves the lake is a major source of electricity for Uganda. Covers 23,146 square miles (59,947 sq. km.).

Victoria Falls. One of the world's most beautiful waterfalls, located in Southern Africa. Part of the ZAMBEZI RIVER, located on the border between ZAMBIA and ZIMBABWE. There, the river is 1

mile (1.6 km.) wide and falls 355 feet (108 meters).

Virunga. Range of volcanic mountains extending across Africa's GREAT RIFT VALLEY, along the border of THE DEMOCRATIC REPUBLIC OF THE CONGO with RWANDA and UGANDA. The highest of its eight volcanic peaks is KARISIMBI, at 14,787 feet (4,507 meters); the others are Mgahinga, MIKENO, Muhavura, NYAMURAGIRA, Nyiragongo, Bisoke, and Sabyinyo.

Virunga National Park. Protected area in east DEMOCRATIC REPUBLIC OF THE CONGO, near the UGANDA border. Created in 1925 and called Albert National Park until 1969. Covers about 1.95 million acres (790,000 hectares) Includes part of the SEMLIKI RIVER basin, the RWENZORI range, the western shores of Lake EDWARD, and the VIRUNGA volcanoes. Variability in elevations results in a diversity of climates and vegetation, including marshes, savannas, lava plains, steppes, and several forest cover types. Some of the mountain peaks are permanently snowcapped. Wildlife is abundant; includes elephants, lions, mountain gorillas, and okapi. Designated as a World Heritage site in 1979 and placed on the list of World Heritage in Danger in 1994. A 2014 documentary titled *Virunga* depicts the political chaos in the country and efforts to fight poachers.

Volta Dam. See AKOSOMBO DAM.

Volta River. River system that drains 75 percent of the West African nation of GHANA. The Black Volta and White Volta flow southward from BURKINA FASO to form the Volta River in Ghana. Along with its tributary, the Oti River, it feeds Lake Volta, which has a surface area of 3,275 square miles (8,482 sq. km.).

Wagadugu. See OUAGADOUGOU.

Western Sahara. Desert region of NORTH AFRICA, with a landmass of 102,700 square miles (266,000 sq. km.). Most of its 2018 population of 567,402 lives in two of the region's cities, EL AAIÚN, just south of the disputed border with MOROCCO, and DAKHLA, two-thirds of the way down the Atlantic coast toward MAURITANIA. Both of these cities are international ports. The ownership of the area directly south of Morocco and east of ALGERIA and Mauritania has long been disputed, although it is generally considered to belong to Morocco, which currently occupies it. Western Sahara was a Spanish protectorate until 1976, when Spain transferred the country to Morocco and Mauritania. In 1979 Mauritania abandoned its claim to the area, ceding it to Morocco. An Algerian-backed guerilla group, the Polisario Front, has demanded independence for the area. Morocco's claim to it remains disputed.

Windhoek. Capital and commercial and transportation center of NAMIBIA in Southern Africa. Population in 2011 was 325,858. Located on a dry plateau in the central part of the country, surrounded by hills, at an altitude of 5,400 feet (1,650 meters). Receives most of its rain between December and March. This area has some of the richest diamond fields in the world. Established by German soldiers in the late 1880s.

Witwatersrand. Mountainous part of SOUTH AFRICA that contains some of the richest gold mines in the world. Gold was discovered there in 1886 and created a major gold rush. The immigrants, mainly British, were known as uitlanders (outlanders), and soon outnumbered the local farmers of Dutch ancestry, called Boers. The Boers' attempts to keep the uitlanders from participating in government led to the Boer War. Estimated to be the source of 40 percent of all the gold ever mined.

Yamoussoukro. Capital of CÔTE D'IVOIRE in West Africa since 1983. Located in the south central part of the country. Birthplace of the country's first president, Félix Houphouët-Boigney. Yamoussoukro Basilica, one of the largest Christian churches in the world, is located here. Population was 361,893 in 2020.

Yaoundé. Capital and commercial center of CAMEROON. Located on the forested plateau between the Nyong and Sanaga rivers in the south central section of the country. Founded in 1888 when the country was under German rule. Population was 2,765,600 in 2015.

Yorubaland. Region of southwestern NIGERIA historically inhabited by the Yoruba-speaking people.

Zaire. See DEMOCRATIC REPUBLIC OF THE CONGO.

Zambezi River. Fourth-longest river in Africa. Begins in ZAMBIA and flows 2,200 miles (3,520 km.) to the INDIAN OCEAN, passing through ANGOLA, BOTSWANA, and MOZAMBIQUE. Partially controlled by the Kariba Dam, which has created Lake KARIBA. Barely navigable because it is shallow and has many rapids that interrupt its flow. Its main tributaries are the Lungwebungu, Luenal, Kafue, Luangwa, and KWANDO rivers. Its best-known feature is the great waterfall known as VICTORIA FALLS. The Zambezi and its tributaries were first explored by David Livingstone, who published a description of it in 1865.

Zambia. Republic in south central Africa. Total area of 290,587 square miles (752,618 sq. km.) with a 2020 population of 17,426,623. Capital is LUSAKA, with 1.7 million residents. Located mostly on an elevated plateau. The ZAMBEZI, Kafue, and Luangwa rivers run through it. Much of the northern part of the country is swampland surrounding Lake Bangweulu. A main feature is the impoundment created by the Kariba Dam on the Zambezi River. The country is mineral-rich, with large deposits of zinc, tin, copper, and cobalt. Copper provides 60 percent of Zambia's export earnings. Most of the country is made up of grassland, and most residents are peasant farmers who grow corn, cassava, and sugarcane. The plateau is broken by the Muchinga Mountains in the northeast. It has a temperate climate, with the hot season lasting only from September to November. Ancient volcanic activity has left the area with rich soil, but only one-third can be cultivated due to mountains, forests, and rough savanna. The GREAT RIFT VALLEY runs the length of the country from north to south, and Lake NYASA fills most of the valley. Zambia was formerly under British rule and a colony of then NORTHERN RHODESIA. It became independent as Zambia in 1964.

Zanzibar. Low-lying island off the east coast of Africa. A British protectorate until gaining independence in 1963, it merged with the mainland nation of TANGANYIKA in 1964, to become part of the UNITED REPUBLIC OF TANZANIA. Situated 20 miles (35 km.) from the mainland; covers 640 square miles (1,660 sq. km.). Long a stop for Arab traders, Zanzibar was a major center for the slave trade before British rule came in 1890. Best known for its clove industry; also produces citrus crops and coconuts. Its population in 2012 was 1,303,569. Chief city, also named Zanzibar, is on the main island and had a population of 206,292 in 2002.

Zimbabwe. Republic just to the north of SOUTH AFRICA. Total area of 150,872 square miles (390,757 sq. km.) with a 2020 population of 14,546,314. Capital and largest city is HARARE (called Salisbury in colonial times), with 1.6 million residents in 2013. The composition of its population has shifted because of the emigration of white residents after independence and the installation of a black majority government in 1980. Economy is largely based on agriculture, although it is a leading supplier of chromium and produces several other minerals. Formerly known as Southern Rhodesia and as Rhodesia. Named after the megalithic Great Zimbabwe Ruins in the southeastern part of the country.

Femi Ferreira; C. James Haug; Grove Koger;
Kikombo Ilunga Ngoy; R. Baird Shuman

APPENDIX

THE EARTH IN SPACE

THE SOLAR SYSTEM

Earth's solar system comprises the Sun and its planets, as well as all the natural satellites, asteroids, meteors, and comets that are captive around it. The solar system formed from an interstellar cloud of dust and gas, or nebula, about 4.6 billion years ago. Gravity drew most of the dust and gas together to make the Sun, a medium-size star with an estimated life span of 10 billion years. Its system is located in the Orion arm of the Milky Way galaxy, about two-thirds of the way out from the center.

During the Sun's first 100 million years, the remaining rock and ice smashed together into increasingly larger chunks, or planetesimals, until the planets, moons, asteroids, and comets reached their present state. The resulting disk-shaped solar system can be divided into four regions—terrestrial planets, giant planets, the Kuiper Belt, and the Oort Cloud—each containing its own types of bodies.

Terrestrial Planets

In the first region are the terrestrial (Earth-like) planets Mercury, Venus, Earth, and Mars. Mercury, the nearest to the Sun, orbits at an average distance of 36 million miles (58 million km.) and Mars, the farthest, at 142 million miles (228 million km.). Astronomers call the distance from the Sun to Earth (93 million miles/150 million km.) an astronomical unit (AU) and use it to measure planetary distances.

Terrestrial planets are rocky and warm and have cores of dense metal. All four planets have volcanoes, which long ago spewed out gases that created atmospheres on all but Mercury, which is too close to the Sun to hold onto an atmosphere. Mercury is heavily cratered, like the earth's moon. Venus has a permanent thick cloud cover and a surface temperature hot enough to melt lead. The air on Mars is very thin and usually cold, made mostly of carbon dioxide. Its dry, rock-strewn surface has many craters. It also has the largest known volcano in the solar system, Olympus Mons, which is 16 miles (25 km.) high.

Average temperatures and air pressures on Earth allow liquid water to collect on the surface, a unique feature among planets within the solar system. Meanwhile, Earth's atmosphere—mostly nitrogen and oxygen—and a strong magnetic field protect the surface from harmful solar radiation. These are the conditions that nurture life, according to scientists. It is widely accepted that Mars had abundant water very early in its history, but all large areas of liquid water have since disappeared. A fraction of this water is retained on modern Mars as both ice and locked into the structure of abundant water-rich materials, including clay minerals (phyllosilicates) and sulfates. Studies of hydrogen isotopic ratios indicate that asteroids and comets from beyond 2.5 AU provide the source of Mars' water. Like Earth, Mars has polar ice caps, although those on Mars are made up mostly of carbon dioxide ice (dry ice), while those on Earth are made up of water ice.

A single natural satellite, the Moon, orbits Earth, probably created by a collision with a huge planetesimal more than 4 billion years ago. Mars has two tiny moons that may have drifted to it from the asteroid belt. A broad ring from 2 to 3.3 AU from the Sun, this belt is composed of space rocks as small as dust grains and as large as 600 miles (1,000 km.) in diameter. Asteroids are made of mineral compounds, especially those containing iron, carbon, and silicon. Although the asteroid belt contains

FORMATION OF THE SOLAR SYSTEM

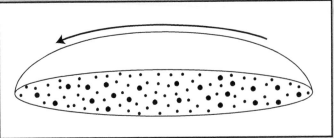

1. The solar system began as a cloud of rotating interstellar gas and dust.

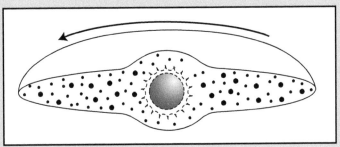

2. Gravity pulled some gases toward the center.

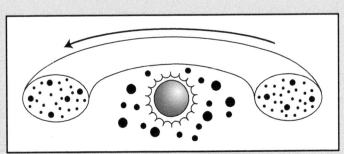

3. Rotation accelerated, and centrifugal force pushed icy, rocky material away from the proto-Sun. Small planetesimals rotate around the Sun in interior orbits.

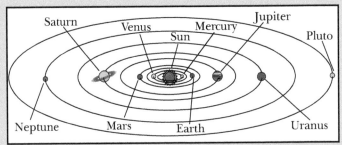

4. The interior, rocky material formed Mercury, Venus, Earth, and Mars. The outer, gaseous material formed Jupiter, Saturn, Uranus, Neptune, and Pluto.

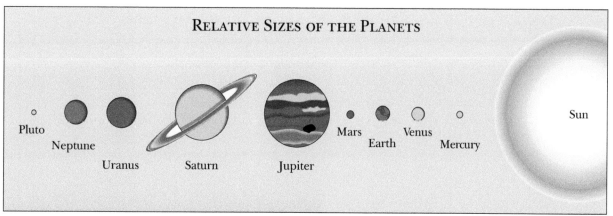

RELATIVE SIZES OF THE PLANETS

Note: The size of the Sun and distances between the planets are not to scale.; Source: Data are from Jet Proulsion Laboratory, California Institute of Technology. The Deep Space Network. Pasadena, Calif.:JPL, 1988, p. 17.

enough material for a planet, one did not form there because Jupiter's gravity prevented the asteroids from crashing together. The belt separates the first region of the solar system from the second.

The Giant Planets

The second region belongs to the gas giants Jupiter, Saturn, Uranus, and Neptune. The closest, Jupiter, is 5.2 AU from the Sun, and the most distant, Neptune, is 30.11 AU. Jupiter is the largest planet in the solar system, its diameter 109 times larger than Earth's. The giant planets have solid cores, but most of their immense size is taken up by hydrogen, helium, and methane gases that grow thicker and thicker until they are like sludge near the core. On Jupiter, Saturn, and Uranus, the gases form wide bands over the surface. The bands sometimes have immense circular storms like hurricanes, but hundreds of times larger. The Great Red Spot of Jupiter is an example. It has winds of up to 250 miles (400 km.) per hour, and is at least a century old.

These planets have such strong gravity that each has attracted many moons to orbit it. In fact, they are like miniature solar systems. Jupiter has the most moons—eighteen—and Neptune has the fewest—eight—but Neptune's moon Triton is the largest of all. Most moons are balls of ice and rock, but Jupiter's Europa and Saturn's Titan may have liquid water below ice-bound surfaces. Several moons appear to have volcanoes, and a wispy atmosphere covers Titan. Additionally, the giant planets have rings of broken rock and ice around them, no more

than 330 feet (100 meters) thick. Saturn's hundreds of rings are the brightest and most famous.

The Kuiper Belt

The third region of the solar system, the Kuiper Belt, contains the dwarf planet, Pluto. Pluto has a single moon, Charon. It does not orbit on the same plane, called the ecliptic, as the rest of the planets do. Instead, its orbit diverges more than seventeen degrees above and below the ecliptic. Its orbit's oval shape brings Pluto within the orbit of Neptune for a large percentage of its long year, which is equal to 248 Earth years. Two-thirds the size of the earth's moon, Pluto has a thin, frigid methane atmosphere. Charon is half Pluto's size and orbits less than 32,000 miles (20,000 km.) from Pluto's surface. Some astronomers consider Pluto and Charon to be a double planet.

The Kuiper Belt holds asteroids and the "short-period" comets that pass by Earth in orbits of 20 to 200 years. These bodies are the remains of

OTHER EARTHS

By the year 2000 astronomers had detected twenty-eight planets circling stars in the Sun's neighborhood of the galaxy. Planets, they think, are common. Those found were all gas giants the size of Saturn or larger. Earth-size planets are much too small to spot at such great distances. Where there are gas giants, there also may be terrestrial dwarfs, as in Earth's solar system. Where there are terrestrial planets, there may be liquid water and, possibly, life.

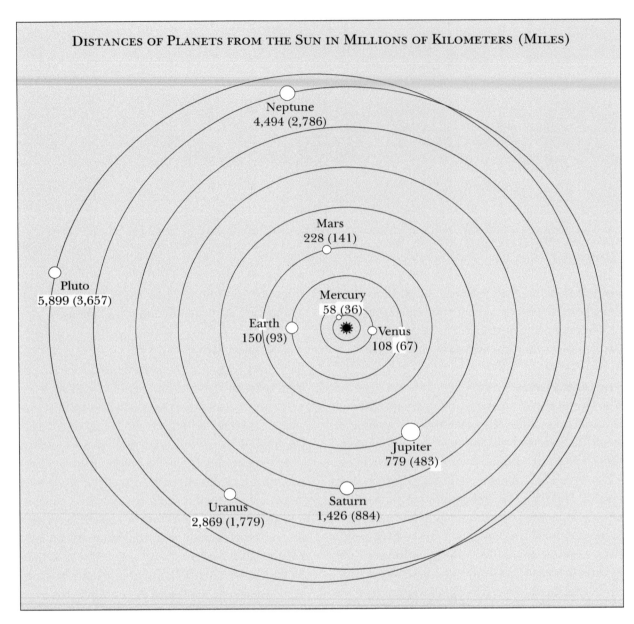

DISTANCES OF PLANETS FROM THE SUN IN MILLIONS OF KILOMETERS (MILES)

Neptune
4,494 (2,786)

Mars
228 (141)

Pluto
5,899 (3,657)

Mercury
58 (36)

Earth
150 (93)

Venus
108 (67)

Jupiter
779 (483)

Uranus
2,869 (1,779)

Saturn
1,426 (884)

planet formation and did not collect into planets because distances between them are too great for many collisions to occur. Most of them are loosely compacted bodies of ice and mineral—"dirty snowballs," as they were termed by the famous astronomer Fred Lawrence Whipple (November 5, 1906–August 30, 2004). An estimated 200 million Kuiper Belt objects orbit within a band of space from 30 to 50 AU from the Sun.

The Oort Cloud

In contrast to the other regions of the solar system, the Oort Cloud is a spherical shell surrounding the entire solar system. It is also a collection of com-

ets—as many as two trillion, scientists calculate. The inner edge of the cloud forms at a distance of about 20,000 AU from the Sun and extends as far out as 100,000 AU. The Oort Cloud thus gives the solar system a theoretical diameter of 200,000 AU—a distance so vast that light needs more than three years to cross it. No astronomer has yet detected an Oort Cloud object, because the cloud is so far away. Occasionally, however, gravity from a nearby star dislodges an object in the cloud, causing it to fall toward the Sun. When observers on Earth see such an object sweep by in a long, cigar-shaped orbit, they call it a long-period comet.

The outer edge of the Oort Cloud marks the farthest reach of the Sun's gravitational power to bind bodies to it. In one respect, the Oort Cloud is part of interstellar space.

In addition to light, the Sun sends out a constant stream of charged particles—atoms and subatomic particles—called the solar wind. The solar wind shields the solar system from the interstellar medium, but it only does so out to about 100 AU, a boundary called the heliopause. That is a small fraction of the distance to the Oort Cloud.

Roger Smith

EARTH'S MOON

The fourth-largest natural satellite in the solar system, Earth's moon has a diameter of 2,159.2 miles (3,475 km.)—less than one-third the diameter of Earth. The Moon's mass is less than one-eightieth that of Earth.

The Moon orbits Earth in an elliptical path. When it is at perigee (when it is closest to Earth), it is 221,473 miles (356,410 km.) distant. When it is at apogee (farthest from Earth), it is 252,722 miles (406,697 km.) distant.

The Moon completes one orbit around Earth every 27.3 Earth days. Because it rotates at about the same rate that it orbits the earth, observers on Earth only see one side of the Moon. The changing angles between Earth, the Sun, and the Moon determine how much of the Moon's illuminated surface can be seen from Earth and cause the Moon's changing phases.

Volcanism

Naked-eye observations of the Moon from Earth reveal dark areas called *maria*, the plural form of the Latin word *mare* for sea. The maria are the remains of ancient lava flows from inside gigantic impact craters; the last eruptions were more than 3 billion years ago. The lava consists of basalt, similar in composition to Earth's oceanic crust and many volcanoes. The maria have names such as Mare Serenitatis (15° to 40°N, 5° to 20°E) and Mare Tranquillitatis (0° to 20°N, 15° to 45°E). Some of the smaller dark areas on the Moon also have names that are water-related: lacus (lake), sinus (bay), and palus (marsh).

Impact Craters

Observing the Moon with an optical aid, such as a telescope or a pair of binoculars, provides a closer view of impact craters. Impact craters of various sizes cover 83 percent of the Moon's surface. More than 33,000 craters have been counted on the Moon.

One of the easiest craters to observe from the Earth is Tycho. Located at 43.3°S, longitude 11.2 degrees west, it is about 50 miles (85 km.) wide. Surrounding Tycho are rays of dusty material, known as

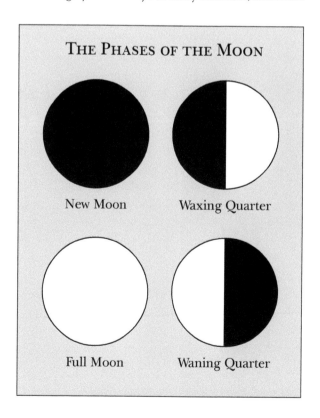

THE PHASES OF THE MOON

New Moon

Waxing Quarter

Full Moon

Waning Quarter

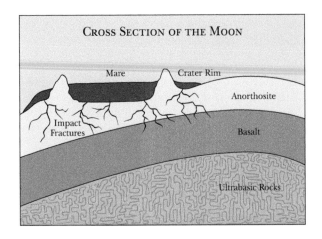

ejecta, that appear to radiate from the crater. When an object from space, such as a meteoroid, slams into the Moon's surface, it is vaporized upon impact. The dust and debris from the interior of the crater fall back onto the lunar surface in a pattern of rays. Because the ejecta is disrupted by subsequent impacts, only the youngest craters still have rays. Sometimes, pieces of the ejecta fall back and create smaller craters called secondary craters. The ejecta rays of Tycho extend to almost 1,865 miles (3,000 km.) beyond the crater's edge.

Other Lunar Features

Near the crater called Archimedes is the Apennines mountain range, which has peaks nearly 20,000 feet (60,000 meters) high—altitudes comparable to South America's Andes.

The Moon also has valleys. Two of the most well known are the Alpine Valley, which is about 115 miles (185 km.) long; and the Rheita Valley, located about 155 miles (250 km.) from the Stevinus crater, which is 238 miles (383 km.) long, 15.5 miles (25 km.) wide, and 2,000 feet (609 meters) deep.

Smaller than valleys and resembling cracks in the lunar surface are features called rilles, which are thought to be places of ancient lava flow. Many rilles can be seen near the Aristarchus crater. Rilles are often up to 3 miles (5 km.) wide and can stretch for more than 104 miles (167 km.).

A wrinkle in the lunar surface is called a ridge. Many ridges are found around the boundaries of the maria. The Serpentine Ridge cuts through Mare Serenitatis.

Exploration of the Moon

Robotic spacecraft were the first visitors to explore the Moon. The Russian spacecraft Luna 1 made the first flyby of the Moon in January, 1959. Eight months later, Luna 2 made the first impact on the Moon's surface. In October, 1959, Luna 3 was the first spacecraft to photograph the side of the Moon not visible from Earth. In 1994 the United States' *Clementine* spacecraft was the first probe to map the Moon's composition and topography globally.

The first humans to land on the Moon were the U.S. astronauts Neil Armstrong and Edwin "Buzz" Aldrin. On July 20, 1969, they landed in the *Eagle* lunar module, during the Apollo 11 mission. Armstrong's famous statement, "That's one small step for man, one giant leap for mankind," was heard around the world by millions of people who watched the first humans set foot on the lunar surface, at the Sea of Tranquillity. The last twentieth century human mission to reach the lunar surface, Apollo 17, landed there in December, 1972. Astronauts Gene Cernan and geologist Jack Schmitt landed in the Taurus-Littrow Valley (20°N, 31°E).

Noreen A. Grice

THE SUN AND THE EARTH

Of all the astronomical phenomena that one can consider, few are more important to the survival of life on Earth than the relationship between Earth and the Sun. With the exception of small amounts of residual (endogenic) energy that have remained inside the earth from the time of its formation some 4.5 billion years ago and which sustain some specialized forms of life along some oceanic rift systems, almost all other forms of life, including human, depend on the exogenic light and energy that the earth receives directly from the Sun.

The enormous variety of ecosystems on Earth are highly dependent on the angles at which the Sun's rays strike Earth's spherical surface. These angles, which vary greatly with latitude and time of year, determine many commonly observed phenomena, such as the height of the Sun above the horizon, the changing lengths of day and night throughout the year, and the rhythm of the seasons. Daily and seasonal changes have profound effects on the many climatic regions and life cycles found on earth.

The Sun

The center of Earth's solar system, the Sun is but one ordinary star among some 100 billion stars in an ordinary cluster of stars called the Milky Way galaxy. There are at least 10 billion galaxies in the universe, each with billions of stars. Statistically, the chances are good that many of these stars have their own solar systems. Late twentieth century astronomical observations discovered the presence of what appear to be planets, large ones similar in size to Jupiter, orbiting other stars.

Earth's Sun is an average star in terms of its physical characteristics. It is a large sphere of incandescent gas that has a diameter more than 100 times that of Earth, a mass more than 300,000 times that of Earth, and a volume 1.3 million times that of Earth. The Sun's surface gravity is thirty-four times that of Earth.

The conversion of hydrogen into helium in the Sun's interior, a process known as nuclear fusion, is the source of the Sun's energy. The amount of mass that is lost in the fusion process is miniscule, as evidenced by the fact that it will take perhaps 15 million years for the Sun to lose one-millionth of its total mass. The Sun is expected to continue shining through another several billion years.

Earth Revolution

The earth moves about the Sun in a slightly elliptical orbit called a revolution. It takes one year for the earth

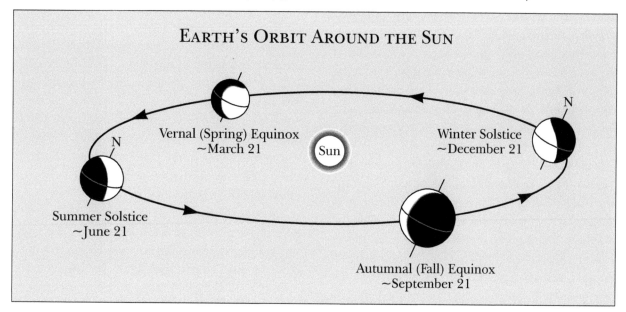

EARTH'S ORBIT AROUND THE SUN

Sun

Vernal (Spring) Equinox ~March 21

Winter Solstice ~December 21

Summer Solstice ~June 21

Autumnal (Fall) Equinox ~September 21

to make one revolution at an average orbital velocity of about 29.6 kilometers per second (18.5 miles per second). Earth-sun relationships are described by a tropical year, which is defined as the period of time (365.25 average solar days) from one vernal equinox to another. To balance the tropical year with the calendar year, a whole day (February 29) is added every fourth year (leap year). Other minor adjustments are necessary so as to balance the system.

Perihelion and Aphelion

The average distance between Earth and the Sun is approximately 93 million miles (150 million km.). At that distance, sunlight, which travels at the speed of light (186,000 miles/300,000 kilometers per second), takes about 8.3 minutes to reach the earth. Since the earth's orbit is an ellipse rather than a circle, the earth is closest to the Sun on about January 3—a distance of 91.5 million miles (147 million km.). This position in space is called perihelion, which comes from the Greek *peri*, meaning "around" or "near," and *helios*, meaning the Sun. Earth is farthest from the Sun on about July 4 at aphelion (Greek *ap*, "away from," and *helios*), with a distance of 152 million kilometers (94.5 million miles).

Axial Inclination

Astronomers call the imaginary surface on which Earth orbits around the Sun the plane of the ecliptic. The earth's axis is inclined 66.5 degrees to the plane of the ecliptic (or 23.5 degrees from the perpendicular to the plane of the ecliptic), and it maintains this orientation with respect to the stars. Thus, the North Pole points in the same direction to Polaris, the North Star, as it revolves about the Sun. Consequently, the Northern Hemisphere tilts away from the Sun during one-half of Earth's orbit and toward the Sun through the other half.

Winter solstice occurs on December 21 or 22, when the tilt of the Northern Hemisphere away from the Sun is at its maximum. The opposite condition occurs during summer solstice on June 21 or 22, when the Northern Hemisphere reaches its maximum tilt toward the Sun. The equinoxes occur midway between the solstices when neither the Southern nor the Northern Hemisphere is tilted toward the Sun. The

---

ECLIPSES

The Sun's diameter is 400 times larger than the moon's; however, the moon is 400 times closer to Earth than the Sun, making the two objects appear nearly the same size in the sky to observers on Earth. As the moon orbits Earth, it crosses the plane of the Earth-Sun orbit twice each month. If one of the orbit-crossing points (called nodes) occurs during a new or full moon phase, a solar or lunar eclipse can occur.

A solar eclipse occurs when the moon and the Sun appear to be in the exact same place in the sky during a new moon phase. When that happens, the moon blocks the light of the Sun for up to seven minutes. Because solar eclipses can be seen only from certain places on Earth, some people travel around the world—sometimes to remote places—to view them.

A lunar eclipse occurs when Earth is positioned between the Sun and the moon and casts its shadow on the moon. In contrast to solar eclipses, lunar eclipses are visible from every place on Earth from which the moon can be seen.

---

vernal and autumnal equinoxes occur on March 20 or 21 and September 22 or 23, respectively.

The axial inclination of 66.6 degrees (or 23.5 degrees from the perpendicular) explains the significance of certain parallels on the earth. The noon sun shines directly overhead on the earth at varying latitudes on different days—between 23.5°S and 23.5°N. The parallels at 23.5°S and 23.5°N are called the Tropics of Capricorn and Cancer, respectively.

During the winter and summer solstices, the area on the earth between the Arctic Circle (at 66.5°N) and the North Pole has twenty-four hours of darkness and daylight, respectively. The same phenomena occurs for the area between the Antarctic Circle (at 66.5°S) and the South Pole, except that the seasons are reversed in the Southern Hemisphere. At the poles, the Sun is below the horizon for six months of the year.

For those living outside the tropics (poleward of 23.5 degrees north and south latitude), the noon sun will never shine directly overhead. Hours of daylight will also vary greatly during the year. For example, daylight will range from approximately nine hours during the winter solstice to fifteen hours during the summer solstice for persons living near 40°N, such as in Philadelphia, Denver, Madrid, and Beijing.

Solar Radiation

Given the size of the earth and its distance from the Sun, it is estimated that this planet receives only about one two-billionth part of the total energy released by the Sun. However, this seemingly small amount is enough to drive the massive oceanic and atmospheric circulation systems and to support all life processes on Earth.

Solar energy is not evenly distributed on Earth. The higher the angle of the Sun in the sky, the greater the duration and intensity of the insolation.

To illustrate this, note how easy it is look at the Sun when it is very low on the horizon—near dawn and sunset. At those times, the Sun's rays have to penetrate much more of the atmosphere, so more of the sunlight is absorbed. When the Sun's rays are coming in at a low angle, the same solar energy is spread over a larger area, thereby leading to less insolation per unit of area. Thus, the equatorial region receives much more solar energy than the polar region. This radiation imbalance would make the earth decidedly less habitable were it not for the atmospheric and oceanic circulation systems (such as the warm Gulf Stream) that move the excess heat from the Tropics to the middle and high latitudes.

Robert M. Hordon

THE SEASONS

Earth's 365-day year is divided into seasons. In most parts of the world, there are four seasons—winter, spring, summer, and fall (also called autumn). In some tropical regions—those close to the equator—there are only two seasons. In areas close to the equator, temperatures change little throughout the year; however, amounts of rainfall vary greatly, resulting in distinct wet and dry seasons. The polar regions of the Arctic and Antarctic also have little variation in temperature, remaining cold throughout the year. Their seasons are light and dark, because the Sun shines almost constantly in the summer and hardly at all in the winter.

The four seasons that occur throughout the northern and southern temperate zones—between the tropics and the polar regions—are climatic seasons, based on temperature and weather changes. Winter is the coldest season; it is the time when days are short and few crops can be grown. It is followed by spring, when the days lengthen and the earth warms; this is the time when planting typically be-

gins, and animals that hibernate (from the French word for winter) during the winter leave their dens.

Summer is the hottest time of the year. In many areas, summer is marked by drought, but other regions experience frequent thunderstorms and humid air. In the fall, the days again become shorter and cooler. This is the time when many crops are harvested. In ancient cultures, the turning of the seasons was marked by festivals, acknowledging the importance of seasonal changes to the community's survival.

Each season is defined as lasting three months. Winter begins at the winter solstice, which is the time when the Sun is farthest from the equator. In the Northern Hemisphere, this occurs on December 21 or 22, when the Sun is directly over the tropic of Capricorn. Summer begins at the other solstice, June 20 or 21 in the Northern Hemisphere, when the Sun is directly over the tropic of Cancer. The winter solstice is the shortest day of the year; the summer solstice is the longest.

Spring and fall begin on the two equinoxes. At an equinox, the Sun is directly above the earth's equator and the lengths of day and night are approximately equal everywhere on Earth. In the Northern Hemisphere, the vernal (spring) equinox occurs on March 21 or 22; in the Southern Hemisphere, it is the autumnal (fall) equinox. The Northern Hemisphere's autumnal equinox (and the Southern Hemisphere's vernal equinox) occurs September 22 or 23.

Seasons and the Hemispheres

The relationship of the seasons to the calendar is opposite in the Northern and Southern Hemispheres. On the day that a summer solstice occurs in the Northern Hemisphere, the winter solstice occurs in the Southern Hemisphere. Thus, when it is summer in the Southern Hemisphere, it is winter in the Northern Hemisphere, and vice versa.

The Sun and the Seasons

The reason why summers and winters differ in the temperate zones is often misunderstood. Many people think that winter happens when the Sun is more distant from Earth than it is in summer. What causes Earth's seasons is not the changing distances between the earth and the Sun, but the tilt of the earth's axis. A line drawn from the North Pole to the South Pole through the center of the earth (the earth's axis) is not perpendicular to the plane of the earth's orbit (the ecliptic). The earth's axis and the perpendicular to the ecliptic make an angle of 23.5 degrees. This tilts the Northern Hemisphere toward the Sun when the earth is on one side of its orbit around the Sun, and tilts the Southern Hemisphere toward the Sun when the earth moves around to the Sun's opposite side. When the Sun appears to be at its highest in the sky, and its rays are most direct, summer occurs. When the Sun appears to be at its lowest, and its rays are indirect, there is winter.

Local Phenomena

Local conditions can have important effects on seasonal weather. At locations near oceans, sea breezes develop during the day, and evenings are characterized by land breezes. Sea breezes bring cooler ocean air in toward land. This results in temperatures at the shore often being 5 to 11 degrees Fahrenheit (3 to 6 degrees Celsius) lower than temperatures a few miles inland.

At night, when land temperatures are lower than ocean temperatures, land breezes move air from the land toward the water. As a result, coastal regions have less seasonal temperature variations than inland areas do. For example, coastal areas seldom become cold enough to have snow in the winter, even though inland areas at the same latitude do.

Hailstorms

Hail usually occurs during the summer, and is associated with towering thunderstorm clouds, called cumulonimbus. Hail is occasionally confused with sleet. Sleet is a wintertime event, and occurs when warmer layers of air sit above freezing layers near the ground. Rain that forms in the warmer, upper layer solidifies into tiny ice pellets in the lower, sub-freezing layer before hitting the ground.

Hail is an entirely different phenomenon. When cold air plows into warmer, moist air—called a cold front boundary—powerful updrafts of rising air can be created. The warm, moist air propelled upward by the heavier cold air can reach velocities approaching 100 miles (160 kilometers) per hour. Ice crystals form above the freezing level in the cumulonimbus clouds and fall into lower, warmer parts of the clouds, where they become coated with water. Picked up by an updraft, the coated ice crystals are carried back to a higher, colder levels where their water coatings freeze. This cycle can repeat many times, producing hailstones that have multiple, concentric layers of ice.

Hailstorms can be very damaging. Hail can ruin crops, dent car bodies, crack windshields, and injure people. The Midwest of the United States is particularly susceptible to hailstorms. There, warm, moist air from the Gulf of Mexico often meets much colder, drier air originating in Canada. This combination produces the extreme atmospheric instability necessary for that kind of weather.

Alvin S. Konigsberg

EARTH'S INTERIOR

EARTH'S INTERNAL STRUCTURE

Earth is one of the nine known planets in the Sun's solar system that formed from a giant cloud of cosmic dust called a nebula. This event is thought to have happened between 4.44 billion years ago (based on the age of the oldest-known Moon rock) and 4.56 billion years ago (the age of meteorite bombardment). After Earth's formation, heat released by colliding particles combined with the heat energy released by the decay of radioactive elements to cause some or all of Earth's interior to melt. This melting began the process of differentiation, which allowed the heavier elements, mainly iron and nickel, to sink toward Earth's center while the lighter, rocky components moved upward, as a result of the contrast in density of the earth's forming elements.

This process of differentiation was probably the most important event of Earth's early history. It changed the planet from a homogeneous mixture with neither continents nor oceans to a planet with three layers: a dense core beginning at 1,800 miles (2,900 km.) deep and ending at Earth's center, 3,977 miles (6,400 km.) below the surface; a mantle beginning between 3 and 44 miles (5-70 km.) deep and ending at Earth's core; and a crust going from Earth's surface to about 3-6 miles (5-10 km.) deep for oceanic crust and 22-44 miles (35-70 km.) deep for continental crust.

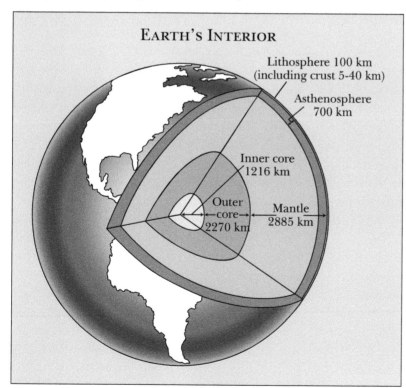

EARTH'S INTERIOR

Lithosphere 100 km
(including crust 5-40 km)

Asthenosphere
700 km

Inner core
1216 km

Outer
core
2270 km

Mantle
2885 km

Layering of the Earth

Earth's layers can be classified either by their composition (the traditional method) or by their mechanical behavior (strength). Compositional classification identifies several distinct concentric layers, each with its own properties. The outermost layer of Earth is the crust or skin. This is divided into continental and oceanic crusts. The continental crust varies in thickness between 22 and 25 miles (35 and 40 km.) under flat continental regions and up to 44 miles (70 km.) under high mountains. The oceanic crust is made up of igneous rocks rich in iron and magnesium, such as basalt and peridotite. The upper continental crust is composed mainly of alumino-silicates. The old-

PROPERTIES OF SEISMIC WAVES

Seismologists use two types of body waves—primary (P-waves) and secondary (S-waves) waves—to estimate seismic velocities of the different layers within the earth. In most rock types P-waves travel between 1.7 and 1.8 times more quickly than S-waves; therefore, P-waves always arrive first at seismographic stations. P-waves travel by a series of compressions and expansions of the material through which they travel. P-waves can travel through solids, liquids, or gases. When P-waves travel in air, they are called sound waves.

The slower S-waves, also called shear waves, move like a wave in a rope. This movement makes the S-wave more destructive to structures like buildings and highway overpasses during earthquakes. Because S-waves can travel only through solids and cannot travel through Earth's outer core, seismologists concluded that Earth's outer core must be liquid or at least must have the properties of a fluid.

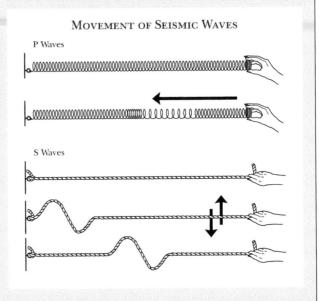

MOVEMENT OF SEISMIC WAVES

P Waves

S Waves

est continental crustal rock exceeds 3.8 billion years, while oceanic crustal rocks are not older than 180 million years. The oceanic crust is heavier than the continental crust.

Earth's next layer is the mantle, which is made up mostly of ferro-magnesium silicates. It is about 1,800 miles (2,900 km.) thick and is separated into the upper and lower mantle. Most of Earth's internal heat is contained within the mantle. Large convective cells in the mantle circulate heat and may drive plate-tectonic processes.

The last layer is the core, which is separated into the liquid outer core and the solid inner core. The outer core is 1,429 miles (2,300 km.) thick, twice as thick as the inner core. The outer core is mainly composed of a nickel-iron alloy, while the inner core is almost entirely composed of iron. Earth's magnetic field is believed to be controlled by the liquid outer core.

In the mechanical layering classification of the earth's interior, the layers are separated based on mechanical properties or strength (resistance to flowing or deformation) in addition to composition. The uppermost layer is the lithosphere (sphere of rock), which comprises the crust and a solid portion of the upper mantle. The lithosphere

is divided into many plates that move in relation to each other due to tectonic forces. The solid lithosphere floats atop a semiliquid layer known as the asthenosphere (weak sphere), which enables the lithosphere to move around.

Exploring Earth's Interior

Volcanic activity provides natural samples of the outer 124 miles (200 km.) of Earth's interior. Meteorites—samples of the solar system that have collided with Earth—also provide clues about Earth's composition and early history. The most ambitious human effort to penetrate Earth's interior was made by the former Soviet Union, which drilled a super-deep research well, named the Kola Well, near Murmansk, Russia. This was an attempt to penetrate the crust and reach the upper mantle. The reported depth of the Kola Well is a little more than 7.5 miles (12 km.). Although impressive, the drilled depth represents less than 0.2 percent of the distance from the earth's surface to its center.

A great deal of knowledge about Earth's composition and structure has been obtained through computer modeling, high-pressure laboratory experiments, and meteorites, but most of what is known about Earth's interior has been acquired by studying seismic waves generated by earthquakes and nuclear explosions. As seismic waves are transmitted, reflected, and refracted through the earth, they carry information to the surface about the materials through which they have traveled. Seismic waves are recorded at receiver stations (seismographic stations) and processed to provide a picturelike image of Earth's interior.

Changes in P- and S-wave velocities within Earth reveal the sequence of layers that make up Earth's interior. P-wave velocity depends on the elasticity, rigidity, and density of the material. By contrast, S-wave velocity depends only on the rigidity and density of the material. There are sharp variations in velocity at different depths, which correspond to boundaries between the different layers of Earth. P-wave velocity within crustal rocks ranges from 3.6-4.2 miles (6-7 km.) per second.

The boundary between the crust and the mantle is called the Mohorovičić discontinuity or Moho. At Moho, P-wave velocity increases from 4.2-4.8 miles (7-8 km.) per second. Beyond the crust-mantle boundary, P-wave velocity increases gradually up to about 8.1 miles (13.5 km.) per second at the core-mantle boundary. At this depth, S-waves are not transmitted and P-wave velocity, decreases from 8.1 to 4.8 miles (13.5 to 8 km.) per second, which strongly supports the concept that the outer core is liquid, since S-waves cannot travel through liquids. As P-waves enter the inner core, their velocity again increases, to about 6.8 miles (11.3 km.) per second.

Earth's interior seems to be characterized by a gradual increase with depth in temperature, pressure, and density. Extensive experimental and modeling work indicates that the temperature at 62 miles (100 km.) is between 1,200 and 1,400 degrees Celsius (2,192 to 2,552 degrees Fahrenheit). The temperature at the core-mantle boundary—about 1,802 miles (2,900 km.) deep—is calculated to be about 8,130 degrees Fahrenheit (4,500 degrees Celsius). At Earth's center the temperature may exceed 12,092 degrees Fahrenheit (6,700 degrees Celsius). Although at Earth's surface, heat energy is slowly but continuously lost as a result of outgassing, such as from volcanic eruptions, its interior remains hot.

Seismic Tomography and Future Exploration
Seismic tomography is one of the newest tools that earth scientists are using to develop three-dimensional velocity images of Earth's interior. In seismic tomography, several crossing seismic waves from different sources (earthquakes and nuclear explosions) are analyzed in much the same way that computerized axial tomography (CAT) scanners are used in medicine to obtain images of human organs. Seismic tomography is providing two- and three-dimensional images from the crust to the core-mantle boundary. Fast P-wave velocities have been correlated to cool material—for example, a piece of sinking lithosphere (cool rigid layer) such as in regions underneath the Andes Mountains (subduction zone); slow P-wave velocities have been correlated with hot materials—for example, rising mantle plumes of hot spots such as the one responsible for volcanic activity in the Hawaiian Islands.

Rubén A. Mazariegos-Alfaro

PLATE TECTONICS

The theory of plate tectonics provides an explanation for the present-day structure of the large landforms that constitute the outer part of the earth. The theory accounts for the global distribution of continents, mountains, hills, valleys, plains, earthquake activity, and volcanism, as well as various associations of igneous, metamorphic, and sedimentary rocks, the formation and location of min-

MAJOR TECTONIC PLATES AND MID-OCEAN RIDGES

Types of Boundaries: Divergent // Convergent ⤳ Transform /

eral resources, and the geology of ocean basins. Everything about the earth is related either directly or indirectly to plate tectonics.

Basic Theory

Plate-tectonic theory is based on an Earth model in which a rigid, outer shell—the lithosphere—lies above a hotter, weaker, partially molten part of the mantle called the asthenosphere. The lithosphere varies in thickness between 6 and 90 miles (10 and 150 km.), and comprises the crust and the underlying, upper mantle. The asthenosphere extends from the base of the lithosphere to a depth of about 420 miles (700 km.). The brittle lithosphere is broken into a pattern of internally rigid plates that move horizontally relative to each other across the earth's surface.

More than a dozen plates have been distinguished, some extending more than 2,500 miles (4,000 km.) across. Exhibiting independent motion, the plates grind and scrape against each other,

similar to chunks of ice in water, or like giant rafts cruising slowly on the asthenosphere. Most of the earth's dynamic activity, including earthquakes and volcanism, occurs along plate boundaries. The global distribution of these tectonic phenomena delineates the boundaries of the plates.

Geological observations, geophysical data, and theoretical models support the existence of three types of plate boundaries. Divergent boundaries occur where adjacent plates move away from each other. Convergent boundaries occur where adjacent plates move toward each other. Transform boundaries occur where plates slip past one another in directions parallel to their common boundaries.

The continents were formed by the movement at plate boundaries, and continental landforms were generated by volcanic eruptions and continental plates colliding with each other. The velocity of plate movement varies from plate to plate and even within portions of the same plate, ranging from 0.8 to 8 inches (2 to 20 centimeters) per year. The rates

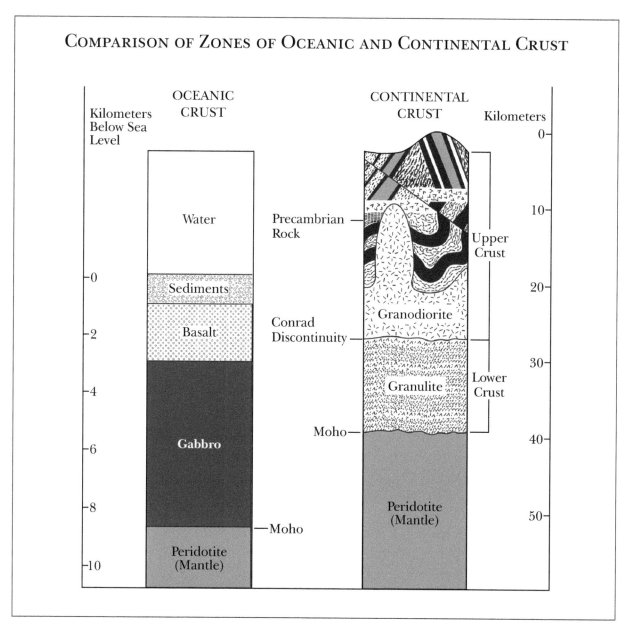

COMPARISON OF ZONES OF OCEANIC AND CONTINENTAL CRUST

are calculated from the distance to the midoceanic ridge crests, along with the age of the sea floor as determined by radioactive dating methods.

Convection currents that are driven by heat from radioactive decay in the mantle are important mechanisms involved in moving the huge plates. Convection currents in the earth's mantle carry magma (molten rock) up from the asthenosphere. Some of this magma escapes to form new lithosphere, but the rest spreads out sideways beneath the lithosphere, slowly cooling in the process. Assisted by gravity, the magma flows outward, dragging the overlying lithosphere with it, thus continu-

ing to open the ridges. When the flowing hot rock cools, it becomes dense enough to sink back into the mantle at convergent boundaries.

A second plate-driving mechanism is the pull of dense, cold, down-flowing lithosphere in a subduction zone on the rest of the trailing plate, further opening up the spreading centers so magma can move upward.

Divergent Plate Boundaries
During the 1950s and 1960s, oceanographic studies revealed that Earth's seafloors were marked by a nearly continuous system of submarine ridges,

THE SUPERCONTINENTS

The theory of plate tectonics explains the present-day distribution of major landforms, seismic and volcanic activity, and physiographic features of ocean basins. Many scientists also use the theory to explain the history of Earth's surface. Evidence indicates that the modern continents once formed a single landmass called Pangaea, meaning "all lands." According to the theory of plate tectonics, approximately 200 million years ago Pangaea began to split into two supercontinents, Laurasia and Gondwanaland. Eventually, as a result of tectonic forces, Laurasia split into North America, Europe, and most of Asia. Gondwanaland broke up into India, South America, Africa, Australia, and Antarctica.

LAURASIA AND GONDWANALAND

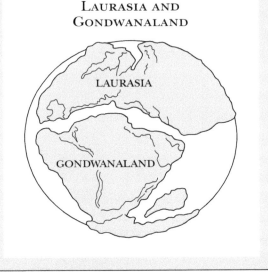

PANGAEA

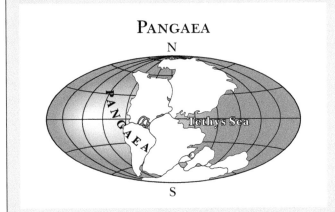

more than 40,000 miles (64,000 km.) in length. Detailed investigations revealed that the midoceanic ridge system has a central rift valley that runs along its length and that the ridge system is associated with volcanic and earthquake activity. The earthquakes are frequent, shallow, and mild.

Magnetic studies of the seafloor indicate that the oceanic lithosphere has been segmented into a series of long magnetic strips that run parallel to the axis of the midoceanic ridges. On either side of the ridge, the ocean floor consists of alternating bands of rock, magnetized either parallel to or exactly opposite of the present-day direction of the earth's magnetic field.

Midoceanic ridges, or divergent plate boundaries, are tensional features representing zones of weakness within the earth's crust, where new seafloor is created by the welling up of mantle material from the asthenosphere into cracks along the ridges. As rifting proceeds, magma ascends to fill in the fissures, creating new oceanic crust. Iron minerals within the magma become aligned to the existing Earth polarity as the rock cools and crystallizes. The oceanic floor slowly moves away from the oce-

anic ridge toward deep ocean trenches, where it descends into the mantle to be melted and recycled to the earth's surface to generate new rocks and landforms.

As the seafloor spreads outward from the rift center, about half of the material is carried to either side of the rift, which is later filled by another influx of molten basalt. When the polarity of the earth changes, the subsequent molten basalt is magnetized in the opposite polarity. The continuation of this process over geologic time leads to the young geologic age of the seafloor and the magnetic symmetry around the midoceanic ridges.

Not all spreading centers are underneath the oceans. An example of continental rifting in its embryonic stage can be observed in the Red Sea, where the Arabian plate has separated from the African plate, creating a new oceanic ridge. Another modern-day example of continental divergent activity is East Africa's Great Rift Valley system. If this rifting continues, it will eventually fragment Africa, producing an ocean that will separate the resulting pieces. Through divergence, large plates are made into smaller ones.

Convergent Plate Boundaries

Because Earth's volume is not changing, the increase in lithosphere created along divergent boundaries must be compensated for by the destruction of lithosphere elsewhere. Otherwise, the radius of Earth would change. The compensation occurs at convergent plate boundaries, where plates are moving together. Three scenarios are possible along convergent boundaries, depending on whether the crust involved is oceanic or continental.

If both converging plates are made of oceanic crust, one will inevitably be older, cooler, and denser than the other. The denser plate eventually subducts beneath the less-dense plate and descends into the asthenosphere. The boundary along the two interacting plates, called a subduction zone, forms a trench. Some trenches are more than 620 miles (1,000 km.) long, 62 miles (100 km.) wide, and 6.8 miles (11 km.) deep. Heated by the hot asthenosphere beneath, the subducted plate becomes hot enough to melt.

Because of buoyancy, some of the melted material rises through fissures and cracks to generate volcanoes along the overlying plate. Over time, other parts of the melted material eventually migrate to a divergent boundary and rise again in cyclic fashion to generate new seafloor. The volcanoes generated along the overriding plate often form a string of islands called island arcs. Japan, the Philippines, the Aleutians, and the Mariannas are good examples of island arcs resulting from subduction of two plates consisting of oceanic lithosphere. Intense earthquakes often occur along subduction zones.

If the leading edge of one of the two convergent plates is oceanic crust and the other is continental crust, the oceanic plate is always the one subducted, because it is always denser. A classic example of this case is the western boundary of South America. On the oceanic side of the boundary, a trench was formed where the oceanic plate plunged underneath the continental plate. On the continental side, a fold mountain belt—the Andes—was formed as the oceanic lithosphere pushed against continental lithosphere.

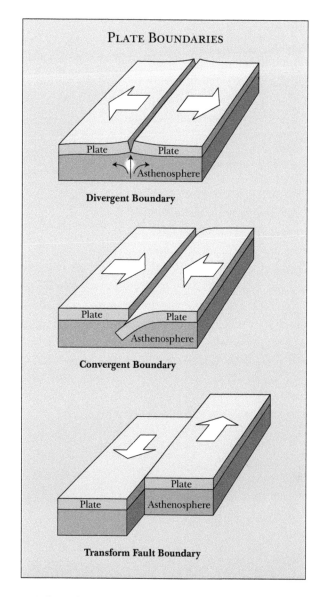

PLATE BOUNDARIES

Divergent Boundary

Convergent Boundary

Transform Fault Boundary

When the oceanic plate descends into the mantle, some of the material melts and works its way up through the mountain belt to produce rather violent volcanoes. The boundary between the plates is a region of earthquake activity. The earthquakes range from shallow to relatively deep, and some are quite severe.

The last type of convergent plate boundary involves the collision of two continental masses of lithosphere, which can result in folding, faulting, metamorphism, and volcanic activity. When the plates collide, neither is dense enough to be forced into the asthenosphere. The collision compresses and thickens the continental edges, twisting and deforming the rocks and uplifting the land to form unusually high fold mountain belts. The prototype example is the collision of India with Asia, resulting in the formation of the Himalayas. In this case, the earthquakes are typically shallow, but frequent and severe.

Transform Plate Boundaries
The actual structure of a seafloor spreading ridge is more complex than a single, straight crack. Instead, ridges comprise many short segments slightly offset from one another. The offsets are a special kind of fault, or break in the lithosphere, known as a transform fault, and their function is to connect segments of a spreading ridge. The opposite sides of a transform fault belong to two different plates that are grinding against each other in opposite directions.

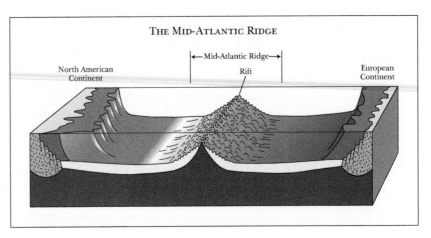

THE MID-ATLANTIC RIDGE

Transform faults form the boundaries that allow the plates to move relative to each another. The classic case of a transform boundary is the San Andreas Fault. It slices off a small piece of western California, which rides on the Pacific plate, from the rest of the state, which resides on the North American plate. As the two plates scrape past each other, stress builds up, eventually being released in earthquakes that can be quite violent.

Mantle Plumes and Hot Spots
Most plate tectonic features are near plate boundaries, but the Hawaiian Islands are not. In the late twentieth century, the only active volcanoes in the Hawaiian Islands were on the island of Hawaii, at the southeast end of the chain. Radiometric dating and examination of states of erosion show that, when proceeding along the chain to the northwest, successive islands are progressively older.

Evidently, the same heat source produced all the volcanoes in the Hawaiian chain. Known as a mantle plume, it has remained stationary while the Pacific plate rides over it, producing a volcanic trail from which absolute motion of the plate can be determined. Since mantle plumes do not move with the plates, the plumes must originate beneath the lithosphere, probably far below it. Resulting volcanoes are called hot spots to distinguish them from subduction-zone volcanoes. Iceland is a good example of a hot spot, as is Yellowstone. At least 100 hot spots are distributed around Earth.

Alvin K. Benson

CALIFORNIA'S SAN ANDREAS FAULT

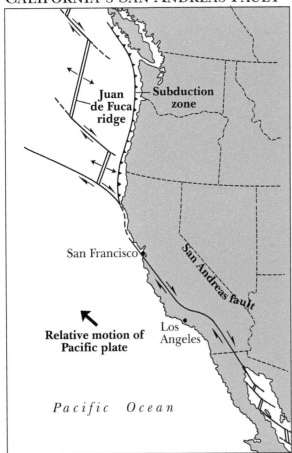

VOLCANOES

Volcanoes form mountains both on land and in the sea and either do it on a grand scale or merely create minute bumps on the seafloor. Volcanoes do not occur in a random pattern, but are found in distinct zones that are related to plate dynamics. Each of the three types of volcanism on Earth is characterized by specific types of eruptions and magma compositions. Molten magma is the rock material below the earth's crust that forms igneous rock as it cools.

Types of Volcanoes

Geologists generally group volcanoes into four main kinds—cinder cones, composite volcanoes, shield volcanoes, and lava domes.

Cinder cones are built from congealed lava ejected from a single vent. As the gas-charged lava is blown into the air, it breaks into small fragments that solidify and fall as *cinders* around the vent to form a circular or oval cone. Most cinder cones have a bowl-shaped *crater* at the summit and rarely rise more than a thousand feet or so above their surroundings. Cinder cones are numerous in western North America and in other volcanic terrains of the world.

Composite volcanoes —sometimes called stratovolcanoes—include some of the Earth's grandest mountains, including Mount Fuji in Japan, Mount Cotopaxi in Ecuador, Mount Shasta in California, Mount Hood in Oregon, and Mount St. Helens and Mount Rainier in Washington. The essential feature of a composite volcano is a conduit system through which magma deep in the Earth's crust rises to the surface. They are typically steep-sided, symmetrical cones of large dimension built of alternating layers of lava flows, volcanic ash, cinders, blocks, and bombs. They may rise as much as 8,000 feet above their bases. Most have a crater at the summit that contains a central vent or a clustered group of vents. Lavas either flow through breaks in the crater wall or fissures on the flanks of the cone.

Shield volcanoes, the third type of volcano, are built almost entirely of fluid lava flows that pour out in all directions from a central vent, or group of vents, building a broad, gently sloping cone. They are built up slowly as thousands of highly fluid lava flows—basalt lava—spread over great distances, and then cool into thin sheets. Some of the largest volcanoes in the world are shield volcanoes. The Hawaiian Islands are composed of linear chains of these volcanoes including Kilauea and Mauna Loa on the island of Hawaii—two of the world's most active volcanoes. The floor of the ocean is more than 15,000 feet deep at the bases of the islands. As Mauna Loa, the largest of the shield volcanoes (and also the world's largest active volcano), projects 13,679 feet above sea level, its top is over 28,000 feet above the deep ocean floor.

Volcanic Composition

Volcanoes in the midocean ridges and plume environments draw most of their magmas from the earth's mantle and produce mainly dark, magnesium-rich basaltic magmas. When basaltic magmas accumulate in the continental crust (for example, at Yellowstone), the large-scale crustal melting leads to rhyolitic volcanism, the volcanic equivalent of granites. Arc magmas cover a wider range of magmatic compositions, ranging from arc basalt to light-colored, silica-rich rhyolites; the latter are commonly erupted in the form of the silica-rich volcanic rock known as pumice, or the black volcanic glass known as obsidian. Andesites, named after the Andes Mountains, are a common volcanic rock in stratovolcanoes, intermediate in composition between basalt and rhyolite.

Magmas form from several processes that lead to partial melting of a solid rock. The simplest is adding heat—for example, plumes carrying heat from deep levels in the mantle to shallower levels, where melting occurs. Decompressional (lowering the pressure) melting of the mantle occurs where the

SOME VOLCANIC HOT SPOTS AROUND THE WORLD

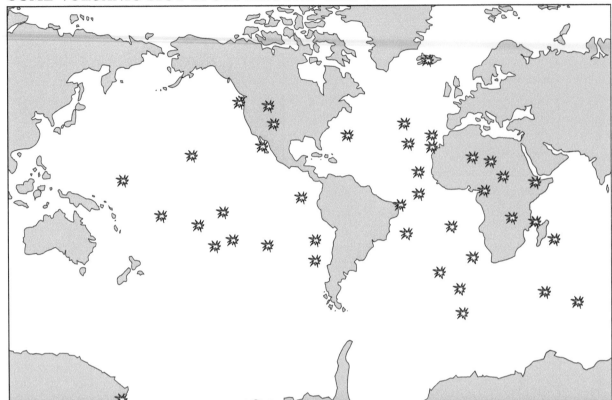

ocean floor is thinned or carried away by seafloor spreading in midocean ridge environments.

Genesis of Magma

Adding a "flux" to a solid mineral mixture may lower the substance's melting point. The most common theory about arc magma genesis invokes the addition of a low-melting-point substance to the arc mantle, a layer of mantle material at about 60 to 90 miles (100 to 150 km.) below the volcanic arc. The relatively dry arc mantle would usually start to melt at about 2,100 to 2,300 degrees Fahrenheit (1,200 to 1,300 degrees Celsius). However, the addition of water and other gases can lower the melting point of the mixture. The water and its dissolved chemicals are supposedly derived from the subducted slab, the former ocean floor that is pushed back into the earth.

The sequence of events is as follows: New basaltic ocean floor forms at midocean ridge volcanoes. The new hot magma interacts with seawater, leading to vents at the seafloor with their mineralized deposits. The seafloor becomes hydrated, and sulfur and chlorine from seawater are locked up in newly formed minerals. During subduction, this altered seafloor with slivers of sediment, including limestone, is gradually warmed up and starts to decompose, adding a flux to the surrounding mantle rocks. The mantle rocks then start to melt, and these magmas with minor inherited oceanic materials start to rise and pond at the bottom of the crust. There the magmas sit and wait for an opportunity to erupt, while cooling and crystallizing. Thus, arc magmas bear a chemical signature of subducted oceanic components while their chemical compositions range from basalt to rhyolite.

Volcanic Eruptions

Volcanic eruptions occur as a result of the rise of magma into the volcano (from depths as great as several miles) and then into the throat of the volcano. In basaltic volcanoes, the magmas have relatively little gas, and the magma simply overflows and forms large lava flows, sometimes associated

VOLCANIC ERUPTION AND CALDERA FORMATION

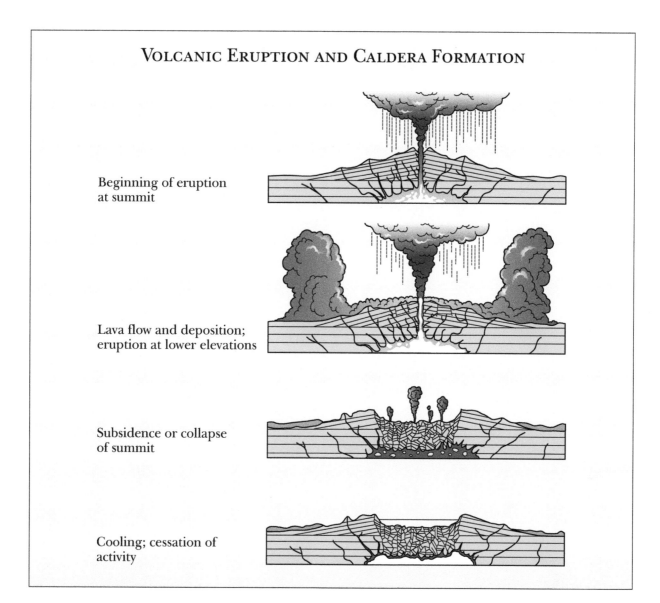

Beginning of eruption
at summit

Lava flow and deposition;
eruption at lower elevations

Subsidence or collapse
of summit

Cooling; cessation of
activity

with fire fountains. Stratovolcanoes can erupt regularly with small explosions or catastrophically after long periods of dormancy. Mount Stromboli, a volcano in Italy, erupts every twenty minutes, with an explosion that creates a column 650 to 980 feet (200 to 300 meters) high. Mount St. Helens in the U.S. state of Washington had a catastrophic eruption in 1980 after about 200 years of dormancy. It emitted an ash plume that reached more than 12 miles (20 km.) into the atmosphere.

After long magma storage periods in the crust, crystallization and melting of crustal material can lead to silica-rich magmas. These are viscous and can have high dissolved water contents—up to 4 to 6 percent by weight. When these magmas break out,

the eruption can be violent and form an eruption column 12 to 35 miles (20 to 55 km.) high. Many cubic miles of magma can be ejected. This leads to so-called plinian ash falls, with showers of pumice and ash over thousands of square miles, with the ash commonly carried around the globe by the high-level winds known as jet streams.

If the volume of ejected magma is large, the volcano empties itself and collapses into the hole, leading to a caldera—a volcanic collapse structure. The caldera at Crater Lake in Oregon is related to a large pumice eruption about 76,000 years ago. Basaltic volcanoes can also form collapse calderas when large volumes of lava have been extruded in a short time. Examples of famous basaltic calderas can be found in Hawaii's Mount Kilauea and the Galapagos Islands.

Volcanic Plumes

The dynamics of volcanic plumes has been studied from eruption photographs, experiments, and theoretical work. The rapidly expanding hot gases force the viscous magma out of the throat of the volcano, where it freezes into pumice. The kinetic energy of the ejected mass carries it 2 to 2.5 miles (3-4 km.) above the volcano. During this phase, air is entrained in the column, diluting the concentration of ash and pumice particles. The hot particles heat the entrained air, the mixture of hot air and solids becomes less dense than the surrounding atmosphere, and a buoyant column rises high into the sky.

The height of an eruption column is not directly proportional to the force of the eruption but is strongly dependent on the rate of heat release of the volcano. If little of the entrained air is heated up, the column will collapse back to the ground and an ash flow forms, which may deposit ash around the volcano. These types of eruptions are among the most devastating, creating glowing ash clouds traveling at speeds up to 60 miles (100 km.) per hour, burning everything in their path. The 1902 eruption of Mount Pelée on Martinique in the Caribbean was such an eruption and killed nearly 30,000 people in a few minutes.

Many volcanoes that are high in elevation are glaciated, and their eruptions lead to large-scale ice melting and possibly mixing of water, magma, and volcanic debris. Massive hot mudflows can race down from the volcano, following river valleys and filling up low areas. The 1980 Mount St. Helens eruption created many mudflows, some of which reached the Pacific Ocean, ninety miles to the west. A catastrophic mudflow event occurred in 1984 at Nevado del Ruiz, a volcano in Colombia, where 20,000 people were buried in mud and perished. When magma intrudes under the ice, meltwater can accumulate and then escape catastrophically, but such meltwater bursts are rare outside Iceland.

Minerals and Gases in Eruptions

The gas-rich character of arc magmas leads to fluid escape at various levels in the volcanoes, and these fluids tend to be rich in chlorine. They can transport metals such as copper, lead, zinc, and gold at high concentrations, and lead to the enrichment of these metals in the fractured volcanic rocks. Many of the world's largest copper ore deposits are associated with older arc volcanism, where erosion has removed most of the volcanic structure and laid the volcano innards bare. Many active volcanoes have modern hydrothermal (hot-water) systems, leading to acid hot springs and crater lakes and the potential to harness geothermal energy. Some areas in Japan, New Zealand, and Central America have an abundance of geothermal energy resources, which are gradually being developed.

Apart from the dangers of eruptions, continuous emissions of large amounts of sulfur dioxide, hydrochloric acid, and hydrofluoric acid present a danger of air pollution and acid rain. Incidences of emphysema and other irritations of the respiratory system are common in people living on the slopes of active volcanoes. The large lava emissions in Iceland in the eighteenth century led to acid fogs all over Europe. Many cattle died in Iceland during this period from the hydrofluoric acid vapors. High levels of fluorine in drinking water can lead to fluorosis, a disease that attacks the bone structure. The discharge of highly acidic fluids from hot springs and crater lakes can cause widespread environmental contamination, which can present a danger for crops gathered from fields irrigated with these waters and for local ecosystems in general.

Johan C. Varekamp

GEOLOGIC TIME SCALE

A major difference between the geosciences (earth sciences) and other sciences is the great enormity of their time scale. One might compare the magnitude of geologic time for geoscientists to the vastness of space for astronomers. Every geological process, such as the movement of crustal plates (plate tectonics), the formation of mountains, and the advance and retreat of glaciers, must be considered within the context of time.

Although certain geologic events, such as floods and earthquakes, seem to occur over short periods of time, the vast majority of observed geological features formed over a great span of time. Consequently, modern geoscientists consider Earth to be exceedingly old. Using radiometric age-dating techniques, they calculate the age of Earth as 4.6 billion years old.

Early miners were probably the first to recognize the need for a scale by which rock and mineral units could be compared over large geographic areas. However, before a time scale—and even geology as a science—could develop, certain principles had to be established. This did not occur until the late eighteenth century when James Hutton, a Scottish naturalist, began his extensive examinations of rock relationships and natural processes at work on the earth. His work was amplified by Charles Lyell in his textbook *Principles of Geology* (1830-1833). After careful observation, Hutton concluded that the natural processes and functions he observed had operated in the same basic manner in the past, and that, in general, natural laws were invari-

able. That idea became known as the principle of uniformitarianism.

The Birth of Stratigraphy

In 1669 Nicholas Steno, a Danish physician working in Italy, recognized that horizontal rock layers contained a chronological record of Earth history and formulated three important principles for interpreting that history. The principle of superposition states that in a succession of undeformed strata, the oldest stratum lies at the bottom, with successively younger ones above. The principle of original horizontality states that because sedimentary particles settle from fluids under gravitational influence, sedimentary rock layers must be horizon-

---

EARTH'S HISTORY COMPRESSED INTO ONE CALENDAR YEAR

One way to visualize events in Earth's history is to compress geologic events into a single calendar year. Earth's birth, 4.6 billion years ago, would occur during the first minute of January 1. The first three-quarters of Earth's history is obscure and would take place from January to mid-October. During this time, Earth gained an oxygenated atmosphere, and the earliest life-forms evolved. The first organisms with hard parts preserved in the fossil record (approximately 570 million years ago) would appear around November 15. The extinction of the dinosaurs (65 million years ago) would occur on Christmas

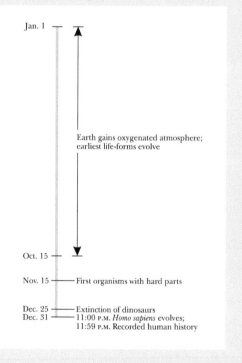

Day. Homo sapiens would first appear at approximately 11 P.M. on December 31, and all of recorded human history would occur in the last few seconds of New Year's Eve.

tal; if not, they have suffered from subsequent disturbance. The principle of original lateral continuity states that strata originally extended in all directions until they thinned to zero or terminated against the edges of the original area of deposition.

In the late eighteenth century, the English surveyor William Smith recognized the wide geographic uniformity of rock layers and discovered the utility of fossils in correlating these layers. By 1815, Smith had completed a geologic map of England and was able to correlate English rock layers with layers exposed across the English Channel in France.

From the need to classify and organize rock layers into an orderly form arose a subdiscipline of modern geology—stratigraphy, the study of rock layers and their age relationships. In 1835 two British geologists, Adam Sedgwick and Roderick Murchison, began organizing rock units into a formal stratigraphic classification. Large divisions, called eras, were based upon well known and characteristic fossils, and included a number of smaller subdivisions, called periods.

The periods are often subdivided into smaller units called epochs. Each period is defined by a representative sequence of rock strata and fossils. For instance, the Devonian period is named for exposures of rock in Devonshire in southern England, while the Jurassic period is defined by strata exposed in the Jura Mountains in northern Switzerland.

Approximately 80 percent of Earth's history is included in the Cryptozoic eon (meaning obscure life). Fossils from the Cryptozoic eon are rare, and the rock record is very incomplete. After the Cryptozoic eon came the Paleozoic (ancient life), Mesozoic (middle life), and Cenozoic (recent life) eras. Most of the life forms that evolved during the Paleozoic and Mesozoic eras are now extinct, whereas 90 percent of the

life-forms that evolved up to the middle Cenozoic era still exist.

The Geologic Time Scale

The geologic time scale is continually in revision as new rock formations are discovered and dated. The ages shown in the table below are in millions of years ago (MYA) before the present and represent the beginning of that particular period. It would be impossible to list all the significant events in Earth's history, but one or two are provided for each period. Note that in the United States, the Carboniferous period has been subdivided into the Mississippian period (older) and the Pennsylvanian period (younger).

The Fossil Record

The word "fossil" comes from the Latin *fossilium*, meaning "dug from beneath the surface of the ground." Fossils are defined as any physical evidence of past life. Fossils can include not only shells, bones, and teeth, but also tracks, trails, and bur-

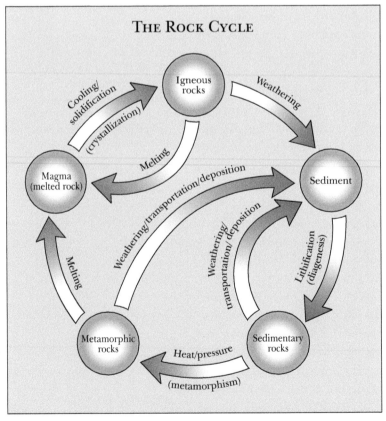

THE ROCK CYCLE

rows. The latter group are referred to as trace fossils. Fossils demonstrate two important truths about life on Earth: First, thousands of species of plants and animals have existed and later became extinct. Second, plants and animals have evolved through time, and the communities of life that have existed on Earth have changed.

Some organisms are slow to evolve and may exist in several geologic time periods, while others evolve quickly and are restricted to small intervals of time within a particular period. The latter, referred to as index fossils, are the most useful to geoscientists for correlating rock layers over wide geographic areas and for recognizing geologic time.

The fossil record is incomplete, because the process of preservation favors organisms with hard parts that are rapidly buried by sediments soon after death. For this reason, the vast majority of fossils are represented by marine invertebrates with exoskeletons, such as clams and snails. Under special circumstances, soft-bodied organism can be preserved, for instance the preservation of insects in amber, made famous by the feature film *Jurassic Park* (1993).

The Rock Cycle

A rock is a naturally formed aggregate of one or more minerals. Three types of rocks exist in the earth's crust, each reflecting a different origin. Igneous rocks have cooled and solidified from molten material either at or beneath Earth's surface. Sedimentary rocks form when preexisting rocks are weathered and broken down into fragments that accumulate and become compacted or cemented together. Fossils are most commonly found in sedimentary rocks. Metamorphic rocks form when heat, pressure, or chemical reactions in Earth's interior change the mineral or chemical composition and structure of any type of preexisting rock.

Over the huge span of geologic time, rocks of any one of these basic types can change into either of the other types or into a different form of the same type. For this reason, older rocks become increasingly more rare. The processes by which the various rock types change over time are illustrated in the rock cycle.

Larry E. Davis

EARTH'S SURFACE

INTERNAL GEOLOGICAL PROCESSES

The earth is layered into a core, a mantle, and a crust. The topmost mantle and the crust make up the lithosphere. Beneath this is a layer called the asthenosphere, which is composed of moldable and partly liquid materials. Heat transference within the asthenosphere sets up convection cells that diverge from hot regions and converge to cold regions. Consequently, the overlying lithosphere is segmented into ridged plates that are moved by the convection process. The hot asthenosphere does not rise along a line. This causes the development of a structure called a transform plate boundary, which is perpendicular to and offsetting the divergent boundary.

The topographic features at Earth's surface, such as mountains, rift valleys, oceans, islands, and ocean trenches, are produced by extension or compression forces that act along divergent, convergent, or transform plate boundaries. The extension and compression forces at Earth's surface are powered by convection within the asthenosphere.

Mountains and Depressions in Zones of Compression

Compression along convergent plate boundaries yields three types of mountain: island arcs that are partly under water; mountains along a continental edge, such as the Andes; and mountains at continental interiors, such as the Alps. At convergent plate boundaries, the denser of the two colliding plates slides down into the asthenosphere and causes volcanic activity to form on the leading edge of the upper plate. Island arcs such as the Aleutians and the Caribbean are formed when an oceanic plate descends beneath another oceanic plate.

Volcanic mountain chains such as the Andes of South America are formed when an oceanic plate descends beneath a continental plate. In both the island arc type and Andean type collisions, a deep depression in the oceans, called a trench, marks the place where neighboring plates are colliding and where the denser plates are pulled downward into the asthensophere. If the colliding plates are of similar density, neither plate will go into the asthenosphere. Instead, the edges of the neighboring plates will be folded and faulted and excess material will be pushed upward to form a block mountain, such as the mountain chain that stretches from the Alps through to the Himalayas. This type of mountain chain is not associated with a trench.

The Appalachians of the eastern United States are an example of the alpine type of mountain belt. When the Appalachians were forming 300 million years ago, rock layers were deformed. The deformation included folding to form ridges and valleys; fracturing along joint sets, with one joint set being parallel to ridges, while the other set is perpendicular; and thrust faulting, in which rock blocks were detached and shoved upward and northwestward.

Millions of years of erosion have reduced the height of the mountains and have produced topographic inversion in the foothills. Topographic inversion occurs because joints create wider fractures at upfolded ridges and narrower fractures at downfolded valleys. Erosion is then accelerated at upfolded ridges, converting ancient ridges into valleys, while ancient valleys stand as ridges. The Valley and Ridge Province of the Appalachians is noted for such topographic inversion.

West of the Valley and Ridge Province of the Appalachians is the Allegheny Plateau, which is bounded by a cliff on its eastern side. In general, plateaus are flat topped because the rock layer that covers the surface is resistant to weathering. The cliff side is formed by erosion along joint or fault surfaces.

The Sierra Nevada range, which formed 70 million years ago, is an example of an Andean type of mountain belt. Millions of years of erosion there has exposed igneous rocks that formed at depth. Over the years, the force of compression that formed the Sierras has evolved to form a zone of extension between the Sierras and the Colorado Plateau.

Mountains and Depressions in Zones of Extension

Extension is a strain that involves an increase in length and causes crustal thinning and faulting. Extension is associated with convergent boundaries, divergent boundaries, and transform boundaries.

Extension Associated with a Convergent Boundary

During the formation of the Sierra Nevada, an oceanic plate that was subducted beneath California declined at a shallow angle eastward toward the Colorado Plateau. Later, the subducted plate peeled off and molten asthenosphere took its place. From the asthenosphere, lava ascended through fractures to form volcanic mountains in Arizona and Utah, and lava flowed and volcanic ash fell as far west as California. The lithosphere has been heated up and has become buoyant, so the Colorado Plateau rises to higher elevations, and rock layers slide westward from it in a zone of extension that characterizes the Basin and Range Province.

In the extension zone, the top rock layers move westward on curved displacement planes that are steep at the surface and nearly horizontal at depth. When rock layers move westward over a curved detachment surface, the trailing edge of the rock layers roll over and are tilted toward the east so they do not leave space in buried rocks. On the other hand, a west-facing slope is left behind on a mountain from which the rock layers were detached. There-fore, movement along one curved detachment surface creates a valley, and movement along several such detachment surfaces forms a series of valleys separated by ridges, as in the Basin and Range Province. The amount of the displacement along the curved surfaces is not uniform. For example, more displacement has created wide zones of valleys such as the Las Vegas valley in Nevada, and Death Valley in California.

Extension Associated with a Divergent Boundary

The longest mountain chain on Earth lies under the Pacific Ocean. It is about 37,500 miles (60,000 km.) long, 31.3 miles (50 km.) wide, and 2 miles (3 km.) high. The central part of this midoceanic ridge is marked by a depression, about 3,000 feet (1,000 meters) deep, and is called a rift valley. A part of the submarine ridge, called the East Pacific Rise, forms the seafloor sector in the Gulf of California and reappears off the coast of northern California, Oregon, and Washington as the Juan de Fuca Ridge. Another part forms the seafloor sector in the Gulf of Aden and Red Sea seafloor, part of which is exposed in the Afar of Ethiopia. From the Afar southward to the southern part of Mozambique is the longest exposed rift valley on land, the East African Great Rift Valley.

A rift valley is the place where old rocks are pushed aside and new rocks are created. Blocks of rock that are detached from the rift walls slide down by a series of normal fault displacements. The ridge adjacent to the central rift is present because hot rocks are less dense and buoyant. If the process of divergences continues from the rifting stage to a drifting stage, as the rocks move farther away from the central rift, the rocks become older, colder, and denser, and push on the underlying asthenosphere to create basins. These basins will be flooded by oceanic water as neighboring continents drift away. However, not all processes of divergence advance from the rifting to the drifting stage.

Extension Associated with Transform Boundary

The best-known example of a transform boundary is the San Andreas Fault that offsets the East Pacific

Rise from the Juan de Fuca Ridge, and is exposed on land from the Gulf of California to San Francisco. Along transform boundaries, there are pull-apart basins that may be filled to form lakes, such as the Salton Sea in Southern California. Another example is the Aqaba transform of the Middle East, along which the Sea of Galilee and the Dead Sea are located.

H. G. Churnet

EXTERNAL PROCESSES

Continuous processes are at work shaping the earth's surface. These include breaking down rocks, moving the pieces, and depositing the pieces in new locations. Weathering breaks down rocks through atmospheric agents. The process of moving weathered pieces of rock by wind, water, ice, or gravity is called erosion. The materials that are deposited by erosion are called sediment.

Mechanical weathering occurs when a rock is broken into smaller pieces but its chemical makeup is not changed. If the rock is broken down by a change in its chemical composition, the process is called chemical weathering.

Mechanical Weathering
Different types of mechanical weathering occur, depending on climatic conditions. In areas with moist climates and fluctuating temperatures, rocks can be broken apart by frost wedging. Water fills in cracks in rocks, then freezes during cold nights. As the ice expands and pushes out on the crack walls, the crack enlarges. During the warm days, the water thaws and flows deeper into the enlarged crack. Over time, the crack grows until the rock is broken apart. This process is active in mountains, producing a pile of rock pieces at the mountain base called talus.

Salt weathering occurs in areas where much salt is available or there is a high evaporation rate, such as along the seashore. Salt crystals form when salty moisture enters rock cracks. Growing crystals settle in the bottom of the crack and apply pressure on the crack walls, enlarging the crack.

Thermal expansion and contraction occur in climates with fluctuating temperatures, such as deserts. All minerals expand during hot days and contract during cold nights, and some minerals expand and contract more than others. This process continues until the rock loosens up and breaks into pieces.

Mechanical exfoliation can happen to a rock body overlain by a thick rock or sediment layer. If the heavy overlying layer over a portion of the rock body is removed, pressure is relieved and the exposed rock surface will expand in response. This expanding surface will break off into sheets parallel to the surface, but the remaining rock body remains under pressure and unchanged.

When plant roots grow into cracks in rocks, they enlarge the cracks and break up the rocks. Finally, abrasion can occur to rock fragments during transport. Either the fragments collide, breaking apart, or fragments are scraped against rocks, breaking off pieces.

Chemical Weathering
Water and oxygen create two common causes of chemical weathering. For example, dissolution occurs when water or another solution dissolves minerals within a rock and carries them away. Hydrolysis can occur when water flows through earth materials. The hydrogen ions or the hydroxide ions of the water may react with minerals in the

299

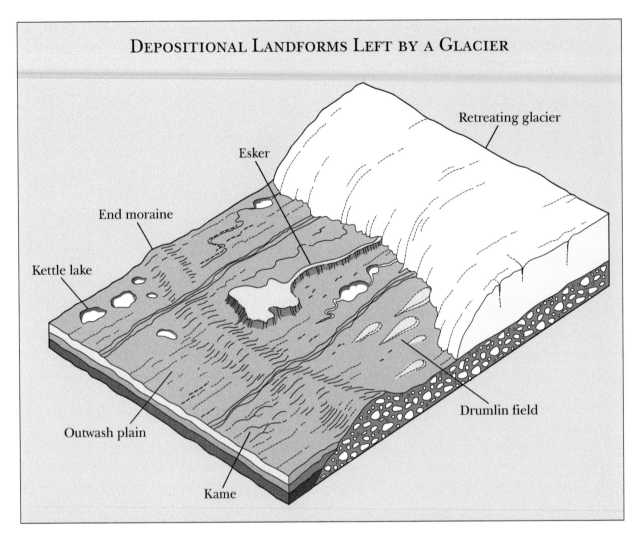

DEPOSITIONAL LANDFORMS LEFT BY A GLACIER

Retreating glacier

Esker

End moraine

Kettle lake

Drumlin field

Outwash plain

Kame

rocks. When this occurs, the chemical composition of the mineral is changed, and a new mineral is formed. Hydrolysis often produces clay minerals.

Some elements in minerals combine with oxygen from the atmosphere, creating a new mineral. This process is called oxidation. Some of these oxidation minerals are commonly referred to as rust.

Mass Movement

Weathered rock pieces (sediments) are transported (eroded) by one or more of four transport processes: water (streams and oceans), wind, ice (glaciers), or gravity. Mass movement transports earth materials down slopes by the pull of gravity. Gravity, constantly working to pull surface materials down, parallel to the slope, is the most important factor affecting mass movement. There is also a force in-

volved perpendicular to the slope that contributes to the effects of friction.

Friction, the second factor, is determined by the earth material type involved. For example, weathering may create cracks in rocks, which form planes of weakness on which the mass movement can occur. Loose sediments always tend to roll downhill.

The third factor is the slope angle. Each earth material has its own angle of repose, which is the steepest slope angle on which the materials remain stable. Beyond this slope angle, earth materials will move downslope.

Water, the fourth factor, affects the stability of the earth material in the slope. Friction is weakened by water between the mineral grains in the rock. For example, water can make clay quite slippery, causing the mass movement.

HOW HYDROLOGY SHAPES GEOGRAPHY

Water and ice sculpt the landscape over time. Fast-flowing rivers erode the soil and rock through which they flow. When rivers slow down in flatter areas, they deposit eroded sediments, creating areas of rich soils and deltas at the mouths of the rivers. Over time this process wears down mountain ranges. The Appalachian Mountain range on the eastern side of the North American continent is hundreds of millions of years older than the Rocky Mountain range on the continent's western side. Although the Appalachians once rivaled the Rockies in size, they have been made smaller by time and erosion.

Canyons are carved by rivers, as the Grand Canyon was carved by the Colorado River, which exposed rocks billions of years old. Ice also changes the landscape. Large ice sheets from past ice ages could have been well over 1 mile (1,600 meters) thick, and they scoured enormous amounts of soil and rock as they slowly moved over the land surface. Terminal moraines are the enormous mounds of soil pushed directly in front of the ice sheets. Long Island, New York, and Cape Cod, Massachusetts, are two examples of enormous terminal moraines that were left behind when the ice sheets retreated.

The rooting system of vegetation, the fifth factor, helps make the surficial materials of the slope stable by binding the loose materials together.

Mass movements can be classified by their speed of movement. Creep and solifluction are the two types of slow mass movement, which are measured in fractions of inches per year. Creep is the slowest mass movement process, where unconsolidated materials at the surface of a slope move slowly downslope. The materials move slightly faster at the surface than below, so evidence of creep appears in the form of slanted telephone poles. During solifluction, the warm sun of the brief summer season in cold regions thaws the upper few feet of the earth. This waterlogged soil flows downslope over the underlying permafrost.

Rapid mass movement processes occur at feet per second or miles per hour. Falls occur when loose rock or sediment is dislodged and drops from a steep slope, such as along sea cliffs where waves erode the cliff base. Topples occur when there is an overturning movement of the mass. A topple can turn into a fall or a slide. A slide is a mass of rock or sediment that becomes dislodged and moves along a plane of weakness, such as a fracture. A slump is a slide that separates along a concave surface. Lateral spreads occur when a fractured earth mass spreads out at the sides.

A flow occurs when a mass of wet or dry rock fragments or sediment moves downslope as a highly viscous fluid. There are several different flow types. A debris flow is a mass of relatively dry, broken pieces of earth material that suddenly has water added. The debris flow occurs on steeper slopes and moves at speeds of 1-25 miles (2-40 km.) per hour. A debris avalanche occurs when an entire area of soil and underlying weathered bedrock becomes detached from the underlying bedrock and moves quickly down the slope. This flow type is often triggered by heavy rains in areas where vegetation has been removed. An earthflow is a dry mass of clayey or silty material that moves relatively slowly down the slope. A mudflow is a mass of earth material mixed with water that moves quickly down the slope.

A quick clay can occur when partially saturated, solid, clayey sediments are subjected to an earthquake, explosion, or loud noise and become liquid instantly.

Sherry L. Eaton

FLUVIAL AND KARST PROCESSES

Earth's landscape has been sculptured into an almost infinite variety of forms. The earth's surface has been modified by various processes for thousands, even hundreds of millions, of years to arrive at the modern configuration of landscapes.

Each process that transforms the surface is classified as either endogenic or exogenic. Endogenic processes are driven by the earth's internal heat and energy and are responsible for major crustal deformation. Endogenic processes are considered constructional, because they build up the earth's surface and create new landforms, such as mountain systems. Conversely, exogenic processes are considered destructional because they result in the wearing away of landforms created by endogenic processes. Exogenic processes are driven by solar energy putting into motion the earth's atmosphere and water, resulting in the lowering of features originally created by endogenic processes.

The most effective exogenic processes for wearing away the landscape are those that involve the action of flowing water, commonly referred to as fluvial processes. Water flows over the surface as runoff, after it evaporates into the atmosphere and infiltrates into the soil. The water that is left over flows down under the influence of gravity and has tremendous energy for sculpturing the earth's surface. Although flowing water is the most effective agent for modifying the landscape, it represents less than 0.01 percent of all the water on Earth's surface. By comparison, nearly 75 percent of the earth's surface water is stored within glaciers.

Drainage Basins

Fluvial processes can be considered from a variety of spatial scales. The largest scale is the drainage basin. A drainage basin is the area defined by topographic divides that diverts all water and material within the basin to a single outlet. Every stream of any size has its own drainage basin, and every portion of the earth's land surfaces are located within a drainage basin. Drainage basins vary tremendously in size, de-pending on the size of the river considered. For example, the largest drainage basin on earth is the Amazon, which drains about 2.25 million square miles (5.83 million sq. km.) of South America.

The Amazon Basin is so large that it could contain nearly the entire continent of Australia. By comparison, the Mississippi River drainage basin, the largest in North America, drains an area of about 1,235,000 square miles (3,200,000 sq. km.). Smaller rivers have much smaller basins, with many draining only an area roughly the size of a football field. While basins vary tremendously in size, they are spatially organized, with larger basins receiving the drainage from smaller basins, and eventually draining into the ocean. Because drainage basins receive water and material from the landscape within the basin, they are sensitive to environmental change that occurs within the basin. For example, during the twentieth century, the Mississippi River was influenced by many human-imposed changes that occurred either within the basin or directly within the channel, such as agriculture, dams and reservoirs, and levees.

Drainage Networks and Surface Erosion

Drainage basins can be subdivided into drainage networks by the arrangement of their valleys and interfluves. Interfluves are the ridges of higher elevation that separate adjacent valleys. Where an interfluve represents a natural boundary between two or more basins, it is referred to as a drainage divide. Valleys contain the larger rivers and are easily distinguished from interfluves by their relatively low, flat surfaces. Interfluves have relatively steep slopes and, for this reason, are eroded by runoff. The term erosion refers to the transport of material, in this case sediment that is dislodged from the surface.

Runoff starts as a broad sheet of slow-moving water that is not very erosive. As it continues to flow downslope, it speeds up and concentrates into rills, which are narrow, fast-moving lines of water. Because the runoff is concentrated within rills, the water travels faster and has more energy for erosion. Thus, rills are responsible for transporting sediment from higher points of elevation within the basin to the valleys, which are at a lower elevation. Rills can become powerful enough to scour deeply into the surface, developing into permanent channels called gullies.

The presence of many gullies indicates significant erosion on the landscape and represents an expensive and long-lasting problem if it is not remedied after initial development. The formations of gullies is often associated with human manipulation of the earth. For example, gullies can develop after improper land management, particularly intensive agricultural and grazing practices. A change in land use from natural vegetation, such as forests or prairie, can result in a type of land cover that is not suited for preventing erosion. Such land surfaces become susceptible to the formation of gullies during heavy, prolonged rains.

At a smaller scale, fluvial processes can be considered from the perspective of the river channel. River channels are located within the valleys of basins, offering a permanent conduit for drainage. Higher in the basin, river channels and valleys are relatively narrow, but grow larger toward the mouth of the basin as they receive drainage from smaller rivers within the basin. River channels may be categorized by their planform pattern, which refers to their overhead appearance, such as would be viewed from the window of an airplane.

The two major types of rivers are meandering and braided. Meandering rivers have a single channel that is sinuous and winding. These rivers are characterized as having orderly and symmetrical bends, causing the river to alternate directions as it flows across its valley. In contrast, braided rivers contain numerous channels divided by small islands, which results in a disorganized pattern. The islands within a braided river channel are not permanent. Instead, they erode and form over the course of a few years, or even during large flood events. Meandering channels usually have narrow and deep channels, but braided river channels are shallow and wide.

Sediment and Floodplains

Another distinction between braided and meandering river channels is the types of sediment they transport. Braided rivers transport a great amount of sediment that is deposited into midchannel islands within the river. Also, because braided rivers are frequently located higher in the drainage basin, they may have larger sediments from the erosion of adjacent slopes. In contrast, meandering river channels are located closer to the mouth of the basin and transport fine-grained sediment that is easily stored within point bars, which results in symmetrical bends within the river.

The sediments of both meandering and braided rivers are deposited within the valleys onto floodplains. Floodplains are wide, flat surfaces formed from the accumulation of alluvium, which is a term for sediment that is deposited by water. Floodplain sediments are deposited with seasonal flooding. When a river floods, it transports a large amount of sediment from the channel to the adjacent floodplain. After the water escapes the channel, it loses energy and can no longer transport the sediment. As a result, the sediment falls out of suspension and is deposited onto the floodplain. Because flooding occurs seasonally, floodplain deposits are layered and may accumulate into very thick alluvial deposits over thousands of years.

Karst Processes and Landforms

A specialized type of exogenic process that is also related to the presence of water is karst. Karst processes and topography are characterized by the solution of limestone by acidic groundwater into a number of distinctive landforms. While fluvial processes lower the landscape from the surface, karst processes lower the landscape from beneath the surface. Because limestone is a very permeable sedimentary rock, it allows for a large amount of groundwater flow. The primary areas for solution of the limestone occur along bedding planes and joints. This creates a positive feedback by increasing the amount of water flowing through the rock, thereby further increasing solution of the limestone. The result is a complex maze of underground conduits and caverns, and a surface with few rivers because of the high degree of infiltration.

The surface topography of karst regions often is characterized as undulating. A closer inspection reveals numerous depressions that lack surface outlets. Where this is best developed, it is referred to as cockpit karst. It occurs in areas underlain by extensive limestone and receiving high amounts of precipitation, for example, southern Illinois and Indiana in the midwestern United States, and in Puerto Rico and Jamaica.

Sinkholes are also common to karstic regions. Sinkholes are circular depressions having steep-sided vertical walls. Sinkholes can form either from the sudden collapse of the ceiling of an underground cavern or as a result of the gradual solution and lowering of the surface. Sinkholes can fill with sediments washed in from surface runoff. This reduces infiltration and results in the development of small circular lakes, particularly common in central Florida. Over time, erosion causes the vertical walls to retreat, resulting in uvalas, which are much larger flat-floored depressions.

Where there are numerous adjacent sinkholes, the retreat and expansion of the depressions causes them to coalesce, resulting in the formation of poljes. Unlike uvalas, poljes have an irregular shape, and the floor of the basin is not flat because of differences between the coalescing sinkholes.

Caves are among the most characteristic features of karst regions, but can only be seen beneath the surface. Caves can traverse the subsurface for miles, developing into a complex network of interconnected passages. Some caves develop spectacular formations as a result of the high amount of dissolved limestone transported by the groundwater. The evaporation of water results in the accumulation of carbonate deposits, which may grow for thousands of years. Some of the most common deposits are stalactites, which grow downward from the ceiling of the cave, and stalagmites, which grow upward and occasionally connect with stalactites to form large vertical columns.

Paul F. Hudson

GLACIATION

In areas where more snow accumulates each winter than can thaw in summer, glaciers form. Glacier ice, called firn, looks like rock but is not as strong as most rocks and is subject to intermittent thawing and freezing. Glacier ice can be brittle and fracture readily into crevasses, while other ice behaves as a plastic substance. A glacier is thickest in the area receiving the most snow, called the zone of accumulation. As the thickness piles up, it settles down and squeezes the limit of the ice outward in all directions. Eventually, the ice reaches a climate where the ice begins to melt and evaporate. This is called the zone of ablation.

Alpine Glaciation

Varied topographic evidence throughout the alpine environment attests to the sculpturing ability of glacial ice. The world's most spectacular mountain scenery has been produced by alpine glaciation, including the Matterhorn, Yosemite Valley, Glacier National Park, Mount Blanc, the Tetons, and Rocky

A FUTURE ICE AGE

If past history is an indicator, some time in the future conditions again will become favorable for the growth of glaciers. As recently as 1300 to 1600 CE, a cold period known at the Little Ice Age settled over Northern Europe and Eastern North America. Viking colonies perished as agriculture became unfeasible, and previously ice-free rivers in Europe froze over.

Another ice age would probably develop rapidly and be impossible to stop. Active mountain glaciers would bury living forests. Great ice caps would again cover Europe and North America, and move at a rate of 100 feet (30 meters) per day. Major cities and populations would shift to the subtropics and the topics.

Mountain National Park, all of which are visited by large numbers of people annually. Although alpine glaciation is still an active process of land sculpture in the high mountain ranges of the world, it is much less active than it was in the Ice Age of the Pleistocene epoch.

The prerequisites for alpine, or mountain, glaciation to become active are a mountainous terrain with Arctic climatic conditions in the higher elevations, and sufficient moisture to help snow and ice develop into glacial ice. As glaciers move out from their points of origin, they erode into the sides of mountains and increase the local relief in the higher elevations. The erosional features produced by alpine glaciation dominate mountain topography and usually are the most visible features on topographic maps. The eroded material is transported downvalley and deposited in a variety of landforms.

One kind of an erosional feature is a cirque, a hollow bowl-shaped depression. The bowl of the cirque commonly contains a small round lake or tarn. A steep-walled mountain ridge called an arête forms between two cirques. A high pyramidal peak, called horn, is formed by the intersecting walls of three or more cirques.

Erosion is particularly rapid at the head of a glacier. In valleys, moving glaciers press rock fragments against the sides, widening and deepening them by abrasion and forming broad U-shaped valleys. When glaciers recede, tributary streams become higher than the floor of the U-shaped valley and waterfalls occur over these hanging valleys. As the ice continues to melt, residual sediments called moraines may be deposited. Moraines are made up of glacier till, a collection of sediment of all sizes. Bands of sediment along the side of a valley glacier are lateral moraines; those crossing the valley are end or recessional moraines; where two glaciers join, a medial moraine is formed. Meltwater may also sort out the finer materials, transport them downvalley, and deposit them in beds as outwash.

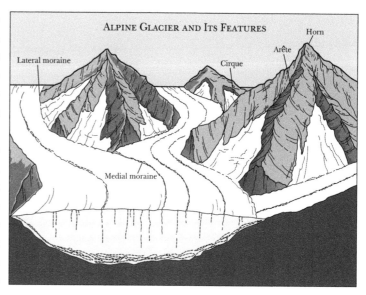

ALPINE GLACIER AND ITS FEATURES

Horn

Arête

Cirque

Lateral moraine

Medial moraine

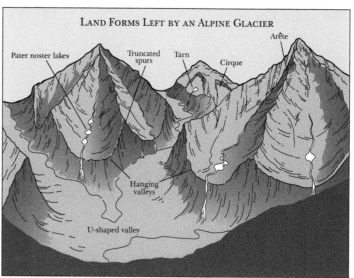

LAND FORMS LEFT BY AN ALPINE GLACIER

Arête

Pater noster lakes

Truncated spurs

Tarn

Cirque

Hanging valleys

U-shaped valley

Continental Glaciation

In the modern world, continental glaciation operates on a large scale only in Greenland and Antarctica. However, its existence in previous geologic ages is evidenced by strata of tillite (a compacted rock formed of glacial deposits) or, more frequently, by surficial deposits of glacial materials.

Much of the geomorphology of the northeastern quadrant of North America and the northwestern portion of Europe was formed during the Ice Age. During that time, great masses of ice accumulated on the continents and moved out from centers near the Hudson Bay and the Fennoscandian Shield, extending over the continents in great advancing and retreating lobes. In North America, the four major

FEATURES OF A CONTINENTAL GLACIER

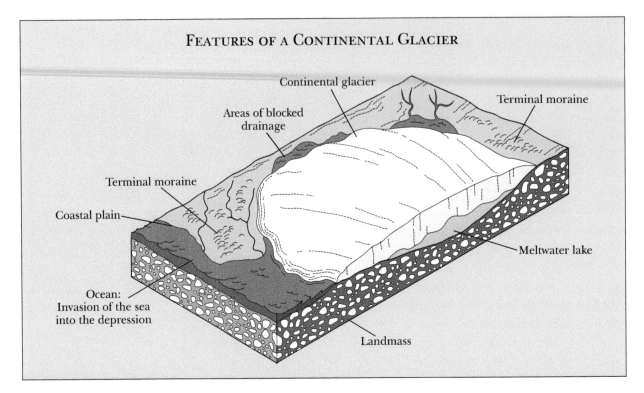

Continental glacier

Areas of blocked drainage

Terminal moraine

Terminal moraine

Coastal plain

Meltwater lake

Ocean:
Invasion of the sea
into the depression

Landmass

stages of lobe advance were the Wisconsin (the most recent), the Illinoian, the Kansan, and the Nebraskan (the oldest). Between each of these major advances were pluvial periods in which the ice melted and great quantities of water rushed over or stood on the continents, creating distinctive features which can still be detected today.

The two major functions of gradation are accomplished by the processes of scour (degradation) in the areas close to the centers and deposition (aggradation) adjacent to the terminal or peripheral areas of the lobes. Thus, the overall effect of continental glaciation is to reduce relief—to scour high areas and fill in lower regions—unlike the changes caused by alpine glaciation.

Although continental glaciation usually does not result in the spectacular scenery of alpine glaciation, it was responsible for creating most of the Great Lakes and the lakes of Wisconsin, Michigan, Minnesota, Finland, and Canada; for gravel deposits; and for the rich agricultural lands of the Midwest, to mention just a few of its effects.

While glaciers were leveling hilly sections of North America and Europe by scraping them bare

of soil and cutting into the ice itself, they acquired a tremendous load of material. As a glacier warms and melts, there is a tremendous outflow of water, and the streams thus formed carry with them the debris of the glacier. The material deposited by glaciers is called drift or outwash. Glaciofluvial drift can be recognized by its separation into layers of finer sands and coarser gravels.

Kettles and kames are the most common features of the end moraines found at the outermost edges of a glacier. A kettle is a depression left when a block of ice, partially or completely buried in deposits of drift, melts away. Most of the lakes in the upper Great Lakes of the United States are kettle lakes. A kame is a round, cone-shaped hill. Kames are produced by deposition from glacial meltwater. Sometimes, the outwash material poured into a long and deep crevasse, rather than a hole. These tunnels have had their courses choked by debris, revealed today by long, narrow ridges, generally referred to as eskers.

Ron Janke

DESERT LANDFORMS

Deserts are often striking in color, form, or both. The underlying lack of water in deserts produces unique features not found in humid regions. Arid lands cover approximately 30 percent of the earth's land surface, an area of about 15.4 million square miles (40 million sq. km.). Arid lands include deserts and surrounding steppes, semiarid regions that act as transition zones between arid and humid lands.

Many of the world's largest and driest deserts are found between 20 and 40 degrees north and south latitude. These include the Mojave and Sonoran Deserts of the United States, the Sahara in northern Africa, and the Great Sandy Desert in Australia. In these deserts, the subtropical high prevents cloud formation and precipitation while increasing rates of surface evaporation.

Some arid lands, like China's Gobi Desert, form because they are far from oceans that are the dominant source for atmospheric water vapor and precipitation. Others, like California's Death Valley, are arid because mountain ranges block moisture from coming from the sea. The combination of mountain barriers and very low elevations makes Death Valley the hottest, driest desert in North America.

Sand Dunes

Many people envision deserts as vast expanses of blowing sand. Although wind plays a more important role in deserts than it does elsewhere, only about 25 percent of arid lands are covered by sand. Broad regions that are covered entirely in sand (such as portions of northwestern Africa, Arabia, and Australia) are referred to as sand seas. Why is wind more effective here than elsewhere?

The lack of soil water and vegetation, both of which act to bind grains together, allows enhanced eolian (wind) erosion. Very small particles are picked up and suspended within the moving air mass, while sand grains bounce along the surface. Removal of material often leaves behind depressions called blowouts or deflation hollows. Moving grains abrade cobbles and boulders at the surface, creating uniquely sculpted and smoothed rocks known as ventifacts. Bedrock outcrops can be streamlined as they are blasted by wind-borne grains to form features called yardangs. As these rocks are ground away, they contribute additional sediment to the wind.

Desert sand dunes are not stationary features—instead, they represent accumulations of moving sand. Wind blows sand along the desert floor. Where it collects, it forms dunes. Typically, dunes have relatively shallow windward faces and steeper slip faces. Sand grains bounce up the windward face and then eventually cascade down the slip face, the movement of individual grains driving movement of the entire dune in a downwind direction.

Four major dune types are found within arid regions. Barchan dunes are crescent-shaped features, with arms that point downwind. They may occur as isolated structures or within fields. They form where winds blow in a single direction and where the supply of sand is limited. With a larger supply of sand, barchan dunes can join with one another to form a transverse dune field.

There, ridges are perpendicular to the predominant wind direction. With quartering winds (that is, winds that vary in direction throughout a range of about 45 degrees) dune ridges form that are parallel to the average wind direction. These so-called longitudinal dunes have no clearly defined windward and slip faces. Where winds blow sand from all directions, star dunes form. Sand collects in the middle of the feature to form a peaked center with arms that spiral outward.

Badlands, Mesas, and Buttes

As scarce as it may be, water is still the dominant force in shaping desert landscapes. Annual precipitation may be low, but the amount of precipitation in a single storm may be a large fraction of the yearly total. An arid landscape that is underlain by poorly cemented rock or sediment, such as that found in western South Dakota, may form badlands as a result of the erosive ability of storm-water run-off. Overall aridity prevents vegetation from establishing the interconnected root system that holds soil particles together in more humid regions.

Cloudbursts cause rapid erosion that forms numerous gullies, deeply incised washes, and hoodoos, which are created when rock or sediment that is more resistant protects underlying material from erosion. Over time, protected sections stand as prominent spires while surrounding material is removed. Landscapes like those found in Badlands National Park in South Dakota are devoid of vegetation and erode rapidly during storms.

Arid regions that are underlain by flat-lying rock units can form mesas and buttes. Water follows fractures and other lines of weakness, forming ever-widening canyons. Over time, these grow into broad valleys. In northern Arizona's Monument Valley, remnants of original bedrock stand as isolated, flat-topped structures. Broad mesas are marked by their flat tops (made of a resistant rock like sandstone or basalt) and steep sides. Buttes are much narrower, with a small resistant cap, but are often as tall and steep as neighboring mesas.

Desert Pavement and Desert Varnish

Much of the desert floor is covered by desert pavement, an accumulation of gravel and cobbles that forms a surface fabric that can interconnect tightly. Fine material has been removed by wind and water, leaving behind larger fragments that inhibit further erosion. In many areas, desert pavements have been stable for long periods of time, as evidenced by their surface patina of desert varnish. Desert varnish is a thin outer coating of wind-deposited clay mixed with iron and manganese oxides. Varying in color from light brown to black, these coatings are thought to adhere to rocks by the action of single-celled microorganisms. Under a microscope, desert varnish can be seen to be made up of very fine layers. A thick, dark patina means that a rock has been exposed for a long time.

Playas

Where neither dunes nor rocky pavements cover the desert floor, one may find an accumulation of saline minerals. A playa is a flat surface that is often blindingly white in color. Playas are usually found in the centers of desert valleys and contain material that mineralized during the evaporation of a lake. Dry lake beds are a common feature of the Great Basin in the western United States. During glacial stages, the last of which occurred about 20,000 years ago, lakes grew in what are now arid, closed valleys. As the climate warmed, these lakes shrank, and many dried completely. As a lake evaporates, minerals that were held in solution crystallize, forming salts, including halite (table salt). These salt deposits frequently are mined for useful household and industrial chemicals.

Richard L. Orndorff

DEATH VALLEY PLAYA

California's Death Valley is the driest desert in the United States, with an average rainfall of only 1.5 inches (38 millimeters) per year at the town of Furnace Creek. It is also consistently one of the hottest places on Earth, with a record high of 134 degrees Fahrenheit (57 degrees Celsius). In the distant past, however, Death Valley held lakes that formed in response to global cooling. Over 120,000 years ago, Death Valley hosted a 295-foot-deep (90 meters) body of water called Lake Manley. Evidence of this lake remains in evaporite deposits that make up the playa in the valley's center, in wave-cut shorelines, and in beach bars.

OCEAN MARGINS

Ocean margins are the areas where land borders the sea. Although often referred to as coastlines or beaches, ocean margins cover far greater territory than beaches. An ocean margin extends from the coastal plain—the fertile farming belt of land along the seacoast—to the edge of the gently sloping land submerged in water, called the continental shelf.

Ocean margin constitutes 8 percent of the world's surface. It is rich in minerals, both above and below water, and is home to 25 percent of Earth's people, along with 90 percent of the marine life. This fringe of land at the border of the ocean is ever changing. Tides wash sediment in and leave it behind, just below sea level. This process, called deposition, builds up land in some areas of the coastline. At the same time, ocean waves, winds, and storms wear away or erode parts of the shoreline. As land is worn away or built up, the amount of land above sea level changes. Factors such as climate, erosion, deposition, changes in sea level, and the effects of humans constantly change the shape of the ocean margin on Earth.

Beach Dynamics

The two types of coasts or land formations at the ocean margin are primary coasts and secondary coasts. Primary coasts are formed by systems on land, such as the melting of glaciers, wind or water erosion, and sediment deposited by rivers. Deltas and fjords are examples of primary coasts. Secondary coasts are formed by ocean patterns, such as erosion by waves or currents, sediment deposition by waves or currents, or changes by marine plants or animals. Beaches, coral reefs, salt marshes, and mangrove swamps are examples of secondary coasts.

Sediment carried by rivers to the sea is deposited to form deltas at the mouths of the rivers. Some of the sediment can wash out to sea, causing formations to build up at a distance from the shore. These formations eventually become barrier islands, which are often little more than 10 feet (3 km.) above sea level. As

a consequence, heavy storms, such as hurricanes, can cause great damage to barrier islands. Barrier islands naturally protect the coastline from erosion, however, especially during heavy coastal storms.

Sea level changes also affect the shape of the coastline. As oceans slowly rise, land is slowly consumed by the ocean. Barrier islands, having low sea levels, may slowly be covered with water. The melting of continental glaciers increased the sea level 0.06 inch (0.15 centimeter) per year during the twentieth century. As ocean waters warm, they expand, eating away at sea levels. Global warming caused by carbon dioxide levels in the atmosphere could cause sea levels to rise as much as 0.24 inch (0.6 centimeter) per year as a result of the warming of the water and glacial melting.

Human Influence

The shape of the ocean margin also changes radically as a result of human influence. According to the United States Geological Survey, 39 percent of the people living in the United States live directly on the the coasts. According to UN Atlas of Oceans, about 44 percent of the world's population lives within 93 miles (150 kilometers) of the coast. Pollution from toxins, dredging, recreational boating, and waste disposal kills plants and animals along the ocean margin. This changes the coastal shape, as mangrove forests, coral reefs, and other coastal lifeforms die.

A greater concern along the coastal fringe, however, is human development. Not only are people drawn to the fertile soil along the coastal zone of the continent, but they also develop islands and coves into resort communities. To protect homes and hotels along the coastal zone from coastal erosion, people build breakwalls, jetties, and sand and stone bars called groins.

These human-made barriers disrupt the natural method by which the ocean carries material along the coast. Longshore drift, a zigzag movement, deposits sediment from one area of the beach farther along the shoreline. Breakwalls, jetties, and groins disrupt this flow. As the ocean smashes against a breakwall, the property behind it may be safe for the present, but the coastline neighboring the breakwall takes a greater beating. The silt and sediment from upshore, which would replace that carried downshore, never arrives. Eventually, the breakwall will break down under the impact of the ocean force. Areas with breakwalls and jetties often suffer greater damage in coastal storms than areas that remain naturally open to the changing forces of the ocean.

To compensate for the destructive nature of human-made barriers, many recreational beaches re-

place lost sand with dredgings or deposit truckloads of sand from inland sources. For example, Virginia Beach in the United States spends between US$2 million and US$3 millon annually to restore beaches for the tourist season in this way.

Despite the changes in the shape of the ocean margin, it continues to provide a stable supply of resources—fish, seafood, minerals, sponges, and other marine plants and animals. Offshore drilling of oil and natural gas often takes place within 200 miles (322 km.) of shorelines.

Lisa A. Wroble

EARTH'S CLIMATES

THE ATMOSPHERE

The thin layer of gases that envelops the earth is the atmosphere. This layer is so thin that if the earth were the size of a desktop globe, more than 99 percent of its atmosphere would be contained within the thickness of an ordinary sheet of paper. Despite its thinness, the atmosphere sustains life on Earth, protecting it from the Sun's searing radiation and regulating the earth's temperature. Storms of the atmosphere carry water to the continents, and weathering by its wind and rain helps shape them.

Composition of the Atmosphere

The earth's atmosphere consists of gases, microscopic particles called aerosol, and clouds consisting of water droplets and ice particles. Its two principal gases are nitrogen and oxygen. In dry air, nitrogen occupies 78 percent, and oxygen 21 percent, of the atmosphere's volume. Argon, neon, xenon, helium, hydrogen, and other trace gases together equal less than 1 percent of the remaining volume.

These gases are distributed homogeneously in a layer called the homosphere, which occurs between the earth's surface and about 50 miles (80 km.) altitude. Above 50 miles altitude, in the heterosphere, the concentration of heavier gases decreases more rapidly than lighter gases.

The atmosphere has no firm top. It simply thins out until the concentration of its gas molecules approaches that of the gases in outer space. The concentration of nitrogen and oxygen remains essentially constant in the atmosphere because a balance exists between the production and removal of these gases at the earth's surface. Decaying organic matter adds nitrogen to the atmosphere, while soil bacteria remove nitrogen. Oxygen enters the atmosphere primarily through photosynthesis and is removed through animal respiration, combustion, and decay of organic material, and by chemical reactions involving the creation of oxides.

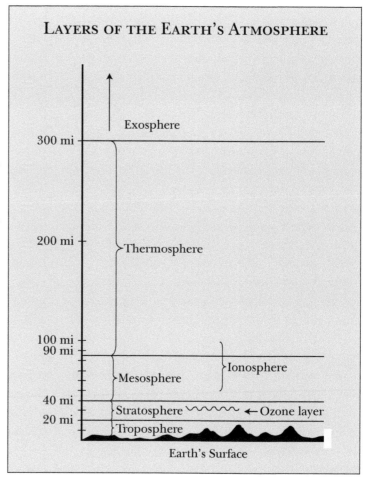

LAYERS OF THE EARTH'S ATMOSPHERE

THE GREENHOUSE EFFECT

Clouds and atmospheric gases such as water vapor, carbon dioxide, methane, and nitrous oxide absorb part of the infrared radiation emitted by the earth's surface and reradiate part of it back to the earth. This process effectively reduces the amount of energy escaping to space and is popularly called the "greenhouse effect" because of its role in warming the lower atmosphere. The greenhouse effect has drawn worldwide attention because increasing concentrations of carbon dioxide from the burning of fossil fuels result in a global warming of the atmosphere.

Scientists know that the greenhouse analogy is incorrect. A greenhouse traps warm air within a glass building where it cannot mix with cooler air outside. In a real greenhouse, the trapping of air is more important in maintaining the temperature than is the trapping of infrared energy. In the atmosphere, air is free to mix and move about.

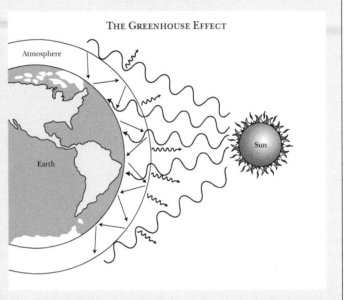

THE GREENHOUSE EFFECT

The atmosphere contains many gases that are present in small, variable concentrations. Three gases—water vapor, carbon dioxide and ozone—are vital to life on Earth. Water vapor enters the atmosphere through evaporation, primarily from the oceans, and through transpiration by plants. It condenses to form clouds, which provide the rain and snow that sustain life outside the oceans. The concentration of water vapor varies from about 4 percent by volume in tropical humid climates to a small fraction of a percent in polar dry climates. Water vapor plays an important role in regulating the temperature of the earth's surface and the atmosphere. Clouds reflect some of the incoming solar radiation, while water vapor and clouds both absorb earth's infrared radiation.

Carbon dioxide also absorbs the earth's infrared radiation. The global average atmospheric carbon dioxide in 2018 was 407.4 parts per million (ppm for short), with a range of uncertainty of plus or minus 0.1 ppm. Carbon dioxide levels today are higher than at any point in at least the past 800,000 years. The annual rate of increase in atmospheric carbon dioxide over the past 60 years is about 100 times faster than previous natural increases, such as those that occurred at the end of the last ice age 11,000–17,000 years ago.

Carbon dioxide enters the atmosphere as the result of decay of organic material, through respiration, during volcanic eruptions, and from the burning of fossil fuels. It is removed during photosynthesis and by dissolving in ocean water, where it is used by organisms and converted to carbonates. The increase in atmospheric carbon dioxide associated with the burning of fossil fuels has raised concerns that the earth's atmosphere may be warming through enhancement of the greenhouse effect.

Ozone, a gas consisting of molecules containing three oxygen atoms, forms in the upper atmosphere when oxygen atoms and oxygen molecules combine. Most ozone exists in the upper atmosphere between 6.2 and 31 miles (10 and 50 km.) in altitude, in concentrations of no more than 0.0015 percent by volume. This small amount of ozone sustains life outside the oceans by absorbing most of the Sun's ultraviolet radiation, thereby shielding the earth's surface from the radiation's harmful effects on living organisms. Paradoxically, ozone is an irritant near the earth's surface and is the major component of photochemical smog. Other gases that contribute to pollution include methane, nitrous oxide, hydrocarbons, and chlorofluorocarbons.

Aerosols represent another component of atmospheric pollution. Aerosols form in the atmosphere during chemical reactions between gases, through mechanical or chemical interactions between the earth, ocean surface, and atmosphere, and during evaporation of droplets containing dissolved or solid material. These microscopic particles are always present in air, with concentrations of about a few hundred per cubic centimeter in clean air to as many as a million per cubic centimeter in polluted air. Aerosols are essential to the formation of rain and snow, because they serve as centers upon which cloud droplets and ice particles form.

Energy Exchange in the Atmosphere

The Sun is the ultimate source of the energy in Earth's atmosphere. Its radiation, called electromagnetic radiation because it propagates as waves with electric and magnetic properties, travels to the surface of the earth's atmosphere at the speed of light. This energy spans many wavelengths, some of which the human eye perceives as colors. Visible wavelengths make up about 44 percent of the Sun's energy. The remainder of the Sun's radiant energy cannot be seen by human eyes. About 7 percent arrives as ultraviolet radiation, and most of the remaining energy is infrared radiation.

The Sun is not the only source of radiation. All objects emit and absorb radiation to some degree. Cooler objects such as the earth emit nearly all their energy at infrared wavelengths. Objects heat when they absorb radiation and cool when they emit radiation. The radiation emitted by the earth and atmosphere is called terrestrial radiation.

The balance between absorption of solar radiation and emission of terrestrial radiation ultimately determines the average temperature of the earth-atmosphere system. The vertical temperature distribution within the atmosphere also depends on the absorption and emission of radiation within the atmosphere, and the transfer of energy by the processes of conduction, convection, and latent heat exchange. Conduction is the direct transfer of heat from molecule to molecule. This process is most important in transferring heat from the earth's surface to the first few centimeters of the at-

---

THE OZONE HOLE

Since the 1970s, balloon-borne and satellite measurements of stratospheric ozone have shown rapidly declining stratospheric ozone concentrations over the continent of Antarctica, termed the "ozone hole." The lowest concentrations occur during the Antarctic spring, in September and October. The decrease in ozone has been associated with an increase in the concentration of chlorine, a gas introduced into the stratosphere through chemical reactions involving sunlight and chlorofluorocarbons—synthetic chemicals used primarily as refrigerants. The ozone hole over Antarctica has raised concern about possible worldwide reduction in the concentration of upper atmospheric ozone.

---

mosphere. Convection, the transfer of heat by rising or sinking air, transports heat energy vertically through the atmosphere.

Latent heat is the energy required to change the state of a substance, for example, from a liquid to a gas. Energy is transferred from the earth's surface to the atmosphere through latent heat exchange when water evaporates from the oceans and condenses to form rain in the atmosphere.

Only 48 percent of the solar energy reaching the top of the earth's atmosphere is absorbed by the earth's surface. The atmosphere absorbs another 23 percent. The remaining 30 percent is scattered back to space by atmospheric gases, clouds and the earth's surface. To understand the importance of terrestrial radiation and the greenhouse effect in the atmosphere's energy balance, consider the solar radiation arriving at the top of the earth to be 100 energy units, with 48 energy units absorbed by the earth's surface and 23 units by the atmosphere.

The earth's surface actually emits 117 units of energy upward as terrestrial radiation, more than twice as much energy as it receives from the Sun. Only 6 of these units are radiated to space—the atmosphere absorbs the remaining energy. Latent heat exchange, conduction, and convection account for another 30 units of energy transferred from the surface to the atmosphere. The atmosphere, in turn, radiates 96 units of energy back to the earth's surface (the greenhouse effect), and 64 units to space. The earth's and atmosphere's energy budget remains in balance, the atmosphere gaining and losing 160 units of energy, and the earth gaining and losing 147 units of energy.

Vertical Structure of the Atmosphere

Temperature decreases rapidly upward away from the earth's surface, to about –60 degrees Farenheit (–51 degrees Celsius) at an altitude of about 7.5 miles (12 km.). Above this altitude, temperature increases with height to about 32 degrees Farenheit (0 degrees Celsius) at an altitude of 31 miles (50 km.). The layer of air in the lower atmosphere where temperature decreases with height is called the troposphere. It contains about 75 percent of the atmosphere's mass. The layer of air above the troposphere, where temperature increases with height, is called the stratosphere. All but 0.1 percent of the remaining mass of the atmosphere resides in the stratosphere.

The stratosphere exists because ozone in the stratosphere absorbs ultraviolet light and converts it to heat. The boundary between the troposphere and stratosphere is called the tropopause. The tropopause is extremely important because it acts as a lid on the earth's weather. Storms can grow vertically in the troposphere, but cannot rise far, if at all, beyond the tropopause. In the polar regions, the tropopause can be as low as 5 miles (8 km.) above the surface, while in the tropics, the tropopause can be as high as 11 miles (18 km.). For this reason, tropical storms can extend to much higher altitudes than storms in cold regions.

The mesosphere extends from the top of the stratosphere, the stratopause, to an altitude of about 56 miles (90 km.). Temperature decreases with height within the mesosphere. The lowest average temperatures in the atmosphere occur at the mesopause, the top of the mesosphere, where the temperature is about –130 degrees Farenheit (–90 degrees Celsius). Only 0.0005 percent of the atmosphere's mass remains above the mesopause. In this uppermost layer, the thermosphere, there are few atoms and molecules. Oxygen molecules in the thermosphere absorb high-energy solar radiation. In this near vacuum, absorption of even small amounts of energy causes a large increase in temperature. As a result, temperature increases rapidly with height in the lower thermosphere, reaching about 1,300 degrees Farenheit (700 degrees Celsius) above 155 miles (250 km.) altitude.

The upper mesosphere and thermosphere also contain ions, electrically charged atoms or molecules. Ions are created in the atmosphere when air molecules collide with high-energy particles arriving from space or absorb high-energy solar radiation. Ions cannot exist very long in the lower atmosphere, because collisions between newly formed ions quickly restore ions to their uncharged state. However, above about 37 miles (60 km.) collisions are less frequent and ions can exist for longer times. This region of the atmosphere, called the ionosphere, is particularly important for amplitude-modulated (AM) radio communication because it reflects standard AM radio waves. At night, the lower ionosphere disappears as ions recombine, allowing AM radio waves to travel longer distances when reflected. For this reason, AM radio station signals can sometimes travel great distances at night.

The top of the atmosphere occurs at about 310 miles (500 km.). At this altitude, the distance between individual molecules is so great that energetic molecules can move into free space without colliding with neighbor molecules. In this uppermost layer, called the exosphere, the earth's atmosphere merges into space.

Robert M. Rauber

GLOBAL CLIMATES

A region's climate is the sum of its long-term weather conditions. Most descriptions of climate emphasize temperature and precipitation characteristics, because these two climatic elements usually exert more impact on environmental conditions and human activities than do other elements, such as wind, humidity, and cloud cover. Climatic descriptions of a region generally cover both mean conditions and extremes. Climatic means are important because they represent average conditions that are frequently experienced; extreme conditions, such as severe storms, excessive heat and cold, and droughts, are important because of their adverse impact.

Important Climate Controls

A region's climate is largely determined by the interaction of six important natural controls: sun angle, elevation, ocean currents, land and water heating and cooling characteristics, air pressure and wind belts, and orographic influence.

Sun angle—the height of the Sun in degrees above the nearest horizon—largely controls the amount of solar heating that a site on Earth receives. It strongly influences the mean temperatures of most of the earth's surface, because the Sun is the ultimate energy source for nearly all the atmosphere's heat. The higher the angle of the Sun in the sky, the greater the concentration of energy, per unit area, on the earth's surface (assuming clear skies). From a global perspective, the Sun's mean angle is highest, on average, at the equator, and becomes progressively lower poleward. This causes a gradual decrease in mean temperatures with increasing latitude.

Sun angles also vary seasonally and daily. Each hemisphere is inclined toward the Sun during spring and summer, and away from the Sun during fall and winter. This changing inclination causes mean sun angles to be higher, and the length of daylight longer, during the spring and summer. Therefore, most locations, especially those outside the tropics, have warmer temperatures during these two seasons. The earth's rotation causes sun angles to be higher during midday than in the early morning and late afternoon, resulting in warmer temperatures at midday. Heating and cooling lags cause both seasonal and daily maximum and minimum temperatures typically to occur somewhat after the periods of maximum and minimum solar energy receipt.

Variations in elevation—the distance above sea level—can cause locations at similar latitudes to vary greatly in temperature. Temperatures decrease an average of about 3.5 degrees Fahrenheit per thousand feet (6.4 degrees Celsius per thousand meters). Therefore, high mountain and plateau stations are much colder than low-elevation stations at the same latitude.

Surface ocean currents can transport masses of warm or cold water great distances from their source regions, affecting both temperature and moisture conditions. Warm currents facilitate the evaporation of copious amounts of water into the atmosphere and add buoyancy to the air by heating it from below. This results in a general increase in precipitation totals. Cold currents evaporate water relatively slowly and chill the overlying air, thus stabilizing it and reducing its potential for precipitation.

The influence of ocean currents on land areas is greatest in coastal regions and decreases inland. The west coasts of continents (except for Europe) generally are paralleled by relatively cold currents, and the east coasts by relatively warm currents. For example, the warm Gulf Stream flows northward off the eastern United States, while the West Coast is cooled by the southward-flowing California Current.

Land can change temperature much more readily than water. As a result, the air over continents typically experiences larger annual temperature ranges (that is, larger temperature differences between summer and winter) and shorter heating and cooling lags than does the air over oceans. This same effect causes continental interiors and the leeward (downwind) coasts of continents typically to have larger temperature ranges than do windward (upwind) coasts. Climates that are dominated by air from landmasses are often described as continental climates. Conversely, climates dominated by air from oceans are described as maritime climates.

The seasonal heating and cooling of continents can also produce a monsoon influence, which has to do with annual shifts of wind patterns. Areas influenced by a monsoon, such as Southeast Asia, tend to have a predominantly onshore flow of moist maritime air during the summer. This often produces heavy rains. An offshore flow of dry air predominates in winter, producing fair weather.

Earth's atmosphere displays a banded, or beltlike, pattern of air pressure and wind systems. High pressure is associated with descending air and dry weather; low pressure is associated with rising air, which produces cloudiness and often precipitation. Wind is produced by differences in air pressure. The air blows outward from high-pressure systems and into low-pressure systems in a constant attempt to equalize air pressures.

The direction and speed of movement of weather systems, such as weather fronts and storms, are controlled by wind patterns, especially those several kilometers above the surface. The seasonal shift of global temperatures caused by the movement of the Sun's vertical rays between the Tropics of Cancer and Capricorn produces a latitudinal migration of both air pressure and wind belts. This shift affects the annual temperature and precipitation patterns of many regions.

Four air-pressure belts exist in each hemisphere. The intertropical convergence zone (ITCZ) is a broad belt of low pressure centered within a few degrees of latitude of the equator. The subtropical highs are high-pressure belts centered between 20 and 40 degrees north and south latitude, which are responsible for many of the world's deserts. The subpolar lows are low-pressure belts centered about 50 or 70 degrees north and south latitude. Finally, the polar highs are high-pressure centers located near the North and South Poles.

The air pressure gradient between these belts produces the earth's major wind belts. The regions between the ITCZ and the subtropical highs are dominated by the trade winds, a broad belt in each hemisphere of easterly (that is, moving east to west) winds. The middle latitudes are mostly situated between the subtropical highs and the subpolar lows and are within the westerly wind belt. This wind belt causes winds, and weather systems, to travel generally from west to east in the United States and Canada. Finally, the high-latitude zones between the subpolar lows and polar highs are situated within the polar easterlies.

The final factor affecting climate— orographic influence—is the lifting effect of mountain peaks or ranges on winds that pass over them. As air approaches a mountain barrier, it rises, typically producing clouds and precipitation on the windward (upwind) side of the mountains. After it crosses the crest, it descends the leeward (downwind) side of the mountains, generally producing dry weather. Most of the world's wettest locations are found on the windward sides of high mountain ranges; some deserts, such as those of the western interior United States, owe their aridity to their location on the leeward sides of orographic barriers.

World Climate Types

The global distribution of the world climate controls is responsible for the development of fourteen widely recognized climate types. In this section, the major characteristics of each of these climates will be briefly described. The climates are discussed in a rough poleward sequence.

Tropical Wet Climate

Sometimes called the tropical rainforest climate, the tropical wet climate exists chiefly in areas lying within 10 degrees of the equator. It is an almost seasonless climate, characterized by year-round warm, humid, rainy conditions that allow land areas to support a dense broadleaf forest cover. The warm temperatures, which for most locations average near 80 degrees Fahrenheit (27 degrees Celsius) throughout the year, result from the constantly high midday sun angles experienced at this low latitude.

WORLD CLIMATE REGIONS

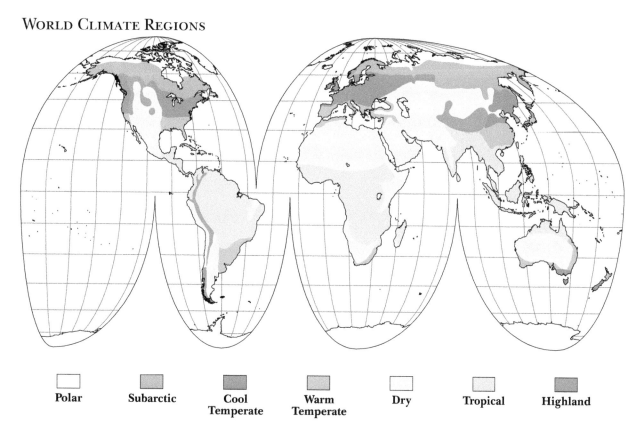

Polar   Subarctic   Cool Temperate   Warm Temperate   Dry   Tropical   Highland

The heavy precipitation totals result from the heating and subsequent rising of the warm moist air to form frequent showers and thunderstorms, especially during the afternoon hours. The dominance of the ITCZ enhances precipitation totals, helping make this climate type one of the world's rainiest.

Tropical Monsoonal Climate

The tropical monsoonal climate occurs in low-latitude areas, such as Southeast Asia, that have a warm, rainy climate with a short dry season. Temperatures are similar to those of the tropical wet climate, with the warmest weather often occurring during the drier period, when sunshine is more abundant. The heavy rainfalls result from the nearness of the ITCZ for much of the year, as well as the dominance of warm, moist air masses derived from tropical oceans. During the brief dry season, however, the ITCZ has usually shifted into the opposite hemisphere, and windflow patterns often have changed so as to bring in somewhat drier air derived from continental sources.

Tropical Savanna Climate

The tropical savanna climate, also referred to as the tropical wet-dry climate, occupies a large portion of the tropics between 5 and 20 degrees latitude in both hemispheres. It experiences a distinctive alternation of wet and dry seasons, caused chiefly by the seasonal shift in latitude of the subtropical highs and ITCZ. Summer is typically the rainy season because of the domination of the ITCZ. In many areas, an onshore windflow associated with the summer monsoon increases rainfalls at this time. In winter, however, the ITCZ shifts into the opposite hemisphere and is replaced by drier and more stable air associated with the subtropical high. In addition, the winter monsoon tendency often produces an outflow of continental air. The long dry season inhibits forest growth, so vegetation usually consists of a cover of drought-resistant shrubs or the tall savanna grasses after which the climate is named.

Subtropical Desert Climate

The subtropical desert climate has hot, arid conditions as a result of the year-round dominance of the subtropical highs. Summertime temperatures in this climate soar to the highest readings found anywhere on earth. The world's record high temperature was 134 degrees Fahrenheit (56.7 degrees Celsius), recorded in Furnace Creek Ranch, California (formerly Greenland Ranch) on July 10, 1913. Rainfall totals in this type of climate are generally less than 10 inches (25 centimeters) per year. What rainfall does occur often arrives as brief, sometimes violent, afternoon thunderstorms. Although summer temperatures are extremely hot, the dry air enables rapid cooling during the winter, so that temperatures are cool to mild at this time of year.

Subtropical Steppe Climate

The subtropical steppe climate is a semiarid climate, found mostly on the margins of the subtropical deserts. Precipitation usually ranges from 10 to 30 inches (25 to 75 centimeters), sufficient for a ground cover of shrubs or short steppe grasses. Areas on the equatorward margins of subtropical deserts typically receive their precipitation during a brief showery period in midsummer, associated with the poleward shift of the ITCZ. Areas on the poleward margins of the subtropical highs receive most of their rainfall during the winter, due to the penetration of cyclonic storms associated with the equatorward shift of the westerly wind belt.

Mediterranean Climate

The Mediterranean climate, also sometimes referred to as the dry summer subtropics, has a distinctive pattern of dry summers and more humid, moderately wet winters. This pattern is caused by the seasonal shift in latitude of the subtropical high and the westerlies. During the summer, the subtropical high shifts poleward into the Mediterranean climate regions, blanketing them with dry, warm, stable air. As winter approaches, this pressure center retreats equatorward, allowing the westerlies, with their eastward-traveling weather fronts and cyclonic storms, to overspread this region. The Mediterranean climate is found on the windward sides of continents, particularly the area surrounding the Mediterranean Sea and much of California. This results in the predomi-

nance of maritime air and relatively mild temperatures throughout the year.

Humid Subtropical Climate

The humid subtropical climate is found on the eastern, or leeward, sides of continents in the lower middle latitudes. The most extensive land area with this climate is the southeastern United States, but it is also seen in large areas in South America, Asia, and Australia. Temperature ranges are moderately large, with warm to hot summers and cool to mild winters. Mean temperatures for a given location are dictated largely by latitude, elevation, and proximity to the coast. Precipitation is moderate. Winter precipitation is usually associated with weather fronts and cyclonic storms that travel eastward within the westerly wind belt. During summer, most precipitation is in the form of brief, heavy afternoon and evening thunderstorms. Some coastal areas are subject to destructive hurricanes during the late summer and autumn.

Midlatitude Desert Climate

This type of climate consists of areas within the western United States, southern South America, and Central Asia that have arid conditions resulting from the moisture-blocking influence of mountain barriers. This climate is highly continental, with warm summers and cold winters. When precipitations occurs, it frequently comes in the form of winter snowfalls associated with weather fronts and cyclonic storms. Rainfall in summer typically occurs as afternoon thunderstorms.

Midlatitude Steppe

The midlatitude steppe climate is located in interior portions of continents in the middle latitudes, particularly in Asia and North America. This climate has semiarid conditions caused by a combination of continentality resulting from the large distance from oceanic moisture sources and the presence of mountain barriers. Like the midlatitude desert climate, this climate has large annual temperature ranges, with cold winters and warm summers. It also receives winter rains and snows chiefly from weather fronts and cyclonic storms; summer rains occur largely from afternoon convectional storms. In the Great Plains of the United States, spring can bring very turbulent conditions, with blizzards in early spring and hailstorms and tornadoes in mid to late spring.

Marine West Coast

This type of climate is typically located on the west coasts of continents just poleward of the Mediterranean climate. Its location in the heart of the westerly wind belt on the windward sides of continents produces highly maritime conditions. As a result, cloudy and humid weather is common, along with frequent periods of rainfall from passing weather fronts and cyclonic storms. These storms are often well developed in winter, resulting in extended periods of wet and windy weather. Precipitation amounts are largely controlled by the presence and strength of the orographic effect; mountainous coasts like the northwestern United States and the west coast of Canada are much wetter than are flatter areas like northern Europe. Temperatures are held at moderate levels by the onshore flow of maritime air. As a consequence, winters are relatively mild and summers relatively cool for the latitude.

Humid Continental Climate

The humid continental climate is found in the northern interiors of Eurasia (Europe and Asia) and North America. It does not occur in the Southern Hemisphere because of the absence of large land masses in the upper midlatitudes of that hemisphere. This climate type is characterized by low to moderate precipitation that is largely frontal and cyclonic in nature. Most precipitation occurs in summer, but cold winter temperatures typically cause the surface to be frozen and snow-covered for much of the late fall, winter, and early spring. Temperature ranges in this climate are the largest in the world. A town in Siberia, Verkhoyansk, holds the Guinness world record for the greatest temperature range at a single location is 221 degrees Fahrenheit (105 degrees Celsius), from -90 degrees Fahrenheit (-68 degrees Celsius) to 99 degrees Fahrenheit (37 degrees Celsius). Winter temperatures in parts of both North America and Siberia can fall well below -49 degrees

Fahrenheit (-45 degrees Celsius), making these the coldest permanently settled sites in the world.

Tundra Climate

The tundra climate is a severely cold climate that exists mostly on the coastal margins of the Arctic Ocean in extreme northern North America and Eurasia, and along the coast of Greenland. The high-latitude location and proximity to icy water cause every month to have average temperatures below 50 degrees Fahrenheit (10 degrees Celsius), although a few months in summer have means above freezing. As a result of the cold temperatures, tundra areas are not forested, but instead typically have a sparse ground cover of grasses, sedges, flowers, and lichens. Even this vegetation is buried by a layer of snow during most of the year. Cold temperatures lower the water vapor holding capacity of the air, causing precipitation totals to be generally light. Most precipitation is associated with weather fronts and cyclonic storms and occurs during the summer half of the year.

Ice Cap Climate

The most poleward and coldest of the world's climates is called the ice cap climate. It is found on the continent of Antarctica, interior Greenland, and some high mountain peaks and plateaus. Because monthly mean temperatures are subfreezing throughout the year, areas with this climate are glaciated and have no permanent human inhabitants.

The coldest temperatures of all occur in interior Antarctica, where a Russian research station named Vostok recorded the world's coldest temperature of -128.6 degrees Fahrenheit (-89.2 degrees Celsius) on July 21, 1983. This climate receives little precipitation because the atmosphere can hold very little water vapor. A major moisture surplus exists, however, because of the lack of snowmelt and evaporation. This causes the build up of a surface snow cover that eventually compacts to form the icecaps that bury the surface. Snowstorms are often accompanied by high winds, producing blizzard conditions.

Global Warming

Though warming has not been uniform across the planet, the upward trend in the globally averaged temperature shows that more areas are warming than cooling. According to the National Oceanic and Atmospheric Administration (NOAA) 2018 Global Climate Summary, the combined land and ocean temperature has increased at an average rate of 0.13°F (0.07°C) per decade since 1880; however, the average rate of increase since 1981 (0.31°F/ 0.17°C) is more than twice as great. It is strongly suspected that human activities that increase the accumulation of greenhouse gases (heat-trapping gases) in the atmosphere may play a key role in the temperature rise.

Levels of carbon dioxide (CO_2) in the atmosphere are higher now than they have been at any time in the past 400,000 years. This gas is responsible for nearly two-thirds of the global-warming potential of all human-released gases. Levels surpassed 407 ppm in 2018 for the first time in recorded history. By comparison, during ice ages, CO_2 levels were around 200 parts per million (ppm), and during the warmer interglacial periods, they hovered around 280 ppm. The recent rise in CO_2 shows a remarkably constant relationship with fossil-fuel burning, which is understandable when one considers that about 60 percent of fossil-fuel emissions stay in the air. Atmospheric carbon dioxide concentrations are also increased by deforestation, which is occurring at a rapid rate in several tropical countries. Deforestation causes carbon dioxide levels to rise because trees remove large quantities of this gas from the atmosphere during the process of photosynthesis.

Research indicates that if atmospheric concentrations of greenhouse gases continue to increase at the 1990s pace, global temperatures could rise an additional 1.8 to 6.3 degrees Fahrenheit (1 to 3.5 degrees Celsius) during the twenty-first century. That level of temperature increase would produce major changes in global climates and plant and animal habitats and would cause sea levels to rise substantially.

Ralph C. Scott

CLOUD FORMATION

Clouds are visible manifestations of water in the air. Cloud patterns can provide even a casual observer with much information about air movements and the processes occurring in the atmosphere. The shapes and heights of the clouds and the directions from which they have come are valuable clues in understanding weather.

Importance of Cooling

Clouds are formed when water vapor in the air is transformed into either water droplets or ice crystals. Sometimes large amounts of moisture are added to the air, producing clouds, but clouds generally are formed when a large amount of air is cooled. The amount of water vapor that air can hold varies with temperature: Cold air can hold less water vapor than warmer air. If air is cooled to the point at which it can hold no more water vapor, the water vapor will condense into water droplets. The temperature at which condensation begins is called the dew point. At below freezing temperatures, the water vapor will turn or deposit into ice crystals.

Cloud droplets do not necessarily form even if the air is fully saturated, that is, holding as much water vapor as possible at a given temperature. Once formed, cloud droplets can evaporate again very easily. Two factors hasten the production and growth of cloud droplets. One is the presence of

CLOUD FORMATION

The hydrologic cycle is the continuous circulation of the earth's waters through evaporation, condensation, and precipitation. The cycle also moves water through runoff, infiltration, and transpiration.

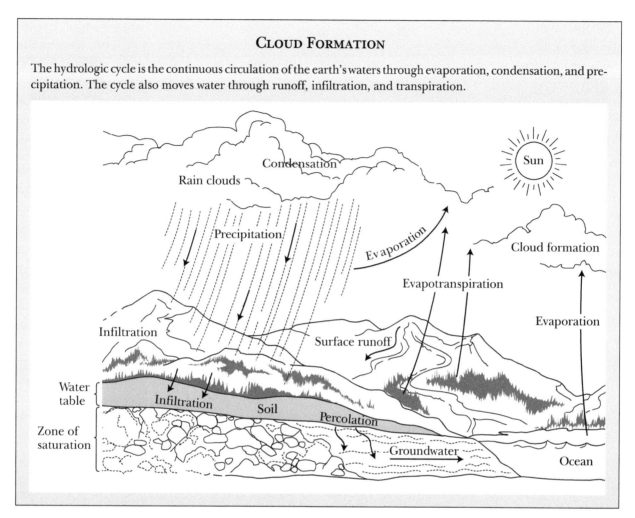

particles in the atmosphere that attract water. These are called hygroscopic particles or condensation nuclei. They include salt, dust, and pollen. Once water vapor condenses on these particles, more condensation can occur. Then the droplets can grow larger and bump into other droplets, growing even larger through this process, called coalescence.

Condensation and cloud droplet growth also is hastened when the air is very cold, at about -40 degrees Farenheit (which is also -40 degrees Celsius). At this temperature ice crystals form, but some water droplets can exist as liquid water. These water droplets are said to be supercooled. The water vapor is more likely to deposit on the ice crystals than on the supercooled water. Thus the ice crystals grow larger and the supercooled water droplets evaporate, resulting in more water vapor to deposit on ice crystals. Whether the cloud droplets start as hygro-

scopic particles or ice crystals, they eventually can grow in size to become a raindrop; around 1 million cloud droplets make one raindrop.

How and Why Rising Air Cools

In order for air to be cooled, it must rise or be lifted. When a volume of air, or an air parcel, is forced to rise through the surrounding air, the parcel expands in size as the pressure of the air around it declines with altitude. Close to the surface, the atmospheric pressure is relatively high because the density of the atmosphere is high. As altitude increases, the atmosphere declines in density, and the still air exerts less pressure. Thus, as an air parcel rises through the atmosphere, the pressure of the surrounding air declines, and the parcel takes up more space as it expands. Since work is done by the parcel as it expands, the parcel cools and its temperature declines.

An alternative explanation of the cooling is that the number of molecules in the air parcel remains the same, but when the volume is larger, the molecules produce less frictional heat because they do not bang into each other as much. The temperature of the air parcel declines, but no heat left the parcel—the change in temperature resulted from internal processes. The process of an air parcel rising, expanding, and cooling is called adiabatic cooling. Adiabatic means that no heat leaves the parcel. If the parcel rises far enough, it will cool sufficiently to reach its dewpoint temperature. With continued cooling, condensation will result—a cloud will be formed. At this height, which is called the lifting condensation level, an invisible parcel of air will turn into a cloud.

Uplift Mechanisms

An initial force is necessary to cause the air parcel to rise and then cool adiabatically. The three major processes are convection, orographic, and frontal or cyclonic.

With certain conditions, convection or vertical movement can cause clouds to form. On a sunny day, usually in the summer, the ground is heated unevenly. Some areas of the ground become warmer and heat the air above, making it warmer and less dense. A stream of air, called a thermal, may rise. As it rises, it cools adiabatically through expansion and may reach its dewpoint temperature. With continued cooling and rising, condensation will occur, forming a cloud. Since the cloud is formed by predominantly vertical motions, the cloud will be cumulus. With continued warming of the surface, the thermals may rise even higher, perhaps producing thunderstorm, or cumulonimbus, clouds. Thus, a sunny summer day can start off without a cloud in the sky, but can be stormy with many thunderstorms by afternoon.

Clouds also can form when air is forced to rise when it meets a mountain or other large vertical barrier. This type of lifting—orographic—is especially prevalent where air moves over the ocean and then is forced to rise up a mountain, as occurs on the west coast of North and South America. As the

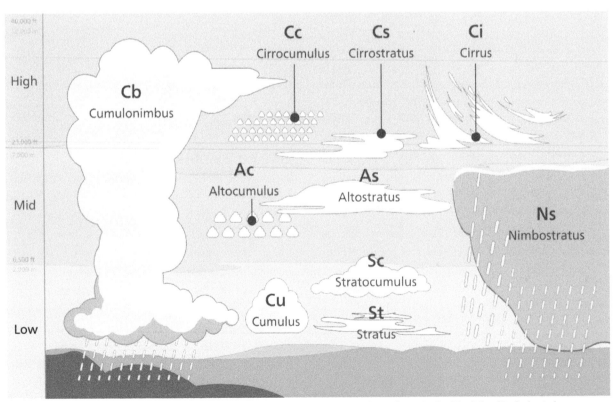

Cloud classification by altitude of occurrence. Multi-level and vertical genus-types not limited to a single altitude level include nimbostratus, cumulonimbus, and some of the larger cumulus species. (Illustration by Valentin de Bruyn)

air rises, it cools adiabatically and eventually becomes so cool that it cannot hold the water vapor. Condensation occurs and clouds form. The air continues to move up the mountain, producing clouds and precipitation on the side of the mountain from which the wind came, the windward side. However, the air eventually must fall down the other side of the mountain, the leeward side. That air is warmed and moisture evaporates, resulting in no clouds.

A third lifting mechanism is frontal, or cyclonic, action. This occurs when a large mass of cold air and a large mass of warm air—often hundreds of miles in area—meet. The warm air mass and the cold air mass will not mix freely, resulting in a border or front between the two air masses. The warm, less dense, air will always rise above the cold, denser, air mass. As the warm air rises, it cools, and when it reaches its dew point, clouds will form. If the warm air displaces the cold air, or a warm front occurs, the warm air will rise gradually, resulting in layered or stratiform clouds. The cloud types will change on an upward diagonal path, with the lowest being

stratus, and nimbostratus if rain occurs, followed by altostratus, then cirrostratus, and cirrus.

On the other hand, if the cold air displaces the warm air, the warm air will be forced to rise much more quickly. The clouds formed will be puffy or cumuliform—cumulus at the lowest levels, altocumulus and cirrocumulus at the highest altitudes. Sometimes cumulonimbus clouds will also form.

Sometimes when a cold front meets a warm front, the whole warm air mass is forced off the ground. This forms a cyclone—an area of low pressure—as the warm air rises. As this air rises, it cools. If it reaches its dew point, condensation and clouds will result. In oceanic tropical areas, a cyclone can form within warm, moist air. This air also will cool and, if it reaches its dew point, will condense and form clouds. Sometimes, these tropical cyclones are the precursors of hurricanes. The clouds associated with cyclones are usually cumulus, including cumulonimbus, as they are formed by rapidly rising air.

Margaret F. Boorstein

STORMS

A storm is an atmospheric disturbance that produces wind, is accompanied by some form of precipitation, and sometimes involves thunder and lightning. Storms that meet certain criteria are given specific names, such as hurricanes, blizzards, and tornadoes.

Stormy weather is associated with low atmospheric pressure, while clear, calm, dry weather is associated with high atmospheric pressure. Because of the way atmospheric pressure and wind direction are related, low-pressure areas are characterized by winds moving cyclonically (in a counterclockwise direction in the Northern Hemisphere; clockwise in the Southern Hemisphere) around the center of the low pressure. Storms of all kinds are associated with

cyclones, but two classes of cyclones—tropical and extratropical—produce most storms.

Tropical Cyclones

These storms develop during the summer and autumn in every tropical ocean except the South Atlantic and eastern South Pacific Oceans. Tropical cyclones that occur in the North Atlantic and eastern North Pacific Oceans are known as hurricanes; in the western North Pacific Ocean, as typhoons; and in the Indian and South Pacific Oceans, as cyclones.

All tropical cyclones develop in three stages. Arising from the formation of the initial atmospheric disturbance that is characterized by a cluster of thunderstorms, the first stage—tropical depres-

ANATOMY OF A HURRICANE

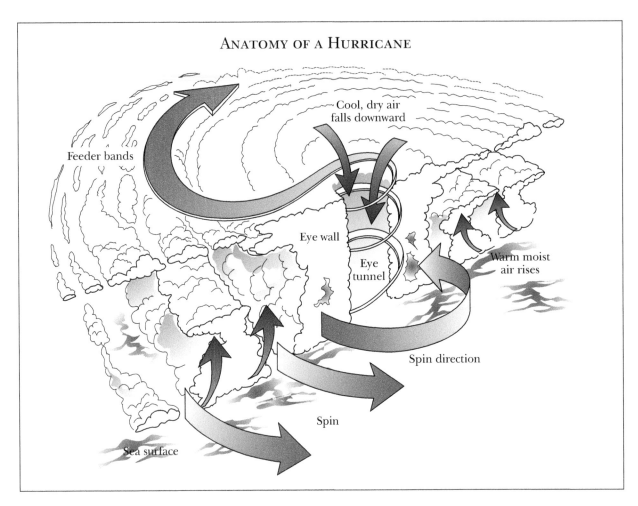

sion—occurs when the maximum sustained surface wind speeds (the average speed over one minute) range from 23–39 miles (37–61 km.) per hour. The second stage—tropical storm—occurs when sustained winds range from 40–73 miles (62–119 km.) per hour. At this stage, the storm is given a name. From 80 to 100 tropical storms develop each year across the world, with about half continuing to the final stage—hurricane—at which sustained wind speeds are 74 miles (120 km.) per hour or greater. Moving over land or into colder oceans initiates the end of the hurricane after a week or so by eliminating the hurricane's fuel—warm water.

A mature hurricane is a symmetrical storm, with the "eye" at the center; the eye develops as winds increase and become circular around the central core of low pressure. Within the eye, it is relatively warm, and there are light winds, no precipitation, and few clouds. This is caused by air descending in the center of the storm. Surrounding the eye is the "eye

wall," a ring of intense thunderstorms that can extend high into the atmosphere. Within the eye wall, the strongest winds and heaviest rainfall are found; this is also where warm, moist air, the hurricane's "fuel," flows into the storm. Spiraling bands of clouds, called "rain bands," surround the eye wall. Precipitation and wind speeds decrease from the eye wall out toward the edge of the rain bands, while atmospheric pressure is lowest in the eye and increases outward.

Hurricanes can be the most damaging storms because of their intensity and size. Damage is caused by high winds and the flying debris they carry, flooding from the tremendous amounts of rain a hurricane can produce, and storm surge. A storm surge, which accounts for most of the coastal property loss and 90 percent of hurricane deaths, is a dome of water that is pushed forward as the storm moves. This wall of water is lifted up onto the coast as the eye wall comes in contact with land. For exam-

NAMING HURRICANES

Hurricanes once were identified by their latitudes and longitudes, but this method of naming became confusing when two or more hurricanes developed at the same time in the same ocean. During World War II hurricanes were identified by radio code letters, such as Able and Baker. In 1953 the National Weather Service began using English female names in an alphabetized list. Male names and French and Spanish names were added in 1978. By 2000 six lists of names were used on a rotating basis. When a hurricane causes much death or destruction, as Hurricane Andrew did in August of 1992—its name is retired for at least ten years.

ple, a 25-foot (8-meter) storm surge created by Hurricane Camille in 1969 destroyed the Richelieu Apartments next to the ocean in Pass Christian, Mississippi. Ignoring advice to evacuate, twenty-five people had gathered there for a hurricane party; all but one was killed.

To help predict the damage that an approaching hurricane can cause, the Saffir- Simpson Scale was developed. A hurricane is rated from 1 (weak) to 5 (devastating), according to its central pressure, sustained wind speed, and storm surge height. Michael was a category 5 storm at the time of landfall on October 10, 2018, near Mexico Beach and Tyndall Air Force Base, Florida. Michael was the first hurricane

to make landfall in the United States as a category 5 since Hurricane Andrew in 1992, and only the fourth on record. The others are the Labor Day Hurricane in 1935 and Hurricane Camille in 1969.

Extratropical Cyclones

Also known as midlatitude cyclones, these storms are traveling low-pressure systems that are seen on newspaper and television daily weather maps. They are created when a mass of moist, warm air from the south contacts a mass of drier, cool air from the north, causing a front to develop. At the front, the warmer air rides up over the colder air. This causes water vapor to condense and produces clouds and rain during most of the year, and snow in the winter.

Thunderstorms

Thunderstorms also develop in stages. During the cumulus stage, strong updrafts of warm air build the storm clouds. The storm moves into the mature stage when updrafts continue to feed the storm, but cool downdrafts are also occurring in a portion of the cloud where precipitation is falling. When the warm updrafts disappear, the storm's fuel is gone and the dissipating stage begins. Eventually, the cloud rains itself out and evaporates.

Thunderstorms can also form away from a frontal system, usually during summer. This formation is related to a relatively small area of warm, moist air

STORM CLASSIFICATIONS

Tropical Classification	Wind speed
Gale-force winds	>15 meters/second
Tropical depression	20-34 knots and a closed circulation
Tropical storm (named)	35-64 knots
Hurricane	65+ knots (74+ mph)
Saffir-Simpson Scale for Hurricanes	
Category 1	63-83 knots (74-95 mph)
Category 2	83-95 knots (96-110 mph)
Category 3	96-113 knots (111-130 mph)
Category 4	114-135 knots (131-155 mph)
Category 5	>135 knots (>155 mph)

Notes: 1 knot = 1 nautical mile/hour = 1.152 miles/hour = 1.85 kilometers/hour.
Source: National Aeronautics and Space Administration, Office of Space Science, Planetary Data System.
http:/atmos.nmsu.edu/jsdap/encyclopediawork.html

rising and creating a thunderstorm that is usually localized and short lived.

Wind, lightning, hail, and flooding from heavy rain are the main destructive forces of a thunderstorm. Lightning occurs in all mature thunderstorms as the positive and negative electrical charges in a cloud attempt to equal out, creating a giant spark. Most lightning stays within the clouds, but some finds its way to the surface. The lightning heats the air around it to incredible temperatures (54,000 degrees Farenheit/30,000 degrees Celsius), which causes the air to expand explosively, creating the shock wave called thunder. Since lightning travels at the speed of light and thunder at the speed of sound, one can estimate how many miles away the lightning is by counting the seconds between the lightning and thunder and dividing by five. People have been killed by lightning while boating, swimming, biking, golfing, standing under a tree, talking on the telephone, and riding on a lawnmower.

Hail is formed in towering cumulonimbus clouds with strong updrafts. It begins as small ice pellets that grow by collecting water droplets that freeze on contact as the pellets fall through the cloud. The strong updrafts push the pellets back into the cloud, where they continue collecting water droplets until they are too heavy to stay aloft and fall as hailstones. The more an ice pellet is pushed back into the cloud, the larger the hailstone becomes. The largest authenticated hailstone in the United States fell near Vivian, South Dakota, on July 23, 2010. It measured 8.0 inches (20 cm) in diameter, 18 1/2 inches (47 cm.) in circumference, and weighed in at 1.9375 pounds (879 grams).

Tornadoes

For reasons not well understood, less than 1 percent of all thunderstorms spawn tornadoes. Called funnel clouds until they touch earth, tornadoes contain the highest wind speeds known.

Although tornadoes can occur anywhere in the world, the United States has the most, with an average of 1000 per year. Tornadoes have occurred in every state, but the greatest number hit a portion of the Great Plains from central Texas to Nebraska, known as "Tornado Alley." There cold Canadian air and warm Gulf Coast air often collide over the flat land, creating the wall cloud from which most tornadoes are spawned. May is the peak month for tornado activity, but they have been spotted in every month.

Because tornado winds cannot be measured directly, the tornado is ranked according to its damage, using the Fujita Intensity Scale. The scale ranges from an F0, with wind speeds less than 72 miles (116 km.) per hour, causing light damage, to an F5, with winds greater than 260 miles (419 km.) per hour, causing incredible damage. Most tornadoes are small, but the larger ones cause much damage and death.

Kay R. S. Williams

Earth's Biological Systems

Biomes

The major recognizable life zones of the continents, biomes are characterized by their plant communities. Temperature, precipitation, soil, and length of day affect the survival and distribution of biome species. Species diversity within a biome may increase its stability and capability to deliver natural services, including enhancing the quality of the atmosphere, forming and protecting the soil, controlling pests, and providing clean water, fuel, food, and drugs. Land biomes are the temperate, tropical, and boreal forests; tundra; desert; grasslands; and chaparral.

Temperate Forest

The temperate forest biome occupies the so-called temperate zones in the midlatitudes (from about 30 to 60 degrees north and south of the equator). Temperate forests are found mainly in Europe, eastern North America, and eastern China, and in narrow zones on the coasts of Australia, New Zealand, Tasmania, and the Pacific coasts of North and South America. Their climates are characterized by high rainfall and temperatures that vary from cold to mild.

Temperate forests contain primarily deciduous trees—including maple, oak, hickory, and beechwood—and, secondarily, evergreen trees—including pine, spruce, fir, and hemlock. Evergreen forests in some parts of the Southern Hemisphere contain eucalyptus trees.

The root systems of forest trees help keep the soil rich. The soil quality and color is due to the action of earthworms. Where these forests are frequently cut, soil runoff pollutes streams, which reduces fisheries because of the loss of spawning habitat.

Racoons, opposums, bats, and squirrels are found in the trees. Deer and black bear roam forest floors. During winter, small animals such as groundhogs and squirrels burrow in the ground.

Tropical Forest

Tropical forests are in frost-free areas between the Tropic of Cancer and the Tropic of Capricorn. Temperatures range from warm to hot year-round, because the Sun's rays shine nearly straight down around midday. These forests are found in northern Australia, the East Indies, southeastern Asia, equatorial Africa, and parts of Central America and northern South America.

Tropical forests have high biological diversity and contain about 15 percent of the world's plant species. Animal life lives at different layers of tropical forests. Nuts and fruits on the trees provide food for birds, monkeys, squirrels, and bats. Monkeys and sloths feed on tree leaves. Roots, seeds, leaves, and fruit on the forest floor feed deer, hogs, tapirs, antelopes, and rodents. The tropical forests produce rubber trees, mahogany, and rosewood. Large animals in these forests include the Asian tiger, the African bongo, the South American tapir, the Central and South American jaguar, the Asian and African leopard, and the Asian axis deer. Deforestation for agriculture and pastures has caused reduction in plant and animal diversity.

Boreal Forest

The boreal forest is a circumpolar Northern Hemisphere biome spread across Russia, Scandinavia, Canada, and Alaska. The region is very cold. Evergreen trees such as white spruce and black spruce

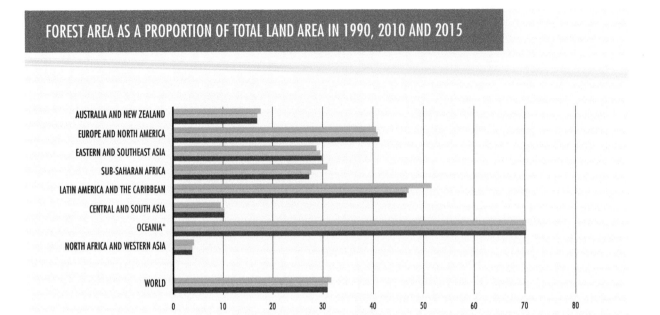

FOREST AREA AS A PROPORTION OF TOTAL LAND AREA IN 1990, 2010 AND 2015

AUSTRALIA AND NEW ZEALAND
EUROPE AND NORTH AMERICA
EASTERN AND SOUTHEAST ASIA
SUB-SAHARAN AFRICA
LATIN AMERICA AND THE CARIBBEAN
CENTRAL AND SOUTH ASIA
OCEANIA*
NORTH AFRICA AND WESTERN ASIA
WORLD

PERCENTAGE

■ 1990   ■ 2010   ■ 2015

NOTE: *Excluding Australia and New Zealand.
SOURCE: Based on UN, 2017a.

dominate this zone, which also contains larch, balsam, pine, and fir, and some deciduous hardwoods such as birch and aspen. The acidic needles from the evergreens make the leaf litter that is changed into soil humus. The acidic soil limits the plants that develop.

Animals in boreal forests include deer, caribou, bear, and wolves. Birds in this zone include goshawks, red-tailed hawks, sapsuckers, grouse, and nuthatches. Relatively few animals emigrate from this habitat during winter. Conifer seeds are the basic winter food. The disappearing aspen habitat of the beaver has decreased their numbers and has reduced the size of wetlands.

Tundra

About 5 percent of the earth's surface is covered with Arctic tundra, and 3 percent with alpine tundra. The Arctic tundra is the area of Europe, Asia, and North America north of the boreal coniferous forest zone, where the soils remain frozen most of the year. Arctic tundra has a permanent frozen subsoil, called permafrost. Deep snow and low temper-

atures slow the soil-forming process. The area is bounded by a 50 degrees Fahrenheit circumpolar isotherm, known as the summer isotherm. The cold temperature north of this line prevents normal tree growth.

The tundra landscape is covered by mosses, lichens, and low shrubs, which are eaten by caribou, reindeer, and musk oxen. Wolves eat these herbivores. Bear, fox, and lemming also live here. The larger mammals, including marine mammals and the overwintering birds, have large fat layers beneath the skin and long dense fur or dense feathers that provide protection. The small mammals burrow beneath the ground to avoid the harsh winter climate. The most common Arctic bird is the old squaw duck. Ptarmigans and eider ducks are also very common. Geese, falcons, and loons are some of the nesting birds of the area.

The alpine tundra, which exists at high altitude in all latitudes, is acted upon by winds, cold temperatures, and snow. The plant growth is mostly cushion and mat-forming plants.

BIOMES OF THE WORLD

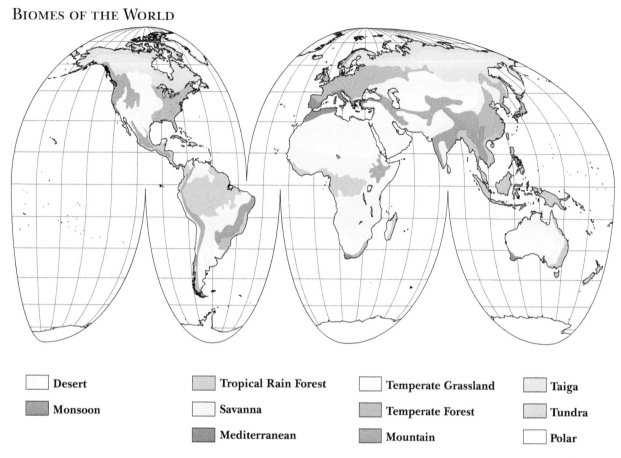

☐ Desert	▨ Tropical Rain Forest	☐ Temperate Grassland	▨ Taiga
▨ Monsoon	☐ Savanna	▨ Temperate Forest	▨ Tundra
	▨ Mediterranean	▨ Mountain	☐ Polar

Desert

The desert biome covers about one-seventh of the earth's surface. Deserts typically receive no more than 10 inches (25 centimeters) of rainfall a year, but evaporation generally exceeds rainfall. Deserts are found around the Tropic of Cancer and the Tropic of Capricorn. As the warm air rises over the equator, it cools and loses its water content. This dry air descends in the two subtropical zones on each side of the equator; as it warms, it picks up moisture, resulting in drying the land.

Rainfall is a key agent in shaping the desert. The lack of sufficient plant cover removes the natural protection that prevents soil erosion during storms. High winds also cut away the ground.

Some desert plants obtain water from deep below the surface, for example, the mesquite tree, which has roots that are 40 feet (13 meters) deep. Other plants, such as the barrel cactus, store large amounts of water in their leaves, roots, or stems. Other plants slow the loss of water by having tiny leaves or shedding their leaves. Desert plants have very short growth periods, because they cannot grow during the long drought periods.

Desert animals protect themselves from the Sun's heat by eating at night, staying in the shade during the day, and digging burrows in the ground. Among the world's large desert animals are the camel, coyote, mule deer, Australian dingo, and Asian saiga. The digestive process of some desert animals produces water. A method used by some animals to conserve water is the reabsorption of water from their feces and urine.

Grassland

Grasslands cover about a quarter of the earth's surface, and can be found between forests and deserts. Treeless grasslands grow in parts of central North America, Central America, and eastern South America that have between 10 and 40 inches (250-1,000 millimeters) of erratic rainfall. The climate has a high rate of evaporation and periodic major droughts. The biome is also subject to fire.

Some grassland plants survive droughts by growing deep roots, while others survive by being dormant. Grass seeds feed the lizards and rodents that become the food for hawks and eagles. Large animals include bison, coyotes, mule deer, and wolves. The grasslands produce more food than any other biome. Poor grazing and agricultural practices and mining destroy the natural stability and fertility of these lands. The reduced carrying capacity of these lands causes an increase in water pollution and erosion of the soil. Diverse natural grasslands appear to be more capable of surviving drought than are simplified manipulated grass systems. This may be due to slower soil mineralization and nitrogen turnover of plant residues in the simplified system.

Savannas are open grasslands containing deciduous trees and shrubs. They are near the equator and are associated with deserts. Grasses grow in clumps and do not form a continuous layer. The northern savanna bushlands are inhabited by oryx and gazelles. The southern savanna supports springbuck and eland. Elephants, antelope, giraffe, zebras, and black rhinoceros are found on the savannas. Lions, leopards, cheetah, and hunting dogs are the primary predators here. Kangaroos are found in the savannas of Australia. Savannas cover South America north and south of the Amazon rainforest, where jaguar and deer can be found.

Chaparral

The chaparral or Mediterranean biome is found in the Mediterranean Basin, California, southern Australia, middle Chile, and Cape Province of South America. This region has a climate of wet winters and summer drought. The plants have tough leathery leaves and may contain thorns. Regional fires clear the area of dense and dead vegetation. Fire, heat, and drought shape the region. The vegetation dwarfing is due to the severe drought and extreme climate changes. The seeds from some plants, such as the California manzanita and South African fire lily, are protected by the soil during a fire and later germinate and rapidly grow to form new plants.

Ocean

The ocean biome covers more than 70 percent of the earth's surface and includes 90 percent of its volume. The ocean has four zones. The intertidal zone is shallow and lies at the land's edge. The continental shelf, which begins where the intertidal zone ends, is a plain that slopes gently seaward. The neritic zone (continental slope) begins at a depth of about 600 feet (180 meters), where the gradual slant of the continental shelf becomes a sharp tilt toward the ocean floor, plunging about 12,000 feet (3,660 meters) to the ocean bottom, which is known as the abyss. The abyssal zone is so deep that it does not have light.

Plankton are animals that float in the ocean. They include algae and copepods, which are microscopic crustaceans. Jellyfish and animal larva are also considered plankton. The nekton are animals that move freely through the water by means of their muscles. These include fish, whales, and squid. The benthos are animals that are attached to or crawl along the ocean's floor. Clams are examples of benthos. Bacteria decompose the dead organic materials on the ocean floor.

The circulation of materials from the ocean's floor to the surface is caused by winds and water temperature. Runoff from the land contains polluting chemicals such as pesticides, nitrogen fertilizers, and animal wastes. Rivers carry loose soil to the ocean, where it builds up the bottom areas. Overfishing has caused fisheries to collapse in every world sector. In some parts of the northwestern Altantic Ocean, there has been a shift from bony fish to cartilaginous fish dominating the fisheries.

Human Impact on Biomes

Human interaction with biomes has increased biotic invasions, reduced the numbers of species, changed the quality of land and water resources, and caused the proliferation of toxic compounds. Managed care of biomes may not be capable of undoing these problems.

Ronald J. Raven

NATURAL RESOURCES

SOILS

Soils are the loose masses of broken and chemically weathered rock mixed with organic matter that cover much of the world's land surface, except in polar regions and most deserts. The two major solid components of soil—minerals and organic matter—occupy about half the volume of a soil. Pore spaces filled with air and water account for the other half. A soil's organic material comes from the remains of dead plants and animals, its minerals from weathered fragments of bedrock. Soil is also an active, dynamic, ever-changing environment. Tiny pores in soil fill with air, water, bacteria, algae, and fungi working to alter the soil's chemistry and

speed up the decay of organic material, making the soil a better living environment for larger plants and animals.

Soil Formation

The natural process of forming new soil is slow. Exactly how long it takes depends on how fast the bedrock below is weathered. This weathering process is a direct result of a region's climate and topography, because these factors influence the rate at which exposed bedrock erodes and vegetation is distributed. Global variations in these factors account for the worldwide differences in soil types.

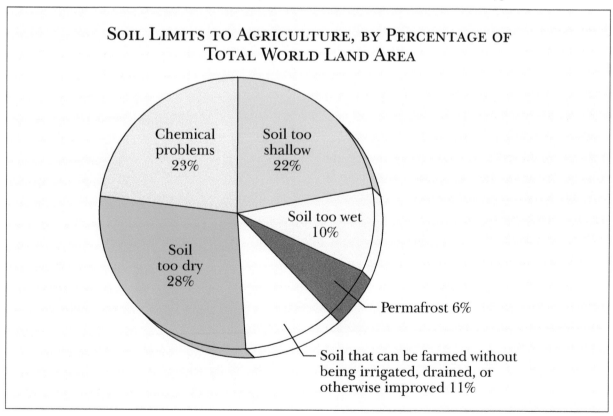

SOIL LIMITS TO AGRICULTURE, BY PERCENTAGE OF TOTAL WORLD LAND AREA

Chemical problems 23%

Soil too shallow 22%

Soil too wet 10%

Soil too dry 28%

Permafrost 6%

Soil that can be farmed without being irrigated, drained, or otherwise improved 11%

Climate is the principal factor in determining the type and rate of soil formation. Temperature and precipitation are the two main climatic factors that influence soil formation, and they vary with elevation and latitude. Water is the main agent of weathering, and the amount of water available depends on how much falls and how much runs off. The amount of precipitation and its distribution during the year influence the kind of soil formed and the rate at which it is formed. Increased precipitation usually results in increased rates of soil formation and deep soils. Temperature and precipitation also determine the kind and amount of vegetation in a region, which determines the amount of available organics.

Topography is a characteristic of the landscape involving slope angle and slope length. Topographic relief governs the amount of water that runs off or enters a soil. On flat or gently sloping land, soil tends to stay in place and may become thick, but as the slope increases so does the potential for erosion. On steep slopes, soil cover may be very thin, possibly only a few inches, because precipitation washes it downhill; on level plains, soil profiles may be several feet thick.

Types of Soil

Typically, bedrock first weathers to form regolith, a protosoil devoid of organic material. Rain, wind, snow, roots growing into cracks, freezing and thawing, uneven heating, abrasion, and shrinking and swelling break large rock particles into smaller ones. Weathered rock particles may range in size from clay to silt, sand, and gravel, with the texture and particle size depending largely on the type of bedrock. For example, shale yields finer-textured soils than sandstone. Soils formed from eroded limestone are rich in base minerals; others tend to be acidic. Generally, rates of soil formation are largely determined by the rates at which silicate minerals in the bedrock weather: the more silicates, the longer the formation time.

In regions where organic materials, such as plant and animal remains, may be deposited on top of regolith, rudimentary soils can begin to form. When waste material is excreted, or a plant or animal dies, the material usually ends up on the earth's surface. Organisms that cause decomposition, such as bacteria and fungi, begin breaking down the remains into a beneficial substance known as humus. Humus restores minerals and nutrients to the soil. It also improves the soil's structure, helping it to retain water. Over time, a skeletal soil of coarse, sandy material with trace amounts of organics gradually forms. Even in a region with good weathering rates and adequate organic material, it can take as long as fifty years to form 12 inches (30 centimeters) of soil. When new soil is formed from weathering bedrock, it can take from 100 to 1,000 years for less than an inch of soil to accumulate.

Water moves continually through most soils, transporting minerals and organics downward by a process called leaching. As these materials travel downward, they are filtered and deposited to form distinct soil horizons. Each soil horizon has its own color, texture, and mineral and humus content. The O-horizon is a thin layer of rotting organics covering the soil. The A-horizon, commonly called topsoil, is rich in humus and minerals. The B-horizon is a subsoil rich in minerals but poor in humus. The C-horizon consists of weathered bedrock; the D-horizon is the bedrock itself.

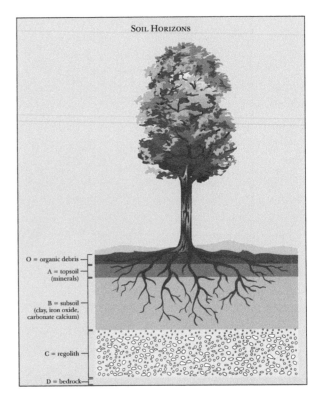

SOIL HORIZONS

O = organic debris

A = topsoil
(minerals)

B = subsoil
(clay, iron oxide,
carbonate calcium)

C = regolith

D = bedrock

Because Earth's surface is made of many different rock types exposed at differing amounts and weathering at different rates at different locations, and because the availability of organic matter varies greatly around the planet due to climatic and seasonal conditions, soil is very diverse and fertile soil is unevenly distributed. Structure and composition are key factors in determining soil fertility. In a fertile soil, plant roots are able to penetrate easily to obtain water and dissolved nutrients. A loam is a naturally fertile soil, consisting of masses of particles from clays (less than 0.002 mm across), through silts (ten times larger) to sands (100 times larger), interspersed with pores, cracks, and crevices.

The Roles of Soil

In any ecosystem, soils play six key roles. First, soil serves as a medium for plant growth by mechanically supporting plant roots and supplying the eighteen nutrients essential for plants to survive. Different types of soil contain differing amounts of these eighteen nutrients; their combination often determines the types of vegetation present in a region, and as a result, influences the number and types of animals the vegetation can support, including humans. Humans rely on soil for crops necessary for food and fiber.

Second, the property of a particular soil is the controlling factor in how the hydrologic system in a region retains and transports water, how contaminants are stored or flushed, and at what rate water is naturally purified. Water enters the soil in the form of precipitation, irrigation, or snowmelt that falls or runs off soil. When it reaches the soil, it will either be surface water, which evaporates or runs into streams, or subsurface water, which soaks into the soil where it is either taken up by plant roots or percolates downward to enter the groundwater system. Passing through soil, organic and inorganic pollutants are filtered out, producing pure groundwater.

Soil also functions as an air-storage facility. Air is pushed into and drawn out of the soil by changes in barometric pressure, high winds, percolating water, and diffusion. Pore spaces within soil provide access to oxygen to organisms living underground as well as to plant roots. Soil pore spaces also contain carbon dioxide, which many bacteria use as a source of carbon.

Soil is nature's recycling system, through which organic waste products and decaying plants and animals are assimilated and their elements made available for reuse. The production and assimilation of humus within soil converts mineral nutrients into forms that can be used by plants and animals, who return carbon to the atmosphere as carbon dioxide. While dead organic matter amounts to only about 1 percent of the soil by weight, it is a vital component as a source of minerals.

Soil provides a habitat for many living things, from insects to burrowing animals, from single microscopic organisms to massive colonies of subterranean fungi. Soils contain much of the earth's genetic diversity, and a handful of soil may contain billions of organisms, belonging to thousands of species. Although living organisms only account for about 0.1 percent of soil by weight, 2.5 acres (one hectare) of good-quality soil can contain at least 300 million small invertebrates—mites, millipedes, insects, and worms. Just 1 ounce (30 grams) of fertile soil can contain 1 million bacteria of a single type, 100 million yeast cells, and 50,000 fungus mycelium. Without these, soil could not convert nitrogen, phosphorus, and sulphur to forms available to plants.

Finally, soil is an important factor in human culture and civilization. Soil is a building material used to make bricks, adobe, plaster, and pottery, and often provides the foundation for roads and buildings. Most important, soil resources are the basis for agriculture, providing people with their dietary needs.

Because the human use of soils has been haphazard and unchecked for millennia, soil resources in many parts of the world have been harmed severely. Human activities, such as overcultivation, inexpert irrigation, overgrazing of livestock, elimination of tree cover, and cultivating steep slopes, have caused natural erosion rates to increase many times over. As a result of mismanaged farm and forest lands, escalated erosional processes wash off or blow away an estimated 75 billion tons of soil annually, eroding away one of civilization's crucial resources.

Randall L. Milstein

WATER

Life on Earth requires water—without it, life on Earth would cease. As human populations grow, the freshwater resources of the world become scarcer and more polluted, while the need for clean water increases. Although nearly three-quarters of Earth's surface is covered with water, only about 0.3 percent of that water is freshwater suitable for consumption and irrigation. This is because more than 97 percent of Earth's water is ocean salt water, and most of the remaining freshwater is frozen in the Antarctic ice cap. Only the small amounts that remain in lakes, rivers, and groundwater is available for human use.

All of earth's water cycles between the ocean, land, atmosphere, plants, and animals over and over. On average, a molecule of surface water cycles from the ocean, to the atmosphere, to the land and back again in less than two weeks. Water consumed by plants or animals takes longer to return to the oceans, but eventually the cycle is completed.

Water's Uses

Water supports the lives of all living creatures. People use for drinking, cooking, cleaning, and bathing. Water also plays a key role in society since humans can travel on it, make electricity with it, fish in it, irrigate crops with it, and use it for recreation. Globally, more than 4 trillion cubic meters of freshwater is used each day. Agriculture accounts for about 70 percent, industry uses 20 percent, and domestic and municipal activities use 10 percent. To produce beef requires between 1,320 and 5,283 gallons (5,000 and 20,000 liters) of water for every 35 ounces (1 kg). A similar amount of wheat requires between 660 and 1056 gallons (2,500 and 4,000 liters) of water. Manufactured goods also require significant amounts of freshwater; a car consumes between 13,000 and 20,000 gallons (49,210 to 75,708 liters), a smartphone has a water footprint of nearly 3,100 gallons (11,734 liters) and a teeshirt uses around 660 gallons (2,498 liters).

The average American family uses more than 300 gallons of drinking quality water per day at home. Roughly 70 percent of this use occurs indoors for drinking, bathing and showering, flushing the toilet, and washing dishes. Outdoor water use for landscape watering, washing cars, and cleaning windows, etc., accounts for 30 percent of household use, although it can be much higher in drier parts of the country. For example, the arid West has some of the highest per capita residential water use because of landscape irrigation.

As the world's population grows, the demand for fresh water will also increase. A study by the World Bank concluded that approximately 80 percent of human illness results from insufficient water supplies and poor water quality caused by lack of sanitation, so careful management of water resources is essential for improving the health of people in the twenty-first century.

Groundwater Supply and Quality

The amount of groundwater in the Earth is seventy times greater than all of the freshwater lakes combined. Groundwater is held within the rocks below the ground surface and is the primary source of water in many parts of the world. In the United States, approximately 50 percent of the population uses some groundwater. However, problems with both groundwater supplies and its quality threaten its future use.

The U.S. Environmental Protection Agency (EPA) found that 45 percent of the large public water systems in the United States that use groundwater were contaminated with synthetic organic chemicals that posed potential health threats. Another major problem occurs when groundwater is used faster than it is replaced by precipitation infiltrating through the ground surface. Many of the arid regions of earth are already suffering from this problem. For example, one-third of the wells in Beijing, China, have gone dry due to overuse. In the United States, the Ogallala Aquifer of the Great Plains, the

THE WORLD AND NORTH AMERICA'S GREATEST RIVERS AND LAKES

Longest river	
Nile (North Africa)	4,130 miles (6,600 km.)
Missouri-Mississippi (United States)	3,740 miles ((6,000 km.)
Largest river by average discharge	
Amazon (South America)	6,181,000 cubic feet/second (175,000 cubic meters/second)
Missouri-Mississippi (United States)	600,440 cubic feet/second (17,000 cubic meters/second)
Largest freshwater lake by volume	
Lake Baikal (Russia)	5,280 cubic miles (22,000 cubic km.)
Lake Superior (United States)	3,000 cubic miles (12,500 cubic km.)

largest in North America, is being severely overused. This aquifer irrigates 30 percent of U.S. farmland, but some areas of the aquifer have declined by up to 60 percent. In one part of Texas, so much water has been pumped out that the aquifer has essentially dried up there. Once depleted, the aquifer will take over 6,000 years to replenish naturally through rainfall.

Surface Water Supply and Quality

Surface water is used for transportation, recreation, electrical generation, and consumption. Ships use rivers and lakes as transport routes, people fish and boat on rivers and lakes, and dams on rivers often are used to generate electricity. The largest river on earth is the Amazon in South America, which has an average flow of 212,500 cubic meters per second, more than twelve times greater than North America's Mississippi River. Earth's largest lake by volume—Lake Baikal in Russia—holds 5,521 cubic miles of water (23,013 cubic kilometers), or approximately 20 percent of Earth's fresh surface water. This is a volume of water approximately equivalent to all five of the North American Great Lakes combined.

Although surface water has more uses, it is more prone to pollution than groundwater. Almost every human activity affects surface water quality. For ex-

ample, water is used to create paper for books, and some of the chemicals used in the paper process are discharged into surface water sources. Most foods are grown with agricultural chemicals, which can contaminate water sources. According to 2018 surveys on national water quality from the U.S. Environmental Protection Agency, nearly half of U.S. rivers and streams and more than one-third of lakes are polluted and unfit for swimming, fishing, and drinking.

Earth's Future Water Supply

Inadequate water supplies and water quality problems threaten the lives of more than 1.5 million people worldwide. The World Health Organization estimates that polluted water causes the death of 361,000 children under five years of age each year and affects the health of 20 percent of Earth's population. As the world's population grows, these problems are likely to worsen.

The United Nations estimates that if current consumption patterns continue, 52 percent of the world's people will live in water-stressed conditions by 2050. Since access to clean freshwater is essential to health and a decent standard of living, efforts must be made to clean up and conserve the planet's freshwater.

Mark M. Van Steeter

EXPLORATION AND TRANSPORTATION

EXPLORATION AND HISTORICAL TRADE ROUTES

The world's exploration was shaped and influenced substantially by economic needs. Lacking certain resources and outlets for trade, many societies built ships, organized caravans, and conducted military expeditions to protect their frontiers and obtain new markets.

Over the last 5,000 years, the world evolved from a cluster of isolated communities into a firmly integrated global community and capitalist world system. By the beginning of the twentieth century, explorers had successfully navigated the oceans, seas, and landmasses and gathered many regional economies into the beginnings of a global economy.

Early Trade Systems

Trade and exploration accompanied the rise of civilization in the Middle East. Egyptian pharaohs, looking for timber for shipbuilding, established trade relations with Mediterranean merchants. Phoenicians probed for new markets off the coast of North Africa and built a permanent settlement at Carthage. By 513 BCE, the Persian Empire stretched from the Indus River in India to the Libyan coast, and it controlled the pivotal trade routes in Iran and Anatolia. A regional economy was taking shape, linking Africa, Asia, and Europe into a blended economic system.

Alexander the Great's victory against the Persian Empire in 330 BCE thrust Greece into a dominant position in the Middle Eastern economy. Trade between the Mediterranean and the Middle East increased, new roads and harbors were constructed, and merchants expanded into sub-Saharan Africa, Arabia, and India. The Romans later benefited from the Greek foundation. Through military and political conquest, Rome consolidated its control over such diverse areas as Arabia and Britain and built a system of roads and highways that facilitated the growth of an expanding world economy. At the apex of Roman power in 200 CE, trade routes provided the empire with Greek marble, Egyptian cloth, seafood from Black Sea fisheries, African slaves, and Chinese silk.

The emergence of a profitable Eurasian trade route linked people, customs, and economies from the South China Sea to the Roman Empire. Although some limited activity occurred during the Hellenistic period, East-West trade flourished following the rise of the Han Dynasty in China. With the opening of the Great Silk Road from 139 BCE to 200 CE, goods and services were exchanged between people from three different continents.

The Great Silk Road was an intricate network of middlemen stretching from China to the Mediterranean Sea. Eastern merchants sold their products at markets in Afghanistan, Iran, and even Syria, and exchanged a variety of commodities through the use of camel caravans. Chinese spices, perfumes, metals, and especially silk were in high demand. The Parthians from central Asia added their own sprinkling of merchandise, introducing both the East and the West to various exotic fruits, rare birds, and ostrich eggs.

Romans peddled glassware, statuettes, and acrobatic performing slaves. Since communication lines were virtually nonexistent during this period, trade routes were the only means by which ideas regarding art, religion, and culture could mix. The contacts and exchanges enacted along the Great Silk Road initiated a process of cultural diffusion among a diversity of cultures and increased each culture's knowledge of the vast frontiers of world geography.

The Atlantic Slave Trade

Beginning in the fifteenth century, European navigators explored the West African coastline seeking gold. Supplies were difficult to procure, because most of the gold mines were located in the interior along the Senegal River and in the Ashanti forests. Because mining required costly investments in time, labor, and security, the Europeans quickly shifted their focus toward the slave trade. Although slavery had existed since antiquity, the Atlantic slave trade generated one of the most significant movements of people in world history. It led to the forced migration of more than 10 million Africans to South America, the Caribbean islands, and North America. It ensured the success of several imperial conquests, and it transformed the demographic, cultural, and political landscape on four continents.

Originally driven by their quest to circumnavigate Africa and open a lucrative trade route with India, the Portuguese initiated a systematic exploration of the West African coastline. The architect of this system, Henry the Navigator, pioneered the use of military force and naval superiority to annex African islands and open up new trade routes, and he increased Portugal's southern frontier with every acquisition. In 1415 his ships captured Ceuta, a prosperous trade center located on the Mediterranean coast overlooking North African trade routes. Over the next four decades, Henry laid claim to the Madeira Islands, the Canary Islands, the Azores, and Cape Verde. After his death, other Portuguese explorers continued his pursuit of circumnavigation of Africa.

Diego Cão reached the Congo River in 1483 and sent several excursions up the river before returning to Lisbon. Two explorers completed the Portuguese mission at the end of the fifteenth century. Vasco da Gama, sailing from 1497 to 1499, and Bartholomeu Dias, from 1498 to 1499, sailed past the southern tip of Africa and eventually reached India. Since Muslims had already created a number of trade links between East Africa, Arabia, and India, Portuguese exploration furthered the integration of various regions into an emerging capitalist world system.

When the Portuguese shifted their trading from gold to slaves, the other European powers followed suit. The Netherlands, Spain, France, and England used their expanding naval technology to explore the Atlantic Ocean and ship millions of slaves across the ocean. A highly efficient and organized trade route quickly materialized. Since the Europeans were unwilling to venture beyond the walls of their coastal fortresses, merchants relied on African sources for slaves, supplying local kings and chiefs with the means to conduct profitable slave-raiding parties in the interior. In both the Congo and the Gold Coast region, many Africans became quite wealthy trading slaves.

In 1750 merchants paid King Tegbessou of Dahomey 250,000 pounds for 9,000 slaves, and his income exceeded the earnings of many in England's merchant and landowning class. After purchasing slaves, dealers sold them in the Americas to work in the mines or on plantations. Commodities such as coffee and sugar were exported back to Europe for home consumption. Merchants then sold alcohol, tobacco, textiles, and firearms to Africans in exchange for more slaves. This practice was abolished by the end of the nineteenth century, but not before more than 10 million Africans had been violently removed from their homeland. The Atlantic slave trade, however, joined port cities from the Gold Coast and Guinea in Africa with Rio de Janeiro, Hispaniola, Havana, Virginia, Charleston, and Liverpool, and constituted a pivotal step toward the rise of a unified global economy.

Magellan and Zheng He

The Portuguese explorer Ferdinand Magellan generated considerable interest in the Asian markets when he led an expedition that sailed around the world from 1519 to 1522. Looking for a quick route to Asia and the Spice Islands, he secured financial backing from the king of Spain. Magellan sailed from Spain in 1519, canvassed the eastern coastline of South America, and visited Argentina. He ultimately traversed the narrow straits along the southern tip of the continent and ventured into the uncharted waters of the Pacific Ocean.

Magellan explored the islands of Guam and the Philippines but was killed in a skirmish on Mactan in 1521. Some of his crew managed to return to Spain in 1522, and one member subsequently published a journal of the expedition that dramatically enhanced the world's understanding of the major sea lanes that connected the continents.

China also opened up new avenues of trade and exploration in Southeast Asia during the fifteenth century. Under the direction of Chinese emperor Yongle, explorer Zheng He organized seven overseas trips from 1405 to 1433 and investigated economic opportunities in Korea, Vietnam, the Indian Ocean, and Egypt. His first voyage consisted of more than 28,000 men and 400 ships and represented the largest naval force assembled prior to World War I.

Zheng's armada carried porcelains, silks, lacquerware, and artifacts to Malacca, the vital port city in Indonesia. He purchased an Arab medical text on drug therapy and had it translated into Chinese. He introduced giraffes and mahogany wood into the mainland's economy, and his efforts helped spread Chinese ideas, customs, diet, calendars, scales and measures, and music throughout the global economy. Zheng He's discoveries, coupled with all the material gathered by the European explorers, provided cartographers and geographers with a credible store of knowledge concerning world geography.

Emerging Global Trade Networks

From 1400 to 1900, several regional economic systems facilitated the exchange of goods and services throughout a growing world system. Building on the triangular relationships produced by the slave trade, the Atlantic region helped spread new food-

stuffs around the globe. Plants and plantation crops provided societies with a plentiful supply of sweet potatoes, squash, beans, and maize. This system, often referred to as the Columbian exchange, also assisted development in other regions by supplying the global economy with an ample money supply in gold and silver. Europeans sent textiles and other manufactures to the Americas. In return, they received minerals from Mexico; sugar and molasses from the Caribbean; money, rum, and tobacco from North America; and foodstuffs from South America. Trade routes also closed the distance between the Pacific coastline in the Americas and the Pacific Rim.

Additional thriving trade routes existed in the African-West Asian region. Linking Europe and Africa with Arabia and India, this area experienced a considerable amount of trade over land and through the sea lanes in the Persian Gulf and Red Sea. Europeans received grains, timber, furs, iron, and hemp from Russia in exchange for wool textiles and silver. Central Asians secured stores of cotton textiles, silk, wheat, rice, and tobacco from India and sold silver, horses, camel, and sheep to the Indians. Ivory, blankets, paper, saltpeter, fruits, dates, incense, coffee, and wine were regularly exchanged among merchants situated along the trade route connecting India, Persia, the Ottoman Empire, and Europe.

Finally, a Russian-Asian-Chinese market provided Russia's ruling czars with arms, sugar, tobacco, and grain, and a sufficient supply of drugs, medicines, livestock, paper money, and silver moved eastward. Overall, this system linked the economies of three continents and guaranteed that a nation could acquire essential foodstuffs, resources, and money from a variety of sources.

Several profitable trade routes existed in the Indian Ocean sector. After Malacca emerged as a key trading port in the sixteenth century, this territory served as an international clearinghouse for the global economy. Indians sent tin, elephants, and wood into Burma and Siam. Rice, silk, and sugar were sold to Bengal. Pepper and other spices were shipped westward across the Arabian Sea, while Ceylon furnished India with vast quantities of jewels, cinnamon, pearls, and elephants. The booming interregional trade routes positioned along the Indian coastline ensured that many of the vast commodities produced in the world system could be obtained in India.

The final region of crucial trade routes was between Southeast Asia and China. While the extent of Asian overseas trade prior to the twentieth century is usually downplayed, an abundance of products flowed across the Bay of Bengal and the South China Sea. Japan procured silver, copper, iron, swords, and sulphur from Cantonese merchants, and Japanese-finished textiles, dyes, tea, lead, and manufactures were in high demand on the mainland. The Chinese also purchased silk and ceramics from the Philippines in exchange for silver. Burma and Siam traded pepper, sappan wood, tin, lead, and saltpeter to China for satin, velvet, thread, and labor. As goods increasingly moved from the Malabar coast in India to the northern boundaries of Korea and Japan, the Pacific Rim played a prominent role in the global economy.

Robert D. Ubriaco, Jr.

ROAD TRANSPORTATION

Roads—the most common surfaces on which people and vehicles move—are a key part of human and economic geography. Transportation activities form part of a nation's economic product: They strengthen regional economy, influence land and natural resource use, facilitate communication and commerce, expand choices, support industry, aid agriculture, and increase human mobility. The need for roads closely correlates with the relative location of centers of population, commerce, industry, and other transportation.

History of Road Making

The great highway systems of modern civilization have their origin in the remote past. The earliest travel was by foot on paths and trails. Later, pack animals and crude sleds were used. The development of the wheel opened new options. As various ancient civilizations reached a higher level, many of them realized the importance of improved roads.

The most advanced highway system of the ancient world was that of the Romans. When Roman civilization was at its peak, a great system of military roads reached to the limits of the empire. The typical Roman road was bold in conception and construction, built in a straight line when possible, with a deep multilayer foundation, perfect for wheeled vehicles.

After the decline of the Roman Empire, rural road building in Europe practically ceased, and roads fell into centuries of disrepair. Commerce traveled by water or on pack trains that could negotiate the badly maintained roads. Eventually, a commercial revival set in, and roads and wheeled vehicles increased.

Interest in the art of road building was revived in Europe in the late eighteenth century. P. Trésaguet, a noted French engineer, developed a new method of lightweight road building. The regime of French dictator Napoleon Bonaparte (1800–1814) encouraged road construction, chiefly for military purposes. At about the same time, two Scottish engineers, Thomas Telford and John McAdam, also developed road-building techniques.

Roads in the United States

Toward the end of the eighteenth century, public demand in the United States led to the improvement of some roads by private enterprise. These improvements generally took the form of toll roads, called "turnpikes" because a pike was rotated in each road to allow entry after the fee was paid, and generally were located in areas adjacent to larger cities. In the early nineteenth century, the federal government paid for an 800-mile-long macadam road from Cumberland, Maryland, to Vandalia, Illinois.

With the development of railroads, interest in road building began to wane. By 1900, however, demand for better roads came from farmers, who wanted to move their agricultural products to market more easily. The bicycle craze of the 1890s and the advent of motorized vehicles also added to the demand for more and better roads. Asphalt and concrete technology was well developed by then; now, the problem was financing. Roads had been primarily a local issue, but the growing demand led to greater state and federal involvement in funding.

The Federal-Aid Highway Act of 1956 was a milestone in the development of highway transportation in the United States; it marked the beginning of the largest peacetime public works program in the history of the world, creating a 41,000-mile National System of Interstate and Defense Highways, built to high standards. Later legislation expanded funding, improved planning, addressed environmental concerns, and provided for more balanced transportation. Other developed countries also developed highway programs but were more restrained in construction.

Roads and Development

Transportation presents a severe challenge for sustainable development. The number of motor vehicles at the end of the second decade of the twenty-first century—estimated at more than 1.2 billion worldwide—is growing almost everywhere at higher rates than either population or the gross domestic product. Overall road traffic grows even more quickly. The tiny nation of San Marino has nearly 1.2 cars per person. Americans own one car for every 1.88 residents. In Great Britain, there is one car for every 5.3 people.

Highways around the world have been built to help strengthen national unity. The Trans-Canada Highway, the world's longest national road, for ex-

---

HIGHWAY CLASSIFICATION

Modern roads can be classified by roadway design or traffic function. The basic type of roadway is the conventional, undivided two-way road. Divided highways have median strips or other physical barriers separating the lanes going in opposite directions.

Another quality of a roadway is its right-of-way control. The least expensive type of system controls most side access and some minor at-grade intersections; the more expensive type has side access fully controlled and no at-grade intersections. The amount of traffic determines the number of lanes. Two or three lanes in each direction is typical, but some roads in Los Angeles have five lanes, while some sections of the Trans-Canada Highway have only one lane. Some highways are paid for entirely from public funds; if users pay directly when they use the road, they are called tollways or turnpikes.

Roads are classified as expressway, arterial, collector, and local in urban areas, with a similar hierarchy in rural areas. The highest level—expressway—is intended for long-distance travel.

---

ample, extends east-west across the breadth of the country. Completed in the 1960s, it had the same goal as the Canadian Pacific Railroad a century before, to improve east-west commerce within Canada.

Sometimes, existing highways need to be upgraded; in less-developed countries, this can simply mean paving a road for all-weather operation. An example of a late-1990s project of this nature was the Brazil-Venezuela Highway project, which had this description: Improve the Brazil-Venezuela highway link by completion of paving along the BR-174, which runs northward from Manaus in the Amazon, through Boa Vista and up to the frontier, so opening a route to the Caribbean. Besides the investment opportunities in building the road itself, the highway would result in investment opportunities in mining, tourism, telecommunications, soy and rice production, trade with Venezuela, manufacturing in the Manaus Free Trade Zone, ecotourism in the Amazon, and energy integration.

Growing road traffic has required increasingly significant national contributions to road construction. Beginning in the 1960s, the World Bank began to finance road construction in several countries. It required that projects be organized to the highest technical and economic standards, with private contracting and international competitive bidding rather than government workers. Still, there were questions as to whether these economic assessments had a road-sector bias and properly incorporated environmental costs. Sustainability was also a question—could the facilities be maintained once they were built?

In the 1990s, the World Bank financed a program to build an asphalt road network in Mozambique. Asphalt makes very smooth roads but is very maintenance-intensive, requiring expensive imported equipment and raw materials. By the end of the decade, the roads required resurfacing but the debt was still outstanding. Alternative materials would have given a rougher road, but it could have been built with local materials and labor.

The European Investment Bank has become a major player in the construction of highways linking Eastern and Western Europe to further European integration. Some of the fastest growth in the world in ownership of autos has been in Eastern Europe. There is a two-way feedback effect between highway construction and auto ownership.

Environment Consequences

Highways and highway vehicles have social, economic, and environmental consequences. Compromise is often necessary to balance transportation needs against these constraints. For example, in Israel, there has been a debate over construction of the Trans-Israel highway, a US$1.3 billon, six-lane highway stretching 180 miles (300 km.) from Galilee to the Negev.

Demand on resources for worldwide road infrastructure far exceeds available funds; governments increasingly are looking to external sources such as tolls. Private toll roads, common in the nineteenth century, are making a comeback. This has spread from the United States to Europe, where private and government-owned highway operators have begun to sell shares on the stock market. Private companies are not only operating and financing roads in Europe, they are also designing and building them. In Eastern Europe, where road construction languished under communism, private financing and toll collecting are seen as the means of supporting badly needed construction.

Industrial development in poor countries is adversely affected by limited transportation. Costs are high-unreliable delivery schedules make it necessary to maintain excessive inventories of raw materials and finished goods. Poor transport limits the radius of trade and makes it difficult for manufacturers to realize the economies of large-scale operations to compete internationally.

In more difficult terrain, roads become more expensive because of a need for cuts and fills, bridges, and tunnels. To save money, such roads often have steeper grades, sharper curves, and reduced width than might be desired. Severe weather changes also damage roads, further increasing maintenance costs.

Stephen B. Dobrow

RAILWAYS

Railroads were the first successful attempts by early industrial societies to develop integrated communication systems. Today, global societies are linked by Internet systems dependent upon communication satellites orbiting around Earth. The speed by which information and ideas can reach remote places breaks down isolation and aids in the developing of a world community. In the nineteenth century, railroads had a similar impact. Railroads were critical for the creation of an urban-industrial society: They linked regions and remote places together, were important contributors in developing nation-states, and revolutionized the way business was conducted through the creation of corporations. Although alternative forms of transportation exist, railroads remain important in the twenty-first century.

The Industrial Revolution and the Railroad

Development of the steam engine gave birth to the railroad. Late in the eighteenth century, James Watt perfected his steam engine in England. Water was superheated by a boiler and vaporized into steam, which was confined to a cylinder behind a piston. Pressure from expanding steam pushes the cylinder forward, causing it to do work if it is attached to wheels. Watt's engine was used in the manufacturing of textiles, thus beginning the Industrial Revolution whereby machine technology mass produced goods for mass consumption. Robert Fulton was the first innovator to commercially apply the steam engine to water transportation. His steamboat *Clermont* made its maiden voyage up the Hudson River in 1807.

Not until the 1820s was a steam engine used for land transportation. Rivers and lakes were natural features where no road needed to be built. Applying steam to land movement required some type of roadbed. In England, George Stephenson ran a locomotive over iron strips attached to wooden rails. Within a short time, England's forges were able to roll rails made completely of iron shaped like an inverted "U."

The amount of profit a manufacturer could make was determined partially by the cost of transportation. The lower the cost of moving cargo and people, the higher the profitability. Several alternatives existed before the emergence of railroads, although there were drawbacks compared to rail transportation. Toll roads were too slow. A loaded wagon pulled by four horses could average 15 miles (25 km.) a day. Canals were more efficient than early railroads, because barges pulled by mules moved faster over waterways. However, canals could not be built everywhere, especially over mountains.

The application of railroad technology, using steam as a power source, made it possible to overcome obstacles in moving goods and people over considerable distances and at profitable costs. Railroads transformed the way goods were purchased by reducing the costs for consumers, thus raising the living standards in industrial societies. Railroads transformed the human landscape by strengthening the link between farm and city, changed commercial cities into industrial centers, and started early forms of suburban growth well before automobiles arrived.

Financing Railroads

Constructing railroads was costly. Tunnels had to be blasted through mountains, and rivers had to be crossed by bridges. Early in the building of U.S. railroads, the nation's iron foundries could not meet the demands for rolled rails. Rails had to be imported from England until local forges developed more efficient technologies. Once a railway was completed, there was a constant need to maintain the right-of-way so that traffic flow would not be disrupted. Accidents were frequent, and it was an early practice to burn damaged cars because salvaging them was too expensive.

In some countries, railroads were built and operated by national governments. In the United States, railroads were privately owned; however, it was impossible for any single individual to finance and operate a rail system with miles of track. Businessmen raised money by selling stocks and bonds. Just as investors buy stocks in modern high-technology companies, investors purchased stocks and bonds in railroads.

Investing in railroads was good as long as they earned profits and returned money to their investors, but not all railroads made sufficient profits to reward their investors. Competition among railroads was heavy in the United States, and some railroads charged artificially low fares to attract as much business as they could. When ambitious investment schemes collapsed, railroads went bankrupt and were taken over by financiers.

Selling shares of common stock and bonds was made possible by creating corporations. Railroads were granted permission from state governments to organize a corporation. Every investor owned a portion of the railroad. Stockholders' interests were served by boards of directors, and all business transactions were opened for public inspection. One important factor of the corporation was that it relieved individuals from the responsibilities associated with accidents. The railroad, as a corporation, was held accountable, and any compensation for claims made against the company came out of corporate funds, not from individual pockets. This had an impact on the law profession, as law schools began specializing in legal matters relevant to railroads and interstate commerce.

The Success of Railroads

Railroads usually began by radiating outward from port cities where merchants engaged in transoceanic trade. A classic example, in the United States, is the country's first regional railroad—the Baltimore and Ohio. Construction commenced from Baltimore in 1828; by 1850, the railroad had crossed the Appalachian Mountains and was on the Ohio River at Wheeling, Virginia.

Once trunk lines were established, rail networks became more intensive as branch lines were built to link smaller cities and towns. Countries with extremely large continental dimensions developed interior articulating cities where railroads from all directions converged. Chicago and Atlanta are two such cities in the United States. Chicago was surrounded by three circular railroads (belts) whose only function was to interchange cars. Railroads from the Pacific Coast converged with lines from the Atlantic Coast as well as routes moving north from the Gulf Coast.

Mechanized farms and heavy industries developed within the network. Railroads made possible the extraction of fossil fuels and metallic ores, the necessary ingredients for industrial growth. Extension of railroads deep into Eastern Europe helped to generate massive waves of immigration into both

North and South America, creating multicultural societies.

Building railroads in Africa and South Asia made it possible for Europe to increase its political control over native populations. The ultimate aim of the colonial railroad was to develop a colony's economy according to the needs of the mother country. Railroads were usually single-line routes transhipping commodities from interior centers to coastal ports for exportation. Nairobi, Kenya, began as a rail hub linking British interests in Uganda with Kenya's port city of Mombasa. Similar examples existed in Malaysia and Indonesia.

Railroads generated conflicts among colonial powers as nations attempted to acquire strategic resources. In 1904–1905 Russia and Japan fought a war in the Chinese province of Manchuria over railroad rights; Imperial Germany attempted to get around British interests in the Middle East by building a railroad linking Berlin with Baghdad to give Germany access to lucrative oil fields. India was a region of loosely connected provinces until British railroads helped establish unification. The resulting sense of national unity led to the termination of British rule in 1947 and independence for India and Pakistan.

In the United States, private railroads discontinued passenger service among cities early in the 1970s and the responsibility was assumed by the federal government (Amtrak). Most Americans riding trains do so as commuters traveling from the suburbs to jobs in the city. The U.S. has no true high-speed trains, aside from sections of Amtrak's Acela line in the Northeast Corridor, where it can reach 150 mph for only 34 miles of its 457-mile span. Passenger service remains popular in Japan and Europe. France, Germany, and Japan operate high-speed luxury trains with speeds averaging above 100 miles (160 km.) per hour.

Railroads are no longer the exclusive means of land transportation as they were early in the twentieth century. Although competition from motor vehicles and air freight provide alternate choices, railroads have remained important. France and England have direct rail linkage beneath the English Channel. In the United States, great railroad mergers and the application of computer technology have reduced operating costs while increasing profits. Transoceanic container traffic has been aided by railroads hauling trailers on flatcars. Railroads began the process of bringing regions within

a nation together in the nineteenth century just as the computer and the World Wide Web began uniting nations throughout the world at the end of the twentieth century.

Sherman E. Silverman

AIR TRANSPORTATION

The movement of goods and people among places is an important field of geographic study. Transportation routes form part of an intricate global network through which commodities flow. Speed and cost determine the nature and volume of the materials transported, so air transportation has both advantages and disadvantages when compared with road, rail, or water transport.

Early Flying Machines

The transport of people and freight by air is less than a century old. Although hot-air balloons were used in the late eighteenth century for military purposes, aerial mapping, and even early photography, they were never commercially important as a means of transportation. In the late nineteenth century, the German count Ferdinand von Zeppelin began experimenting with dirigibles, which added self-propulsion to lighter-than-air craft. These aircraft were used for military purposes, such as the bombing of Paris in World War I. However, by the 1920s zeppelins had become a successful means of passenger transportation. They carried thousands of passengers on trips in Europe or across the Atlantic Ocean and also were used for exploration. Nevertheless, they had major problems and were soon superseded by flying machines heavier than air. The early term for such a machine was an aeroplane, which is still the word used for airplane in Great Britain.

Following pioneering advances with the internal combustion engine and in aerodynamic theory using gliders, the development of powered flight in a

heavier-than-air machine was achieved by Wilbur and Orville Wright in December, 1903. From that time, the United States moved to the forefront of aviation, with Great Britain and Germany also making significant contributions to air transport. World War I saw the further development of aviation for military purposes, evidenced by the infamous bombing of Guernica.

Early Commercial Service

Two decades after the Wright brothers' brief flight, the world's first commercial air service began, covering the short distance from Tampa to St. Petersburg in Florida. The introduction of airmail service by the U.S. Post Office provided a new, regular source of income for commercial airlines in the United States, and from these beginnings arose the modern Boeing Company, United Airlines, and American Airlines. Europe, however, was the home of the world's first commercial airlines. These include the Deutsche Luftreederie in Germany, which connected Berlin, Leipzig, and Weimar in 1919; Farman in France, which flew from Paris to London; and KLM in the Netherlands (Amsterdam to London), followed by Qantas—the Queensland and Northern Territory Aerial Services, Limited—in Australia. The last two are the world's oldest still operating airlines.

Aircraft played a vital role in World War II, as a means of attacking enemy territory, defending territory, and transporting people and equipment. A humanitarian use of air power was the Berlin Airlift of 1948, when Western nations used airplanes to deliver food and medical supplies to the people of West Berlin, which the Soviet Union briefly blockaded on the ground.

Cargo and Passenger Service

The jet engine was developed and used for fighter aircraft during World War II by the Germans, the British, and the United States. Further research led to civil jet transport, and by the 1970s, jet planes accounted for most of the world's air transportation. Air travel in the early days was extremely expensive, but technological advances enabled longer flights with heavier loads, so commercial air travel became both faster and more economical.

Most air travel is made for business purposes. Of business trips between 750 and 1,500 miles (1,207 and 2,414 km.), air captures almost 85 percent, and of trips more than 1,500 miles (2,414 km.), 90 percent are made by air. The United States had 5,087 public airports in 2018, a slight decrease from the 5,145 public airports operating in 2014. Conversely, the number of private airports increased over this period from 13,863 to 14,549.

The biggest air cargo carriers in 2019 were Federal Express, which carried more than 15.71 billion freight tonne kilometres (FTK), Emirates Skycargo (12.27 billion FTK), and United Parcel Service (11.26 billion FTK).

The first commercial supersonic airliner, the British-French Concorde, which could fly at more than twice the speed of sound, began regular service in early 1976. However, the fleet was grounded after a Concorde crash in France in mid-2000. The first space shuttle flew in 1981. There have been 135 shuttle missions since then, ending with the successful landing of Space Shuttle Orbiter Atlantis on July 21, 2011. The shuttles have transported 600 people and 3 million pounds (1.36 million kilograms) of cargo into space.

Health Problems Transported by Air

The high speed of intercontinental air travel and the increasing numbers of air travelers have increased the risk of exotic diseases being carried into destination countries, thereby globalizing diseases previously restricted to certain parts of the world. Passengers traveling by air might be unaware that they are carrying infections or viruses. The worldwide spread of HIV/AIDS after the 1980s was accelerated by international air travel.

Disease vectors such as flies or mosquitoes can also make air journeys unnoticed inside airplanes. At some airports, both airplane interiors and passengers are subjected to spraying with insecticide upon arrival and before deplaning. The West Nile virus (West Nile encephalitis) was previously found only in Africa, Eastern Europe, and West Asia, but in the 1990s it appeared in the northeastern United States, transported there by birds, mosquitos, or people.

It was feared in the mid-1990s that the highly infectious and deadly Ebola virus, which originated in tropical Africa, might spread to Europe and the United States, by air passengers or through the importing of monkeys. The devastation of native bird communities on the island of Guam has been traced to the emergence there of a large population of brown tree snakes, whose ancestors are thought to have arrived as accidental stowaways on a military airplane in the late 1940s.

Ray Sumner

ENERGY AND ENGINEERING

ENERGY SOURCES

Energy is essential for powering the processes of modern industrial society: refining ores, manufacturing products, moving vehicles, heating buildings, and powering appliances. In 1999 energy costs were half a trillion dollars in the United States alone. All technological progress has been based on harnessing more energy and using it more effectively. Energy use has been shaped by geography and also has shaped economic and political geography.

Ancient to Modern Energy

Energy use in traditional tribal societies illustrates all aspects of energy use that apply in modern human societies. Early Stone Age peoples had only their own muscle power, fueled by meat and raw vegetable matter. Warmth for living came from tropical or subtropical climates. Then a new energy source, fire, came into use. It made cold climates livable. It enabled the cooking of roots, grains, and heavy animal bones, vastly increasing the edible food supply. Its heat also hardened wood tools, cured pottery, and eventually allowed metalworking.

Nearly as important as fire was the domestication of animals, which multiplied available muscle energy. Domestic animals carried and pulled heavy loads. Domesticated horses could move as fast as the game to be hunted or large animals to be herded.

Increased energy efficiency was as important as new energy sources in making tribal societies more successful. Cured animal hides and woven cloth were additional factors enabling people to move to cooler climates. Cooking fires also allowed drying meat into jerky to preserve it against times of limited supply. Fire-cured pottery helped protect food against pests and kept water close by. However, energy benefits had costs. Fire drives for hunting may have caused major animal extinctions. Periodic burning of areas for primitive agriculture caused erosion. Trees became scarce near the best campsites because they had been used for camp fires—the first fuel shortage.

Energy Fundamentals

Human use of energy revolves about four interrelated factors: energy sources, methods of harnessing the sources, means of transporting or storing energy, and methods of using energy. The potential energies and energy flows that might be harnessed are many times greater than present use.

The Sun is the primary source of most energy on Earth. Sunlight warms the planet. Plants use photosynthesis to transform water and carbon dioxide into the sugars that power their growth and indirectly power plant-eating and meat-eating animals. Many other energies come indirectly from the Sun. Remains of plants and animals become fossil fuels. Solar heat evaporates water, which then falls as rain, causing water flow in rivers. Regional differences in the amount of sunlight received and reflected cause temperature differences that generate winds, ocean currents, and temperature differences between different ocean layers. Food for muscle power of humans and animals is the most basic energy system.

Energy Sources

Biomass—wood or other vegetable matter that can be burned—is still the most important energy source in much of the world. Its basic use is to provide heat for cooking and warmth. Biomass fuels are often agricultural or forestry wastes. The advantage of biomass is that it is grown, so it can be replaced. However, it has several limitations. Its low energy content per unit volume and unit mass makes it unprofitable to ship, so its use is limited to the amount nearby. Collecting and processing biomass fuels costs energy, so the net energy is less. Biomass energy production may compete with food production, since both come from the soil. Finally, other fuels can be cheaper.

Greater concentration of biomass energy or more efficient use would enable it to better compete against other energy sources. For example, fermenting sugars into fuel alcohol is one means of concentrating energy, but energy losses in processing make it expensive.

Fossil fuels have more concentrated chemical energy than biomass. Underground heat and pressure compacts trees and swampy brush into the progressively more energy-concentrated peat, lignite coal, bituminous coal, and anthracite or black coal, which is mostly carbon. Industrializing regions turned to coal when they had exhausted their firewood. Like wood, coal could be stored and shoveled into the fire box as needed. Large deposits of coal are still available, but growth in the use of coal slowed by the mid-twentieth century because of two competing fossil fuels, petroleum and natural gas.

Petroleum includes gasoline, diesel fuel, and fuel oil. It forms from remains of one-celled plants and animals in the ocean that decompose from sugars into simpler hydrogen and carbon compounds (hydrocarbons). Petroleum yields more energy per unit than coal, and it is pumped rather than shoveled. These advantages mean that an oil-fired vehicle can be cheaper and have greater range than a coal-fired vehicle.

There are also hydrocarbon gases associated with petroleum and coal. The most common is the natural gas methane. Methane does not have the energy density of hydrocarbon liquids, but it burns cleanly and is a fuel of choice for end uses such as homes and businesses.

Petroleum and natural gas deposits are widely scattered throughout the world, but the greatest known deposits are in an area extending from Saudi Arabia north through the Caucasus Mountains. Deposits extend out to sea in areas such as the Persian Gulf, the North Sea, and the Gulf of Mexico. Other sources, such as oil tar sands and shale oil, are currently seen as a potentially important source of energy, but controversies surrounding the extraction, refining, and delivery processes make these energy sources a matter of significant debate and concern.

Heat engines transform the potential of chemical energies. James Watt's steam engine (1782) takes heat from burning wood or coal (external combustion), boils water to steam, and expands it through pistons to make mechanical motion. In the twentieth century, propeller-like steam turbines were developed to increase efficiency and decrease complexity. Auto and diesel engines burn fuel inside the engine (internal combustion), and the hot gases expand through pistons to make mechanical motion. Expanding them through a gas turbine is a jet engine. Heat engines can create energy from other sources, such as concentrated sunlight, nuclear fission, or nuclear fusion. The electrical generator transforms mechanical motion into electricity that can move by wire to uses far away. Such transportation (or wheeling) of electricity means that one power plant can serve many customers in different locations.

Flowing water and wind are two of the oldest sources of industrial power. The Industrial Revolution began with water power and wind power, but they could only be used in certain locations, and they were not as dependable as steam engines. In the early twentieth century, electricity made river power practical again. Large dams along river valleys with adequate water and steep enough slopes enabled areas like the Tennessee Valley to be industrial centers. In the 1970s wind power began to be used again, this time for generating electricity.

Solar energy can be tapped directly for heat or to make electricity. Although sunlight is free, it is not concentrated energy, so getting usable energy requires more equipment cost. Consequently, fossil-fueled heat is cheaper than solar heat, and power from the conventional utility grid has been much less expensive than solar-generated electricity. However, prices of solar equipment continue to drop as technologies improve.

Future Energy Sources

Possible future energy sources are nuclear fission, nuclear fusion, geothermal heat, and tides. Fission reactors contain a critical mass of radioactive heavy elements that sustains a chain reaction of atoms splitting (fissioning) into lighter elements—releasing heat to run a steam turbine. Tremendous amounts of fission energy are available, but reactor costs and safety issues have kept nuclear prices higher than that of coal.

Nuclear fusion involves the same reaction that powers the Sun: four hydrogen atoms fusing into one helium atom. However, duplicating the Sun's heat in a small area without damaging the surrounding reactor may be too expensive to allow profitable fusion reactors.

Geothermal power plants, tapping heat energy from within the earth, have operated since 1904, but widespread use depends on cheaper drilling to make them practical in more than highly volcanic areas. Tidal power is limited to the few bays that concentrate tidal energy.

Energy and Warfare

Much of ancient energy use revolved about herding animals and conducting warfare. Horse riders moved faster and hit harder than warriors on foot. The bow and arrow did not change appreciably for thousands of years. Herders on the plains rode horses and used the bow and arrow as part of tending their flocks, and the small amounts of metal needed for weapons was easily acquired. Consequently, the herders could invade and plunder much more advanced peoples. From Scythians to Parthians to Mongols, these people consistently destroyed the more advanced civilizations.

The geographical effect was that ancient civilizations generally developed only if they had physical barriers separating them from the flat plains of herding peoples. Egypt had deserts and seas. The Greeks and Romans lived on mountainous peninsulas, safe from easy attack. The Chinese built the Great Wall along their northern frontier to block invasions.

Nomadic riders dominated until the advent of an energy system of gunpowder and steel barrels began delivering lead bullets. With them, the Russians broke the power of the Tartars in Eurasia in the late fifteenth century, and various peoples from Europe conquered most of the world. Energy and industrial might became progressively more important in war with automatic weapons, high explosives, aircraft, rockets, and nuclear weapons.

By World War II, oil had become a reason for war and a crucial input for war. The Germans attempted to seize petroleum fields around Baku on the Caspian. Later in the war, major Allied attacks targeted oil fields in Romania and plants in Germany synthesizing liquid fuels. During the Arab-Israeli War of 1973, Arabs countered Western support of Israel with an oil boycott that rocked Western economies. In 1990 Iraq attempted to solve a border dispute with its oil-rich neighbor, Kuwait, by seizing all of Kuwait. An alliance, led by the United States, ejected the Iraqis.

Other wars occur over petroleum deposits that extend out to sea. European nations bordering on the North Sea negotiated a complete demarcation of economic rights throughout that body. Tensions between China and other Asian countries continue to mount over rights to the resources available in the South China Sea. Current estimates suggest that there may be 90 trillion cubic feet of natural gas and 11 billion barrels of oil in proved and probable reserves, with much more potentially undiscovered. The area is claimed by China, Vietnam, Malaysia, and the Philippines. Turkey and Greece have not resolved ownership division of Aegean waters that might have oil deposits.

Energy, Development, and Energy Efficiency

Ancient civilizations tended to grow and use locally available food and firewood. Soils and wood supplies often were depleted at the same time, which often coincided with declines in those civilizations.

The Industrial Revolution caused development to concentrate in new wooded areas where rivers suitable for power, iron ore, and coal were close together, for example, England, Silesia, and the Pittsburgh area. The iron ore of Alsace in France combined with nearby coal from the Ruhr in Germany fueled tremendous growth, not always peacefully.

By the late nineteenth century, the development of Birmingham, Alabama, demonstrated that railroads enabled a wider spread between coal deposits, iron ore deposits, and existing population centers. By the 1920s, the Soviet Union developed entirely new cities to connect with resources. By the 1970s, unit trains and ore-carrying ships transported coal from the thick coal beds in Montana and Wyoming to the United States' East Coast and to countries in Asia.

The mechanized transport of electrical distribution and distribution of natural gas in pipelines also changed settlement patterns. Trains and subway trains allowed cities to spread along rail corridors in the late nineteenth century and early twentieth century. By the 1940s, cars and trucks enabled cities such as Los Angeles and Phoenix to spread into suburbs. The trend continues with independent solar power that allows houses to be sited anywhere.

Advances in technology have allowed people to get more while using less energy. For example, early peoples stampeded herds of animals over cliffs for food, which was mostly wasted. Horseback hunting was vastly more efficient. Likewise, fireplaces in colonial North America were inefficient, sending most of their heat up the chimney. In the late eighteenth century, inventor and statesman Benjamin Franklin developed a metallic cylinder radiating heat in all directions, which saved firewood.

The ancient Greeks and others pioneered the use of passive solar energy and efficiency after they exhausted available firewood. They sited buildings to absorb as much low winter sun as possible and constructed overhanging roofs to shade buildings from the high summer sun. That siting was augmented by heavy masonry building materials that buffered the buildings from extremes of heat and cold. Later,

metal pipes and glass meant that solar energy could be used for water and space heating.

The first seven decades of the twentieth century saw major declines in energy prices, and cars and appliances became less efficient. That changed abruptly with the energy crises and high prices of the 1970s. Since then, countries such as Japan, with few local energy resources, have worked to increase efficiency so they will be less sensitive to energy shocks and be able to thrive with minimal energy inputs. This trend could lead eventually to economies functioning on only solar and biomass inputs.

Solid-state electronics, use of light emitting diode (LED) or compact fluorescent lamps (CFLs) rather than incandescent bulbs, and fuel cells, which convert fuel directly into electricity more efficiently than combustion engines, all could lead to less energy use. The speed of their adoption depends on the price of competing energies. According to the U.S. Energy Information Administration's (EIA) International Energy Outlook 2019 (IEO2019), the global supply of crude oil, other liquid hydrocarbons, and biofuels is expected to be adequate to meet world demand through 2050. However, many have noted that continuing to burn fossil fuels at our current rate is not sustainable, not because reserves will disappear, but because the damage to the climate would be unacceptable.

Energy and Environment
Energy affects the environment in three major ways. First, firewood gathering in underdeveloped countries contributes to deforestation and resulting erosion. Although more efficient stoves and small solar cookers have been designed, efficiency increases are competing against population increases.

Energy production also frequently causes toxic pollutant by-products. Sulfur dioxide (from sulfur impurities in coal and oil) and nitrogen oxides (from nitrogen being formed during combustion) damage lungs and corrode the surfaces of buildings. Lead additives in gasoline make internal combustion engines run more efficiently, but they cause low-grade lead poisoning. Spent radioactive fuel from nuclear fission reactors is so poisonous that it must be guarded for centuries.

Finally, carbon dioxide from the burning of fossil fuels may be accelerating the greenhouse effect, whereby atmospheric carbon dioxide slows the planetary loss of heat. If the effect is as strong as some research suggests, global temperatures may increase several degrees on average in the twenty-first century, with unknown effects on climate and sea level.

Roger V. Carlson

ALTERNATIVE ENERGIES

The energy that lights homes and powers industry is indispensable in modern societies. This energy usually comes from mechanical energy that is converted into electrical energy by means of generators—complex machines that harness basic energy captured when such sources as coal, oil, or wood are burned under controlled conditions. This energy, in turn, provides the thermal energy used for heating, cooling, and lighting and for powering automobiles, locomotives, steamships, and airplanes. Because such natural resources as coal, oil, and wood are being used up, it is vital that these nonrenewable sources of energy be replaced by sources that are renewable and abundant. It is also desirable that alternative sources of energy be developed in order to cut down on the pollution that results from the combustion of the hydrocarbons that make the nonrenewable fuels burn.

The Sun as an Energy Source

Energy is heat. The Sun provides the heat that makes Earth habitable. As today's commonly used fuel resources are used less, solar energy will be used increasingly to provide the power that societies need in order to function and flourish.

There are two forms of solar energy: passive and active. Humankind has long employed passive solar energy, which requires no special equipment. Ancient cave dwellers soon realized that if they inhabited caves that faced the Sun, those caves would be warmer than those that faced away from the Sun. They also observed that dark surfaces retained heat and that dark rocks heated by the Sun would radiate the heat they contained after the Sun had set. Modern builders often capitalize on this same knowledge by constructing structures that face south in the Northern Hemisphere and north in the Southern Hemisphere. The windows that face the Sun are often large and unobstructed by draperies and curtains. Sunlight beats through the glass and, in passive solar houses, usually heats a dark stone or brick floor that will emit heat during the hours when there is no sunlight. Just as an automobile parked in the sunlight will become hot and retain its heat, so do passive solar buildings become hot and retain their heat.

Active solar energy is derived by placing specially designed panels so that they face the Sun. These panels, called flat plate collectors, have a flat glass top beneath which is a panel, often made of copper with a black overlay of paint, that retains heat. These panels are constructed so that heat cannot escape from them easily. When water circulated through pipes in the panels becomes hot, it is either pumped into tanks where it can be stored or circulated through a central heating system.

Some active solar devices are quite complex and best suited to industrial use. Among these is the focusing collector, a saucer-shaped mirror that centers the Sun's rays on a small area that becomes extremely hot. A power plant at Odeillo in the French Pyrenees Mountains uses such a system to concentrate the Sun's rays on a concave mirror. The mirror directs its incredible heat to an enormous, confined body of water that the heat turns to steam, which is then used to generate electricity.

Another active solar device is the solar or photo-voltaic cell, which gathers heat from the Sun and turns it into energy directly. Such cells help to power spacecraft that cannot carry enough conventional fuel to sustain them through long missions in outer space.

Geothermal Heating

The earth's core is incredibly hot. Its heat extends far into the lower surfaces of the planet, at times causing eruptions in the form of geysers or volca-noes. Many places on Earth have springs that are warmed by heat from the earth's core.

In some countries, such as Iceland, warm springs are so abundant that people throughout the coun-try bathe in them through the coldest of winters. In Iceland, geothermal energy is used to heat and light homes, making the use of fossil fuels unnecessary.

Hot areas exist beneath every acre of land on Earth. When such areas are near the surface, it is easy to use them to produce the energy that humans require. As dependence on fossil fuels decreases, means will increasingly be found of drawing on Earth's subterranean heat as a major source of energy.

Wind Power

Anyone who has watched a sailboat move effort-lessly through the water has observed how the wind can be used as a source of kinetic energy—the kind of energy that involves motion—whose movement is transferred to objects that it touches. Wind power has been used throughout human history. In its more refined aspects, it has been employed to power windmills that cause turbines to rotate, pro-viding generators with the power they require to produce electricity.

Windmills typically have from two to twenty blades made of wood or of heavy cloth such as can-vas. Windmills are most effective when they are lo-cated in places where the wind regularly blows with considerable velocity. As their blades turn, they cause the shafts of turbines to rotate, thus powering generators. The electricity created is usually trans-

mitted over metal cables for immediate use or for storage.

Modern vertical-axis wind turbines have two or three strips of curved metal that are attached at both ends to a vertical pole. They can operate effi-ciently even if they are not turned toward the wind. These windmills are a great improvement over the old horizontal axis windmills that have been in use for many years. From 2000 to 2015, cumulative wind capacity around the world increased from 17,000 megawatts to more than 430,000 mega-watts. In 2015, China also surpassed the EU in the number of installed wind turbines and continues to lead installation efforts. Production of wind elec-tricity in 2016 accounted for 16 percent of the elec-tricity generated by renewables.

Oceans as Energy Sources

Seventy percent of the earth's surface is covered by oceans. Their tides, which rise and fall with predict-able regularity twice a day, would offer a ready source of energy once it becomes economically fea-sible to harness them and store the electrical energy they can provide. The most promising spots to build facilities to create electrical energy from the tides are places where the tides are regularly quite dramatic, such as Nova Scotia's Bay of Fundy, where the difference between high and low tides averages about 55 feet (17 meters).

Some tidal power stations that currently exist were created by building dams across estuaries. The sluices of these dams are opened when the tide co-mes in and closed after the resulting reservoir fills. The water captured in the reservoir is held for sev-eral hours until the tide is low enough to create a considerable difference between the level of the wa-

---

OCEAN ENERGY

The oceans have tremendous untapped energy flows in currents and tremendous potential energy in the temperature differences between warmer tropical surface waters and colder deep waters, known as ocean thermal energy conversion. In both cases, the insurmountable cost has been in trans-porting energy to users on shore.

ter in the reservoir and that outside it. Then the sluice gates are opened and, as the water rushes out at a high rate of speed, it turns turbines that generate electricity.

The world's first large-scale tidal power plant was the Rance Tidal Power Station in France, which became operational in 1966. It was the largest tidal power station in terms of output until Sihwa Lake Tidal Power Station opened in South Korea in August 2011.

Future of Renewable Energy

As pollution becomes a huge problem throughout the world, the race to find nonpolluting sources of energy is accelerating rapidly. Scientists are working on unlocking the potential of the electricity generated by microbes as a fuel source, for example. New technologies are making renewable energy sources economically practical. As supplies of fossil fuels have diminished, pressure to become less dependent on them has grown worldwide. Alternative energy sources are the wave of the future.

R. Baird Shuman

ENGINEERING PROJECTS

Human beings attempt to overcome the physical landscape by building forms and structures on the earth. Most structures are small-scale, like houses, telephone poles, and schools. Other structures are great engineering works, such as hydroelectric projects, dams, canals, tunnels, bridges, and buildings.

Hydroelectric Projects

The potential for hydroelectricity generation is greatest in rapidly flowing rivers in mountainous or hilly terrain. The moving water turns turbines that, in turn, generate electricity. Hydroelectric power projects also can be built on escarpments and fall lines, where there is tremendous untapped energy in the falling water.

Most of the potential for hydroelectricity remains untapped. Only about one-sixth of the suitable rivers and falls are used for hydroelectric power. Certain areas of the world have used more of their potential than others. The percent of potential hydropower capacity that has not been developed is 71 percent in Europe, 75 percent in North America, 79 percent in South America, 95 percent in Africa, 95 percent in the Middle East, and 82 percent in Asia-Pacific. China, Brazil, Canada, and the United States currently produce the most hydroelectric power.

In Africa, only Zambia, Zimbabwe, and Ghana produce significant hydroelectricity. The region's total generating capacity needs to increase by 6 percent per year to 2040 from the current total of 125 GW to keep pace with rising electricity demand. In Southeast Asia, countries continue to grapple with the need to build up their hydroelectric plants without causing harm to the rivers that are used to supply food, water, and transportation.

Dams

Dams serve several purposes. One purpose is the generation of hydroelectric power, as discussed above. Dams also provide flood control and irrigation. Rivers in their natural state tend to rise and fall with the seasons. This can cause serious problems for people living in downstream valleys. Flood-control dams also can be used to regulate the flow of water used for irrigation and other projects. A final reason to build dams is to reduce swampland, in order to control insects and the diseases they carry.

Famous dams are found in all regions of the world. In North America, two of the most notable dams are Hoover Dam, completed in 1936, on the Colorado River between Arizona and Nevada; and the Grand Coulee Dam, completed in 1942, on the Columbia River in Washington State.

In South America, the most famous dam is the Itaipu Dam, completed in 1983, on the Paraná River between Brazil and Paraguay. In Africa, the Aswan High Dam was completed in 1970, on the Nile River in Egypt, and the Kariba Dam was completed in 1958, on the Zambezi River between Zambia and Zimbabwe. In Asia, the Three Gorges Dam spans the Yangtze River by the town of Sandouping in Hubei province, China. The Three Gorges Dam has been the world's largest power station in terms of installed capacity (22,500 MW) since 2012.

Bridges

Bridges are built to span low-lying land between two high places. Most commonly, there is a river or other body of water in the way, but other features that might be spanned include ravines, deep valleys and trenches, and swamps. A related engineering

---

ENGINEERING WORKS AND ENVIRONMENTAL PROBLEMS

Although engineering allows humans to overcome natural obstacles, works of engineering often have unintended consequences. Many engineering projects have caused unanticipated environmental problems.

Dams, for instance, create large lakes behind them by trapping water that is released slowly. This water typically contains silt and other material that eventually would have formed soil downstream had the water been allowed to flow naturally. Instead, the silt builds up behind the dam, eventually diminishing the lake's usefulness. As an additional consequence, there is less silt available for soil-building downstream.

Canals also can cause environmental harm by diverting water from its natural course. The river from which water is diverted may dry up, negatively affecting fish, animals, and the people who live downstream.

The benefits of engineering works must be weighed against the damage they do to the environment. They may be worthwhile, but they are neither all good nor all bad: There are benefits and drawbacks in building any engineering project.

---

project is the causeway, in which land in a low-lying area is built up and a road is then constructed on it.

The longest bridge in the world is the Akashi Kaikyo in Japan near Osaka. It was built in 1998 and spans 6,529 feet (1,990 meters), connecting the island of Hōnshū to the small island of Awaji. The Storebælt Bridge in Denmark, also completed in 1998, spans 5,328 feet (1,624 meters), connecting the island of Sjaelland, on which Copenhagen is situated, with the rest of Denmark. Another bridge spanning more than 5,300 feet is the Osman Gazi Bridge in Turkey. The bridge was opened on 1 July 1, 2016, ad to become the longest bridge in Turkey and the fourth-longest suspension bridge in the world by the length of its central span. The length of the bridge is expected to be surpassed by the Çanakkale 1915 Bridge, which is currently under construction across the Dardanelles strait.

Other long bridges can be found across the Humber River in Hull, England; across the Chiang Jiang (Yangtze River) in China; in Hong Kong, Norway, Sweden, and Turkey and elsewhere in Japan.

The longest bridge in the United States is the Lake Pontchartrain Causeway, Louisiana, which spans 24 miles (38.5 km.), the Verrazano-Narrows Bridge in New York City between Staten Island and Brooklyn was once the longest suspension bridge in the world. Completed in 1964, its main span measures 4,260 feet (1,298 meters).

Canals

Moving goods and people by water is generally cheaper and easier, if a bit slower, than moving them by land. Before the twentieth century, that cost savings overwhelmed the advantages of land travel—speed and versatility. Therefore, human beings have wanted to move things by water whenever possible. To do so, they had two choices: locate factories and people near water, such as rivers, lakes, and oceans, or bring water to where the factories and people are, by digging canals.

One of the most famous canals in the world is the Erie Canal, which runs from Albany to Buffalo in New York State. Built in 1825 and running a length of 363 miles (584 km.), the Erie Canal opened up the Great Lakes region of North America to development and led to the rise of New York City as one of the world's dominant cities.

Two other important canals in world history are the Panama Canal and the Suez Canal. The Panama Canal connects the Atlantic and Pacific Oceans over a length of 50.7 miles (81.6 km.) on the isthmus of Panama in Central America. Completed in 1914, the Panama Canal eliminated the long and dangerous sea journey around the tip of South America. The Suez Canal in Egypt, which runs for 100 miles (162 km.) and was completed in 1856, eliminates a similar journey around the Cape of Good Hope in South Africa.

The longest canal in the world is the Grand Canal in China, which was built in the seventh century and stretches a length of 1,085 miles (2,904 km.). It connects Tianjin, near Beijing in the north of China, with Nanjing on the Chang Jiang (Yangtze River) in Central China. The Karakum Canal runs across the Central Asian desert in Turkmenistan from the Amu Darya River westward to Ashkhabad. It was begun in the 1954, and completed in 1988 and is navigable over much of its 854-mile (1,375-km.) length. The Karakum Canal and carries 13 cubic kilometres (3.1 cu mi) of water annually from the Amu-Darya River across the Karakum Desert to irrigate the dry lands of Turkmenistan.

Many canals are found in Europe, particularly in England, France, Belgium, the Netherlands, and Germany, and in the United States and Canada, especially connecting the Great Lakes to each other and to the Ohio and Mississippi Rivers.

Tunnels

Tunnels connect two places separated by physical features that would make it extremely difficult, if not impossible, for them to be connected without cutting directly through them. Tunnels can be used in place of bridges over water bodies so that water traffic is not impeded by a bridge span. Tunnels of this type are often found in port cities, and cities with them include Montreal, Quebec; New York City; Hampton Roads, Virginia; Liverpool, England; or Rio de Janeiro, Brazil.

Tunnels are often used to go through mountains that might be too tall to climb over. Trains especially are sensitive to changes in slope, and train tunnels are found all over the world. Less common are automobile and truck tunnels, although these are also found in many places. Train and automotive tunnels through mountains are common in the Appalachian Mountains in Pennsylvania, the Rockies in the United States and Canada, Japan, and the Alps in Italy, France, Switzerland, and Austria.

The Chunnel

Arguably the most famous—and one of the most ambitious—tunnels in the world goes by the name Chunnel. Completed in 1994, it connects Dover, England, to Calais, France, and runs 31 miles (50 km.). "Chunnel" is short for the Channel Tunnel, named for the English Channel, the body of water that it goes under. It was built as a train tunnel, but cars and trucks can be carried through it on trains. In the year 2000 plans were underway to cut a second tunnel, to carry automobiles and trucks, that would run parallel to the first Chunnel.

The Seikan Tunnel in Japan, connects the large island of Hōnshū with the northern island of Hōkkaidō. The Seikan Tunnel is nearly 2.4 miles (4 km.) longer than Europe's Chunnel; however, the undersea portion of the tunnel is not as long as that of the Chunnel.

Buildings

Historically, North America has been home to the tallest buildings in the world. Chicago has been called the birthplace of the skyscraper and was at one time home to the world's tallest building. In 1998, however, the two Petronas Towers (each 1,483 feet/452 meters tall) were completed in Kuala Lumpur, Malaysia, surpassing the height of the world's tallest building, Chicago's Sears Tower (1,450 feet/442 meters), which had been completed in 1974. In 2019, the tallest completed building in the world is the 2,717-foot (828-metre) tall Burj Khalifa in Dubai, the tallest building since 2008.

Of the twenty tallest buildings standing in the year 2019, China is home to ten (Shanghai Tower, Ping An Finance Center, Goldin Finance 117, Guangzhou CTF Finance Center, Tianjin CFT Finance Center, China Zun, Shanghai World Financial Center, International Commerce Center, Wuhan Greenland Center, Changsha); Malaysia (the Petronas towers) and the United States (One World Trade Center and Central Park Towers) boast two each; Vietnam has one (Landmark 81 in Ho Chi Minh City), as does Russia (Lakhta Center), Taiwan (Taipei 101), South Korea (Lotte World Trade Center), Saudi Aragia (Abraj Al-Bait Clock Tower in Mecca).

Timothy C. Pitts

INDUSTRY AND TRADE

MANUFACTURING

Manufacturing is the process by which value is added to materials by changing their physical form—shape, function, or composition. For example, an automobile is manufactured by piecing together thousands of different component parts, such as seats, bumpers, and tires. The component parts in unassembled form have little or no utility, but pieced together to produce a fully functional automobile, the resulting product has significant utility. The more utility something has, the greater its value. In other words, the value of the component parts increases when they are combined with the other parts to produce a useful product.

Employment in Manufacturing

On a global scale, 28 percent of the world's working population had jobs in the manufacturing sector in the third decade of the century. The rest worked in agriculture (28 percent) and services (49 percent). The importance of each of these sectors varies from country to country and from time period to time period. High-income countries have a higher percentage of their labor force employed in manufacturing than low-income countries do. For example, in the United States 19 percent of the labor force worked in manufacturing by 2019, whereas the African country of Tanzania had only 7 percent of its labor force employed in the manufacturing sector at that time.

At the end of the twentieth century, the vast majority of the U.S. labor force (74 percent) worked in services, a sector that includes jobs such as computer programmers, lawyers, and teachers. By the end of the second decade of the twenty-first century, the percentage has risen to slightly more than 79 percent. Only 1 percent worked in agriculture and mining. This employment structure is typical for a high-income country. In low-income countries, in contrast, the majority of the labor force have agricultural jobs. In Tanzania, for example, 66 percent of the labor force worked in agriculture, while services accounted for 27 percent of the jobs.

The importance of manufacturing as an employer changes over time. In 1950 manufacturing accounted for 38 percent of all jobs in the United States. The percentage of jobs accounted for by the manufacturing sector in high-income countries has decreased in the post-World War II period. The decreasing share of manufacturing jobs in high-income countries is partly attributable to the fact that many manufacturing companies have replaced people with machines on assembly lines. Because one machine can do the work of many people, manufacturing has become less labor-intensive (uses fewer people to perform a particular task) and more capital-intensive (uses machines to perform tasks formerly done by people). In the future, manufacturing in high-income countries is expected to become increasingly capital-intensive. It is not inconceivable that manufacturing's share of the U.S. labor force could fall below 10 percent over the course of the twenty-first century.

Geography of Manufacturing

Every country produces manufactured goods, but the vast bulk of manufacturing activity is concentrated geographically. Four countries—China, the United States, Japan, and Germany—produce almost 60 percent of the world's manufactured goods. The concentration of manufacturing activity in a small number of regions means that there are other regions where very little manufacturing occurs. Africa is a prime example of a region with little manufacturing.

Different countries tend to specialize in the production of different products. For example, 50 percent of the automobiles that were produced in that late 1990s were produced in three countries—Germany, Japan, and the United States. In the production of television sets, the top three countries were China, Japan, and South Korea, which together produced 48 percent of the world's television sets. It is important to note that these patterns change over time. For example, in 1960 the top three automobile-producing countries were Germany, the United Kingdom, and the United States, which together produced 76 percent of the world's automobiles.

Multinational Corporations

A multinational corporation is a corporation that is headquartered in one country but owns business facilities, for example, manufacturing plants, in other countries. Some examples of multinational corporations from the manufacturing sector include the automobile maker Ford, whose headquarters are the in the United States, the pharmaceutical company Bayer, whose headquarters are in Germany, and the candy manufacturer Nestlé, whose headquarters are in Switzerland. Since the end of World War II, multinational corporations have become increasingly important in the world economy. Most multinational corporations are headquartered in high-income countries, such as Japan, the United Kingdom, and the United States.

Companies open manufacturing plants in other countries for a variety of reasons. One of the most common reasons is that it allows them to circumvent barriers to trade that are imposed by foreign governments, especially tariffs and quotas. A tariff is an import tax that is imposed upon foreign-manufactured goods as they enter a country. A quota is a limitation imposed on the volume of a particular good that a particular country can export to another country. The net effect of tariffs and quotas is

to increase the cost of imported goods for consumers.

Governments impose tariffs and quotas partly to raise revenue and partly to encourage consumers to purchase goods manufactured in their own country. Foreign manufacturers faced with tariffs and quotas often begin manufacturing their product in the country imposing the tariffs and quotas. As tariffs and quotas apply to imported goods only, producing in the country imposing the quotas or tariffs effectively makes these trade barriers obsolete.

Companies also open manufacturing plants in other countries because of differences in labor costs among countries. While most manufacturing takes place in high-income countries, some low-income countries have become increasingly attractive as production locations because their workers can be hired much more cheaply than in high-income countries. For example, in late 2019, the average manufacturing job in the United States paid more than US$22.50 per hour. By comparison, manufacturing employees in the Philippines earned a few cents more than US$2.50 per hour.

This dramatic differences in labor costs have prompted some companies to close down their manufacturing plants in high-income countries and open up new plants in low-income countries. This has resulted in high-income countries purchasing more manufactured goods from low-income countries.

More than half the clothing imported into the United States came from Asian countries, for example, China, Taiwan, and South Korea, where labor costs were much lower than in the United States. Much of this clothing was made in factories where workers were paid by companies headquartered in the United States. For example, most of the Nike sports shoes that were sold in the United States were made in China, Indonesia, Vietnam, and Pakistan.

Transportation and Communications Technology

The ability of companies to have manufacturing plants in other countries stems from the fact that the world has a sophisticated and efficient transportation and communications system. An advanced transportation and communications system makes it relatively easy and relatively cheap to transfer information and goods between geographically distant locations. Thus, Nike can manufacture soccer balls in Pakistan and transport them quickly and cheaply to customers in the United States.

The extent to which transportation and communications systems have improved during the last two centuries can be illustrated by a few simple examples. In 1800, when the stagecoach was the primary method of overland transportation, it took twenty hours to travel the ninety miles from Lansing, Michigan, to Detroit, Michigan. Today, with the automobile, the same journey takes approximately ninety minutes. In 1800 sailing ships traveling at an average speed of ten miles per hour were used to transport people and goods between geographically distant countries. In the year 2019 jet-engine aircraft could traverse the globe at speeds in excess of 600 miles per hour. Communications technology has also improved over time.

In 1930 a three-minute telephone call between New York and London, England, cost more than US$385 in 2018 dollars. In the year 2019 the same telephone call could be made for less than a dime.

In addition to modern telephones, there are fax machines, email, videoconferencing capabilities, and a host of other technologies that make communication with other parts of the world both inexpensive and swift.

Future Prospects

The global economy of the twenty-first century presents a wide variety of opportunities and challenges. Sophisticated communications and transportation networks provide increasing numbers of manufacturing companies with more choices as to where to locate their factories. However, high-income countries like the United States are increasingly in competition with other countries (both high- and low-income) to maintain existing and manufacturing investments and attract new ones. Persuading existing companies to keep their U.S. factories open and not move overseas has been a major challenge. Likewise, making the United States as an attractive place for foreign companies to locate their manufacturing plants is an equally challenging task.

Neil Reid

GLOBALIZATION OF MANUFACTURING AND TRADE

Why are most of the patents issued worldwide assigned to Asian corporations? How did a Taiwanese earthquake prevent millions of Americans from purchasing memory upgrades for their computers? Why have personal incomes in Beijing nearly doubled in less than a decade?

Answers to these questions can be found in the geography of globalization. Globalization is an economic, political, and social process characterized by the integration of the world's many systems of manufacturing and trade into a single and increasingly seamless marketplace. The result: a new world geography.

This new geography is associated with the expansion of manufacturing and trade as capitalist principles replace old ideologies and state-controlled economies. With expanded free markets, the process of manufacturing and trading is constantly changing. Globalization delivers economic growth through improved manufacturing processes, newly developed goods, foreign investment in overseas manufacturing, and expanded employment.

The economies of developing countries are slowly transitioning from agricultural to industrial activities. Nevertheless, more than 65 percent of workers in these countries continue to work in agriculture. Meanwhile, developed countries, such as Australia and Germany, are experiencing high-technology service sector growth and reduced manufacturing employment. In the United States, nearly 30 percent of all workers were employed in manufacturing during the 1950s, but by 2019, less than 8.5 percent were.

In between these extremes, former state-controlled economies, like Romania, are adopting more efficient economic development strategies. Other nations and economic models, such as Indonesia and China, are pulled into the global marketplace by the growth and expansion of market economies. Despite the different economic paths of developing, transitioning, and developed nations, manufacturing and trade link all nations together and represent an economic convergence with important implications for political, business, and labor leaders—as well as all the world's citizens.

The geographies of manufacturing and trade can be examined as the distribution and location of economic activities in response to technological change and political and economic change.

Distribution and Location

Questions about where people live, work, and spend their money can be answered by reading product labels in any shopping mall, supermarket, or automobile dealership. They reveal the fact that manufacturing is a multistage process of component fabrication and final product assembly that can occur continents apart. For example, a shirt may be designed in New Jersey, assembled in Costa Rica from North Carolina fabric, and sold in British Columbia. To understand how goods produced in faraway locations are sold at neighborhood stores, geographers investigate the spatial, or geographic, distribution of natural resources, manufacturing plants, trading patterns, and consumption.

Historically, the geography of manufacturing and trade has been closely linked to the distribution of raw materials, workers, and buyers. In earlier times, this meant that manufacturing and trade were highly localized functions. In the eighteenth century, every North American town had cobblers or blacksmiths who produced goods from local resources for sale in local markets. By the start of the Industrial Revolution, improved transportation and manufacturing techniques had significantly enlarged the geography of manufacturing and trade. As distances increased, new manufacturing and trading centers developed. The location of these centers was contingent upon site and situation. Site and situation refer to a physical location, or site, relative to needed materials, transportation networks, and markets. For example, Pittsburgh, Pennsylvania, became the site of a major steel industry because it was near coal and iron resources. Pittsburgh also benefited from its historical role as a port town on a major river system that provided access to both western and eastern markets.

While relative location and transportation costs continue to be important factors, the geographic distribution of production and movement of goods across space is more complex than the simple calculus of site and situation. New global and local geographies of manufacturing and trade have been fueled by two major factors: technology and political change.

Technological Change

The old saying that time is money partially explains where goods are manufactured and traded. By compressing time and space, technology has enabled people, goods, and information to go farther more quickly. In the process, technology has reduced interaction costs, such as telecommunications. Just as steel enabled railroads to push farther westward, new technologies reduce the distance between places and people.

By increasing physical and virtual access to people, places, and things, technology has eliminated many barriers to global trade. However, improved telecommunications and transportation are only part of technology's contribution to globalization. If time is money, new efficient manufacturing processes also have reduced costs and facilitated globalization.

Armed with more efficient production processes, reliable telecommunications infrastructures, and transportation improvements, businesses can increase profits and remain competitive by seeking out lower-cost labor markets thousands of miles from consumers. As trade and manufacturing are increasingly spatially separate activities, the geographic distribution of manufacturing promotes an uneven distribution of income. The global distribution of manufacturing plants is closely related to industry-specific skill and wage requirements. For example, low-wage and low-skill jobs tend to concentrate in the developing regions of Asia, South America, and Africa. Alternately, high-technology and high-wage manufacturing activities concentrate in more developed regions.

In some cases, high wages and global competition force corporations to move their manufacturing plants to save costs and remain competitive. During the early 1990s, this byproduct of globalization was a major issue during the U.S. and Canadian debates to ratify the North American Free Trade Agreement (NAFTA). Focusing on primarily U.S. and Canadian companies that moved jobs to Mexico, the debate contributed to growing anxiety over job security as plants relocate to low-cost labor markets in South America and around the world.

As global competition increases, the geography of manufacturing and trade is increasingly global and rapidly changing. One company that has adapted to the shifting nature of global trade and manufacturing is Nike. Based in Beaverton, Oregon, Nike designs and develops new products at its Oregon world headquarters. However, Nike has internationalized much of its manufacturing capacity to compete in an aggressive athletic apparel industry. Over the last twenty-five years, Nike's strategy has meant shifts in production from high-wage U.S. locations to numerous low-wage labor markets around Pacific Rim.

Political and Economic Change: A New World Order

In order for companies such as Nike to successfully adapt to changing global dynamics, a stable international, or multilateral, trading system must be in place. In 1948 the General Agreement on Tariffs and Trade (GATT) was the first major step toward developing this stable global trading infrastructure. During that same period, the World Bank and International Monetary Fund were created to stabilize and standardize financial markets and practices. However, Cold War politics postponed complete economic integration for nearly half a century. Since the collapse of communism, globalization has accelerated as economies coalesce around the principles of free markets and capitalism. These important changes have become institutionalized through multilateral trade agreements and international trading organizations.

International trading organizations try to minimize or eliminate barriers to free and fair trade between nations. Trade barriers include tariffs (taxes levied on imported goods), product quotas, government subsidies to domestic industry, domestic content rules, and other regulations. Barriers prevent competitive access to domestic markets by artificially raising the prices of imported goods too high or preventing foreign firms from achieving economies of scale. In some cases, tariffs can also be used to promote fair trade by effectively leveling the playing field.

Because tariffs can be used both to promote fair trade and to unfairly protect markets, trading organizations are responsible for distinguishing between the two. For example, the Asian Pacific Economic Cooperation (APEC) forum has established guidelines to promote fair trade and attract foreign investment. APEC initiatives include a public Web-based database of member state tariff schedules and related links. Through programs such as the APEC information-sharing project, trading organizations are streamlining the international business process and promoting the overall stability of international markets.

The Future

As the globalization of manufacturing and trade continues, a new world geography is emerging. Unlike the Cold War's east-west geography and politics of ideology, an economic politics divides the developed and developing world along a north-south axis. While the types of conflicts associated with these new politics and the rules of engagement are unclear, it is evident that a new hierarchy of nations is emerging.

Globalization will raise the economic standard of living in most nations, but it has also widened the gap between richer and poorer countries. A small group of nations generates and controls most of the world's wealth. Conversely, the poorest countries account for roughly two-thirds of the world's population and less than 10 percent of its wealth.

This fundamental question of economic justice was a motive behind globalization's first major political clash. During the 1999 World Trade Organization (WTO) meetings in Seattle, Washington, approximately 50,000 environmentalists, labor unionists, and human and animal rights activists protested against numerous issues, including cultural intolerance, economic injustice, environmental degradation, political repression, and unfair labor practices they attribute to free trade. While the protesters managed to cancel the opening ceremonies, the United Nations secretary-general, Kofi Annan, expressed the general sentiment of most WTO member states. Agreeing that the protesters' concerns were important, Annan also asserted that the globalization of manufacturing and trade should not be used as a scapegoat for domestic failures to protect individual rights. More important, the secretary-general feared that those issues could

be little more than a pretext for a return to unilateral trade policies, or protectionism.

Like the Seattle protesters, supporters of multilateral trade advocate political and economic reforms. Proponents emphasize that open markets promote open societies. Free traders earnestly believe economic engagement encourages rogue nations to improve poor human rights, environmental, and labor records. It is argued that economic engagement raises the expectations of citizens, thereby promoting change.

Conclusion

Technological and political change have made global labor and consumer markets more accessible and established an economic world hierarchy. At the top, one-fifth of the world's population consumes the vast majority of produced goods and controls more than 80 percent of the wealth. At the bottom of this hierarchy, poor nations are industrializing but possess less than 10 percent of the world's wealth. In political, social, and cultural terms, this global economic reality defines the contours and cleavages of a changing world geography. Whether geographers calculate the economic and political costs of a widening gap between rich and poor or chart the flow of funds from Tokyo to Toronto, the globalization of manufacturing and trade will remain central to the study of geography well into the twenty-first century.

Jay D. Gatrell

MODERN WORLD TRADE PATTERNS

Trade, its routes, and its patterns are an integral part of modern society. Trade is primarily based on need. People trade the goods that they have, including money, to obtain the goods that they don't have. Some nations are very rich in agriculture or natural resources, while others are centers of industrial or technical activity. Because nations' needs change

only slowly, trade routes and trading patterns develop that last for long periods of time.

Types of Trade

The movement of goods can occur among neighboring countries, such as the United States and Mexico, or across the globe, as between Japan and

Italy. Some trade routes are well established with regularly scheduled service connecting points. Such service is called liner service. Liners may also serve intermediate points along a trade route to increase their revenue.

Some trade occurs only seasonally, such as the movement of fresh fruits from Chile to California. Some trade occurs only when certain goods are demanded, such as special orders of industrial goods. This type of service is provided by operators called tramps. They go where the business of trade takes them, rather than along fixed liner schedules and routes.

Many people think of international trade as being carried on great ships plying the oceans of the world. Such trade is important; however, a considerable amount of trade is carried by other modes of transportation. Ships and airplanes carry large volumes of freight over large distances, while trucks, trains, barges, and even animal transport are used to move goods over trade routes among neighboring or landlocked countries.

Trade Routes

Through much of human history, trade routes were limited. Shipping trade carried on sailing vessels, for example, was limited by the prevailing winds that powered the ships. Land routes were limited by the location of water, mountain ranges, and the slow development of roads through thick forests and difficult terrain. The mechanization of transportation eventually freed ships and other forms of transport to follow more direct trade routes. Also, the development of canals and transcontinental highway systems allowed trade routes to develop based solely upon economic requirements.

Other changes in trade routes have occurred with industrialization of transport systems. The world began to have a great need for coal. Trade routes ran to the countries in which coal was mined. Ships and trains delivered coal to the power industry worldwide. Later, trade shifted to locations where oil (petroleum) was drilled. Now, oil is delivered to those same powerplants and industrial sites around the world.

Noneconomic Factors

Some trade is not purely economic in nature. Political relationships among countries can play an important part in their trade relations. For example, many national governments try to protect their countries' automobile and electronics industries from outside competition by not allowing foreign goods to be imported easily. Governments control imports by assessing duties, or tariffs, on selected imports.

Some national governments use the concept of cabotage to protect their home transportation industries by requiring that certain percentages of imported and exported trade goods be carried by their own carriers. For example, the U.S. government might require that 50 percent of its trade use American ships, planes, or trucks. The government might also require that all American carriers employ only American citizens.

Nations also can exert pressure on their trading partners by limiting access to port or airport facilities. Stronger nations may force weaker nations into accepting unequal trade agreements. For example, the United States once had an agreement with Germany concerning air passenger service between the two countries. The agreement allowed United States carriers to carry 80 percent of the passengers, while German carriers were permitted to carry only 20 percent of the passengers.

Multilateral Trade

In situations in which pairs of trading nations do not have direct diplomatic contact with each other, they make their trade arrangements through other nations. Such trade is referred to as multilateral. Certain carriers cater to this type of trade. They operate their ships or planes in around-the-world service. They literally travel around the globe picking up and depositing cargo along the way for a variety of nations.

Trade Patterns

For many years, world populations were coast centered. This means that most of the people in the country lived close to the coast. This was due primarily to the availability of water transportation systems to move both goods and people. At this time, major railroad, highway and airline systems did not exist. As railway and highway systems pushed into the interiors of nations, the population followed, and goods were needed as well as produced in these areas. Thus, over the years many inland population centers have developed that require transportation systems to move goods into and away from this area.

In these cases, international trade to these inland centers required the use of a number of different modes of transportation. Each of the different modes required additional paperwork and time for repackaging and securing of the cargo. For example, cargo coming off ships from overseas was unloaded and placed in warehouse storage. At some later time, it was loaded onto trucks that carried it to railyards. There it would be unloaded, stored, and then loaded onto railcars. At the destination, the cargo would once again be shifted to trucks for the final delivery. During the course of the trip, the cargo would have been handled a number of times, with the possibility of damage or loss occurring each time.

Containerization

As more goods began to move in international trade, the systems for packaging and securing of cargo became more standardized. In the 1960s, shipments began to move in containers. These are highway truck trailers which have been removed from the chassis leaving only the box. Container packaging has become the standard for most cargos moving today in both domestic and international trade. With the advent of containerization of cargo in international trade, cargo movements could quickly move intermodally. Intermodal shipping involves the movement of cargo by using more than a single mode of transportation.

Land, water, and air carriers have attempted to make the intermodal movement of cargo in international trade as seamless as possible. They have not only standardized the box for carrying cargo, but they have also standardized the handling equipment, so that containers move quickly from one mode to another. Advances in communications and

---

THE WORLD TRADE ORGANIZATION AND GLOBAL TRADING

In 1998 domestic political pressures and an expected domestic surplus of rice prompted the Japanese government to unilaterally implement a 355-percent tariff on foreign rice, violating the United Nations' General Agreement on Tariffs and Trade (GATT). On April 1, 1999, Japan agreed to return to GATT import levels and imposed new over-quota tariffs. While domestic Japanese politics could have prompted a trade war with rice-exporting countries, the crisis demonstrates how multilateral trading initiatives promote stability. Without an agreement, rice exporters might not have gained access to Japanese markets. By returning to GATT minimum quotas and implementing over-quota taxes, the compromise addressed the interests of both domestic and foreign rice growers.

---

electronic banking allow the paperwork and payments also to be completed and transferred rapidly.

As the demands for products have grown and as the size of industrial plants has grown, the size of movements of raw materials and containerized cargo has also grown. Thus, the sizes of the ships and trains required to move these large volumes of cargo have also increased.

The development of VLCC's (very large crude carriers) has allowed shippers to move large volumes of oil products. The development of large bulk carriers has allowed for the carriage of large volumes of dry raw materials such as grains or iron ore. These large vessels take advantage of what is known as economies of scale. Goods can be moved more cheaply when large volumes of them are moved at the same time. This is because the doubling of the volume of cargo moved does not double the cost to build or operate the vessels in which it is carried. This savings reduces the cost to move large volumes of cargo.

Intermodal Transportation

Intermodal transportation has allowed cargo to move seamlessly across both international boundaries and through different modes of transportation. This seamless movement has changed ocean trade routes over recent years.

The development of the Pacific Rim nations created a demand for trade between East Asia and both the United States and Europe. This trade has usually taken the all-water routes between Asia and Europe. Ships moving from East Asia across the Pacific Ocean pass through the Panama Canal and cross the Atlantic Ocean to reach Western Europe. This journey is in excess of 10,000 miles (16,000 km.) and usually takes about thirty days for most ships to complete. The all-water route from Asia to New York is similar. The distance is almost as great as that to Europe and requires about twenty-one to twenty-four days to complete.

Intermodal transportation has given shippers alternatives to all-water routes. A great volume of Asian goods is now shipped to such western U.S. ports as Seattle, Oakland, and Los Angeles, from which these goods are carried by trains across the United States to New York. The overall lengths of these routes to New York are only about 7,400 miles (12,000 km.) and take between only fifteen and nineteen days to complete. Cargos continuing to Europe are put back on ships in New York and complete their journeys in an additional seven to ten days. Such intermodal shipping can save as much as a week in delivery time.

Airfreight

Another changing trend in trade patterns is the development of airfreight as an international competitor. Modern aircraft have improved dramatically both in their ability to lift large weights of cargo as well as their ability to carry cargos over long distances. Because of the speed at which aircraft travel in comparison to other modes of transportation, goods can be moved quickly over large distances. Thus, high-value cargos or very fragile cargos can move very quickly by aircraft.

The drawback to airfreight movement of cargo is that it is more expensive than other modes of travel. However, for businesses that need to move perishable commodities, such as flowers of the Netherlands, or expensive commodities, such as Paris fashions or Singapore-made computer chips, airfreight has become both economic and essential.

Robert J. Stewart

POLITICAL GEOGRAPHY

FORMS OF GOVERNMENT

Philosophers and political scientists have studied forms of government for many centuries. Ancient Greek philosophers such as Plato and Aristotle wrote about what they believed to be good and bad forms of government. According to Plato's famous work, *The Republic*, the best form of government was one ruled by philosopher-kings. Aristotle wrote that good governments, whether headed by one person (a kingship), a few people (an aristocracy), or many people (a polity), were those that ruled for the benefit of all. Those that were based on narrow, selfish interests were considered bad forms of government, whether ruled by an individual (a tyranny), a few people (an oligarchy), or many people (a democracy). Thus, democracy was not always considered a good form of government.

Constitutions and Political Institutions

All governments have certain things in common: institutions that carry out legislative, executive, and judicial functions. How these institutions are supposed to function is usually spelled out in a country's constitution, which is a guide to organizing a country's political system. Most, but not all, countries have written constitutions. Great Britain, for example, has an unwritten constitution based on documents such as the Magna Carta, the English Bill of Rights, and the Treaty of Rome, and on unwritten codes of behavior expected of politicians and members of the royal family.

The world's oldest written constitution still in use is that of the United States. All countries have written or unwritten constitutions, and most follow them most of the time. Some countries do not follow their constitutions—for example, the Soviet Union did not; other countries, for example France, change their constitutions frequently.

Constitutions usually first specify if the country is to be a monarchy or a republic. Few countries still have monarchies, and those that do usually grant the monarch only ceremonial powers and duties. Countries with monarchies at the beginning of the twenty-first century included Spain, Great Britain, Lesotho, Swaziland, Sweden, Saudi Arabia, and Jordan. Most countries that do not have monarchies are republics.

Constitutions also specify if power is to be concentrated in the hands of a strong national government, which is a unitary system; if it is to be divided between a national and various subnational governments such as states, provinces, or territories, which is a federal system; or if it is to be spread among various subnational governments that might delegate some power to a weak national government, which is a confederate system.

Examples of countries with unitary systems include Great Britain, France, and China; federal systems include the United States, Germany, Russia, Canada, India, and Brazil. There were no confederate systems by the third decade of the twenty-first century, although there are examples from history as well as confederations of various groups and nations. The United States under its eighteenth-century Articles of Confederation and the nineteenth-century Confederate States of America, made up of the rebelling Southern states, were confederate systems. Switzerland was a confederation for much of the nineteenth century. The concept of dividing power between the national and subnational governments is called the vertical axis of power.

MONARCHIES OF THE WORLD

Realm/Kingdom	Monarch	Type
Principality of Andorra	Co-Prince Emmanuel Macron; Co-Prince Archbishop Joan Enric Vives Sicília	Constitutional
Antigua and Barbuda	Queen Elizabeth II	Constitutional
Commonwealth of Australia	Queen Elizabeth II	Constitutional
Commonwealth of the Bahamas	Queen Elizabeth II	Constitutional
Barbados	Queen Elizabeth II	Constitutional
Belize	Queen Elizabeth II	Constitutional
Canada	Queen Elizabeth II	Constitutional
Grenada	Queen Elizabeth II	Constitutional
Jamaica	Queen Elizabeth II	Constitutional
New Zealand	Queen Elizabeth II	Constitutional
Independent State of Papua New Guinea	Queen Elizabeth II	Constitutional
Federation of Saint Kitts and Nevis	Queen Elizabeth II	Constitutional
Saint Lucia	Queen Elizabeth II	Constitutional
Saint Vincent and the Grenadines	Queen Elizabeth II	Constitutional
Solomon Islands	Queen Elizabeth II	Constitutional
Tuvalu	Queen Elizabeth II	Constitutional
United Kingdom of Great Britain and Northern Ireland	Queen Elizabeth II	Constitutional
Kingdom of Bahrain	King Hamad bin Isa	Mixed
Kingdom of Belgium	King Philippe	Constitutional
Kingdom of Bhutan	King Jigme Khesar Namgyel	Constitutional
Brunei Darussalam	Sultan Hassanal Bolkiah	Absolute
Kingdom of Cambodia	King Norodom Sihamoni	Constitutional
Kingdom of Denmark	Queen Margrethe II	Constitutional
Kingdom of Eswatini	King Mswati III	Absolute
Japan	Emperor Naruhito	Constitutional
Hashemite Kingdom of Jordan	King Abdullah II	Constitutional
State of Kuwait	Emir Sabah al-Ahmad	Constitutional
Kingdom of Lesotho	King Letsie III	Constitutional
Principality of Liechtenstein	Prince Regnant Hans-Adam II (Regent: The Hereditary Prince Alois)	Constitutional
Grand Duchy of Luxembourg	Grand Duke Henri	Constitutional
Malaysia	Yang di-Pertuan Agong Abdullah	Constitutional
Principality of Monaco	Sovereign Prince Albert II	Constitutional

MONARCHIES OF THE WORLD *(continued)*

Realm/Kingdom	Monarch	Type
Kingdom of Morocco	King Mohammed VI	Constitutional
Kingdom of the Netherlands	King Willem-Alexander	Constitutional
Kingdom of Norway	King Harald V	Constitutional
Sultanate of Oman	Sultan Haitham bin Tariq	Absolute
State of Qatar	Emir Tamim bin Hamad	Mixed
Kingdom of Saudi Arabia	King Salman bin Abdulaziz	Absolute theocracy
Kingdom of Spain	King Felipe VI	Constitutional
Kingdom of Sweden	King Carl XVI Gustaf	Constitutional
Kingdom of Thailand	King Vajiralongkorn	Constitutional
Kingdom of Tonga	King Tupou VI	Constitutional
United Arab Emirates	President Khalifa bin Zayed	Mixed
Vatican City State	Pope Francis	Absolute theocracy

Whether governments share power with subnational governments or not, there must be institutions to make laws, enforce laws, and interpret laws: the legislative, executive, and judicial branches of government. How these branches interact is what determines whether governments are parliamentary, presidential, or mixed parliamentary-presidential. In a presidential system, such as in the United States, the three branches—legislative, executive, and judicial—are separate, independent, and designed to check and balance each other according to a constitution. In a parliamentary system, the three branches are not entirely separate, and the legislative branch is much more powerful than the executive and judicial branches.

Great Britain is a good example of a parliamentary system. Some countries, such as France and Russia, have created a mixed parliamentary-presidential system, wherein the three branches are separate but are not designed to check and balance each other. In a mixed parliamentary-presidential system, the executive (led by a president) is the most powerful branch of government.

Looking at political systems in this way—how the legislative, executive, and judicial branches of government interact—is to examine the horizontal axis of power. All governments are unitary, federal, or confederate, and all are parliamentary, presidential, or mixed parliamentary-presidential. One can find examples of different combinations. Great Britain is unitary and parliamentary. Germany is federal and parliamentary. The United States is federal and presidential. France is unitary and mixed parliamentary-presidential. Russia is federal and mixed parliamentary-presidential. Furthermore, virtually all countries are either republics or monarchies.

Types of Government

Constitutions describe how the country's political institutions are supposed to interact and provide a guide to the relationship between the government and its citizens. Thus, while governments may have similar political institutions—for example, Germany and India are both federal, parliamentary republics—how the leaders treat their citizens can vary widely. However, governments may have political systems that function similarly although they have different forms of constitutions and institutions. For example, Great Britain, a unitary, parliamentary monarchy with an unwritten constitution, treats its citizens very similarly to the United States, which is a federal, presidential republic with a written constitution.

The three most common terms used to describe the relationships between those who govern and those who are governed are democratic, authoritarian, and totalitarian. Characteristics of democracies are free, fair, and meaningfully contested elections; majority rule and respect for minority rights and opinions; a willingness to hand power to the opposition after an election; the rule of law; and civil rights and liberties, including freedom of speech and press, freedom of association, and freedom to travel. The United States, Canada, Japan, and most European countries are democratic.

An authoritarian system is one that curtails some or all of the characteristics of a democratic regime. For example, authoritarian regimes might permit token electoral opposition by allowing other political parties to run in elections, but they do not allow the opposition to win those elections. If the opposition did win, the authoritarian regime would not hand over power. Authoritarian regimes do not respect the rule of law, the rights of minorities to dissent, or freedom of the press, speech, or association. Authoritarian governments use the police, courts, prisons, and the military to intimidate and threaten their citizens, thus preventing people from uniting to challenge the existing political rulers. Afghanistan, Cuba, Iran, Uzbekistan, Saudi Arabia, Chad, Syria, Libya, Sudan, Belarus, and China are examples of countries with authoritarian regimes.

Totalitarian regimes are similar to authoritarian regimes but are even more extreme. Under a totalitarian regime, there is no legal opposition, no freedom of speech, and no rule of law whatsoever. Totalitarian regimes attempt to control totally all members of the society to the point where everyone always must actively demonstrate their loyalty to and support for the regime. Nazi Germany under Adolf Hitler's rule (1933-1945) and the Soviet Union under Joseph Stalin's rule (1928-1953) are examples of totalitarian regimes. As of 2019, only Eritrea and North Korea are the still have governments classified as totalitarian dictatorships.

Forms of Government: Putting it All Together

In *The Republic*, Plato asserts that people have varied dispositions, and, therefore, there are various types of governments. In recent years, regimes have been created that some call mafiacracies (rule by criminal mafias), narcocracies (rule by narcotics gangs), gerontocracies (rule by very old people), theocracies (rule by religious leaders), and so forth. Such variations show the ingenuity of the human mind in devising forms of government.

Whatever labels that are given to a political system, there remain basic questions to be asked about that regime: Is it a monarchy or a republic? Is all power concentrated in the hands of a national government, or is power shared between a national government and the states or provinces? Are its institutions those of a parliamentary, presidential, or mixed parliamentary-presidential system? Is it democratic, authoritarian, or totalitarian? Finally, does it live up to its constitution, both in terms of how power is supposed to be distributed among institutions and in its relationship between the government and the people? To paraphrase Aristotle, how many rulers are there, and in whose interests do they rule?

Nathaniel Richmond

POLITICAL GEOGRAPHY

Students of politics have been aware that there is a significant relationship between physical and political geography since the time of ancient Greece.

The ancient Greek philosopher Plato argued that a *polis* (politically organized society) must be of limited geographical size and limited population or it would lack cohesion. The ideal *polis* would be only as geographically large as required to feed about 5,000 people, its maximum population.

Plato's illustrious pupil, Aristotle, agreed that stable states must be small. "One can build a wall around the Hellespont," the main territory of ancient Greece, he wrote in his treatise *Politics*, "but that will not make it a polis." Today human ideas differ about the maximum area of a successful state or nation-state, but the close influence of physical geography on political geography and their profound mutual effects on politics itself are not in question.

Geographical Influences on Politics

The physical shape and contours of states may be called their physical geography; the political shape and contours of states, starting with their basic structure as unified state, federation, or confederation, are primary features of their political geography. The idea of "political geography" also can refer to variations in a population's political attitudes and behavior that are influenced by geographical features. Thus, the combination of plentiful land and sparse population tend toward an independent spirit, especially where the economy is agriculturally based. This has historically been the case in the western United States; in the Pampas region of Argentina, where cattle are raised by independent-minded gauchos (cowboys); and on the Brazilian frontier, where government regulation is routinely resisted.

Likewise, where physical geography presents significant difficulties for inhabitants in earning a living or associating, as where there is rough terrain and poor soil or inhospitable climate, the populace is likely to exhibit a hardy, self-reliant character that strongly influences political preferences. Thus, physical geography helps to shape national character, including aspects of a nation's politics.

Furthermore, it is well known that where physical geography isolates one part of a country's population from the rest, political radicalism may take root. This tendency is found in coastal cities and remote regions, where labor union radicalism has often been pronounced. Populations in coastal loca-

tions with access to foreign trade often show a more liberal, tolerant, and outgoing spirit, as reflected in their political opinions. In ancient Greece, the coastal access enjoyed by Athens through a nearby port in the fifth century BCE had a strong influence on its liberal and democratic political order. In modern times, China's coastal cities, such as Tientsin, and North American cities such as San Francisco, show similar influences.

The Geographical Imperative

In many instances, political geography is shaped by what may be called the "geographical imperative." Physical geography in these instances demands, or at least strongly suggests, that political geography follow its course. The numerous valleys of mountainous Greece strongly influenced the emergence of the small, often fiercely independent, polis of ancient times. The formation and borders of Asian states such as Bhutan, Nepal, and Tibet have been strongly influenced by the Himalaya Mountains, and the Alps, which shape Switzerland.

As another example, physical geography demands that the land between the Pacific Ocean and the Andes Mountains along the western edge of South America be organized as a separate country—Chile. Island geography often plays a decisive role in its political geography. The qualified political unity of Great Britain can be directly traced to its insular status. Small islands often find themselves combined into larger units, such as the Hawaiian Islands.

The absence of the geographical imperative, however, leaves political geography an open question. For example, Indonesia comprises some 1,300 islands stretching 3,000 miles in bodies of water such as the Indian Ocean and the Celebes Sea. With so many islands, Indonesia lacks a geographical imperative to be a unified state. It also lacks the imperative of ethnic and cultural homogeneity and cohesion, a circumstance mirrored in its political life, since it has remained unified only through military force. As control by the military waned after the fall of the authoritarian General Suharto in 1998, conflicts among the nation's diverse peoples have threatened its breakup. No such threat, however, confronts Australia, an immense island continent where a European majority dominates a fragmented and primitive aboriginal minority. In Australia, the geographical imperative suggests a unity supported by the cultural unity of the majority.

As many examples show, the geographical imperative is not absolute. For example, mountainous Greece is politically united in the twenty-first century. Although long shielded geographically, Tibet lost its political independence after it was successfully invaded by China. The formerly independent Himalayan state Sikkim was taken over by India. Thus, political will trumps physical geography.

The frequency of exceptions to the geographical imperative illustrates that human freedom, while not unlimited, often plays a key role in shaping political geography. As one example, the Baltic Republics—Lithuania, Latvia, and Estonia—historically have been dominated, or largely swallowed up, by neighboring Russia. By the start of the twenty-first century, however, they had regained their independence through the political will to self-rule and the drive for cultural survival.

Strategically Significant Locations

Locations of great economic or military significance become focal points of political attention and, potentially, of military conflict. There are innumerable such places in the world, but several stand out as models of how important physical geography can be for political geography in the context of international politics.

One significant example is the Panama Canal, without which ships must sail around South America. The Suez Canal, which connects European and Asian shipping, is a similar waterway, saving passage around Africa. The canal's significance was reduced after 1956, however, when its blockage after the Arab-Israeli war of that year led to the building of supertankers too large to traverse it. Another example is Gibraltar, whose fortifications command the entrance to the Mediterranean Sea from the Atlantic Ocean. A final example is the Bosporus, the tiny entrance from the Black Sea to waters leading to the Mediterranean Sea. It is the only warm-water route to and from Eastern Russia and therefore is of great military and economic importance for regional and world power politics.

Charles F. Bahmueller

GEOPOLITICS

Geopolitics is a concept pertaining to the role of purely geographical features in the relations among states in international politics. Geopolitics is especially concerned with the geographical locations of the states in relationship to one another. Geopolitical relationships incorporate social, economic, political, and historical features of the states that interact with purely geographical elements to influence the strategic thinking and behavior of nations in the international sphere.

Coined in 1899 by the Swedish theorist Rudolf Kjellen, the term "geopolitics" combines the logic of the search for security and competition for dominance among states with geographical methodology. *Geopolitics* must not, however, be confused with *political geography*, which focuses on individual states' territorial sizes, boundaries, resources, internal political relations, and relations with other states.

Geopolitical is a term frequently used by military and political strategists, politicians and diplomats, political scientists, journalists, statesmen, and a variety of other government officials, such as policy planners and intelligence analysts.

CAPITALS AND MAJOR CITIES OF THE WORLD

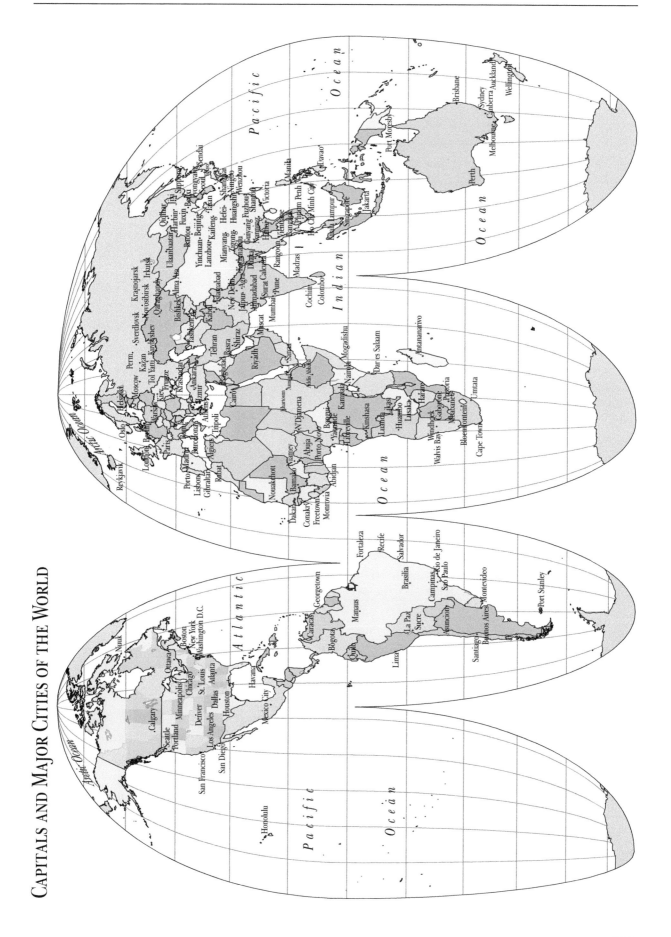

Power Struggles Among States

The idea of geopolitics arises in the course of what might be considered the universal struggle for power among the world's most powerful nations, which compete for political and military leadership. How one state can threaten another, for example, is often influenced by geographical factors in combination with technological, social, economic and other factors. The extent to which individual states can threaten each other depends in no small measure on purely geographical considerations.

By the close of twentieth century the Cold War that had dominated world security concerns was over. Nevertheless, the United States still worried about the danger of being attacked by nuclear missiles fired, not by the former Soviet Union, but by irresponsible, fanatical, or suicidal states. American political leaders and military planners were concerned with the geographical position of so-called "rogue states." or "states of concern." In 1994, North Korea, Cuba, Iran, Libya under Muammar Gaddafi, and Ba'athist Iraq were listed as states of concern. By 2019, a list of state sponsors of terrorism included Iran, North Korea, Sudan, and Syria.

Geographical factors play prominent roles in assessments of the different threats that those states presented to American interests. How far those states are located from American territory determines whether their missiles might pose a serious threat. A missile may be able to reach only the periphery of U.S. soil, or it might be able to carry only a small payload. Similar considerations determine the threat such states pose for U.S. forces stationed abroad, as well as for such important U.S. allies as Japan, Western Europe, or Israel. Such questions are thus said to constitute geopolitical, or geostrategic, considerations.

There are many examples of the influence of geopolitical factors on international relations among nations in the past. For example, the Bosporus, the narrow sea lane linking the Black Sea and the Mediterranean where Istanbul is situated, has long been considered of great strategic importance. In the nineteenth century, the Bosporus was the only direct route through which the Russian

---

A PEACEFULLY RESOLVED BORDER DISPUTE

The peaceful resolution of the border dispute between the Southern African states of Botswana and Namibia was hailed by observers of African politics. Instead of resorting to the armed warfare that so often has marked similar disputes on the continent, the two states chose a different course in 1996, when they found negotiations stalemated. They submitted their claims to the International Court of Justice in The Hague and agreed to accept the court's ruling. Late in 1999, by an eleven-to-four vote, the court ruled for Botswana, and Namibia kept its word to embrace the decision. At issue was a tiny island in the Chobe River on Botswana's northern border. An 1890 treaty between colonial rulers Great Britain and Germany had described the border at the disputed point vaguely, as the river's "main channel." The court took the course of the deepest channel to mark the agreed boundary, giving Botswana title to the 1.4-square-mile (3.5-sq. km.) territory.

---

navy could reach southern Europe and the Mediterranean Sea.

Because of Russia's nineteenth century history of expansionism and its integration into the pre-World War I European state system, with its networks of competing military alliances, the Bosporus took on added geopolitical meaning. It was the congested (and therefore vulnerable) space through which Russian naval power had to pass to reach the Mediterranean.

Historical Origins of Geopolitics

Although political geography was a well-established field by the late nineteenth century, geopolitics was just beginning to emerge as a field of study and political analysis at the end of the century. In 1896 the German theorist Friedrich Ratzel published his *Political Geography*, which put forward the idea of the state as territory occupied by a people bound together by an idea of the state. Ratzel's theory embraced Social Darwinist notions that justified the current boundaries of nations. Ratzel viewed the state as a biological organism in competition for land with other states. The ethical implication of his theory seemed to be that "might makes right."

That theme set the stage for later German geopolitical thought, especially the notion of the need for *Lebensraum* (living room)—space into which the people of a nation could expand. German dictator Adolf Hitler justified his attack on Russia during World War II partly upon his claim that the German people needed more *Lebensraum* to the east. To some modern geographers, the use of geopolitical theories to serve German fascism and to justify other instances of military aggression tarnished geopolitics itself as a field of study.

Historical Development of Geopolitics

Modern geopolitics has further origins in the work of the Scottish geographer Sir Halford John Mackinder. In 1904 he published a seminal article, "The Geographical Pivot of History," in which he argued that the world is made up of a Eurasian "heartland" and a secondary hinterland (the remainder of the world), which he called the "marginal crescent." According to his theory, international politics is the struggle to gain control of the heartland. Any state that managed that feat would dominate the world.

A major proposition of Mackinder's theory was that geographical factors are not merely causative factors, but coercive. He tried to describe the physical features of the world that he believed directed human actions. In his view, "Man and not nature initiates, but nature in large measure controls." Geopolitical factors were therefore to a great extent determinants of the behavior of states. If this were true, geopolitics as a science could have deep relevance and corresponding influence among governments.

After Mackinder's time, the concept of geopolitics had a double significance. On the one hand it was a purely descriptive theory of geographic causation in history. On the other hand, its purveyors also believed, as Mackinder argued in 1904, that geopolitics has "a practical value as setting into perspective some of the competing forces in current international politics." Mackinder sought to promote this field of study as a companion to British state-

craft, a tool to further Britain's national interest. By extension, geopolitical theory could assist any government in forming its political/military strategy.

As applied to the early twentieth-century world of international politics, however, Mackinder's theory had major weaknesses. Among his most glaring oversights were his failure to appreciate the rise of the United States, which attained considerable naval power after the turn of the century. Also, he failed to foresee the crucial strategic role that air power would play in warfare—and with it the immense change that air power could make in geopolitical considerations. Air power moves continents closer together, revolutionizing their geopolitical relationships.

One of Mackinder's chief critics was Nicolas John Spykman. Spykman argued that Mackinder had overvalued the potential economic, and therefore political, power of the Eurasian heartland, which could never reach its full potential because it could not overcome the obstacles to internal transportation. Moreover, the weaknesses of the remainder of the world—in effect, northern, western and southern Europe—could be overcome through forging alliances.

The dark side of geopolitical thought as handmaiden to political and military strategy became apparent in the Germany of the 1920s. At that time German theorists sought the resurrection of a German state broken by failure in World War I, the harsh terms of the Versailles Treaty that ended the war, and the hyperinflation that followed, wiping out the German middle class. In his 1925 article "Why Geopolitik?" Karl Haushofer urged the practical applications of *Geopolitik*. He urged that this form of analysis had not only "come to stay" but could also form important services for German political leaders, who should use all available tools "to carry on the fight for Germany's existence."

Haushofer ominously suggested that the "struggle" for German existence was becoming increasingly difficult because of the growth of the country's population. A people, he wrote, should study the living spaces of other nations so it could be prepared to "seize any possibility to recover lost ground." This discussion clearly implied that, from geopolitical necessity, Germany should seek additional territory to feed itself—a view carried into effect by Hitler in his quest for *Lebensraum* in attacking the Soviet Union, including its wheat-producing breadbasket, the Ukraine.

After World War II, a chastened Haushofer sought to soft-pedal both the direction and influence of his prewar writings. However, Hitler's morally heinous use of *Geopolitik* left geopolitical theorizing permanently tainted, in some eyes. Nevertheless, there is no necessary connection between geopolitics as a purely analytic description and geopolitics as the basis for a selfish search for power and advantage.

Geopolitics in the Twenty-first Century

Geopolitical considerations were unquestionably of profound relevance to the principal states of the post-World War II Cold War period. After the fall of the Berlin Wall in 1989, however, some theorists thought that the age of geopolitics had passed. In 1990 American strategic theorist Edward N. Luttwak, for example, argued that the importance of military power in international affairs had declined precipitously with the winding down of the Cold War. Military power had been overtaken in significance by economic prowess. Consequently, geopolitics had been eclipsed by what Luttwak called "geoeconomics," the waging of geopolitical struggle by economic means.

The view of Luttwak and various geographers of the declining significance of military power and geopolitical analysis, however, was soon proved to be overdrawn by events. As early as the first months of 1991, before the Soviet Union was officially dismantled, military power asserted itself as a key determinant on the international scene. Led by the United States, a far-flung alliance of nations participated in a war to remove Iraqi dictator Saddam Hussein's forces from neighboring Kuwait, which Iraq had illegally occupied. The decisive and successful use of military power in that war dramatically disproved assertions of its growing irrelevance.

Similarly, in the first three decades of the twenty-first century, military power retained its pre-

eminence in the dynamics of international politics, even as economic forces were seen to gather momentum. To states throughout Asia and the West (especially Western Europe and the United States), the relative military capability of potential adversaries, and therefore geopolitics, remained a vital feature of the international order. Central to this view of the world scene is the growing military rivalry of the United States and China in East Asia. As China modernizes and expands its nuclear and conventional forces, it may feel itself capable of challenging America's predominant military power and prestige in East Asia. This possibility heightens the use of geopolitical thinking, giving it currency in analyzing this emerging situation.

Geopolitics as Civilizational Clash

A sometimes controversial expression of geopolitical analysis has been offered by Samuel Huntington of Harvard University. In his *The Clash of Civilizations and the Remaking of World Order* (1996) Huntington constructs a theory to explain certain tendencies of international behavior. He divides the world into a number of cultural groupings, or "civilizations," and argues that the character of various international conflicts can best be explained as conflicts or clashes of civilizations. In his view, Western civilization differs from the civilization of Orthodox Christianity, with a variety of conflicts erupting between the two. An example is the attack by the North Atlantic Treaty Organization (NATO), the bastion of the West, on Serbia, which is part of the Orthodox East.

Huntington's other civilizations include Islamic, Jewish, Eastern Caribbean, Hindu, Sinic (Chinese), and Japanese. The clash between Israel and its neighbors, the struggle between Pakistan and India over Kashmir, the rivalries between the United States and China and between China and India, for example, can be viewed as civilizational conflicts. Huntington has stated, however, that his theory is not intended to explain all of the historical past, and he does not expect it to remain valid long into the future.

Charles F. Bahmueller

NATIONAL PARK SYSTEMS

The world's first national parks were established as a response to the exploitation of natural resources, disappearance of wildlife, and destruction of natural landscapes that took place during the late nineteenth century. Government efforts to preserve natural areas as parks began with the establishment of Yellowstone National Park in the United States in 1872 and were soon adopted in other countries, including Australia, Canada, and New Zealand.

While the preservation of nature continues to be an important benefit provided by national parks, worldwide increases in population and the pressures of urban living have raised public interest in setting aside places that provide opportunities for solitude and interaction with nature.

Because national parks have been established by nations with diverse cultural values, land resources, and management philosophies, there is no single definition of what constitutes a national park. In some countries, areas used principally for recreational purposes are designated as national parks; other countries emphasize preservation of outstanding scenic, geologic, or biological resources. The terminology used for national parks also varies among countries. For example, protected areas that are similar to national parks may be called reserves, preserves, or sanctuaries.

Diverse landscapes are protected within national parks, including swamps, river deltas, dune areas, mountains, prairies, tropical rainforests, temperate forests, arid lands, and marine environments. Individual parks within nations form networks that vary with respect to size, accessibility, function, and the type of natural landscapes preserved. Some national park areas are isolated and sparsely populated, such as Greenland National Park; others, such as Peak District National Park in Great Britain, contain numerous small towns and are easily accessible to urban populations.

The functions of national parks include the preservation of scenic landscapes, geological features, wilderness, and plants and animals within their natural habitats. National parks also serve as outdoor laboratories for education and scientific research and as reservoirs for genetic information. Many are components of the United Nations International Biosphere Reserve Program.

National parks also play important roles in preserving cultures, by protecting archaeological, cultural, and historical sites. The United Nations recognizes several national parks that possess important cultural attributes as World Heritage Sites. Tourism to national parks has become important to the economies of many developing nations, especially in Eastern and Southern Africa, India, Nepal, Ecuador, and Indonesia. Parks are sources of local employment and can stimulate improvements to transportation and other types of infrastructure while encouraging productive use of lands that are of marginal agricultural use.

The International Union for Conservation of Nature has developed a system for classifying the world's protected areas, with Category II areas designated as national parks. Using this definition, there are 3,044 national parks in the world, with a mean average size of 457 square miles (1,183 sq. km.) each. Together, they cover an area of about 1.5 million square miles (4 million sq. km.), accounting for about 2.7 percent of the total land area on Earth.

STEPHEN T. MATHER AND THE U.S. NATIONAL PARK SERVICE

In 1914 businessman and conservationist Stephen T. Mather wrote to Secretary of the Interior Franklin K. Lane about the poor condition of California's Yosemite and Sequoia National Parks. Lane wrote back, "if you don't like the way the national parks are being run, come on down to Washington and run them yourself." Mather accepted the challenge and became an assistant to Lane and later the first director of the U.S. National Park Service, from 1917 to 1929.

North America

In 1916 management of U.S. national parks and monuments was shifted from the U.S. Army to the newly established National Park Service (NPS). The system has since grown in size to protect sixty-one national parks, as well as other natural areas including national monuments, seashores, and preserves.

North America's second-largest system of national parks is Parks Canada, created in 1930. Among the best-known Canadian parks is Banff, established in southern Alberta in 1885. Preserved within this area are glacially carved valleys, evergreen forests, and turquoise lakes. Parks Canada has the goal of protecting representative examples of each of Canada's vegetation and physiographic regions.

Mexico began providing protection for natural areas in the late nineteenth century. Among its system of sixty-seven national parks is Dzibilchaltún, an important Mayan archaeological site on the Yucatán Peninsula. With fewer resources available for park management, the emphasis in Mexico remains the preservation of scenic beauty for public use.

South America

Two of South America's best-known national parks are located within Argentina's park system. Nahuel Huapi National Park preserves two rare deer species of the Andes, while Iguazú National Park, located on the border with Brazil, is home to tapir, ocelot, and jaguar.

Located on a plateau of the western slope of the Andes Mountains in Chile, Lauca National Park is one of the world's highest parks, with an average elevation of more than 14,000 feet (4,267 meters)-an altitude nearly as high as the tallest mountains in the continental United States. Huascarán, another mountain park located in western Peru, boasts twenty peaks that exceed 19,000 feet (5,791 meters) in elevation. The volcanic islands of Galapagos Islands National Park, managed by Ecuador, have been of interest to biologists since British naturalist Charles Darwin studied variation and adaptation in animal species there in 1835.

Australia and New Zealand

Established in 1886, Royal was Australia's first national park. Perhaps better known to tourists, Uluru National Park in Australia's Northern Territory protects two rock domes, Ayer's Rock and Mount Olga, that rise above the plains 15 miles (40 km.) apart.

Along with Australia and other former colonies of Great Britain, New Zealand was a leader in establishing early national parks. The first of these was Tongagiro, created in 1887 to protect sacred lands of the Maori people on the North Island. New Zealand's South Island features several national parks including Fiordland, created in 1904 to preserve high mountains, forests, rivers, waterfalls, and other spectacular features of glacial origin.

Africa

Game poaching continues to be a severe problem in Africa, where animals are slaughtered for ivory, meat, and hides. Many African national parks were established to protect large game. South Africa's national park system began in 1926, when the Sabie Game Preserve of the eastern Transvaal region became Kruger National Park. Among South Africa's greatest attractions to foreign visitors, Kruger is famous for its population of lions and elephants.

East Africa is also known for outstanding game sanctuaries, such as Serengeti National Park, created prior to Tanzania's independence from Great Britain. Another national park in Tanzania, Kilimanjaro, protects Africa's highest and best-known mountain. Other African countries with well-developed park systems include Kenya, the Democratic Republic of the Congo (formerly Zaire), and Zambia. Although there is now a network of national parks in Africa that protects a wide range of habitats in various regions, there remains a need to protect additional areas in the arid northern part of the continent that includes the Sahara Desert.

Europe

In comparison with the United States, the national park concept spread more slowly within Europe. In 1910 Germany set aside Luneburger Heide National Park near the Elbe River, and in 1913, Sweden established Sarek, Stora Sjöfallet, Peljekasje, and Abisko National Parks. Swiss National Park was founded in Switzerland in 1914, in the Lower Engadine region. Great Britain has several national parks, including Lake District, a favorite recreation destination for English poet William Wordsworth. Spain's Doñana National Park, located on its southwestern coast, preserves the largest dune area on the European continent.

Asia

The system of land tenure and rural economy in many Asian countries has made it difficult for national governments to set aside large areas free from human exploitation. Many national parks established by colonial powers prior to World War II were maintained or expanded by countries following independence. For example, Kaziranga National Park is a refuge for the largest heard of rhinocerous in India. Established in 1962, Thailand's Khao Yai National Park protects a sample of the country's wildlife, while Indonesia's Komodo Island National Park preserves the habitat for the large lizards known as Komodo dragons.

In Japan, high population density has made it difficult to limit human activities within large areas. Some Japanese national parks are principally recreation areas rather than wildlife sanctuaries and may contain cultural features such as Shinto shrines. One of the best known national parks in Japan is Fuji-Hakone-Izu, which contains world-famous Mount Fuji, a volcano with a nearly symmetrical shape.

The Future

National parks serve as relatively undisturbed enclaves that protect examples of the world's most outstanding natural and cultural resources. The movement to establish these areas is a relatively recent attempt to achieve an improved balance between human activities and the earth. In recent years, rising incomes and lower costs for international travel have improved the accessibility of national parks to a larger number of persons, meaning that park visitation is likely to continue to rise.

Thomas A. Wikle

BOUNDARIES AND TIME ZONES

INTERNATIONAL BOUNDARIES

International boundaries are the marked or imaginary lines traversing natural terrain of land or water that mark off the territory of one politically organized society—a state or nation-state—from other states. In addition, states claim "air boundaries." While satellites circumnavigate the earth without nations' permission, airplanes and other air vessels that fly much lower must gain the permission of states over whose territory they travel.

The existence of international boundaries is a consequence of the "territoriality" that is a feature of modern human societies. All politically organized societies, except for nomadic tribes, claim to rule some exactly defined geographical territory. International boundaries provide the limits that define this territory.

International boundaries have ancient origins. For example, the oldest sections of the Great Wall of China date back to the Ch'in Dynasty of the second century BCE. The Roman Empire also maintained boundaries to its territories, such as Hadrian's Wall in the north of England, built by the Romans in 122 CE as a defensive barrier against marauders. In these and other ancient instances, however, there was little thought that borders must be exact.

The existence of precisely drawn boundaries among states is relatively recent. The modern state has existed for no more than a few hundred years. In addition, means to determine many boundaries have come into existence only in the nineteenth and twentieth centuries, with the invention of scientific methods and instruments, along with accompanying vocabulary, for determining exact boundaries. The most basic terms of this vocabulary begin with "latitude" and "longitude" and their subdivi-

sions into the "minutes" and "seconds" used in determining boundaries. In modern times, a new attitude toward states' territory was born, especially with the nineteenth century forms of nationalism, which tend to regard every acre of territory as sacred.

Types of Boundaries
There are several types of international boundaries. Some are geographical features, including rivers, lakes, oceans, and seas. Thus boundaries of the United States include the Great Lakes, which border Canada to the north; the Rio Grande, a river that forms part of the U.S. boundary with Mexico to the south; the Atlantic and Pacific Oceans, to the east and west, respectively; and the Gulf of Mexico, to the south. In Africa, Lake Victoria bounds parts of Tanzania, Uganda, and Kenya; and rivers, such as sections of the Congo and the Zambezi, form natural boundaries among many of the continent's states.

Other geographical features, such as mountains, often form international boundaries. The Pyrenees, for example, separate France and Spain and cradle the tiny state of Andorra. In South America, the Andes frequently serve as a boundary, such as between Argentina and Chile. The Himalayas in South Central Asia create a number of borders, such as between India, China, and Tibet and between Nepal, Butan, and their neighbors. When there are no clear geographical barriers between states, boundaries must be decided by mutual consent or the threat of force. In the 2016 presidential campaign, Donald Trump repeatedly called for a wall to be built between the United States and Mexico, claim-

Current Political Boundaries of the World

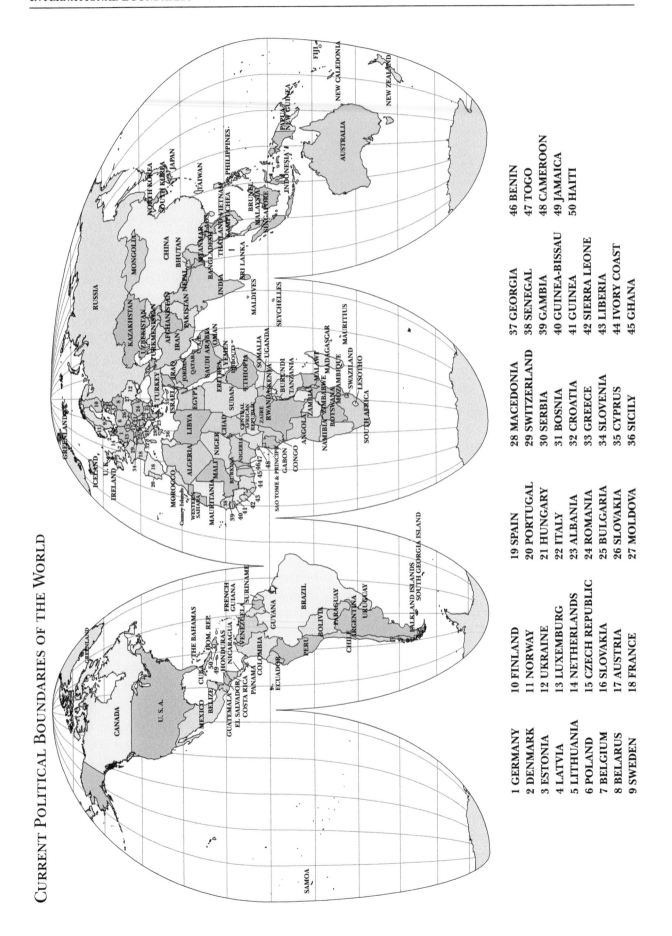

1 GERMANY
2 DENMARK
3 ESTONIA
4 LATVIA
5 LITHUANIA
6 POLAND
7 BELGIUM
8 BELARUS
9 SWEDEN

10 FINLAND
11 NORWAY
12 UKRAINE
13 LUXEMBURG
14 NETHERLANDS
15 CZECH REPUBLIC
16 SLOVAKIA
17 AUSTRIA
18 FRANCE

19 SPAIN
20 PORTUGAL
21 HUNGARY
22 ITALY
23 ALBANIA
24 ROMANIA
25 BULGARIA
26 SLOVAKIA
27 MOLDOVA

28 MACEDONIA
29 SWITZERLAND
30 SERBIA
31 BOSNIA
32 CROATIA
33 GREECE
34 SLOVENIA
35 CYPRUS
36 SICILY

37 GEORGIA
38 SENEGAL
39 GAMBIA
40 GUINEA-BISSAU
41 GUINEA
42 SIERRA LEONE
43 LIBERIA
44 IVORY COAST
45 GHANA

46 BENIN
47 TOGO
48 CAMEROON
49 JAMAICA
50 HAITI

ing that Mexico would pay for it. As of 2019, the wall has not been completed, however.

Creation and Change of International Boundaries

War and conquest often have been used to determine borders. Such wars, however, historically have created hostility among losers. Political pressures to recover lost lands build up among aggrieved losers, and such irredentist claims provide fuel for future wars. A classic example is the fate of the regions of Alsace and Lorraine between France and Germany. Although natural resources in the form of coal played a substantial role in the dispute over this area, national pride was also a potent element.

Whether boundaries are fixed through compelling geographical imperatives or in their absence, states typically sign treaties agreeing to their location. These may be treaties that conclude wars, or boundary commissions set up by those involved may draw up borders to which states give formal agreement. In 1846, for example, negotiators for Great Britain and the United States settled on the forty-ninth parallel as the boundary between the western United States and Canada, although in the United States, "Fifty-four [degrees latitude] Forty [minutes] or Fight" had been a popular motto in the presidential election campaign of 1844.

Sometimes no accepted borders exist because of chronic hostility between states. Thus, maps of the Kashmir region between India and Pakistan, claimed by both countries, show only a "line of control" or cease-fire line to divide the two warring states. Similarly, only a cease-fire line, drawn at the armistice of the Korean War of 1950-1953, divides North and South Korea; a mutually agreed-upon border remains unfixed.

In rare instances, no true boundary exists to mark where a state's territory begins and ends. Classic cases are found on the Arabian Peninsula, where the land borders of principalities, known as the Gulf Sheikdoms, are vague lines in the sand. Such circumstances usually create no difficulties where nothing is at stake, but when oil is discovered, states must come to agreement or risk coming to blows.

In other instances, negotiations and international arbitration have been effective for determining borders. Perhaps the most important principle for determining the borders of newly created states is found in the Latin phrase, *Uti possidetis iurus.* This principle is used when states become independent after having been colonies or constituent parts of a larger state that has broken up. The principle holds that states shall respect the borders in place when they were colonies. *Uti possidetis* was first extensively used in South America in the nineteenth century, when European colonial powers withdrew, leaving several newly born states to determine their own boundaries. The principle may be used as a basis for border agreements among the fifteen states of the former Soviet Union.

Besides war and negotiation, purchase has sometimes been a means of creating international boundaries. For example, in 1853 the United States purchased territory from Mexico in the southwest; in 1867, it purchased Alaska from Russia.

In rare cases, natural boundaries may change naturally or be changed deliberately by one side, incurring resentment among victims. An example occurred in 1997, when Vietnam complained that China had built an embankment on a border river embankment that caused the river to change its course; China countered that Vietnam had built a dam altering the river's course.

Other border difficulties among states include conflicts over water that flows from one country to another. In the 1990s, for example, Mexico complained of excessive U.S. use of Colorado River waters and demanded adjustment.

Border Disputes

Border disputes among states in the past two centuries have been numerous and lethal. In the twentieth century, numerous such controversies degenerated into violence. In Asia, India and Pakistan fought over Kashmir, beginning in 1947-1949 and recurring in 1965 and 1999. China has been involved in violent border disputes with India, especially in 1962; Vietnam in 1979; and Russia in 1969. In South America, border wars between Ecuador and Peru broke out in 1941, 1981, and 1995. This dispute was settled by negotiation in 1998. In Africa, among numerous recent armed conflicts, the bloody border conflict between Eritrea and Ethiopia in the 1990s was notable.

Other recent disputes have ended peacefully. Eritrea avoided violence with Yemen over several Red Sea islands by accepting arbitration by an international tribunal. In 1995 Saudi Arabia and the United Arab Emirates negotiated a peaceful agreement to their border dispute involving oil rights.

As of 2019, there are four ongoing border conflicts: Israeli-Syrian ceasefire line incidents during the Syrian Civil War, the War in Donbass, India-Pakistan military confrontation, and the 2019 Turkish offensive into north-eastern Syria, code-named by Turkey as Operation Peace Spring. Many unresolved boundary disputes might yet lead to conflicts. Among the most complex is the multinational dispute over the 600 tiny Spratly Islands in the South China Sea. Uninhabited but potentially valuable because of oil, the Spratlys are claimed by China, Brunei, Malaysia, Indonesia, the Philippines, Taiwan, and Vietnam.

Border Policies

Problems with international borders are not limited to territorial disputes. Policies regarding how borders should be operated—including the key questions of who and what should be allowed entrance and exit under what conditions—can be expected to continue as long as independent states exist. While the members of the European Union have agreed to allow free passage of people and goods among themselves, this policy does not extent to nonmembers.

The most important purpose of states is to protect the lives and property of their citizens. One of the principal purposes of international boundaries is to further this purpose. Most states insist on controlling their borders, although borders seem increasingly porous. Given the imperatives of control and the increasing difficulties of maintaining it, issues surrounding international borders are expected to continue indefinitely in the twenty-first century.

Charles F. Bahmueller

GLOBAL TIME AND TIME ZONES

Before the nineteenth century, people kept time by local reckoning of the position of the Sun; consequently, thousands of local times existed. In medieval Europe, "hours" varied in length, depending upon the seasons: Each hour was determined by the Roman Catholic Church. In the sixteenth century, Holy Roman emperor Charles V was the first secular ruler to decree hours to be of equal length. As the industrial and scientific revolutions swept Europe, North America, and other areas, some form of time standardization became necessary as communities and regions increasingly interacted. In 1780 Geneva, Switzerland, was the first locality known to employ a standard time, set by the town-hall clockkeeper, throughout the town and its immediate vicinity.

The growth and expansion of railroads, providing the first relatively fast movement of people and goods from city to city, underscored the need for a standard system in Great Britain. As early as 1828, Sir John Herschel, Astronomer Royal, called for a national standard time system based on instruments at the Royal Observatory at Greenwich. That practice began in 1852, when the British telegraph system had developed sufficiently for the Greenwich time signals to be sent instantly to any point in the country.

As railroads expanded through North America, they exposed a problem of local time variation similar to that in Great Britain but on a far larger scale, since the distances between the East and West Coasts were much greater than in Great Britain. In order for long-distance train schedules to work, different parts of the country had to coordinate their clocks. The first to suggest a standard time framework for the United States was Charles F. Dowd, president of Temple Grove Seminary for Women in Saratoga Springs, New York. Initially, Dowd proposed putting all U.S. railroads on a single standard time, based on the time in Washington, D.C. When he realized that the time in California would be behind such a standard by almost four hours, he produced a revised system, establishing four time zones in the United States. Dowd's plan, published in 1870, included the first known map of a time zone system for the country.

Not everyone was happy with the designation of Washington, D.C., as the administrative center of time in the United States. Northeastern railroad executives urged that New York, the commercial capital of the nation, be used instead: Many cities and towns in the region already had standardized to New York time out of practical necessity. Dowd proposed a compromise: to set the entire national time zone system in the United States using the Greenwich prime meridian, already in use in many parts of the world for maritime and scientific purposes. In 1873 the American Association of Railways (AAR) flatly rejected the proposal.

In the end, Dowd proved to be a visionary. In 1878 Sandford Fleming, chief engineer of the government of Canada, proposed a worldwide system of twenty-four time zones, each fifteen degrees of longitude in width, and each bisected by a meridian, beginning with the prime meridian of Greenwich. William F. Allen, general secretary of the AAR and armed with a deep knowledge of railroad practices and politics, took up the crusade and persuaded the railroads to agree to a system. At noon on Sunday, November 18, 1883, most of the more than six hundred U.S. railroad lines dropped the fifty-three arbitrary times they had been using and adopted Greenwich-indexed meridians that defined the times in each of four times zones: eastern,

central, mountain, and Pacific. Most major cities in the United States and Canada followed suit.

Time System for the World
Almost at the same time that American railroads adopted a standard time zone system, the State Department, authorized by the United States Congress, invited governments from around the world to assemble delegates in Washington, D.C., to adopt a global system. The International Meridian Conference assembled in the autumn of 1884, attended by representatives of twenty-five countries. Led by Great Britain and the United States, most favored adoption of Greenwich as the official prime meridian and Greenwich mean time as universal time.

There were other contenders: The French wanted the prime meridian to be set in Paris, and the Germans wanted it in Berlin; others proposed a mountaintop in the Azores or the tip of the Great Pyramid in Egypt. Greenwich won handily. The conference also agreed officially to start the universal day at midnight, rather than at noon or at sunrise, as practiced in many parts of the world. Each time zone in the world eventually came to have a local name, although technically, each goes by a letter in the alphabet in order eastward from Greenwich.

Once a global system was in place, there was a new issue: Many jurisdictions wanted to adjust their clocks for part of the year to account for differences in the number of hours of daylight between summer and winter months. In 1918 Congress decreed a system of daylight saving time for the United States but almost immediately abolished it, leaving state governments and communities to their local options. Daylight saving time, or a form of it, returned in the United States and many Allied nations during World War II. In the Uniform Time Act of 1966, Congress finally established a national system of daylight saving time, although with an option for states to abstain.

To the extent that it indicates how human communities want to manipulate time for social, political, or economic reasons, the issue of daylight saving time, rather than the establishment of a system of world time zones, is a better clue to the geographical issues involved in time administration. Both the history and the present format of the world time zone system show that the mathematically precise arrangement envisioned by many of the pioneers of time zones is not as important as things on the ground.

In the United States, the railroad time system adopted in 1883 drew the boundary between eastern time and central time more or less between the thirteen original states and the trans-Appalachian West: The entire Midwest, including Ohio, Indiana, and Michigan, fell in the central time zone. As the center of population migrated westward, train speeds increased, highways developed, and New York emerged as the center of mass media in the United States, the boundary between the eastern and central time zones marched steadily westward. In 1918 it ran down the middle of Ohio; by the 1960s, it was at the outskirts of Chicago.

One of the principal reasons for the popularity of Greenwich as the site of the prime meridian (zero degrees longitude), is that it places the international date line (180 degrees longitude)—where, in effect, time has to move forward to the next day rather than the next hour—far out in the Pacific Ocean where few people are affected by what otherwise would be an awkward arrangement. However, even this line is somewhat irregular, to avoid placing a small section of eastern Russia and some of the Aleutian Islands of the United States in different days.

Coordinated Universal Time Coordinated Universal Time (or UTC) is the primary time standard by which the world regulates clocks and time. It is within about 1 second of mean solar time at 0° longitude, and is not adjusted for daylight saving time. In some countries, the term Greenwich Mean Time is used. The co-ordination of time and frequency transmissions around the world began on January 1, 1960. UTC was first officially adopted as CCIR Recommendation 374, Standard-Frequency and Time-Signal Emissions, in 1963, but the official abbreviation of UTC and the official English name of Coordinated Universal Time (along with the French equivalent) were not adopted until 1967. UTC uses a *slightly* different second called the *SI second*. That is based on *atomic clocks*. Atomic clocks are more regular than the slightly variable Earth's rotation period. Hence, the essential difference between GMT and UTC is that they use different definitions of exactly how long one second of time is.

By 1950 most nations had adopted the universal time zone system, although a few followed later: Saudi Arabia in 1962, Liberia in 1972. Despite adhering to the system in principle, many nations take considerable liberties with the zones, especially if their territory spans several. All of Western Europe, despite covering an area equivalent to two zones, remains on a single standard. The People's Republic of China, which stretches across five different time zones, arbitrarily sets the entire country officially on Beijing time, eight hours behind Greenwich. Iran, Afghanistan, India, and Myanmar, each of which straddle time zone boundaries, operate on half-hour compromise systems as their time standards (as does Newfoundland). As late as 1978, Guyana's standard time was three hours, forty-five minutes in advance of Greenwich.

It can be argued that adoption of a worldwide system of time zones in the late nineteenth century was one of the earliest manifestations of the emergence of a global economy and society, and has been a crucial factor in the unfolding of this process throughout the twentieth century and beyond.

Ronald W. Davis

GLOBAL EDUCATION

THEMES AND STANDARDS IN GEOGRAPHY EDUCATION

Many people believe that the study of geography consists of little more than knowing the locations of places. Indeed, in the past, whole generations of students grew up memorizing states, capitals, rivers, seas, mountains, and countries. Most students found that approach boring and irrelevant. During the 1990s, however, geography education in the United States underwent a remarkable transformation.

While it remains important to know the locations of places, geography educators know that place name recognition is just the beginning of geographic understanding. Geography classes now place greater emphasis on understanding the characteristics of and the connections between places. Three things have led to the renewal of geography education: the five themes of geography, the national geography standards, and the establishment of a network of geographic alliances.

The Five Themes of Geography

One of the first efforts to move geography education beyond simple memorization was the National Geographic Society's publication of five themes of geography in 1984: location, place, human-environment interactions, movement, and regions. Not intended to be a checklist or recipe for understanding the world, these themes merely provided a framework for teachers—many of whom did not have a background in the subject—to incorporate geography throughout a social studies curriculum. The five themes were promoted widely by the National Geographic Society and are still used by some teachers to organize their classes.

Location is about knowing where things are. Both the absolute location (where a place is on earth's surface) and relative location (the connections between places) are important. The concept of place involves the physical and human characteristics that distinguish one place from another. The theme of human/environment interaction recognizes that people have relationships within defined places and are influenced by their surroundings. For example, many different types of housing have been created as adaptations to the world's diverse climates. The theme of movement involves the flow of people, goods, and ideas around the world. Finally, regions are human creations to help organize and understand Earth, and geography studies how they form and change.

The National Geography Standards

Geography was one of six subjects identified by President George H. W. Bush and the governors of the U.S. states when they formulated the National Education Goals in 1989. While the goals themselves foundered amid the political debate that followed their adoption, one tangible result of the initiative was the creation of Geography for Life: The National Geography Standards. More than 1,000 teachers, professors, business people, and government officials were involved in the writing of Geography for Life. The project wassupported by four geography organizations: the American Geographical Society, the Association of American Geographers, the National Council for Geographic Education, and the National Geographic Society. The resulting book defines what every U.S. student

GEOGRAPHY STANDARDS

The geographically informed person knows and understands the following:

- how to use maps and other geographic representations, tools, and technologies to acquire, process, and report information from a spatial perspective;
- how to use mental maps to organize information about people, places, and environments in a spatial context;
- how to analyze the spatial organization of people, places, and environments on Earth's surface;
- the physical and human characteristics of places;
- that people create regions to interpret Earth's complexity;
- how culture and experience influence people's perceptions of places and regions;
- the physical processes that shape the patterns of Earth's surface;
- the characteristics and spatial distribution of ecosystems on Earth's surface;
- the characteristics, distribution, and migration of human populations on Earth's surface;
- the characteristics, distribution, and complexity of Earth's cultural mosaics;
- the patterns and networks of economic interdependence on Earth's surface;
- the processes, patterns, and functions of human settlement;
- how the forces of cooperation and conflict among people influence the division and control of Earth's surface;
- how human actions modify the physical environment;
- how physical systems affect human systems;
- the changes that occur in the meaning, use, distribution, and importance of resources;
- how to apply geography to interpret the past;
- how to apply geography to interpret the present and plan for the future.

Source: National Geography Standards Project. Geography for Life: National Geography Standards, Second Edition. Washington, D.C.: National Geographics Research and Exploration, 2012.

should know and be able to accomplish in geography.

Each of the eighteen standards is designed to develop students' geographic skills, including asking geographic questions; acquiring, organizing, and analyzing geographic information; and answering the questions. Each standard features explanations, examples, and specific requirements for students in grades four, eight, and twelve.

Geography Alliances and the Future of Geography Education

To publicize efforts in geography education, a network of geography alliances was established between 1986 and 1993. Today, each U.S. state has a geography alliance that links teachers and organizations such as the National Geographic Society and the National Council for Geographic Education to sponsor workshops, teacher training sessions, field experiences, and other ways of sharing the best in geographic teaching and learning.

A 2013 executive summary prepared by the National Geographic Society for the *Road Map for 21st Century Geography Education Project* continues to champion the goal of better geography education in K–12 schools. The Road Map Project represents the collaborative effort of four national organizations: the American Geographical Society (AGS), the Association of American Geographers (AAG), the National Council for Geographic Education (NCGE), and the National Geographic Society (NGS). The project partners share belief that geography education is essential for student success in all aspects of their adult lives—careers, civic lives, and personal decision making. It also is essential for the education of specialists who can help society addressing critical issues in the areas of social welfare, economic stability, environmental health, and international relations.

Eric J. Fournier

Global Data

World Gazetteer of Oceans and Continents

Places whose names are printed in SMALL CAPS *are subjects of their own entries in this gazetteer.*

Aden, Gulf of. Deep-water area between the RED and ARABIAN SEAS, bounded by Somalia, Africa, on the south and Yemen on the north. Water is warmer and saltier in the Gulf of Aden than in the Red and Arabian Seas, because little water enters from rain or land runoff.

Africa. Second-largest continent, connected to ASIA by the narrow isthmus of Suez. Bounded on the east by the INDIAN OCEAN and on the west by the ATLANTIC OCEAN. Countries of Africa are Algeria, Angola, Benin, Botswana, Burkina Faso, Burundi, Cameroon, Central African Republic, Chad, Congo, Côte d'Ivoire (Ivory Coast), the Democratic Republic of Congo, Egypt, Ethiopia, Gabon, Gambia, Ghana, Guinea, Kenya, Liberia, Libya, Madagascar, Malawi, Mali, Mauritania, Morocco, Mozambique, Namibia, Niger, Nigeria, Rio Muni (Mbini), Rwanda, Senegal, Sierra Leone, Somalia, South Africa, Sudan, Tanzania, Togo, Tunisia, Uganda, Western Sahara, Zambia, and Zimbabwe. Climate ranges from hot and rainy near the equator, to hot and dry in the huge Sahara Desert in the north and the Kalahari Desert in the south, to warm and fairly mild at the northern and southern extremes. Paleontological evidence indicates that humans originally evolved in Africa.

Agulhas Current. Warm, swift ocean current moving south along East AFRICA's coast. Part moves between AFRICA and MADAGASCAR to form the Mozambique Current. The warm water of the

Oceans and Continents

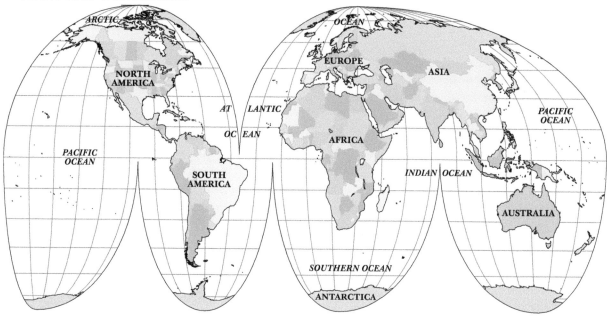

389

Agulhas Current increases the average temperatures in the eastern part of South Africa.

Agulhas Plateau. Relatively small ocean-bottom plateau that lies south of South AFRICA, at the area where the INDIAN and ATLANTIC OCEANS meet.

Aleutian Islands. Chain of volcanic islands that extends 1,100 miles (1,770 km.) from the tip of the Alaska Peninsula to the Kamchatka Peninsula in Russia and forms the boundary between the North PACIFIC OCEAN and the BERING SEA. The area is hazardous to navigation and has been called the "Home of Storms."

Aleutian Trench. Located on the northern margin of the PACIFIC OCEAN, stretching 3,666 miles (5,900 km.) from the western edge of the Aleutian Island chain to Prince William Sound, Alaska. Depth is 25,263 feet (7,700 meters).

American Highlands. Elevated region on the ANTARCTIC coast between Enderby Land and Wilkes Land, located far south of India. The Lambert and Fisher glaciers originate in the American Highlands and move down to feed the AMERY ICE SHELF.

Amery Ice Shelf. Year-round shelf of relatively flat ice in a bay of ANTARCTICA, located at approximately longitude 70 degrees east, between MAC. ROBERTSON LAND and the AMERICAN HIGHLANDS. The ice shelf is fed by the Lambert and Fisher glaciers.

Amundsen Sea. Portion of the southernmost PACIFIC OCEAN off the Wahlgreen Coast of ANTARCTICA, approximately longitude 100 to 120 degrees west. Named for the Norwegian explorer Roald Amundsen, who became the first person to reach the SOUTH POLE in 1911.

Antarctic Circle. Latitude of 66.3 degrees south. South of this line, the Sun does not set on the day of the summer solstice, about December 22 in the SOUTHERN HEMISPHERE, and does not rise on the day of the winter solstice, about June 21.

Antarctic Circumpolar Current. Eastward-flowing current that circles ANTARCTICA and extends from the surface to the deep ocean floor. The largest-volume current in the oceans. Extends northward to approximately 40 degrees south latitude and is driven by westerly winds.

Antarctic Convergence. Meeting place where cold Antarctic water sinks below the warmer sub-Antarctic water.

Antarctic Ocean. See SOUTHERN OCEAN.

Antarctica. Fifth-largest continent, located at the southernmost part of the world. There are two major regions; western Antarctica, which includes the mountainous Antarctic peninsula, and eastern Antarctica, which is mostly a low continental shield area. An ice cap up to 13,000 feet (4,000 meters) thick covers 95 percent of the continent's surface. Temperatures in the austral summer (December, January, and February) rarely rise above 0 degrees Fahrenheit (-18 degrees Celsius) except on the peninsula. By international treaty, the continent is not owned by any single country, and human access is largely regulated. There has never been a self-supporting human habitation on Antarctica.

Arabian Sea. Portion of the INDIAN OCEAN bounded by India on the east, Pakistan on the north, and Oman and Yemen of the Arabian Peninsula on the west.

Arctic Circle. Latitude of 66.3 degrees north. North of this line, the Sun does not set on the day of the summer solstice, about June 21 in the NORTHERN HEMISPHERE, and does not rise on the day of the winter solstice, about December 22.

Arctic Ocean. World's smallest ocean. It centers on the geographic NORTH POLE and connects to the PACIFIC OCEAN through the BERING SEA, and to the ATLANTIC OCEAN through the GREENLAND SEA. The Arctic Ocean is covered with ice up to 13 feet (4 meters) thick all year, except at its edges. Norwegian explorers on the ship *Fram* stayed locked in the icepack from 1893 to 1896, in order to study the movement of polar ice. They drifted in the ice a total of 1,028 miles (1,658 km.), from the Bering Sea to the Greenland Sea, proving that there was no land mass under the Arctic ice at the top of the world. Also known as Arctic Sea or Arctic Mediterranean Sea.

Argentine Basin. Basin on the floor of the western ATLANTIC OCEAN, off the coast of Argentina in SOUTH AMERICA. Among ocean basins, this one is unusually circular.

Ascension Island. Isolated volcanic island in the South ATLANTIC OCEAN, about midway between SOUTH AMERICA and AFRICA. One of the islands visited by British biologist Charles Darwin during his five-year voyage on the *Beagle*.

Asia. Largest continent; joins with EUROPE to form the great Eurasian landmass. Asia is bounded by the ARCTIC OCEAN on the north, the western PACIFIC OCEAN on the east, and the INDIAN OCEAN on the south. Its countries include Afghanistan, Bahrain, Bangladesh, Bhutan, Cambodia, China, India, Iran, Iraq, Irian Jaya, Israel, Japan, Jordan, Kalimantan, Kazakhstan, North and South Korea, Kyrgyzstan, Laos, Lebanon, Malaysia, Myanmar, Mongolia, Nepal, Oman, Pakistan, the Philippines, Russia, Sarawak, Saudi Arabia, Sri Lanka, Sumatra, Syria, Tajikistan, Thailand, Asian Turkey, Turkmenistan, United Arab Emirates, Uzbekistan, Vietnam, and Yemen. Climates include virtually all types on earth, from arctic to tropical, desert to rainforest. Asia has the highest (Mount Everest) and lowest (Dead Sea) surface points in the world. Nearly 60 percent of the world's people live in Asia.

Atlantic Ocean. Second-largest body of water in the world, covering more than 25 percent of Earth's surface. Bordered by NORTH and SOUTH AMERICA on the west, and EUROPE and East AFRICA on the east. The widest part (5,500 miles/8,800 km.) lies between West AFRICA and Mexico, along 20 degrees latitude. Scientists disagree on the north-south boundaries of the Atlantic; if one includes the ARCTIC OCEAN and the SOUTHERN OCEAN, the Atlantic Ocean extends about 13,300 miles (21,400 km.). The deepest spot (28,374 feet/8,648 meters) is found in the PUERTO RICO TRENCH. The Atlantic Ocean has been a major route for trade and communications, especially between North America and

Europe, for hundreds of years. This is because of its relatively narrow size and favorable currents, such as the GULF STREAM.

Australasia. Loosely defined term for the region, which, at the least, includes AUSTRALIA and New Zealand; at the most, it also includes other South Pacific Islands in the region.

Australia. Smallest continent, sometimes called the "island continent." Located between the INDIAN and PACIFIC OCEANS. It is the only continent occupied by a single nation, the Commonwealth of Australia. Australia is the flattest and driest continent; two-thirds is either desert or semiarid. Geologically, it is the oldest and most isolated continent. Unlike any other place on Earth, large mammals never evolved in Australia. Marsupials (pouched, warm-blooded animals) and unusual birds developed in their place.

Azores. Archipelago (group of islands) in the eastern ATLANTIC OCEAN lying about 994 miles (1,600 km.) west of Portugal. The islands are of volcanic origin and have been known, fought over, and used by the Europeans since before the fourteenth century. Spanish explorer Christopher Columbus stopped in the Azores to wait for favorable winds before his first trip across the ATLANTIC OCEAN.

Barents Sea. Partially enclosed section of the ARCTIC OCEAN, bounded by Russia and Norway on the south and the Russian island of Navaya Zemlaya on the east. The Barents Sea was important in World War II because Allied convoys had to cross it, through storms and submarine patrols, to deliver war supplies to Murmansk, the only ice-free port in western Russia. It was named for the Dutch explorer Willem Barents.

Bays. See under individual names.

Beaufort Sea. Area of the ARCTIC OCEAN located off the northern coast of Alaska and western Canada. It is usually frozen over and has no islands. Named for British admiral Sir Francis Beaufort, who devised the Beaufort Wind Scale as a means of classifying wind force at sea.

Bengal, Bay of. Northeast arm of the INDIAN OCEAN, bounded by India on the west and Myanmar on the east. The Ganges River empties into the Bay of Bengal. The great ports of Calcutta and Madras in India, and Rangoon in Myanmar lie in the bay, making it a busy and important area for shipping for centuries.

Benguela Current. Northward-flowing current along the western coast of Southern AFRICA. Normally, the Benguela Current carries cold, rich water that wells up from the ocean depths and supports a large fishing industry. A change in winds can reduce the oxygen supply and kill huge numbers of fish, similar to what may happen off the coast of Peru during El Niño weather conditions.

Bering Sea. Portion of the northernmost PACIFIC OCEAN that is bounded by the state of Alaska on the east, Russia and the Kamchatka Peninsula on the west, and the BERING STRAIT on the north. It is a valuable fishing ground, rich in shrimp, crabs, and fish. Whales, fur seals, sea otters, and walrus are also found there.

Bering Strait. Narrowest point of connection between the BERING SEA and the ARCTIC OCEAN, located between the easternmost point of Siberia on the west and Alaska on the east. The Bering Strait is 52 miles (84 km.) wide. During the Ice Age, when the sea level was lower, humans and animals were able to walk from the Asian continent across a land bridge—now known as Beringia—to the North American continent across the frozen strait, providing the first human access to the Americas.

Bikini Atoll. Small atoll in the Marshall Islands group in the western PACIFIC OCEAN. In the 1940s, the United States began testing nuclear bombs on Bikini and neighboring atolls. The U.S. Army removed the inhabitants of Bikini, and testing occurred from 1946 to 1958. The Bikini inhabitants were allowed to return in 1969, then removed again in 1978 when high levels of radioactivity were found to remain.

Black Sea. Large inland sea situated where southeastern EUROPE meets ASIA; connected to the MEDITERRANEAN SEA through Turkey's Bosporus strait. The sea covers an area of about 178,000 square miles (461,000 sq. km.), with a maximum depth of more than 7,250 feet (2,210 meters).

Brazil Current. Extension of part of the warm, westward-flowing South EQUATORIAL CURRENT, which turns south to the coast of Brazil. The Brazil Current has very salty water because of its long flow across the equator. It joins the WEST WIND DRIFT and moves eastward across the South ATLANTIC OCEAN as part of the SOUTH ATLANTIC GYRE.

California, Gulf of. Branch of the eastern PACIFIC OCEAN that separates Baja California from mainland Mexico. Warm, nutrient-rich water supports a variety of fish, oysters, and sponges. California gray whales migrate to the gulf to give birth and breed, January through March. Fisheries and tourism are important industries in the Gulf of California. Also known as the Sea of Cortés.

California Current. Cool water that flows southeast along the western coast of NORTH AMERICA from Washington State to Baja California. The eastern portion of the NORTH PACIFIC GYRE.

Canada Basin. Part of the ocean floor that lies north of northeastern Canada and Alaska. The BEAUFORT SEA lies above the Canada Basin.

Cape Horn. Southernmost tip of SOUTH AMERICA. It is the site of notoriously severe storms and is hazardous to shipping.

Cape Verde Plateau. ATLANTIC OCEAN plateau lying off the western bulge of the AFRICAN continent. The volcanic Cape Verde Islands lie on the plateau.

Caribbean Sea. Portion of the western ATLANTIC OCEAN bounded by CENTRAL and SOUTH AMERICA to the west and south, and the islands of the Antilles chain on the north and east. Mostly tropical in climate, the Caribbean Sea supports a large variety of plant and animal life. Its islands, including Puerto Rico, the Cayman Islands, and the Virgin Islands, are popular tourist sites.

Caspian Sea. World's largest inland sea. Located east of the Caucasus Mountains at EUROPE's southeasternmost extremity, it dominates the expanses of western Central ASIA. Its basin is 750 miles (1,200 kilometers) long, and its average width is 200 miles (320 kilometers). It covers 149,200 square miles (386,400 sq. km.).

Central America. Region generally understood to constitute the irregularly shaped neck of land linking NORTH and SOUTH AMERICA, containing Belize, Guatemala, Honduras, El Salvador, Nicaragua, Costa Rica, and Panama.

Chukchi Sea. Portion of the ARCTIC OCEAN, bounded by the BERING STRAIT on the south, Siberia on the southwest, and Alaska on the southeast. The Chukchi Sea is the area of exchange between waters and sea life of the PACIFIC and ARCTIC OCEANS, and so is an area of interest to oceanographers and fishermen.

Clarion Fracture Zone. East-west-running fracture zone that begins off the west coast of Mexico and extends approximately 2.500 miles (4,023 km.) to the southwest.

Cocos Basin. Relatively small ocean basin located off the west coast of Sumatra in the northeast INDIAN OCEAN.

Coral Sea. Area of the PACIFIC OCEAN off the northeast coast of AUSTRALIA, between Australia on the southwest, Papua New Guinea and the Solomon Islands on the northeast, and New Caledonia on the east. Site of a naval battle in 1942 that prevented the Japanese invasion of Australia.

Cortés, Sea of. See CALIFORNIA, GULF OF.

Denmark Strait. Channel that separates GREENLAND and ICELAND and connects the North Atlantic Ocean with the ARCTIC OCEAN.

Dover, Strait of. Body of water between England and the European continent, separating the NORTH SEA from the ENGLISH CHANNEL. It is 33 miles (53 km.) wide at its narrowest point. The tunnel between England and France (known as the "Chunnel") was cut into the rock under the Strait of Dover.

Drake Passage. Narrow part of the SOUTHERN OCEAN that connects the ATLANTIC and PACIFIC OCEANS between the southern tip of SOUTH AMERICA and the ANTARCTIC peninsula. Named for sixteenth-century English navigator and explorer Sir Francis Drake, who discovered the passage when his ship was blown into it during a violent storm. Also called Drake Strait.

East China Sea. Area of the western PACIFIC OCEAN bounded by China on the west, the YELLOW SEA on the north, and Japan on the northeast. Large oil deposits were found under the East China Sea floor in the 1980s.

East Pacific Rise. Broad, nearly continuous undersea mountain range that extends from the southern end of Baja California southward, then curves east near ANTARCTICA. It is formed along the southeast side of the Pacific Plate and is part of the RING OF FIRE, a nearly continuous ring of volcanic and tectonic activity around the rim of the Pacific Ocean. Also called East Pacific Ridge.

East Siberian Sea. Portion of the ARCTIC OCEAN bounded by the CHUKCHI SEA on the east, Siberia on the south, and the LAPTEV SEA on the west. Much of the East Siberian Sea is covered with ice year-round.

Eastern Hemisphere. The half of the earth containing EUROPE, ASIA, and AFRICA; generally understood to fall between longitudes 20 degrees west and 160 degrees east.

El Niño. Conditions—also known as El Niño-Southern Oscillation (ENSO) events—that occur every two to ten years and cause weather and ocean temperature changes off the coast of Ecuador and Peru. Most of the time, the PERU CURRENT causes cold, nutrient-rich water to well up off the coast of Ecuador and Peru. During ENSO years, the cold upwelling is replaced by warmer surface water that does not support plankton and fish. Fisheries decline and seabirds starve. Climatic changes of El Niño can bring floods to normally dry areas and drought to wet areas. Effects can extend across NORTH and SOUTH AMERICA, and to the western PACIFIC OCEAN. During the 1990s, the ENSO event fluctuated but did not go completely away, which caused tremendous damage to fisheries and agriculture, storms and droughts in North America, and numerous hurricanes.

Emperor Seamount Chain. Largest known example of submerged underwater volcanic ridges, located in the northern PACIFIC OCEAN and extending southward from the Kamchatka Peninsula in Russia for about 2,500 miles (4,023 km.).

Enderby Land. Section of ANTARCTICA that lies between the INDIAN OCEAN and the South Polar Plateau, east of QUEEN MAUD LAND. Enderby Land lies between approximately longitude 45 and 60 degrees east.

English Channel. Strait water separating continental France from Great Britain. Runs for roughly 350 miles (560 km.), from the ATLANTIC OCEAN in the west to the Strait of Dover in the east.

Equatorial Current. Currents just north and south of the equator that flow from east to west. Equatorial currents are found in the PACIFIC and ATLANTIC OCEANS. The equatorial currents and the trade winds, which move in the same direction, greatly aid oceangoing traffic.

Eurasia. Term for the combined landmass of EUROPE and ASIA.

Europe. Sixth-largest continent, actually a large peninsula of the Eurasian landmass. Europe is densely populated and includes the countries of Albania, Andorra, Austria, Belarus, Belgium, Bulgaria, Bosnia-Herzegovina, Croatia, the Czech Republic, Denmark, Estonia, Finland, France, Germany, Greece, Hungary, Iceland, Ireland, Italy, Latvia, Lithuania, Macedonia, Malta, Moldova, Monaco, the Netherlands, Norway, Poland, Portugal, Romania, Slovakia, Spain, Switzerland, Turkey, and the United Kingdom (England, Northern Ireland, Scotland, and Wales). Climate ranges from near arctic in the north, to temperate and Mediterranean in the south.

Florida Current. Water moving northward along the east coast of Florida to Cape Hatteras, North Carolina, where it joins the GULF STREAM.

Fundy, Bay of. Large inlet on the North American Atlantic coast, northwest of Maine, separating New Brunswick and Nova Scotia in Canada. Renowned for having the largest tidal change in the world, more than 56 feet (17 meters).

Galápagos Islands. Located directly on the equator, 600 miles (965 km.) west of Ecuador. The islands are volcanic in origin and sit directly in the cold PERU CURRENT, which cools the islands and

creates unusual microclimates and fogs. The extreme isolation of the islands allowed unique species to develop. Biologist Charles Darwin visited the Galápagos in the 1830s, and the unusual organisms he observed helped him to conceive the theory of evolution.

Galápagos Rift. Divergent plate boundary extending between the GALÁPAGOS ISLANDS and SOUTH AMERICA. The first hydrothermal vent community was discovered in 1977 in the Galápagos Rift. This unusual type of biological habitat is based on energy from bacteria that use heat and chemicals to make food, instead of sunlight.

Grand Banks. Portion of the northwest ATLANTIC OCEAN southeast of Nova Scotia and Newfoundland. The Grand Banks are extremely rich fishing grounds, although in the 1980s and 1990s catches of cod, flounder, and many other fish dropped dramatically due to overfishing and pollution.

Great Barrier Reef. Largest coral reef in the world, lying in the CORAL SEA off the east coast of AUSTRALIA. The reef system and its small islands stretch for more than 1,100 miles (1,750 km.) and is difficult to navigate through. The reefs are home to an incredible variety of tropical marine life, including large numbers of sharks.

Greenland. Largest island in the world that is not rated as a continent; lies between the northernmost part of the ATLANTIC OCEAN and the ARCTIC OCEAN, northeast of the North American continent. About 90 percent of Greenland is permanently covered with an ice sheet and glaciers. Residents engage in limited agriculture, growing potatoes, turnips, and cabbages. Most people live along the southwest coast, where the climate is warmed by the NORTH ATLANTIC CURRENT.

Greenland Sea. Body of water bounded by GREENLAND on the west, ICELAND on the north, and Spitsbergen on the east. It is often ice-covered.

Guinea, Gulf of. Arm of the North ATLANTIC OCEAN below the great bulge of West AFRICA.

Gulf Stream. Current of westward-moving warm water originating along the equator in the ATLANTIC OCEAN. The mass of water moves along the east coast of Florida as the Florida Current, then turns in a northeasterly direction off North Carolina to become the Gulf Stream. The Gulf Stream flows northeast past Newfoundland and the western edge of the British Isles. The warmer water of the Gulf Stream moderates the climate of northwestern Europe, causing temperatures in winter to be several degrees warmer than in areas of North America at the same latitudes. The Gulf Stream decreases the time required for ships to travel from North America to Europe. This was an important factor in trade and communication in American Colonial times and has continued to be significant.

Gulfs. See under individual names.

Hatteras Abyssal Plain. Part of the floor of the northwest Atlantic Ocean Basin, east of North Carolina. It rises to form shallow sandbars around Cape Hatteras, which are a notorious navigational hazard. In the seventeenth and eighteenth centuries, so many ships were lost in the area that Cape Hatteras became known as "The Graveyard of the Atlantic."

horse latitudes. Latitude belts between 30 and 35 degrees north and south latitude, where winds are usually light and variable and the climate mostly hot and dry.

Humboldt Current. See Peru Current

Iceland. Island country bounded by the Greenland Sea on the north, the Norwegian Sea on the east, and the Atlantic Ocean on the south and west. Total area of 39,768 square miles (103,000 sq. km.). The nearest land mass is Greenland, 200 miles (320 km.) to the northwest. Situated on top of the northern part of the Atlantic Mid-Oceanic Ridge, it is characterized by major volcanic activities, geothermal springs, and glaciers.

Idzu-Bonin Trench. Ocean trench in the western Pacific Ocean, about 6,082 miles (9,810 km.) long and 2,624 feet (800 meters) deep.

Indian Ocean. Third-largest of the world's oceans, bounded by the continents of Africa to the west, Asia to the north, Australia to the east, and Antarctica to the south. Most of the Indian Ocean lies below the equator. It has an approxi-mate area of 33 million square miles (76 million sq. km.) and an average depth of about 13,120 feet (4,000 meters). Its deepest point is 24,442 feet (7,450 meters), in the Java Trench. The Indian Ocean was the first major ocean to be used as a trade route, particularly by the Egyptians. About 600 BCE, the Egyptian ruler Necho sent an expedition into the Indian Ocean, and the ship circumnavigated Africa, probably the first time this feat was accomplished. Warm winds blowing over the northern part of the Indian Ocean from May to September pick up huge amounts of moisture, which falls on India and Sri Lanka as monsoons. Fishing is important and mostly is done by small, family boats. About 40 percent of the world's offshore oil production comes from the Indian Ocean.

Indonesian Trench. See Java Trench.

Intracoastal Waterway. Series of bays, sounds, and channels, part natural and part human-made, that extends along the eastern coast of the United States from the Delaware River in New Jersey, south to the tip of Florida, then around the west coast of Florida. It extends around the Gulf Coast to the Rio Grande in Texas. It runs 2,455 miles (3,951 km.) and is an important, protected route for commercial and pleasure boat traffic.

Japan, Sea of. Marginal sea of the western Pacific Ocean that is bounded by Japan on the east and the Russian mainland on the west. Its surface area is approximately 377,600 square miles (978,000 sq. km.). It has an average depth of 5,750 feet (1,750 meters) and a maximum depth of 12,276 feet (3,742 meters).

Japan Trench. Ocean trench approximately 497 miles (800 km.) long, beginning at the eastern edge of the Japanese islands and stretching southward toward the Mariana Trench. Depth is 27,560 feet (8,400 meters).

Java Sea. Portion of the western Pacific Ocean between the islands of Java and Borneo. The sea has a total surface area of 167,000 square miles (433,000 sq. km.) and a comparatively shallow average depth of 151 feet (46 meters).

Java Trench. Ocean trench in the INDIAN OCEAN, 2,790 miles (4,500 km.) long and 24,443 feet (7,450 meters) deep. Also called the Indonesian Trench.

Kermadec Trench. Ocean trench approximately 930 miles (1,500 km.) long, located in the southwest PACIFIC OCEAN, beginning northeast of New Zealand. It has a depth of 32,800 feet (10,000 meters). Its northern end connects with the TONGA TRENCH.

Kurile Trench. Ocean trench approximately 1,367 miles (2,200 km.) long along the northeast rim of the PACIFIC OCEAN, beginning at the north end of the Japanese island chain and extending northeastward. Depth is 34,451 feet (10,500 meters).

Labrador Current. Cold current that begins in Baffin Bay between GREENLAND and northeastern Canada and flows southward. The Labrador Current sometimes carries icebergs into North Atlantic shipping channels; such an iceberg caused the famous sinking of the great passenger ship *Titanic* in 1912.

Laptev Sea. Marginal sea of the ARCTIC OCEAN off the coast of northern Siberia. The Taymyr Peninsula bounds it on the west and the New Siberian Islands on the east. Its area is about 276,000 square miles (714,000 sq. km.). Its average depth is 1,896 feet (578 meters), and the greatest depth is 9,774 feet (2,980 meters).

Lord Howe Rise. Elevation of the floor of the western PACIFIC OCEAN that lies between AUSTRALIA and New Guinea and under the TASMAN SEA.

Mac. Robertson Land. Land near the coast of ANTARCTICA, located between the INDIAN OCEAN and the south Polar Plateau, east of ENDERBY LAND. Mac.Robertson Land lies between approximately longitude 60 and 65 degrees east.

Macronesia. Loose grouping of islands in the ATLANTIC OCEAN that includes the Azores, Madeira, the Canary Islands and Cape Verde. The term is derived from Greek words meaning "large" and "island" and should not be confused with MICRONESIA, small islands in the central and North PACIFIC OCEAN.

Madagascar. Large island nation, officially called the Malagasy Republic, located in the INDIAN OCEAN about 200 miles from the southeast coast of AFRICA. Although geographically tied to the African continent, it has a culture more closely tied to those of France and Southeast Asia. Area is 226,657 square miles (587,042 sq. km.).

Magellan, Strait of. Waterway connecting the south ATLANTIC OCEAN with the South Pacific. Ships passing through the strait, north of Tierra del Fuego Island, avoid some of the world's roughest seas around CAPE HORN.

magnetic poles. The two points on the earth, one in the NORTHERN HEMISPHERE and one in the SOUTHERN HEMISPHERE, which are defined by the internal magnetism of the earth. Each point attracts one end of a compass needle and repels the opposite end.

Malacca, Strait of. Relatively narrow passage (200 miles/322 kilometers wide) bordered by Malaysia and Sumatra and linking the SOUTH CHINA SEA and the JAVA SEA. It is one of the most heavily traveled waterways in the world, with more than one thousand ships every week.

Mariana Trench. Lowest point on Earth's surface, with a maximum depth of 36,150 feet (11,022 meters) in the Challenger Deep. The Mariana Trench is located on the western margin of the PACIFIC OCEAN southeast of Japan, and is approximately 1,584 miles (2,550 km.) long.

Marie Byrd Land. Section of ANTARCTICA located at the base of the Antarctic peninsula and shaped like a large peninsula itself. It is bounded at its base by the ROSS ICE SHELF and the Ronne Ice Shelf.

Mediterranean Sea. Large sea that separates the continents of EUROPE, AFRICA, and ASIA. It takes its name from Latin words meaning "in the middle of land"—a reference to its nearly land-locked nature. Covers about 969,100 square miles (2.5 million sq. km.) and extends 2,200 miles (3,540 km.) from west to east and about 1,000 miles (1,600 km.) from north to south at its widest. Its greatest depth is 16,897 feet (5,150 meters).

Melanesia. One of three divisions of the Pacific Islands, along with MICRONESIA and POLYNESIA; located in the western Pacific. The name Melanesia, for "dark islands," was given to the area because of its inhabitants' dark skins

Mexico, Gulf of. Nearly enclosed arm of the western ATLANTIC OCEAN, bounded by the states of Florida, Alabama, Mississippi, Louisiana, and Texas, and Mexico and the Yucatan Peninsula. Cuba is located in the gap between the Yucatan Peninsula and Florida. Most ocean water enters through the Yucatan passage and exits the Gulf of Mexico around the tip of Florida, becoming the FLORIDA CURRENT. Fisheries, tourism, and oil production are important activities.

Micronesia. One of three divisions of the Pacific Islands, along with MELANESIA and POLYNESIA. Micronesia means "small islands." Micronesia's islands are mostly atolls and coral islands, but some are of volcanic origin. The more than 2,000 islands of Micronesia are located in the Pacific Ocean east of the Philippines, mostly north of the EQUATOR.

Mid-Atlantic Ridge. Steep-sided, underwater mountain range running down the middle of the ATLANTIC OCEAN. Formed by the divergent boundaries, or region where tectonic plates are separating.

Mozambique Current. See AGULHAS CURRENT.

New Britain Trench. Ocean trench in the southwest PACIFIC OCEAN, about 5,158 miles (8,320 km.) long and 2,460 feet (750 meters) deep.

New Hebrides Basin. Part of the CORAL SEA, located east of AUSTRALIA and west of the New Hebrides island chain. The basin contains volcanic islands, both old and recent.

New Hebrides Trench. Ocean trench in the southwest PACIFIC OCEAN, about 5,682 miles (9,165 km.) long and 3,936 feet (1,200 meters) deep.

North America. Third-largest continent, usually considered to contain all land and nearby islands in the WESTERN HEMISPHERE north of the Isthmus of Panama, which connects it to SOUTH AMERICA. The major mainland countries are Canada, the United States, Mexico, Guatemala, El Salvador, Honduras, Nicaragua, Costa Rica, and Panama. Island countries include the islands of the CARIBBEAN SEA and GREENLAND. Climate ranges from arctic to tropical.

North Atlantic Current. Continuation of the GULF STREAM, originating near the GRAND BANKS off Newfoundland. It curves eastward and divides into a northern branch, which flows into the NORWEGIAN SEA, a southern branch, which flows eastward, and a branch that forms the Canary Current and flows south along the coast of EUROPE.

North Atlantic Gyre. Large mass of water, located in the ATLANTIC OCEAN in the NORTHERN HEMISPHERE, that rotates clockwise. Warm water moves toward the pole and cold water moves toward the equator.

North Pacific Current. Eastward flow of water in the PACIFIC OCEAN in the NORTHERN HEMISPHERE. It originates as the Kuroshio Current and moves from Japan toward NORTH AMERICA.

North Pacific Gyre. Large mass of water, located in the PACIFIC OCEAN in the NORTHERN HEMISPHERE, that rotates clockwise. Warm water moves toward the pole and cold water moves toward the equator.

North Pole. Northern end of the earth's geographic axis, located at 90 degrees north latitude and longitude zero degrees. The North Pole itself is located on the Polar Abyssal Plain, about 14,000 feet (4,000 meters) deep in the ARCTIC OCEAN. U.S. explorer Robert Edwin is credited with being the first person to reach the North Pole, in 1909, although there is historical dispute over the claim. The North Pole is different from the North MAGNETIC POLE.

North Sea. Arm of the northeastern ATLANTIC OCEAN, bounded by Great Britain on the west and Norway, Denmark, and Germany on the east and south. The North Sea is one of the great fishing areas of the world and an important source of oil.

Northern Hemisphere. The half of the earth above the equator.

Norwegian Sea. Section of the North Atlantic Ocean. Norway borders it on the east and Iceland on the west. A submarine ridge linking

Greenland, Iceland, the Faroe Islands, and northern Scotland separates the Norwegian Sea from the open Atlantic Ocean. Cut by the Arctic Circle, the sea is often associated with the Arctic Ocean to the north. Reaches a maximum depth of about 13,020 feet (3,970 meters).

Oceania. Loosely applied term for the large island groups of the central and South Pacific; sometimes used to include Australia and New Zealand.

Okhotsk, Sea of. Nearly enclosed area of the northwestern Pacific Ocean bounded by Russia's Kamchatka Peninsula on the east and Siberia on the west. It is open to the Pacific Ocean on the south side only through Japan and the Kuril Islands, a string of islands belonging to Russia.

Pacific Ocean. Largest body of water in the world, covering more than one-third of Earth's surface—an area of about 70 million square miles (181 million sq. km.), more than the entire land area of the world. At its widest point, between Panama in Central America and the Philippines, it stretches 10,700 miles (17,200 km.). It runs 9,600 miles (15,450 km.) from the Bering Strait in the north to Antarctica in the south. Bordered by North and South America in the east, and Asia and Australia in the west. The average depth is about 12,900 feet (3,900 meters). It contains the deepest point on Earth (36,150 feet/11,022 meters), in the Challenger Deep of the Mariana Trench, southwest of Japan. The Pacific Ocean bottom is more geologically varied than the Indian or Atlantic Oceans; it has more volcanoes, ridges, trenches, seamounts, and islands. The vast size of the Pacific Ocean was a formidable barrier to travel, communications, and trade well into the nineteenth century. However, evidence shows that people crossed the Pacific Ocean in rafts or canoes as early as 3,000 BCE.

Pacific Rim. Modern term for the nations of Asia and North and South America that border, or are in, the Pacific Ocean. Used mostly in discussions of economic growth.

Palau Trench. Ocean trench in the western Pacific Ocean, about 250 miles (400 km.) long and 26,425 feet (8,054 meters) deep.

Palmer Land. Section of Antarctica that occupies the base of the Antarctic peninsula.

Panama, Isthmus of. Narrow neck of land that joins Central and South America. In 1914 the Panama Canal was opened through the isthmus, creating a direct sea link between the Pacific Ocean and the Caribbean Sea. The canal stretches about 50 miles (80 km.) from Panama City on the Pacific to Colón on the Caribbean. More than 12,000 ships pass through the canal annually.

Persian Gulf. Large extension of the Arabian Sea that separates Iran from the Arabian Peninsula in the Middle East. It covers about 88,000 square miles (226,000 sq. km.) and is about 620 miles (1,000 km.) long and 125–185 miles (200–300 km.) wide.

Peru-Chile Trench. Ocean trench that runs along the eastern boundary of the Pacific Ocean, off the western edge of South America. It is 3,666 miles (5,900 km.) long and 26,576 feet (8,100 meters) deep.

Peru Current. Cold, broad current that originates in the southernmost part of the South Pacific Gyre and flows up the west coast of South America. Off the coast of Peru, prevailing winds usually push the warmer surface water to the west. This causes the nutrient-rich, colder water of the Peru Current to well up to the surface, which provides excellent feeding for fish. At times, the upwelling ceases and biological, economic, and climatic catastrophe can result in El Niño weather conditions. Also known as the Humboldt Current.

Philippine Trench. Ocean trench located on the western rim of the Pacific Ocean, at the eastern margin of the Philippine islands. It is about 870 miles (1,400 km.) long and 34,451 feet (10,500 meters) deep.

Polynesia. One of three main divisions of the Pacific Islands, along with Melanesia and Micronesia. The islands are spread through the central and South Pacific. Polynesia means "many islands." Mostly small, the islands are predominantly coral atolls, but some are of volcanic origin.

Puerto Rico Trench. Ocean trench in the western ATLANTIC OCEAN, about 27,500 feet (8,385 meters) deep and 963 miles (1,550 km.) long.

Queen Maud Land. Section of ANTARCTICA that lies between the ATLANTIC OCEAN and the south Polar Plateau, between approximately longitude 15 and 45 degrees east.

Red Sea. Narrow arm of water separating AFRICA from the ARABIAN PENINSULA. One of the saltiest bodies of ocean water on Earth, as a result of high evaporation and little freshwater input. It was used as a trade route for Mediterranean, Indian, and Chinese peoples for centuries before the Europeans discovered it in the fifteenth century. The Suez Canal was opened in 1869 between the MEDITERRANEAN SEA and the Red Sea, cutting the distance from the northern INDIAN OCEAN to northern EUROPE by about 5,590 miles (9,000 km.). This greatly increased the economic and military importance of the Red Sea.

Ring of Fire. Nearly continuous ring of volcanic and tectonic activity around the margins of the PACIFIC OCEAN.

Ross Ice Shelf. Thick layer of ice in the Ross SEA off the coast of ANTARCTICA. The relatively flat ice is attached to and nourished by a continental glacier.

Ross Sea. Bay in the SOUTHERN OCEAN off the coast of ANTARCTICA, located south of New Zealand. Named for English explorer James Clark Ross, the first person to break through the Antarctic ice pack in a ship, in 1841.

St. Peter and St. Paul. Cluster of rocks showing above the surface of the ATLANTIC OCEAN between Brazil and West AFRICA. Important landmarks in the days of slave ships.

Sargasso Sea. Warm, salty area of water located in the ATLANTIC OCEAN south and east of Bermuda, formed from water that circulates around the center of the NORTH ATLANTIC GYRE. Named for the seaweed, *Sargassum*, that floats on the surface in large amounts.

Scotia Sea. Area of the southernmost ATLANTIC OCEAN between the southern tip of SOUTH AMERICA and the ANTARCTIC peninsula. The area is known for severe storms.

Seas. See under individual names.

Siam, Gulf of. See THAILAND, GULF OF.

South America. Fourth-largest continent, usually considered to contain all land and nearby islands in the Western Hemisphere south of the Isthmus of Panama, which connects it to NORTH AMERICA. Countries are Argentina, Bolivia, Brazil, Chile, Colombia, Ecuador, French Guiana, Guyana, Paraguay, Peru, Suriname, Uruguay, and Venezuela. Climate ranges from tropical to cold, nearly sub-Antarctic.

South Atlantic Gyre. Large mass of water, located in the ATLANTIC OCEAN in the SOUTHERN HEMISPHERE, that rotates counterclockwise. Warm water moves toward the pole and cold water moves toward the equator.

South China Sea. Portion of the western PACIFIC OCEAN that lies along the east coast of China, Vietnam, and the southeastern part of the Gulf of Thailand. The eastern and southern edges are defined by the Philippine and Indonesian Islands.

South Equatorial Current. Part of the SOUTH ATLANTIC GYRE that is split in two by the eastern prominence of Brazil. One part moves along the northeastern coast of SOUTH AMERICA toward the CARIBBEAN SEA and the North ATLANTIC OCEAN; the other turns southward and forms the BRAZIL CURRENT.

South Pacific Gyre. Large mass of water, located in the PACIFIC OCEAN in the SOUTHERN HEMISPHERE, that rotates counterclockwise. Warm water moves toward the pole and cold water moves toward the equator.

South Pole. Southern end of the earth's geographic axis, located at 90 degrees south latitude and longitude zero degrees. The first person to reach the South Pole was Norwegian explorer Roald Amundsen, in 1911. The South Pole is different from the South MAGNETIC POLE.

Southeastern Pacific Plateau. Portion of the PACIFIC OCEAN floor closest to SOUTH AMERICA.

Southern Hemisphere. The half of the earth below the equator.

Southern Ocean. Not officially recognized as one of the major oceans, but a commonly used term for water surrounding ANTARCTICA and extending northward to 50 degrees south latitude. Also known as the Antarctic Ocean.

Straits. See under individual names.

Sunda Shelf. One of the largest continental shelves in the world, nearly 772,000 square miles (2 million sq. km.). Located in the JAVA SEA, SOUTH CHINA SEA, and Gulf of THAILAND. The area was above water in the Quaternary period, enabling large animals such as elephants and rhinoceros to migrate to Sumatra, Java, and Borneo.

Surtsey Island. Island formed by a volcanic explosion off the coast of ICELAND in 1963. It is valuable to scientists studying how island flora and fauna develop and is a popular tourist site.

Tashima Current. See TSUSHIMA CURRENT.

Tasman Sea. Area of the PACIFIC OCEAN off the southeast coast of AUSTRALIA, between Australia and Tasmania on the west and New Zealand on the east. First crossed by the Moriois people sometime before 1300 CE Also called the Tasmanian Sea.

Tasmanian Sea. See TASMAN SEA.

Thailand, Gulf of. Also known as the Gulf of Siam, inlet of the South China Sea, located between the Malay Archipelago and the Southeast Asian mainland. Bounded by Thailand, Cambodia, and Vietnam.

Tonga Trench. Ocean trench in the PACIFIC OCEAN, northeast of New Zealand. It stretches for 870 miles (1,400 km.), beginning at the northern end of the KERMADEC TRENCH. Depth is 32,810 feet (10,000 meters).

Tsushima Current. Warm current in the western PACIFIC OCEAN that flows out of the YELLOW SEA into the Sea of JAPAN in the spring and summer. Also called Tashima Current.

Walvis Ridge (Walfisch Ridge). Long, narrow undersea elevation near the southwestern coast of AFRICA, which extends about 1,900 miles (3,000 km.) in a southwesterly direction under the ATLANTIC OCEAN.

Weddell Sea. Bay in the SOUTHERN OCEAN bounded by the ANTARCTIC peninsula on the west and a northward bulge of ANTARCTICA on the east, stretching from approximately longitude 60 to 10 degrees west. One of the harshest environments on Earth; surface water temperatures stay near 32 degrees Fahrenheit (0 degrees Celsius) all year. The Weddell Sea was the site of much whaling and seal hunting in the nineteenth and twentieth centuries.

West Caroline Trench. See YAP TRENCH.

West Wind Drift. Surface portion of the ANTARCTIC CIRCUMPOLAR CURRENT, driven by westerly winds. Often extremely rough; seas as high as 98 feet (30 meters) have been reported.

Western Hemisphere. The half of the earth containing NORTH and SOUTH AMERICA; generally understood to fall between longitudes 160 degrees east and 20 degrees west.

Wilkes Land. Broad section near the coast of ANTARCTICA, which lies south of AUSTRALIA and east of the AMERICAN HIGHLANDS. Wilkes Land is the nearest landmass to the South MAGNETIC POLE.

Yap Trench. Ocean trench in the western PACIFIC OCEAN, about 435 miles (700 km.) long and 27,900 feet (8,527 meters) deep. Also called the West Caroline Trench.

Yellow Sea. Area of the PACIFIC OCEAN bounded by China on the north and west and Korea on the east. Named for the large amounts of yellow dust carried into it from central China by winds and by the Yangtze, Yalu, and Yellow Rivers. Parts of the sea often show a yellow color from the dust.

Kelly Howard

WORLD'S OCEANS AND SEAS

Name	Approximate Area		Average Depth	
	Sq. Miles	Sq. Km.	Feet	Meters
Pacific Ocean	64,000,000	165,760,000	13,215	4,028
Atlantic Ocean	31,815,000	82,400,000	12,880	3,926
Indian Ocean	25,300,000	65,526,700	13,002	3,963
Arctic Ocean	5,440,200	14,090,000	3,953	1,205
Mediterranean & Black Seas	1,145,100	2,965,800	4,688	1,429
Caribbean Sea	1,049,500	2,718,200	8,685	2,647
South China Sea	895,400	2,319,000	5,419	1,652
Bering Sea	884,900	2,291,900	5,075	1,547
Gulf of Mexico	615,000	1,592,800	4,874	1,486
Okhotsk Sea	613,800	1,589,700	2,749	838
East China Sea	482,300	1,249,200	617	188
Hudson Bay	475,800	1,232,300	420	128
Japan Sea	389,100	1,007,800	4,429	1,350
Andaman Sea	308,100	797,700	2,854	870
North Sea	222,100	575,200	308	94
Red Sea	169,100	438,000	1,611	491
Baltic Sea	163,000	422,200	180	55

MAJOR LAND AREAS OF THE WORLD

Area	Approximate Land Area		Percent of World Total
	Sq. Mi.	Sq. Km.	
World	57,308,738	148,429,000	100.0
Asia (including Middle East)	17,212,041	44,579,000	30.0
Africa	11,608,156	30,065,000	20.3
North America	9,365,290	24,256,000	16.3
Central America, South America, & Caribbean	6,879,952	17,819,000	8.9
Antarctica	5,100,021	13,209,000	8.9
Europe	3,837,082	9,938,000	6.7
Oceania, including Australia	2,967,966	7,687,000	5.2

Major Islands of the World

Island	Location	Area	
		Sq. Mi.	Sq. Km.
Greenland	North Atlantic Ocean	839,999	2,175,597
New Guinea	Western Pacific Ocean	316,615	820,033
Borneo	Western Pacific Ocean	286,914	743,107
Madagascar	Western Indian Ocean	226,657	587,042
Baffin	Canada, North Atlantic Ocean	183,810	476,068
Sumatra	Indonesia, northeast Indian Ocean	182,859	473,605
Hōnshū	Japan, western Pacific Ocean	88,925	230,316
Great Britain	North Atlantic Ocean	88,758	229,883
Ellesmere	Canada, Arctic Ocean	82,119	212,688
Victoria	Canada, Arctic Ocean	81,930	212,199
Sulawesi (Celebes)	Indonesia, western Pacific Ocean	72,986	189,034
South Island	New Zealand, South Pacific Ocean	58,093	150,461
Java	Indonesia, Indian Ocean	48,990	126,884
North Island	New Zealand, South Pacific Ocean	44,281	114,688
Cuba	Caribbean Sea	44,218	114,525
Newfoundland	Canada, North Atlantic Ocean	42,734	110,681
Luzon	Philippines, western Pacific Ocean	40,420	104,688
Iceland	North Atlantic Ocean	39,768	102,999
Mindanao	Philippines, western Pacific Ocean	36,537	94,631
Ireland	North Atlantic Ocean	32,597	84,426
Hōkkaidō	Japan, western Pacific Ocean	30,372	78,663
Hispaniola	Caribbean Sea	29,355	76,029
Tasmania	Australia, South Pacific Ocean	26,215	67,897
Sri Lanka	Indian Ocean	25,332	65,610
Sakhalin (Karafuto)	Russia, western Pacific Ocean	24,560	63,610
Banks	Canada, Arctic Ocean	23,230	60,166
Devon	Canada, Arctic Ocean	20,861	54,030
Tierra del Fuego	Southern tip of South America	18,605	48,187
Kyūshū	Japan, western Pacific Ocean	16,223	42,018
Melville	Canada, Arctic Ocean	16,141	41,805
Axel Heiberg	Canada, Arctic Ocean	15,779	40,868
Southampton	Hudson Bay, Canada	15,700	40,663

COUNTRIES OF THE WORLD

Country	Region	Population	Area Square Miles	Area Square Kilometers	Population Density Persons/ Sq. Mi.	Population Density Persons/ Sq. Km.
Afghanistan	Asia	31,575,018	249,347	645,807	127	49
Albania	Europe	2,862,427	11,082	28,703	259	100
Algeria	Africa	42,545,964	919,595	2,381,741	47	18
Andorra	Europe	76,177	179	464	425	164
Angola	Africa	29,250,009	481,354	1,246,700	60	23
Antigua and Barbuda	Caribbean	104,084	171	442	609	235
Argentina	South America	44,938,712	1,073,518	2,780,400	41	16
Armenia	Europe	2,962,100	11,484	29,743	259	100
Australia	Australia	25,576,880	2,969,907	7,692,024	9	3
Austria	Europe	8,877,036	32,386	83,879	275	106
Azerbaijan	Asia	10,027,874	33,436	86,600	300	116
Bahamas	Caribbean	386,870	5,382	13,940	73	28
Bahrain	Asia	1,543,300	300	778	5,136	1,983
Bangladesh	Asia	167,888,084	55,598	143,998	3,020	1,166
Barbados	Caribbean	287,025	166	430	1,730	668
Belarus	Europe	9,465,300	80,155	207,600	119	46
Belgium	Europe	11,515,793	11,787	30,528	976	377
Belize	Central America	398,050	8,867	22,965	44	17
Benin	Africa	11,733,059	43,484	112,622	269	104
Bhutan	Asia	821,592	14,824	38,394	55	21
Bolivia	South America	11,307,314	424,164	1,098,581	26	10
Bosnia and Herzegovina	Europe	3,511,372	19,772	51,209	179	69
Botswana	Africa	2,302,878	224,607	581,730	10.4	4
Brazil	South America	210,951,255	3,287,956	8,515,767	64	25
Brunei	Asia	421,300	2,226	5,765	189	73
Bulgaria	Europe	7,000,039	42,858	111,002	163	63
Burkina Faso	Africa	20,244,080	104,543	270,764	194	75
Burundi	Africa	10,953,317	10,740	27,816	1,020	394
Cambodia	Asia	16,289,270	69,898	181,035	233	90
Cameroon	Africa	24,348,251	179,943	466,050	135	52
Canada	North America	37,878,499	3,855,103	9,984,670	10	4
Cape Verde	Africa	550,483	1,557	4,033	352	136
Central African Republic	Africa	4,737,423	240,324	622,436	21	8
Chad	Africa	15,353,184	495,755	1,284,000	31	12
Chile	South America	17,373,831	291,930	756,096	60	23
China, People's Republic of	Asia	1,400,781,440	3,722,342	9,640,821	376	145
Colombia	South America	46,103,400	440,831	1,141,748	105	40
Comoros	Africa	873,724	719	1,861	1,215	469

Country	Region	Population	Area Square Miles	Area Square Kilometers	Population Density Persons/ Sq. Mi.	Population Density Persons/ Sq. Km.
Costa Rica	Central America	5,058,007	19,730	51,100	256	99
Côte d'Ivoire (Ivory Coast)	Africa	25,823,071	124,680	322,921	207	80
Croatia	Europe	4,087,843	21,831	56,542	186	72
Cuba	Caribbean	11,209,628	42,426	109,884	264	102
Cyprus	Europe	864,200	2,276	5,896	381	147
Czech Republic	Europe	10,681,161	30,451	78,867	350	135
Dem. Republic of the Congo	Africa	86,790,567	905,446	2,345,095	96	37
Denmark	Europe	5,814,461	16,640	43,098	350	135
Djibouti	Africa	1,078,373	8,880	23,000	122	47
Dominica	Caribbean	71,808	285	739	251	97
Dominican Republic	Caribbean	10,358,320	18,485	47,875	559	216
Ecuador	South America	17,398,588	106,889	276,841	163	63
Egypt	Africa	99,873,587	387,048	1,002,450	258	100
El Salvador	Central America	6,704,864	8,124	21,040	826	319
Equatorial Guinea	Africa	1,358,276	10,831	28,051	124	48
Eritrea	Africa	3,497,117	46,757	121,100	75	29
Estonia	Europe	1,324,820	17,505	45,339	75	29
Eswatini (Swaziland)	Africa	1,159,250	6,704	17,364	174	67
Ethiopia	Africa	107,534,882	410,678	1,063,652	262	101
Fed. States of Micronesia	Pacific Islands	105,300	271	701	388	150
Fiji	Pacific Islands	884,887	7,078	18,333	124	48
Finland	Europe	5,527,405	130,666	338,424	41	16
France	Europe	67,022,000	210,026	543,965	319	123
Gabon	Africa	2,067,561	103,347	267,667	21	8
Gambia	Africa	2,228,075	4,127	10,690	539	208
Georgia	Europe	3,729,600	26,911	69,700	140	54
Germany	Europe	83,073,100	137,903	357,168	603	233
Ghana	Africa	30,280,811	92,098	238,533	329	127
Greece	Europe	10,724,599	50,949	131,957	210	81
Grenada	Caribbean	108,825	133	344	818	316
Guatemala	Central America	17,679,735	42,042	108,889	420	162
Guinea	Africa	12,218,357	94,926	245,857	129	50
Guinea-Bissau	Africa	1,604,528	13,948	36,125	114	44
Guyana	South America	782,225	83,012	214,999	9.3	3.6
Haiti	Caribbean	11,263,077	10,450	27,065	1,077	416
Honduras	Central America	9,158,345	43,433	112,492	210	81
Hungary	Europe	9,764,000	35,919	93,029	272	105
Iceland	Europe	360,390	39,682	102,775	9.1	3.5
India	Asia	1,357,041,500	1,269,211	3,287,240	1,069	413
Indonesia	Asia	268,074,600	735,358	1,904,569	365	141
Iran	Asia	83,096,438	636,372	1,648,195	131	50

Country	Region	Population	Area Square Miles	Area Square Kilometers	Population Density Persons/ Sq. Mi.	Population Density Persons/ Sq. Km.
Iraq	Asia	39,309,783	169,235	438,317	233	90
Ireland	Europe	4,921,500	27,133	70,273	181	70
Israel	Asia	9,141,680	8,522	22,072	1,073	414
Italy	Europe	60,252,824	116,336	301,308	518	200
Jamaica	Caribbean	2,726,667	4,244	10,991	642	248
Japan	Asia	126,140,000	145,937	377,975	865	334
Jordan	Asia	10,587,132	34,495	89,342	307	119
Kazakhstan	Asia	18,592,700	1,052,090	2,724,900	18	7
Kenya	Africa	47,564,296	224,647	581,834	212	82
Kiribati	Pacific Islands	120,100	313	811	383	148
Kuwait	Asia	4,420,110	6,880	17,818	642	248
Kyrgyzstan	Asia	6,309,300	77,199	199,945	83	32
Laos	Asia	6,492,400	91,429	236,800	70	27
Latvia	Europe	1,910,400	24,928	64,562	78	30
Lebanon	Asia	6,855,713	4,036	10,452	1,740	672
Lesotho	Africa	2,263,010	11,720	30,355	194	75
Liberia	Africa	4,475,353	37,466	97,036	119	46
Libya	Africa	6,470,956	683,424	1,770,060	9.6	3.7
Liechtenstein	Europe	38,380	62	160	622	240
Lithuania	Europe	2,793,466	25,212	65,300	111	43
Luxembourg	Europe	613,894	998	2,586	614	237
Madagascar	Africa	25,680,342	226,658	587,041	114	44
Malawi	Africa	17,563,749	45,747	118,484	383	148
Malaysia	Asia	32,715,210	127,724	330,803	256	99
Maldives	Asia	378,114	115	298	3,287	1,269
Mali	Africa	19,107,706	482,077	1,248,574	39	15
Malta	Europe	493,559	122	315	3,911	1,510
Marshall Islands	Pacific Islands	55,500	70	181	795	307
Mauritania	Africa	3,984,233	397,955	1,030,700	10.4	4
Mauritius	Africa	1,265,577	788	2,040	1,606	620
Mexico	North America	126,577,691	759,516	1,967,138	166	64
Moldova	Europe	2,681,735	13,067	33,843	205	79
Monaco	Europe	38,300	0.78	2.02	49,106	18,960
Mongolia	Asia	3,000,000	603,902	1,564,100	4.9	1.9
Montenegro	Europe	622,182	5,333	13,812	117	45
Morocco	Africa	35,773,773	172,414	446,550	207	80
Mozambique	Africa	28,571,310	308,642	799,380	93	36
Myanmar (Burma)	Asia	54,339,766	261,228	676,577	207	80
Namibia	Africa	2,413,643	318,580	825,118	7.5	2.9
Nauru	Pacific Islands	11,000	8	21	1,357	524
Nepal	Asia	29,609,623	56,827	147,181	521	201

Country	Region	Population	Area Square Miles	Area Square Kilometers	Population Density Persons/ Sq. Mi.	Population Density Persons/ Sq. Km.
Netherlands	Europe	17,370,348	16,033	41,526	1,083	418
New Zealand	Pacific Islands	4,952,186	104,428	270,467	47	18
Nicaragua	Central America	6,393,824	46,884	121,428	137	53
Niger	Africa	21,466,863	458,075	1,186,408	47	18
Nigeria	Africa	200,962,000	356,669	923,768	565	218
North Korea	Asia	25,450,000	46,541	120,540	546	211
North Macedonia	Europe	2,077,132	9,928	25,713	210	81
Norway	Europe	5,328,212	125,013	323,782	41	16
Oman	Asia	4,183,841	119,499	309,500	36	14
Pakistan	Asia	218,198,000	310,403	803,940	703	271
Palau	Pacific Islands	17,900	171	444	104	40
Panama	Central America	4,158,783	28,640	74,177	145	56
Papua New Guinea	Pacific Islands	8,558,800	178,704	462,840	47	18
Paraguay	South America	7,052,983	157,048	406,752	44	17
Peru	South America	32,162,184	496,225	1,285,216	65	25
Philippines	Asia	108,785,760	115,831	300,000	939	363
Poland	Europe	38,386,000	120,728	312,685	319	123
Portugal	Europe	10,276,617	35,556	92,090	290	112
Qatar	Asia	2,740,479	4,468	11,571	614	237
Republic of the Congo	Africa	5,399,895	132,047	342,000	41	16
Romania	Europe	19,405,156	92,043	238,391	210	81
Russia[1]	Europe/Asia	146,877,088	6,612,093	17,125,242	23	9
Rwanda	Africa	12,374,397	10,169	26,338	1,217	470
St Kitts and Nevis	Caribbean	56,345	104	270	541	209
St Lucia	Caribbean	180,454	238	617	756	292
St Vincent and Grenadines	Caribbean	110,520	150	389	736	284
Samoa	Pacific Islands	199,052	1,093	2,831	181	70
San Marino	Europe	34,641	24	61	1,471	568
Sahrawi Arab Dem. Rep.[2]	Africa	567,421	97,344	252,120	6	2.3
São Tomé and Príncipe	Africa	201,784	386	1,001	523	202
Saudi Arabia	Asia	34,218,169	830,000	2,149,690	41	16
Senegal	Africa	16,209,125	75,955	196,722	212	82
Serbia	Europe	6,901,188	29,913	77,474	231	89
Seychelles	Africa	96,762	176	455	552	213
Sierra Leone	Africa	7,901,454	27,699	71,740	285	110
Singapore	Asia	5,638,700	279	722.5	20,212	7,804
Slovakia	Europe	5,450,421	18,933	49,036	287	111
Slovenia	Europe	2,084,301	7,827	20,273	267	103
Solomon Islands	Pacific Islands	682,500	10,954	28,370	62	24
Somalia	Africa	15,181,925	246,201	637,657	62	24
South Africa	Africa	58,775,022	471,359	1,220,813	124	48

Country	Region	Population	Area Square Miles	Area Square Kilometers	Population Density Persons/ Sq. Mi.	Population Density Persons/ Sq. Km.
South Korea	Asia	51,811,167	38,691	100,210	1,339	517
South Sudan	Africa	12,575,714	248,777	644,329	52	20
Spain	Europe	46,934,632	195,364	505,990	241	93
Sri Lanka	Asia	21,803,000	25,332	65,610	860	332
Sudan	Africa	40,782,742	710,251	1,839,542	57	22
Suriname	South America	568,301	63,251	163,820	9.1	3.5
Sweden	Europe	10,344,405	173,860	450,295	59	23
Switzerland	Europe	8,586,550	15,940	41,285	539	208
Syria	Asia	17,070,135	71,498	185,180	238	92
Taiwan	Asia	23,596,266	13,976	36,197	1,689	652
Tajikistan	Asia	9,127,000	55,251	143,100	166	64
Tanzania	Africa	55,890,747	364,900	945,087	153	59
Thailand	Asia	66,455,280	198,117	513,120	335	130
Timor-Leste	Asia	1,167,242	5,760	14,919	202	78
Togo	Africa	7,538,000	21,853	56,600	344	133
Tonga	Pacific Islands	100,651	278	720	362	140
Trinidad and Tobago	Caribbean	1,359,193	1,990	5,155	683	264
Tunisia	Africa	11,551,448	63,170	163,610	183	71
Turkey	Europe/Asia	82,003,882	302,535	783,562	271	105
Turkmenistan	Asia	5,851,466	189,657	491,210	31	12
Tuvalu	Pacific Islands	10,200	10	26	1,020	392
Uganda	Africa	40,006,700	93,263	241,551	429	166
Ukraine[3]	Europe	41,990,278	232,820	603,000	180	70
United Arab Emirates	Asia	9,770,529	32,278	83,600	303	117
United Kingdom	Europe	66,435,600	93,788	242,910	708	273
United States	North America	330,546,475	3,796,742	9,833,517	87	34
Uruguay	South America	3,518,553	68,037	176,215	52	20
Uzbekistan	Asia	32,653,900	172,742	447,400	189	73
Vanuatu	Pacific Islands	304,500	4,742	12,281	65	25
Vatican City	Europe	1,000	0.17	0.44	5,887	2,273
Venezuela	South America	32,219,521	353,841	916,445	91	35
Vietnam	Asia	96,208,984	127,882	331,212	751	290
Yemen	Asia	28,915,284	175,676	455,000	166	64
Zambia	Africa	16,405,229	290,585	752,612	57	22
Zimbabwe	Africa	15,159,624	150,872	390,757	101	39

Notes: (1) Including the population and area of Autonomous Republic of Crimea and City of Sevastopol, Ukraine's administrative areas on the Crimean Peninsula, which are claimed by Russia; (2) Administration is split between Morocco and the Sahrawi Arab Democratic Republic (Western Sahara), both of which claim the entire territory; (3) Excludes Crimea.
Source: U.S. Census Bureau, International Data Base

PAST AND PROJECTED WORLD POPULATION GROWTH, 1950-2050

Year	Approximate World Population	Ten-Year Growth Rate (%)
1950	2,556,000,053	18.9
1960	3,039,451,023	22.0
1970	3,706,618,163	20.2
1980	4,453,831,714	18.5
1990	5,278,639,789	15.2
2000	6,082,966,429	12.6
2010	6,848,932,929	10.7
2020	7,584,821,144	8.7
2030	8,246,619,341	7.3
2040	8,850,045,889	5.6
2050	9,346,399,468	—

Note: The listed years are the baselines for the estimated ten-year growth rate figures; for example, the rate for 1950-1960 was 18.9%.
Source: U.S. Census Bureau, International Data Base

WORLD'S LARGEST COUNTRIES BY AREA

| | | | Area | |
Rank	Country	Region	Sq. Miles	Sq. Km.
1	Russia	Europe/Asia	6,612,093	17,125,242
2	Canada	North America	3,855,103	9,984,670
3	United States	North America	3,796,742	9,833,517
4	China	Asia	3,722,342	9,640,821
5	Brazil	South America	3,287,956	8,515,767
6	Australia	Australia	2,969,907	7,692,024
7	India	Asia	1,269,211	3,287,240
8	Argentina	South America	1,073,518	2,780,400
9	Kazakhstan	Asia	1,052,090	2,724,900
10	Algeria	Africa	919,595	2,381,741
11	Democratic Rep. of the Congo	Africa	905,446	2,345,095
12	Saudi Arabia	Asia	830,000	2,149,690
13	Mexico	North America	759,516	1,967,138
14	Indonesia	Asia	735,358	1,904,569
15	Sudan	Africa	710,251	1,839,542
16	Libya	Africa	683,424	1,770,060
17	Iran	Asia	636,372	1,648,195
18	Mongolia	Asia	603,902	1,564,100
19	Peru	South America	496,225	1,285,216
20	Chad	Africa	495,755	1,284,000
21	Mali	Africa	482,077	1,248,574
22	Angola	Africa	481,354	1,246,700
23	South Africa	Africa	471,359	1,220,813
24	Niger	Africa	458,075	1,186,408
25	Colombia	South America	440,831	1,141,748
26	Bolivia	South America	424,164	1,098,581
27	Ethiopia	Africa	410,678	1,063,652
28	Mauritania	Africa	397,955	1,030,700
29	Egypt	Africa	387,048	1,002,450
30	Tanzania	Africa	364,900	945,087

Source: U.S. Census Bureau, International Data Base

WORLD'S SMALLEST COUNTRIES BY AREA

Rank	Country	Region	Area Sq. Miles	Sq. Km.
1	Vatican City*	Europe	0.17	0.44
2	Monaco*	Europe	0.78	2.02
3	Nauru	Pacific Islands	8	21
4	Tuvalu	Pacific Islands	10	26
5	San Marino*	Europe	24	61
6	Liechtenstein*	Europe	62	160
7	Marshall Islands	Pacific Islands	70	181
8	Saint Kitts and Nevis	Central America	104	270
9	Maldives	Asia	115	298
10	Malta	Europe	122	315
11	Grenada	Caribbean	133	344
12	Saint Vincent and the Grenadines	Central America	150	389
13	Barbados	Caribbean	166	430
14	Antigua and Barbuda	Caribbean	171	442
15	Palau	Pacific Islands	171	444
16	Seychelles	Africa	176	455
17	Andorra*	Europe	179	464
18	Saint Lucia	Central America	238	617
19	Federated States of Micronesia	Pacific Islands	271	701
20	Tonga	Pacific Islands	278	720

Note: Asterisks (*) denote countries on continents; all other countries are islands or island groups.
Source: U.S. Census Bureau, International Data Base

World's Largest Countries by Population

Rank	Country	Region	Population
1	China	Asia	1,401,028,280
2	India	Asia	1,357,746,150
3	United States	North America	329,229,067
4	Indonesia	Asia	268,074,600
5	Pakistan	Asia	218,385,000
6	Brazil	South America	211,032,216
7	Nigeria	Africa	200,962,000
8	Bangladesh	Asia	167,983,726
9	Russia	Europe/Asia	146,877,088
10	Mexico	North America	126,577,691
11	Japan	Asia	126,020,000
12	Philippines	Asia	108,210,625
13	Ethiopia	Africa	107,534,882
14	Egypt	Africa	99,930,038
15	Vietnam	Asia	96,208,984
16	Democratic Rep. of the Congo	Africa	86,790,567
17	Germany	Europe	83,149,300
18	Iran	Asia	83,142,818
19	Turkey	Europe/Asia	82,003,882
20	France	Europe	67,060,000
21	Thailand	Asia	66,461,867
22	United Kingdom	Europe	66,435,600
23	Italy	Europe	60,252,824
24	South Africa	Africa	58,775,022
25	Tanzania	Africa	55,890,747
26	Myanmar	Asia	54,339,766
27	South Korea	Asia	51,811,167
28	Kenya	Africa	47,564,296
29	Spain	Europe	46,934,632
30	Colombia	South America	46,127,200

Source: U.S. Census Bureau, International Data Base

WORLD'S SMALLEST COUNTRIES BY POPULATION

Rank	Country	Region	Population
1	Vatican City	Europe	1,000
2	Tuvalu	Pacific Islands	10,200
3	Nauru	Pacific Islands	11,000
4	Palau	Pacific Islands	17,900
5	San Marino	Europe	34,641
6	Monaco	Europe	38,300
7	Liechtenstein	Europe	38,380
8	Marshall Islands	Pacific Islands	55,500
9	Saint Kitts and Nevis	Central America	56,345
10	Dominica	Caribbean	71,808
11	Andorra	Europe	76,177
12	Seychelles	Africa	96,762
13	Tonga	Pacific Islands	100,651
14	Antigua and Barbuda	Caribbean	104,084
15	Federated States of Micronesia	Pacific Islands	105,300
16	Grenada	Caribbean	108,825
17	Saint Vincent and the Grenadines	Central America	110,520
18	Kiribati	Pacific Islands	120,100
19	Saint Lucia	Central America	180,454
20	Samoa	Pacific Islands	199,052
21	São Tomé and Príncipe	Africa	201,784
22	Barbados	Caribbean	287,025
23	Vanuatu	Pacific Islands	304,500
24	Iceland	Europe	360,390
25	Maldives	Asia	378,114
26	Bahamas	Caribbean	386,870
27	Belize	Central America	398,050
28	Brunei	Asia	421,300
29	Malta	Europe	493,559
30	Cape Verde	Africa	550,483

Source: U.S. Census Bureau, International Data Base.

WORLD'S MOST DENSELY POPULATED COUNTRIES

Rank	Country	Region	Population	*Area* Sq. Miles	Sq. Km.	*Persons Per Square* Mile	Km.
1	Monaco	Europe	38,300	0.78	2.02	49,106	18,960
2	Singapore	Asia	5,638,700	279	722.5	20,212	7,804
3	Vatican City	Europe	1,000	0.17	0.44	5,887	2,273
4	Bahrain	Asia	1,543,300	300	778	5,136	1,983
5	Malta	Europe	493,559	122	315	3,911	1,510
6	Maldives	Asia	378,114	115	298	3,287	1,269
7	Bangladesh	Asia	167,983,726	55,598	143,998	3,021	1,167
8	Lebanon	Asia	6,855,713	4,036	10,452	1,740	672
9	Barbados	Caribbean	287,025	166	430	1,730	668
10	Taiwan	Asia	23,596,266	13,976	36,197	1,689	652
11	Mauritius	Africa	1,265,577	788	2,040	1,606	620
12	San Marino	Europe	34,641	24	61	1,471	568
13	Nauru	Pacific Islands	11,000	8	21	1,344	519
14	South Korea	Asia	51,811,167	38,691	100,210	1,339	517
15	Rwanda	Africa	12,374,397	10,169	26,338	1,217	470
16	Comoros	Africa	873,724	719	1,861	1,215	469
17	Netherlands	Europe	17,426,881	16,033	41,526	1,087	420
18	Haiti	Caribbean	11,263,077	10,450	27,065	1,077	416
19	Israel	Asia	9,149,500	8,522	22,072	1,074	415
20	India	Asia	1,357,746,150	1,269,211	3,287,240	1,070	413
21	Burundi	Africa	10,953,317	10,740	27,816	1,020	394
22	Tuvalu	Pacific Islands	10,200	10	26	1,015	392
23	Belgium	Europe	11,515,793	11,849	30,689	974	376
24	Philippines	Asia	108,210,625	115,831	300,000	934	361
25	Japan	Asia	126,020,000	145,937	377,975	862	333
26	Sri Lanka	Asia	21,803,000	25,332	65,610	860	332
27	El Salvador	Central America	6,704,864	8,124	21,040	826	319
28	Grenada	Caribbean	108,825	133	344	818	316
29	Marshall Islands	Pacific Islands	55,500	70	181	795	307
30	Saint Lucia	Central America	180,454	238	617	756	292

Source: U.S. Census Bureau, International Data Base

WORLD'S LEAST DENSELY POPULATED COUNTRIES

Rank	Country	Region	Population	Area Sq. Miles	Area Sq. Km.	Persons Per Square Mile	Persons Per Square Km.
1	Mongolia	Asia	3,000,000	603,902	1,564,100	4.9	1.9
2	Western Sahara	Africa	567,421	97,344	252,120	6	2.3
3	Namibia	Africa	2,413,643	318,580	825,118	7.5	2.9
4	Australia	Australia	25,594,366	2,969,907	7,692,024	9	3
5	Suriname	South America	568,301	63,251	163,820	9.1	3.5
6	Iceland	Europe	360,390	39,682	102,775	9.1	3.5
7	Guyana	South America	782,225	83,012	214,999	9.3	3.6
8	Libya	Africa	6,470,956	683,424	1,770,060	9.6	3.7
9	Canada	North America	37,898,384	3,855,103	9,984,670	10	4
10	Mauritania	Africa	3,984,233	397,955	1,030,700	10.4	4
11	Botswana	Africa	2,302,878	224,607	581,730	10.4	4
12	Kazakhstan	Asia	18,592,700	1,052,090	2,724,900	18	7
13	Central African Republic	Africa	4,737,423	240,324	622,436	21	8
14	Gabon	Africa	2,067,561	103,347	267,667	21	8
15	Russia	Europe/Asia	146,877,088	6,612,093	17,125,242	23	9
16	Bolivia	South America	11,307,314	424,164	1,098,581	26	10
17	Chad	Africa	15,353,184	495,755	1,284,000	31	12
18	Turkmenistan	Asia	5,851,466	189,657	491,210	31	12
19	Oman	Asia	4,183,841	119,499	309,500	36	14
20	Mali	Africa	19,107,706	482,077	1,248,574	39	15
21	Argentina	South America	44,938,712	1,073,518	2,780,400	41	16
22	Saudi Arabia	Asia	34,218,169	830,000	2,149,690	41	16
23	Republic of the Congo	Africa	5,399,895	132,047	342,000	41	16
24	Finland	Europe	5,527,405	130,666	338,424	41	16
25	Norway	Europe	5,328,212	125,013	323,782	41	16
26	Paraguay	South America	7,052,983	157,048	406,752	44	17
27	Belize	Central America	398,050	8,867	22,965	44	17
28	Algeria	Africa	42,545,964	919,595	2,381,741	47	18
29	Niger	Africa	21,466,863	458,075	1,186,408	47	18
30	Papua New Guinea	Pacific Islands	8,558,800	178,704	462,840	47	18

Source: U.S. Census Bureau, International Data Base

WORLD'S MOST POPULOUS CITIES

Rank	City	Country	Region	Population
1	Chongqing	China	Asia	30,484,300
2	Shanghai	China	Asia	24,256,800
3	Beijing	China	Asia	21,516,000
4	Chengdu	China	Asia	16,044,700
5	Karachi	Pakistan	Asia	14,910,352
6	Guangzhou	China	Asia	14,043,500
7	Istanbul	Turkey	Europe	14,025,000
8	Tokyo	Japan	Asia	13,839,910
9	Tianjin	China	Asia	12,784,000
10	Mumbai	India	Asia	12,478,447
11	São Paulo	Brazil	South America	12,252,023
12	Moscow	Russia	Europe/Asia	12,197,596
13	Kinshasa	Dem. Rep. of Congo	Africa	11,855,000
14	Baoding	China	Asia	11,194,372
15	Lahore	Pakistan	Asia	11,126,285
16	Wuhan	China	Asia	11,081,000
17	Delhi	India	Asia	11,034,555
18	Harbin	China	Asia	10,635,971
19	Suzhou	China	Asia	10,459,890
20	Cairo	Egypt	Africa	10,230,350
21	Seoul	South Korea	Asia	10,197,604
22	Jakarta	Indonesia	Asia	10,075,310
23	Lima	Peru	South America	9,174,855
24	Mexico City	Mexico	North America	9,041,395
25	Ho Chi Minh City	Vietnam	Asia	8,993,082
26	Dhaka	Bangladesh	Africa	8,906,039
27	London	United Kingdom	Europe	8,825,001
28	Bangkok	Thailand	Asia	8,750,600
29	Xi'an	China	Asia	8,705,600
30	New York	United States	North America	8,622,698
31	Bangalore	India	Asia	8,425,970
32	Shenzhen	China	Asia	8,378,900
33	Nanjing	China	Asia	8,230,000
34	Tehran	Iran	Asia	8,154,051
35	Rio de Janeiro	Brazil	South America	6,718,903
36	Shantou	China	Asia	5,391,028
37	Kolkata	India	Asia	4,486,679
38	Shijiazhuang	China	Asia	4,303,700
39	Los Angeles	United States	North America	3,884,307
40	Buenos Aires	Argentina	South America	3,054,300

MAJOR LAKES OF THE WORLD

Lake	Location	Surface Area		Maximum Depth	
		Sq. Mi.	Sq. Km.	Feet	Meters
Caspian Sea	Central Asia	152,239	394,299	3,104	946
Superior	North America	31,820	82,414	1,333	406
Victoria	East Africa	26,828	69,485	270	82
Huron	North America	23,010	59,596	750	229
Michigan	North America	22,400	58,016	923	281
Aral	Central Asia	13,000	33,800	223	68
Tanganyika	East Africa	12,700	32,893	4,708	1,435
Baikal	Russia	12,162	31,500	5,712	1,741
Great Bear	North America	12,000	31,080	270	82
Nyasa	East Africa	11,600	30,044	2,316	706
Great Slave	North America	11,170	28,930	2,015	614
Chad	West Africa	9,946	25,760	23	7
Erie	North America	9,930	25,719	210	64
Winnipeg	North America	9,094	23,553	204	62
Ontario	North America	7,520	19,477	778	237
Balkhash	Central Asia	7,115	18,428	87	27
Ladoga	Russia	7,000	18,130	738	225
Onega	Russia	3,819	9,891	361	110
Titicaca	South America	3,141	8,135	1,214	370
Nicaragua	Central America	3,089	8,001	230	70
Athabasca	North America	3,058	7,920	407	124
Rudolf	Kenya, East Africa	2,473	6,405	—	—
Reindeer	North America	2,444	6,330	—	—
Eyre	South Australia	2,400	6,216	varies	varies
Issyk-Kul	Central Asia	2,394	6,200	2,297	700
Urmia	Southwest Asia	2,317	6,001	49	15
Torrens	Australia	2,200	5,698	—	—
Vänern	Sweden	2,141	5,545	322	98
Winnipegosis	North America	2,086	5,403	59	18
Mobutu Sese Seko	East Africa	2,046	5,299	180	55
Nettilling	North America	1,950	5,051	—	—

Note: The sizes of some lakes vary with the seasons.

MAJOR RIVERS OF THE WORLD

River	Region	Source	Outflow	Approximate Length Miles	Km.
Nile	N. Africa	Tributaries of Lake Victoria	Mediterranean Sea	4,180	6,690
Mississippi-Missouri-Red Rock	N. America	Montana	Gulf of Mexico	3,710	5,970
Yangtze Kiang	East Asia	Tibetan Plateau	China Sea	3,602	5,797
Ob	Russia	Altai Mountains	Gulf of Ob	3,459	5,567
Yellow (Huang He)	East Asia	Kunlun Mountains, west China	Gulf of Chihli	2,900	4,667
Yenisei	Russia	Tannu-Ola Mountains, western Tuva, Russia	Arctic Ocean	2,800	4,506
Paraná	S. America	Confluence of Paranaiba and Grande Rivers	Río de la Plata	2,795	4,498
Irtysh	Russia	Altai Mountains, Russia	Ob River	2,758	4,438
Congo	Africa	Confluence of Lualaba and Luapula Rivers, Congo	Atlantic Ocean	2,716	4,371
Heilong (Amur)	East Asia	Confluence of Shilka and Argun Rivers	Tatar Strait	2,704	4,352
Lena	Russia	Baikal Mountains, Russia	Arctic Ocean	2,652	4,268
Mackenzie	N. America	Head of Finlay River, British Columbia, Canada	Beaufort Sea	2,635	4,241
Niger	West Africa	Guinea	Gulf of Guinea	2,600	4,184
Mekong	Asia	Tibetan Plateau	South China Sea	2,500	4,023
Mississippi	N. America	Lake Itasca, Minnesota	Gulf of Mexico	2,348	3,779
Missouri	N. America	Confluence of Jefferson, Gallatin, and Madison Rivers, Montana	Mississippi River	2,315	3,726
Volga	Russia	Valdai Plateau, Russia	Caspian Sea	2,291	3,687
Madeira	S. America	Confluence of Beni and Maumoré Rivers, Bolivia-Brazil boundary	Amazon River	2,012	3,238
Purus	S. America	Peruvian Andes	Amazon River	1,993	3,207
São Francisco	S. America	S.W. Minas Gerais, Brazil	Atlantic Ocean	1,987	3,198

River	Region	Source	Outflow	Approximate Length	
				Miles	Km.
Yukon	N. America	Junction of Lewes and Pelly Rivers, Yukon Terr., Canada	Bering Sea	1,979	3,185
St. Lawrence	N. America	Lake Ontario	Gulf of St. Lawrence	1,900	3,058
Rio Grande	N. America	San Juan Mountains, Colorado	Gulf of Mexico	1,885	3,034
Brahmaputra	Asia	Himalayas	Ganges River	1,800	2,897
Indus	Asia	Himalayas	Arabian Sea	1,800	2,897
Danube	Europe	Black Forest, Germany	Black Sea	1,766	2,842
Euphrates	Asia	Confluence of Murat Nehri and Kara Su Rivers, Turkey	Shatt-al-Arab	1,739	2,799
Darling	Australia	Eastern Highlands, Australia	Murray River	1,702	2,739
Zambezi	Africa	Western Zambia	Mozambique Channel	1,700	2,736
Tocantins	S. America	Goiás, Brazil	Pará River	1,677	2,699
Murray	Australia	Australian Alps, New S. Wales	Indian Ocean	1,609	2,589
Nelson	N. America	Head of Bow River, western Alberta, Canada	Hudson Bay	1,600	2,575
Paraguay	S. America	Mato Grosso, Brazil	Paraná River	1,584	2,549
Ural	Russia	Southern Ural Mountains, Russia	Caspian Sea	1,574	2,533
Ganges	Asia	Himalayas	Bay of Bengal	1,557	2,506
Amu Darya (Oxus)	Asia	Nicholas Range, Pamir Mountains, Turkmenistan	Aral Sea	1,500	2,414
Japurá	S. America	Andes, Colombia	Amazon River	1,500	2,414
Salween	Asia	Tibet, south of Kunlun Mountains	Gulf of Martaban	1,500	2,414
Arkansas	N. America	Central Colorado	Mississippi River	1,459	2,348
Colorado	N. America	Grand County, Colorado	Gulf of California	1,450	2,333
Dnieper	Russia	Valdai Hills, Russia	Black Sea	1,419	2,284
Ohio-Allegheny	N. America	Potter County, Pennsylvania	Mississippi River	1,306	2,102
Irrawaddy	Asia	Confluence of Nmai and Mali rivers, northeast Burma	Bay of Bengal	1,300	2,092
Orange	Africa	Lesotho	Atlantic Ocean	1,300	2,092

River	Region	Source	Outflow	Approximate Length Miles	Km.
Orinoco	S. America	Serra Parima Mountains, Venezuela	Atlantic Ocean	1,281	2,062
Pilcomayo	S. America	Andes Mountains, Bolivia	Paraguay River	1,242	1,999
Xi Jiang	East Asia	Eastern Yunnan Province, China	China Sea	1,236	1,989
Columbia	N. America	Columbia Lake, British Columbia, Canada	Pacific Ocean	1,232	1,983
Don	Russia	Tula, Russia	Sea of Azov	1,223	1,968
Sungari	East Asia	China-North Korea boundary	Amur River	1,215	1,955
Saskatchewan	N. America	Canadian Rocky Mountains	Lake Winnipeg	1,205	1,939
Peace	N. America	Stikine Mountains, British Columbia, Canada	Great Slave River	1,195	1,923
Tigris	Asia	Taurus Mountains, Turkey	Shatt-al-Arab	1,180	1,899

HIGHEST PEAKS IN EACH CONTINENT

Continent	Mountain	Location	Height Feet	Height Meters
Asia	Everest	Tibet & Nepal	29,028	8,848
South America	Aconcagua	Argentina	22,834	6,960
North America	McKinley	Alaska	20,320	6,194
Africa	Kilimanjaro	Tanzania	19,340	5,895
Europe	Elbrus	Russia & Georgia	18,510	5,642
Antarctica	Vinson Massif	Ellsworth Mountains	16,066	4,897
Australia	Kosciusko	New South Wales	7,316	2,228

Note: The world's highest sixty-six mountains are all in Asia.

MAJOR DESERTS OF THE WORLD

MAJOR DESERTS OF THE WORLD

Desert	Location	Approximate Area		Type
		Sq. Miles	Sq. Km.	
Antarctic	Antarctica	5,400,000	14,002,200	polar
Sahara	North Africa	3,500,000	9,075,500	subtropical
Arabian	Southwest Asia	1,000,000	2,593,000	subtropical
Great Western (Gibson, Great Sandy, and Great Victoria)	Australia	520,000	1,348,360	subtropical
Gobi	East Asia	500,000	1,296,500	cold winter
Patagonian	Argentina, South America	260,000	674,180	cold winter
Kalahari	Southern Africa	220,000	570,460	subtropical
Great Basin	Western United States	190,000	492,670	cold winter
Thar	South Asia	175,000	453,775	subtropical
Chihuahuan	Mexico	175,000	453,775	subtropical
Karakum	Central Asia	135,000	350,055	cold winter
Colorado Plateau	Southwestern United States	130,000	337,090	cold winter
Sonoran	United States and Mexico	120,000	311,160	subtropical
Kyzylkum	Central Asia	115,000	298,195	cold winter
Taklimakan	China	105,000	272,265	cold winter
Iranian	Iran	100,000	259,300	cold winter
Simpson	Eastern Australia	56,000	145,208	subtropical
Mojave	Western United States	54,000	140,022	subtropical
Atacama	Chile, South America	54,000	140,022	cold coastal
Namib	Southern Africa	13,000	33,709	cold coastal
Arctic	Arctic Circle			polar

HIGHEST WATERFALLS OF THE WORLD

Waterfall	Location	Source	Height Feet	Meters
Angel	Canaima National Park, Venezuela	Rio Caroni	3,212	979
Tugela	Natal National Park, South Africa	Tugela River	3,110	948
Utigord	Norway	glacier	2,625	800
Monge	Marstein, Norway	Mongebeck	2,540	774
Mutarazi	Nyanga National Park, Zimbabwe	Mutarazi River	2,499	762
Yosemite	Yosemite National Park, California, U.S.	Yosemite Creek	2,425	739
Espelands	Hardanger Fjord, Norway	Opo River	2,307	703
Lower Mar Valley	Eikesdal, Norway	Mardals Stream	2,151	655
Tyssestrengene	Odda, Norway	Tyssa River	2,123	647
Cuquenan	Kukenan Tepuy, Venezuela	Cuquenan River	2,000	610
Sutherland	Milford Sound, New Zealand	Arthur River	1,904	580
Kjell	Gudvanger, Norway	Gudvangen Glacier	1,841	561
Takkakaw	Yoho Natl Park, British Columbia, Canada	Takkakaw Creek	1,650	503
Ribbon	Yosemite National Park, California, U.S.	Ribbon Stream	1,612	491
Upper Mar Valley	near Eikesdal, Norway	Mardals Stream	1,536	468
Gavarnie	near Lourdes, France	Gave de Pau	1,388	423
Vettis	Jotunheimen, Norway	Utla River	1,215	370
Hunlen	British Columbia, Canada	Hunlen River	1,198	365
Tin Mine	Kosciusko National Park, Australia	Tin Mine Creek	1,182	360
Silver Strand	Yosemite National Park, California, U.S.	Silver Strand Creek	1,170	357
Basaseachic	Baranca del Cobre, Mexico	Piedra Volada Creek	1,120	311
Spray Stream	Lauterburnnental, Switzerland	Staubbach Brook	985	300
Fachoda	Tahiti, French Polynesia	Fautaua River	985	300
King Edward VIII	Guyana	Courantyne River	850	259
Wallaman	near Ingham, Australia	Wallaman Creek	844	257
Gersoppa	Western Ghats, India	Sharavati River	828	253
Kaieteur	Guyana	Rio Potaro	822	251
Montezuma	near Rosebery, Tasmania	Montezuma River	800	240
Wollomombi	near Armidale, Australia	Wollomombi River	722	2203

Source: Fifth Continent Australia Pty Limited

Glossary

Places whose names are printed in SMALL CAPS *are subjects of their own entries in this glossary.*

Ablation. Loss of ice volume or mass by a GLACIER. Ablation includes melting of ice, SUBLIMATION, DEFLATION (removal by WIND), EVAPORATION, and CALVING. Ablation occurs in the lower portions of glaciers.

Abrasion. Wearing away of ROCKS in STREAMS by grinding, especially when rocks and SEDIMENT are carried along by stream water. The STREAMBED and VALLEY are carved out and eroded, and the rocks become rounded and smoothed by abrasion.

Absolute location. Position of any PLACE on the earth's surface. The absolute location can be given precisely in terms of DEGREES, MINUTES, and SECONDS of LATITUDE (0 to 90 degrees north or south) and of LONGITUDE (0 to 180 degrees east or west). The EQUATOR is 0 degrees latitude; the PRIME MERIDIAN, which runs through Greenwich in England, is 0 degrees longitude.

Abyss. Deepest part of the OCEAN. Modern TECHNOLOGY—especially sonar—has enabled accurate mapping of the ocean floors, showing that there are MOUNTAIN CHAINS, or RIDGES, in all the oceans, as well as deep CANYONS or TRENCHES closer to the edges of the oceans.

Acid rain. PRECIPITATION containing high levels of nitric or sulfuric acid; a major environmental problem in parts of North America, Europe, and Asia. Natural precipitation is slightly acidic (about 5.6 on the pH SCALE), because CARBON DIOXIDE—which occurs naturally in the ATMOSPHERE—is dissolved to form a weak carbonic acid.

Adiabatic. Change of TEMPERATURE within the ATMOSPHERE that is caused by compression or expansion without addition or loss of heat.

Advection. Horizontal movement of AIR from one PLACE to another in the ATMOSPHERE, associated with WINDS.

Advection fog. FOG that forms when a moist AIR mass moves over a colder surface. Commonly, warm moist air moves over a cool OCEAN CURRENT, so the air cools to SATURATION POINT and fog forms. This phenomenon, known as sea fog, occurs along subtropical west COASTS.

Aerosol. Substances held in SUSPENSION in the ATMOSPHERE, as solid particles or liquid droplets.

Aftershock. EARTHQUAKE that follows a larger earthquake and originates at or near the focus of the latter; many aftershocks may follow a major earthquake, decreasing in frequency and magnitude with time.

Agglomerate. Type of ROCK composed of volcanic fragments, usually of different sizes and rough or angular.

Aggradation. Accumulation of SEDIMENT in a STREAMBED. Aggradation often results from reduced flow in the channel during dry periods. It also occurs when the STREAM's load (BEDLOAD and SUSPENDED LOAD) is greater than the stream capacity. A BRAIDED STREAM pattern often results.

Air current. Air currents are caused by differential heating of the earth's surface, which causes heated air to rise. This causes WINDS at the surface as well as higher in the earth's ATMOSPHERE.

Air mass. Large body of air with distinctive homogeneous characteristics of TEMPERATURE, HUMIDITY, and stability. It forms when air remains stationary over a source REGION for a period of time, taking on the conditions of that region. An air mass can extend over a million square miles with a depth of more than a mile. Air masses are classified according to moisture content (*m* for maritime or *c* for continental) and temperature

(*A* for ARCTIC, *P* for polar, *T* for tropical, or *E* for equatorial). The air masses affecting North America are mP, cP, and mT. The interaction of AIR masses produces WEATHER. The line along which air masses meet is a FRONT.

Albedo. Measure of the reflective properties of a surface; the ratio of reflected ENERGY (INSOLATION) to the total incoming energy, expressed as a percentage. The albedo of Earth is 33 percent.

Alienation (land). Land alienation is the appropriation of land from its original owners by a more powerful force. In preindustrial societies, the ownership of agricultural land is of prime importance to subsistence farmers.

Alkali flat. Dry LAKEBED in an arid REGION, covered with a layer of SALTS. A well-known example is the Alkali Flat area of White Sands National Monument in New Mexico; it is the bed of a large lake that formed when the GLACIERS were melting. It is covered with a form of gypsum crystals called selenite. This material is blown off the surface into large SAND DUNES. Also called a salina. See also BITTER LAKE.

Allogenic sediment. SEDIMENT that originates outside the PLACE where it is finally deposited; SAND, SILT, and CLAY carried by a STREAM into a LAKE are examples.

Alluvial fan. Common LANDFORM at the mouth of a CANYON in arid REGIONS. Water flowing in a narrow canyon immediately slows as it leaves the canyon for the wider VALLEY floor, depositing the SEDIMENTS it was transporting. These spread out into a fan shape, usually with a BRAIDED STREAM pattern on its surface. When several alluvial fans grow side by side, they can merge into one continuous sloping surface between the HILLS and the valley. This is known by the Spanish word *bajada*, which means "slope."

Alluvial system. Any of various depositional systems, excluding DELTAS, that form from the activity of RIVERS and STREAMS. Much alluvial SEDIMENT is deposited when rivers top their BANKS and FLOOD the surrounding countryside. Buried alluvial sediments may be important water-bearing RESERVOIRS or may contain PETROLEUM.

Alluvium. Material deposited by running water. This includes not only fertile SOILS, but also CLAY, SILT, or SAND deposits resulting from FLUVIAL processes. FLOODPLAINS are covered in a thick layer of alluvium.

Altimeter. Instrument for measuring ALTITUDE, or height above the earth's surface, commonly used in airplanes. An altimeter is a type of ANEROID BAROMETER.

Altitudinal zonation. Existence of different ECOSYSTEMS at various ELEVATIONS above SEA LEVEL, due to TEMPERATURE and moisture differences. This is especially pronounced in Central America and South America. The hot and humid COASTAL PLAINS, where bananas and sugarcane thrive, is the *tierra caliente*. From about 2,500 to 6,000 feet (750–1,800 meters) is the *tierra templada*; crops grown here include coffee, wheat, and corn, and major cities are situated in this zone. From about 6,000 to 12,000 feet (1,800–3,600 meters) is the *tierra fria*; here only hardy crops such as potatoes and barley are grown, and large numbers of animals are kept. From about 12,000 to 15,000 feet (3,600 to 4,500 meters) lies the *tierra helada*, where hardy animals such as sheep and alpaca graze. Above 15,000 feet (4,500 meters) is the frozen *tierra nevada*; no permanent life is possible in the permanent SNOW and ICE FIELDS there.

Angle of repose. Maximum angle of steepness that a pile of loose materials such as SAND or ROCK can assume and remain stable; the angle varies with the size, shape, moisture, and angularity of the material.

Antecedent river. STREAM that was flowing before the land was uplifted and was able to erode at the pace of UPLIFT, thus creating a deep CANYON. Most deep canyons are attributed to antecedent rivers. In the Davisian CYCLE OF EROSION, this process was called REJUVENATION.

Anthropogeography. Branch of GEOGRAPHY founded in the late nineteenth century by German geographer Friedrich Ratzel. The field is closely related to human ECOLOGY—the study of humans, their DISTRIBUTION over the earth, and their interaction with their physical ENVIRONMENT.

Anticline. Area where land has been UPFOLDED symmetrically. Its center contains stratigraphically older ROCKS. See also SYNCLINE.

Anticyclone. High-pressure system of rotating WINDS, descending and diverging, shown on a WEATHER chart by a series of closed ISOBARS, with a high in the center. In the NORTHERN HEMISPHERE, the rotation is CLOCKWISE; in the SOUTHERN HEMISPHERE, the rotation is COUNTERCLOCKWISE. An anticyclone brings warm weather.

Antidune. Undulatory upstream-moving bed form produced in free-surface flow of water over a SAND bed in a certain RANGE of high flow speeds and shallow flow depths.

Antipodes. TEMPERATE ZONE of the SOUTHERN HEMISPHERE. The term is now usually applied to the countries of Australia and New Zealand. The ancient Greeks had believed that if humans existed there, they must walk upside down. This idea was supported by the Christian Church in the Middle Ages.

Antitrade winds. WINDS in the upper ATMOSPHERE, or GEOSTROPHIC winds, that blow in the opposite direction to the TRADE WINDS. Antitrade winds blow toward the northeast in the NORTHERN HEMISPHERE and toward the southeast in the SOUTHERN HEMISPHERE.

Aperiodic. Irregularly occurring interval, such as found in most WEATHER CYCLES, rendering them virtually unpredictable.

Aphelion. Point in the earth's 365-DAY REVOLUTION when it is at its greatest distance from the SUN. This is caused by Earth's elliptical ORBIT around the Sun. The distance at aphelion is 94,555,000 miles (152,171,500 km.) and usually falls on July 4. The opposite of PERIHELION.

Aposelene. Earth's farthest point from the MOON.

Aquifer. Underground body of POROUS ROCK that contains water and allows water PERCOLATION through it. The largest aquifer in the United States is the Ogallala Aquifer, which extends south from South Dakota to Texas.

Arête. Serrated or saw-toothed ridge, produced in glaciated MOUNTAIN areas by CIRQUES eroding on either side of a RIDGE or mountain RANGE. From the French word for knife-edge.

Arroyo. Spanish word for a dry STREAMBED in an arid area. Called a WADI in Arabic and a WASH in English.

Artesian well. WELL from which GROUNDWATER flows without mechanical pumping, because the water comes from a CONFINED AQUIFER, and is therefore under pressure. The Great Artesian Basin of Australia has hundreds of artesian wells, called BORES, that provide drinking water for sheep and cattle. The name comes from the Artois REGION of France, where the phenomenon is common. A subartesian well is sunk into an UNCONFINED AQUIFER and requires a pump to raise water to the surface.

Asteroid belt. REGION between the ORBITS of Mars and Jupiter containing the majority of ASTEROIDS.

Asthenosphere. Part of the earth's UPPER MANTLE, beneath the LITHOSPHERE, in which PLATE movement takes place. Also known as the low-velocity zone.

Astrobleme. Remnant of a large IMPACT CRATER on Earth.

Astronomical unit (AU). Unit of measure used by astronomers that is equivalent to the average distance from the SUN to Earth (93 million miles/150 million km.).

Atmospheric pressure. Weight of the earth's ATMOSPHERE, equally distributed over earth's surface and pressing down as a result of GRAVITY. On average, the atmosphere has a force of 14.7 pounds per square inch (1 kilogram per centimeter) squared at SEA LEVEL, also expressed as 1013.2 millibars. Variations in atmospheric pressure, high or low, cause WINDS and WEATHER changes that affect CLIMATE. Pressure decreases rapidly with ALTITUDE or distance from the surface: Half of the total atmosphere is found below 18,000 feet (5,500 meters); more than 99 percent of the atmosphere is below 30 miles (50 km.) of the surface. Atmospheric pressure is measured with a BAROMETER.

Atoll. Ring-shaped growth of CORAL REEF, with a LAGOON in the middle. Charles Darwin, who observed many Pacific atolls during his voyage on the *Beagle* in the nineteenth century, suggested that they were created from FRINGING REEFS around volcanic ISLANDS. As such islands sank beneath the water (or as SEA LEVELS rose), the coral continued growing upward. SAND resting atop an atoll enables plants to grow, and small human societies have arisen on some atolls. The world's largest atoll, Kwajalein in the Marshall Islands, measures about 40 by 18 miles (65 by 30 km.), but perhaps the most famous atoll is Bikini Atoll—the SITE of nuclear-bomb testing during the 1950s.

Aurora. Glowing and shimmering displays of colored lights in the upper ATMOSPHERE, caused by interaction of the SOLAR WIND and the charged particles of the IONOSPHERE. Auroras occur at high LATITUDES. Near the North Pole they are called aurora borealis or northern lights; near the South Pole, aurora australis or southern lights.

Austral. Referring to an object or occurrence that is located in the SOUTHERN HEMISPHERE or related to Australia.

Australopithecines. Erect-walking early human ancestors with a cranial capacity and body size within the RANGE of modern apes rather than of humans.

Avalanche. Mass of SNOW and ice falling suddenly down a MOUNTAIN slope, often taking with it earth, ROCKS, and trees.

Bank. Elevated area of land beneath the surface of the OCEAN. The term is also used for elevated ground lining a body of water.

Bar (climate). Measure of ATMOSPHERIC PRESSURE per unit surface area of 1 million dynes per square centimeter. Millibars (thousandths of a bar) are the MEASUREMENT used in the United States. Other countries use kilopascals (kPa); one kilopascal is ten millibars.

Bar (land). RIDGE or long deposit of SAND or gravel formed by DEPOSITION in a RIVER or at the COAST. Offshore bars and baymouth bars are common coastal features.

Barometer. Instrument used for measuring ATMOSPHERIC PRESSURE. In the seventeenth century, Evangelista Torricelli devised the first barometer—a glass tube sealed at one end, filled with mercury, and upended into a bowl of mercury. He noticed how the height of the mercury column changed and realized this was a result of the pressure of air on the mercury in the bowl. Early MEASUREMENTS of atmospheric pressure were, therefore, expressed as centimeters of mercury, with average pressure at SEA LEVEL being 29.92 inches (760 millimeters). This cumbersome barometer was replaced with the ANEROID BAROMETER—a sealed and partially evacuated box connected to a needle and dial, which shows changes in atmospheric pressure. See also ALTIMETER.

Barrier island. Long chain of SAND islands that forms offshore, close to the COAST. LAGOONS or shallower MARSHES separate the barrier islands from the mainland. Such LOCATIONS are hazardous for SETTLEMENTS because they are easily swept away in STORMS and HURRICANES.

Basalt. IGNEOUS EXTRUSIVE ROCK formed when LAVA cools; often black in color. Sometimes basalt occurs in tall hexagonal columns, such as the Giant's Causeway in Ireland, or the Devils Postpile at Mammoth, California.

Basement. Crystalline, usually PRECAMBRIAN, IGNEOUS and METAMORPHIC ROCKS that occur beneath the SEDIMENTARY ROCK on the CONTINENTS.

Basin order. Approximate measure of the size of a STREAM BASIN, based on a numbering scheme applied to RIVER channels as they join together in their progress downstream.

Batholith. Large LANDFORM produced by IGNEOUS INTRUSION, composed of CRYSTALLINE ROCK, such as GRANITE; a large PLUTON with a surface area greater than 40 square miles (100 sq. km.). Most mountain RANGES have a batholith underneath.

Bathymetric contour. Line on a MAP of the OCEAN floor that connects points of equal depth.

Beaufort scale. SCALE that measures WIND force, expressed in numbers from 0 to 12. The original Beaufort scale was based on descriptions of the state of the SEA. It was adapted to land conditions, using descriptions of chimney smoke, leaves of trees, and similar factors. The scale was devised in the early nineteenth century by Sir Francis Beaufort, a British naval officer.

Belt. Geographical REGION that is distinctive in some way.

Bergeron process. PRECIPITATION formation in COLD CLOUDS whereby ice crystals grow at the expense of supercooled water droplets.

Bight. Wide or open BAY formed by a curve in the COASTLINE, such as the Great Australian Bight.

Biogenic sediment. SEDIMENT particles formed from skeletons or shells of microscopic plants and animals living in seawater.

Biostratigraphy. Identification and organization of STRATA based on their FOSSIL content and the use of fossils in stratigraphic correlation.

Bitter lake. Saline or BRACKISH LAKE in an arid area, which may dry up in the summer or in periods of DROUGHT. The water is not suitable for drinking. Another name for this feature is "salina." See also ALKALI FLAT.

Block lava. LAVA flows whose surfaces are composed of large, angular blocks; these blocks are generally larger than those of AA flows and have smooth, not jagged, faces.

Block mountain. MOUNTAIN or mountain RANGE with one side having a gentle slope to the crest, while the other slope, which is the exposed FAULT SCARP, is quite steep. It is formed when a large block of the earth's CRUST is thrust upward on one side only, while the opposite side remains in place. The Sierra Nevada in California are a good example of block mountains. Also known as fault-block mountain.

Blowhole. SEA CAVE or tunnel formed on some rocky, rugged COASTLINES. The pressure of the seawater rushing into the opening can force a jet of seawater to rise or spout through an opening in the roof of the cave. Blowholes are found in Scotland, Tasmania, and Mexico, and on the Hawaiian ISLANDS of Kauai and Maui.

Bluff. Steep slope that marks the farthest edge of a FLOODPLAIN.

Bog. Damp, spongy ground surface covered with decayed or decaying VEGETATION. Bogs usually are formed in cool CLIMATES through the in-filling, or silting up, of a LAKE. Moss and other plants grow outward toward the edge of the lake, which gradually becomes shallower, until the surface is completely covered. Bogs also can form on cold, damp MOUNTAIN surfaces. Many bogs are filled with PEAT.

Bore. Standing WAVE, or wall, of water created in a narrow ESTUARY when the strong incoming, or FLOOD, TIDE meets the RIVER water flowing outward; it moves upstream with the advancing tide, and downstream with the EBB TIDE. South America's Amazon River and Asia's Mekong River have large bores. In North America, the bore in the Bay of Fundy is visited by many tourists each year. Its St. Andrew's wharf is designed to handle changes in water level of as much as 53 feet (15 meters) in one DAY.

Boreal. Alluding to an item or event that is in the NORTHERN HEMISPHERE.

Bottom current. Deep-sea current that flows parallel to BATHYMETRIC CONTOURS.

Brackish water. Water with SALT content between that of SALT WATER and FRESH WATER; it is common in arid areas on the surface, in coastal MARSHES, and in salt-contaminated GROUNDWATER.

Braided stream. STREAM having a CHANNEL consisting of a maze of interconnected small channels within a broader STREAMBED. Braiding occurs when the stream's load exceeds its capacity, usually because of reduced flow.

Breaker. WAVE that becomes oversteepened as it approaches the SHORE, reaching a point at which it cannot maintain its vertical shape. It then breaks, and the water washes toward the shore.

Breakwater. Large structure, usually of ROCK, built offshore and parallel to the COAST, to absorb WAVE ENERGY and thus protect the SHORE. Between the breakwater and the shore is an area of calm water, often used as a boat anchorage or

HARBOR. A similar but smaller structure is a seawall.

Breeze. Gentle WIND with a speed of 4 to 31 miles (6 to 50 km.) per hour. On the BEAUFORT SCALE, the numbers 2 through 6 represent breezes of increasing strength.

Butte. Flat-topped HILL, smaller than a MESA, found in arid REGIONS.

Caldera. Large circular depression with steep sides, formed when a VOLCANO explodes, blowing away its top. The ERUPTION of Mount St. Helens produced a caldera. Crater Lake in Oregon is a caldera that has filled with water. From the Spanish word for kettle.

Calms of Cancer. Subtropical BELT of high pressure and light WINDS, located over the OCEAN near 25 DEGREES north LATITUDE. Also known as the HORSE LATITUDES.

Calms of Capricorn. Subtropical BELT of high pressure and light WINDS, located over the OCEAN near 25 DEGREES south LATITUDE.

Calving. Loss of glacial mass when GLACIERS reach the SEA and large blocks of ice break off, forming ICEBERGS.

Cancer, tropic of. PARALLEL of LATITUDE at 23.5 DEGREES north; this line is the latitude farthest north on the earth where the noon SUN is ever directly overhead. The REGION between it and the tropic of CAPRICORN is known as the TROPICS.

Capricorn, tropic of. Line of LATITUDE at 23.5 DEGREES south; this line is the latitude farthest south on the earth where the noon SUN is ever directly overhead. The REGION between it and the tropic of CANCER is known as the TROPICS.

Carbon dating. Method employed by physicists to determine the age of organic matter—such as a piece of wood or animal tissue—to determine the age of an archaeological or paleontological SITE. The method works on the principle that the amount of radioactive carbon in living matter diminishes at a steady, and measurable, rate after the matter dies. Technique is also known as carbon-14 dating, after the radioactive carbon-14 isotope it uses. Also known as radiocarbon dating.

Carrying capacity. Number of animals that a given area of land can support, without additional feed being necessary. Lush GRASSLAND may have a carrying capacity of twenty sheep per acre, while more arid, SEMIDESERT land may support only two sheep per acre. The term sometimes is used to refer to the number of humans who can be supported in a given area.

Catastrophism. Theory, popular in the eighteenth and nineteenth centuries, that explained the shape of LANDFORMS and CONTINENTS and the EXTINCTION of species as the results of intense or catastrophic events. The biblical FLOOD of Noah was one such event, which supposedly explained many extinctions. Catastrophism is linked closely to the belief that the earth is only about 6,000 years old, and therefore tremendous forces must have acted swiftly to create present LANDSCAPES. An alternative or contrasting theory is UNIFORMITARIANISM.

Catchment basin. Area of land receiving the PRECIPITATION that flows into a STREAM. Also called catchment or catchment area.

Central place theory. Theory that explains why some SETTLEMENTS remain small while others grow to be middle-sized TOWNS, and a few become large cities or METROPOLISES. The explanation is based on the provision of goods and services and how far people will travel to acquire these. The German geographer Walter Christaller developed this theory in the 1930s.

Centrality. Measure of the number of functions, or services, offered by any CITY in a hierarchy of cities within a COUNTRY or a REGION. See also CENTRAL PLACE THEORY.

Chain, mountain. Another term for mountain RANGE.

Chemical farming. Application of artificial FERTILIZERS to the SOIL and the use of chemical products such as insecticides, fungicides, and herbicides to ensure crop success. Chemical farming is practiced mainly in high-income countries, because the cost of the chemical products is high. Farmers in low-income economies rely more on natural organic fertilizers such as animal waste.

Chemical weathering. Chemical decomposition of solid ROCK by processes involving water that change its original materials into new chemical combinations.

Chlorofluorocarbons (CFCs). Manufactured compounds, not occurring in nature, consisting of chlorine, fluorine, and carbon. CFCs are stable and have heat-absorbing properties, so they have been used extensively for cooling in refrigeration and air-conditioning units. Previously, they were used as propellants for aerosol products. CFCs rise into the STRATOSPHERE where ULTRAVIOLET RADIATION causes them to react with OZONE, changing it to oxygen and exposing the earth to higher levels of ultraviolet (UV) radiation. Therefore, the manufacture and use of CFCs was banned in many countries. The commercial name for CFCs is Freon.

Chorology. Description or mapping of a REGION. Also known as chorography.

Chronometer. Highly accurate CLOCK or timekeeping device. The first accurate and effective chronometers were constructed in the mid-eighteenth century by John Harrison, who realized that accurate timekeeping was the secret to NAVIGATION at SEA.

Cinder cone. Small conical HILL produced by PYROCLASTIC materials from a VOLCANO. The material of the cone is loose SCORIA.

Circle of illumination. Line separating the sunlit part of the earth from the part in darkness. The circle of illumination moves around the earth once in every approximately 24 hours. At the VERNAL and autumnal EQUINOXES, the circle of illumination passes through the POLES.

Cirque. Circular BASIN at the head of an ALPINE GLACIER, shaped like an armchair. Many cirques can be seen in MOUNTAIN areas where glaciers have completely melted since the last ICE AGE.

City Beautiful movement. Planning and architectural movement that was at its height from around 1890 to the 1920s in the United States. It was believed that classical architecture, wide and carefully laid-out streets, parks, and urban monuments would reflect the higher values of the society and be a civilizing, even uplifting, experience for the citizens of such cities. Civic pride was fostered through remodeling or modernizing older URBAN AREAS. Chicago, Illinois, and Pasadena, California, are cities where the planners of the City Beautiful movement left their imprint.

Clastic. ROCK or sedimentary matter formed from fragments of older rocks.

Climatology. Study of Earth CLIMATES by analysis of long-term WEATHER patterns over a minimum of thirty years of statistical records. Climatologists—scientists who study climate—seek similarities to enable grouping into climatic REGIONS. Climate patterns are closely related to natural VEGETATION. Computer TECHNOLOGY has enabled investigation of phenomena such as the EL NIÑO effect and global climate change. The KÖPPEN CLIMATE CLASSIFICATION system is the most commonly used scheme for climate classification.

Climograph. Graph that plots TEMPERATURE and PRECIPITATION for a selected LOCATION. The most commonly used climographs plot monthly temperatures and monthly precipitation, as used in the KÖPPEN CLIMATE CLASSIFICATION. Also spelled "climagraph." The term climagram is rarely used.

Clinometer. Instrument used by surveyors to measure the ELEVATION of land or the inclination (slope) of the land surface.

Cloud seeding. Injection of CLOUD-nucleating particles into likely clouds to enhance PRECIPITATION.

Cloudburst. Heavy rain that falls suddenly.

Coal. One of the FOSSIL FUELS. Coal was formed from fossilized plant material, which was originally FOREST. It was then buried and compacted, which led to chemical changes. Most coal was formed during the CARBONIFEROUS PERIOD (286 million to 360 million years ago) when the earth's CLIMATE was wetter and warmer than at present.

Coastal plain. Large area of flat land near the OCEAN. Coastal plains can form in various ways, but FLUVIAL DEPOSITION is an important process. In the United States, the coastal plain extends from Texas to North Carolina.

Coastal wetlands. Shallow, wet, or flooded shelves that extend back from the freshwater-saltwater interface and may consist of MARSHES, BAYS, LAGOONS, tidal flats, or MANGROVE SWAMPS.

Cognitive map. Mental image that each person has of the world, which includes LOCATIONS and connections. These maps expand as children mature, from plans of their rooms, to their houses, to their neighborhoods. Adults know certain parts of the CITY and the streets connecting them.

Coke. Type of fuel produced by heating COAL.

Col. Lower section of a RIDGE, usually formed by the headward EROSION of two CIRQUE GLACIERS at an ARÊTE. Sometimes called a saddle.

Colonialism. Control of one COUNTRY over another STATE and its people. Many European countries have created colonial empires, including Great Britain, France, Spain, Portugal, the Netherlands, and Russia.

Columbian exchange. Interaction that occurred between the Americas and Europe after the voyages of Christopher Columbus. Food crops from the New World transformed the diet of many European countries.

Comet. Small body in the SOLAR SYSTEM, consisting of a solid head with a long gaseous tail. The elliptical ORBIT of a comet causes it to range from very close to the SUN to very far away. In ancient times, the appearance of a comet in the sky was thought to be an omen of great events or changes, such as war or the death of a king.

Comfort index. Number that expresses the combined effects of TEMPERATURE and HUMIDITY on human bodily comfort. The index number is obtained by measuring ambient conditions and comparing these to a chart.

Commodity chain. Network linking labor, production, delivery, and sale for any product. The chain begins with the production of the raw material, such as the extraction of MINERALS by miners, and extends to the acquisition of the finished product by a consumer.

Complex crater. IMPACT CRATER of large diameter and low depth-to-diameter ratio caused by the presence of a central UPLIFT or ring structure.

Composite cone. Cone or VOLCANO formed by volcanic explosions in which the LAVA is of different composition, sometimes fluid, sometimes PYROCLASTS such as cinders. The alternation of layers allows a concave shape for the cone. These are generally regarded as the world's most beautiful volcanoes. Composite volcanoes are sometimes called STRATOVOLCANOES.

Condensation nuclei. Microscopic particles that may have originated as DUST, soot, ASH from fires or VOLCANOES, or even SEA SALT; an essential part of CLOUD formation. When AIR rises and cools to the DEW POINT (saturation), the moisture droplets condense around the nuclei, leading to the creation of raindrops or snowflakes. A typical air mass might contain 10 billion condensation nuclei in a single cubic yard (1 cubic meter) of air.

Cone of depression. Cone-shaped depression produced in the WATER TABLE by pumping from a WELL.

Confined aquifer. AQUIFER that is completely filled with water and whose upper BOUNDARY is a CONFINING BED; it is also called an artesian aquifer.

Confining bed. Impermeable layer in the earth that inhibits vertical water movement.

Confluence. PLACE where two STREAMS or RIVERS flow together and join. The smaller of the two streams is called a TRIBUTARY.

Conglomerate. Type of SEDIMENTARY ROCK consisting of smaller rounded fragments naturally cemented together by another MINERAL. If the cemented fragments are jagged or angular, the rock is called breccia.

Conical projection. MAP PROJECTION that can be imagined as a cone of paper resting like a witch's hat on a globe with a light source at its center; the images of the CONTINENTS would be projected onto the paper. In reality, maps are constructed mathematically. A conic projection can show only part of one HEMISPHERE. This projection is suitable for constructing a MAP of the United States, as a good EQUAL-AREA representation can be achieved. Also called conic projection.

Consequent river. RIVER that flows across a LANDSCAPE because of GRAVITY. Its direction is determined by the original slope of the land. TRIBUTARY streams, which develop later as EROSION proceeds, are called subsequent streams.

Continental climate. CLIMATE experienced over the central REGIONS of large LANDMASSES; drier and subject to greater seasonal extremes of TEMPERATURE than at the CONTINENTAL MARGINS.

Continental rift zones. Continental rift zones are PLACES where the CONTINENTAL CRUST is stretched and thinned. Distinctive features include active VOLCANOES and long, straight VALLEY systems formed by normal FAULTS. Continental rifting in some cases has evolved into the breaking apart of a CONTINENT by SEAFLOOR SPREADING to form a new OCEAN.

Continental shelf. Shallow, gently sloping part of the seafloor adjacent to the mainland. The continental shelf is geologically part of the CONTINENT and is made of CONTINENTAL CRUST, whereas the OCEAN floor is OCEANIC CRUST. Although continental shelves vary greatly in width, on average they are about 45 miles (75 km.) wide and have slopes of 7 minutes (about one-tenth of a degree). The average depth of a continental shelf is about 200 feet (60 meters). The outer edge of the continental shelf is marked by a sharp change in angle where the CONTINENTAL SLOPE begins. Most continental shelves were exposed above current SEA LEVEL during the PLEISTOCENE EPOCH and have been submerged by rising sea levels over the past 18,000 years.

Continental shield. Area of a CONTINENT that contains the oldest ROCKS on Earth, called CRATONS. These are areas of granitic rocks, part of the CONTINENTAL CRUST, where there are ancient MOUNTAINS. The Canadian Shield in North America is an example.

Convectional rain. Type of PRECIPITATION caused when AIR over a warm surface is warmed and rises, leading to ADIABATIC cooling, CONDENSATION, and, if the air is moist enough, rain.

Convective overturn. Renewal of the bottom waters caused by the sinking of SURFACE WATERS that have become denser, usually because of decreased TEMPERATURE.

Convergence (climate). AIR flowing in toward a central point.

Convergence (physiography). Process that occurs during the second half of a SUPERCONTINENT CYCLE, whereby crustal PLATES collide and intervening OCEANS disappear as a result of plate SUBDUCTION.

Convergent plate boundary. Compressional PLATE BOUNDARY at which an oceanic PLATE is subducted or two continental plates collide.

Convergent plate margin. Area where the earth's LITHOSPHERE is returned to the MANTLE at a SUBDUCTION ZONE, forming volcanic "ISLAND ARCS" and associated HYDROTHERMAL activity.

Conveyor belt current. Large CYCLE of water movement that carries warm water from the north Pacific westward across the Indian Ocean, around Southern Africa, and into the Atlantic, where it warms the ATMOSPHERE, then returns at a deeper OCEAN level to rise and begin the process again.

Coordinated universal time (UTC). International basis of time, introduced to the world in 1964. The basis for UTC is a small number of ATOMIC CLOCKS. Leap seconds are occasionally added to UTC to keep it synchronized with universal time.

Core-mantle boundary. SEISMIC discontinuity 1,790 miles (2,890 km.) below the earth's surface that separates the MANTLE from the OUTER CORE.

Core region. Area, generally around a COUNTRY's CAPITAL CITY, that has a large, dense POPULATION and is the center of TRADE, financial services, and production. The rest of the country is referred to as the PERIPHERY. On a larger scale, the CONTINENT of Europe has a core region, which includes London, Paris, and Berlin; Iceland, Portugal, and Greece are peripheral LOCATIONS.

Coriolis effect. Apparent deflection of moving objects above the earth because of the earth's ROTA-

TION. The deflection is to the right in the NORTHERN HEMISPHERE and to the left in the SOUTHERN HEMISPHERE. The deflection is inversely proportional to the speed of the earth's rotation, being negligible at the EQUATOR but at its maximum near the POLES. The Coriolis effect is a major influence on the direction of surface WINDS. Sometimes called Coriolis force.

Corrasion. EROSION and lowering of a STREAMBED by FLUVIAL action, especially by ABRASION of the bedload (material transported by the STREAM) but also including SOLUTION by the water.

Cosmogony. Study of the origin and nature of the SOLAR SYSTEM.

Cotton Belt. Part of the United States extending from South Carolina through Georgia, Alabama, Mississippi, Tennessee, Louisiana, Arkansas, Texas, and Oklahoma, where cotton was grown on PLANTATIONS using slave labor before the Civil War. After that war, the South stagnated for almost a century. Racial SEGREGATION contributed to cultural isolation from the rest of the United States. Cotton is still produced in this REGION, but California has overtaken the Southern STATES as a cotton producer, and other agricultural products, such as soybeans and poultry, have become dominant crops in the old Cotton Belt. In-migration, due to the SUN BELT attraction, has led to rapid urban growth.

Counterurbanization. Out-migration of people from URBAN AREAS to smaller TOWNS or RURAL areas. As large modern cities are perceived to be overcrowded, stressful, polluted, and dangerous, many of their residents move to areas they regard as more favorable. Such moves are often related to individuals' retirements; however, younger workers and families are also part of counterurbanization.

Crater morphology. Structure or form of CRATERS and the related processes that developed them.

Craton. Large, geologically old, relatively stable CORE of a continental LITHOSPHERIC PLATE, sometimes termed a CONTINENTAL SHIELD.

Creep. Slow, gradual downslope movement of SOIL materials under gravitational stress. Creep tests are experiments conducted to assess the effects of time on ROCK properties, in which environmental conditions (surrounding pressure, TEMPERATURE) and the deforming stress are held constant.

Crestal plane. Plane or surface that goes through the highest points of all beds in a fold; it is coincident with the axial plane when the axial plane is vertical.

Cross-bedding. Layers of ROCK or SAND that lie at an angle to horizontal bedding or to the ground.

Crown land. Land belonging to a NATION's MONARCHY.

Crude oil. Unrefined OIL, as it occurs naturally. Also called PETROLEUM.

Crustal movements. PLATE TECTONICS theorizes that Earth's CRUST is not a single rigid shell, but comprises a number of large pieces that are in motion, separating or colliding. There are two types of crust—the older continental and the much younger OCEANIC CRUST. When PLATES diverge, at SEAFLOOR SPREADING zones, new (oceanic) crust is created from the MAGMA that flows out at the MID-OCEAN RIDGES. When plates converge and collide, denser oceanic crust is SUBDUCTED under the lighter CONTINENTAL CRUST. The boundaries at the areas where plates slide laterally, neither diverging nor converging, are called TRANSFORM FAULTS. The San Andreas Fault represents the world's best-known transform BOUNDARY. As a result of crustal movements, the earth can be deformed in several ways. Where PLATE BOUNDARIES converge, compression can occur, leading to FOLDING and the creation of SYNCLINES and ANTICLINES. Other stresses of the crust can lead to fracture, or faulting, and accompanying EARTHQUAKES. LANDFORMS created in this way include HORSTS, GRABEN, and BLOCK MOUNTAINS.

Culture hearth. LOCATION in which a CULTURE has developed; a CORE REGION from which the culture later spread or diffused outward through a larger REGION. Mesopotamia, the Nile Valley, and the Peruvian ALTIPLANO are examples of culture hearths.

Curie point. TEMPERATURE at which a magnetic MINERAL locks in its magnetization. Also known as Curie temperature.

Cycle of erosion. Influential MODEL of LANDSCAPE change proposed by William Morris Davis near the end of the nineteenth century. The UPLIFT of a relatively flat surface, or PLAIN, in an area of moderate RAINFALL and TEMPERATURE, led to gradual EROSION of the initial surface in a sequence Davis categorized as Youth, Maturity, and Old Age. The final landscape was called PENEPLAIN. Davis also recognized the stage of REJUVENATION, when a new uplift could give new ENERGY to the cycle, leading to further downcutting and erosion. The model also was used to explain the sequence of LANDFORMS developed in REGIONS of ALPINE GLACIERS. The model has been criticized as misleading, since CRUSTAL MOVEMENT is continuous and more frequent than Davis perhaps envisaged, but remained useful as a description of TOPOGRAPHY. Also known as the Davisian cycle or geomorphic cycle.

Cyclonic rain. In the NORTHERN HEMISPHERE winter, two low-pressure systems or CYCLONES—the Aleutian Low and the Icelandic Low—develop over the OCEAN near 60 DEGREES north LATITUDE. The polar FRONT forms where the cold and relatively dry ARCTIC AIR meets the warmer, moist air carried by westerly WINDS. The warm air is forced upward, cools, and condenses. These cyclonic STORMS often move south, bringing winter PRECIPITATION to North America, especially to the STATES of Washington and Oregon.

Cylindrical projection. MAP PROJECTION that represents the earth's surface as a rectangle. It can be imagined as a cylinder of paper wrapped around a globe with a light source at its center; the images of the CONTINENTS would be projected onto the paper. In reality, MAPS are constructed mathematically. It is impossible to show the North Pole or South Pole on a cylindrical projection. Although the map is conformal, distortion of area is extreme beyond 50 DEGREES north and south LATITUDES. The Mercator projection, developed in the sixteenth century by the Flemish cartographer Gerardus Mercator, is the best-known cylindrical projection. It has been popular with seamen because the shortest route between two PORTS (the GREAT CIRCLE route) can be plotted as straight lines that show the COMPASS direction that should be followed. Use of this projection for other purposes, however, can lead to misunderstandings about size; for example, compare Greenland on a globe and on a Mercator map.

Datum level. Baseline or level from which other heights are measured, above or below. MEAN SEA LEVEL is the datum commonly used in surveying and in the construction of TOPOGRAPHIC MAPS.

Daylight saving time. System of seasonal adjustments in CLOCK settings designed to increase hours of evening sunlight during summer months. In the spring, clocks are set ahead one hour; in the fall, they are put back to standard time. In North America, these changes are made on the first Sunday in April and the last Sunday in October. The U.S. Congress standardized daylight saving time in 1966; however, parts of Arizona, Indiana, and Hawaii do not follow the system.

Débâcle. In a scientific context, this French word means the sudden breaking up of ice in a RIVER in the spring, which can lead to serious, sudden flooding.

Debris avalanche. Large mass of SOIL and ROCK that falls and then slides on a cushion of AIR downhill rapidly as a unit.

Debris flow. Flowing mass consisting of water and a high concentration of SEDIMENT with a wide RANGE of size, from fine muds to coarse gravels.

Declination, magnetic. Measure of the difference, in DEGREES, between the earth's NORTH MAGNETIC POLE and the North Pole on a MAP; this difference changes slightly each year. The needle of a magnetic COMPASS points to the earth's geomagnetic pole, which is not exactly the same as the North Pole of the geographic GRID or the set of lines of LATITUDE and LONGITUDE. The geomagnetic poles, north and south, mark the ends of the AXIS of the earth's MAGNETIC FIELD, but this field is not stationary. In fact, the geomagnetic poles have completely reversed hundreds of times throughout earth history. Lines of equal magnetic declination are called ISOGONIC LINES.

Declination of the Sun. LATITUDE of the SUBSOLAR POINT, the PLACE on the earth's surface where the SUN is directly overhead. In the course of a year, the declination of the Sun migrates from 23.5 DEGREES north LATITUDE, at the (northern) summer SOLSTICE, to 23.5 degrees south latitude, at the (northern) WINTER SOLSTICE. Hawaii is the only part of the United States that experiences the Sun directly overhead twice a year.

Deep-focus earthquakes. EARTHQUAKES occurring at depths ranging from 40 to 400 miles (70–700 km.) below the earth's surface. This RANGE of depths represents the zone from the base of the earth's CRUST to approximately one-quarter of the distance into Earth's MANTLE. Deep-focus earthquakes provide scientists information about the PLANET's interior structure, its composition, and SEISMICITY. Observation of deep-focus earthquakes has played a fundamental role in the discovery and understanding of PLATE TECTONICS.

Deep-ocean currents. Deep-ocean currents involve significant vertical and horizontal movements of seawater. They distribute oxygen- and nutrient-rich waters throughout the world's OCEANS, thereby enhancing biological productivity.

Defile. Narrow MOUNTAIN PASS or GORGE through which troops could march only in single file.

Deflation. EROSION by WIND, resulting in the removal of fine particles. The LANDFORM that typically results is a deflation hollow.

Deforestation. Removal or destruction of FORESTS. In the late twentieth century, there was widespread concern about tropical deforestation—destruction of the tropical RAINFOREST—especially that of Brazil. Forest clearing in the TROPICS is uneconomic because of low SOIL fertility. Deforestation causes severe EROSION and environmental damage; it also destroys habitat, which leads to the EXTINCTION of both plant and animal species.

Degradation. Process of CRATER EROSION from all processes, including WIND and other meteorological mechanisms.

Degree (geography). Unit of LATITUDE or LONGITUDE in the geographic GRID, used to determine ABSOLUTE LOCATION. One degree of latitude is about 69 miles (111 km.) on the earth's surface. It is not exactly the same everywhere, because the earth is not a perfect sphere. One degree of longitude varies greatly in length, because the MERIDIANS converge at the POLES. At the EQUATOR, it is 69 miles (111 km.), but at the North or South Pole it is zero.

Degree (temperature). Unit of MEASUREMENT of TEMPERATURE, based on the CELSIUS SCALE, except in the United States, which uses the FAHRENHEIT SCALE. On the Celsius scale, one degree is one-hundredth of the difference between the freezing point of water and the boiling point of water.

Demographic measure. Statistical data relating to POPULATION.

Demographic transition. MODEL of POPULATION change that fits the experience of many European countries, showing changes in birth and death rates. In the first stage, in preindustrial countries, population size was stable because both BIRTH RATES and DEATH RATES were high. Agricultural reforms, together with the INDUSTRIAL REVOLUTION and subsequent medical advances, led to a rapid fall in the death rate, so that the second and third stages of the model were periods of rapid population growth, often called the POPULATION EXPLOSION. In the fourth stage of the model, birth rates fall markedly, leading again to stable population size.

Dendritic drainage. Most common pattern of STREAMS and their TRIBUTARIES, occurring in areas of uniform ROCK type and regular slope. A MAP, or aerial photograph, shows a pattern like the veins on a leaf—smaller streams join the main stream at an acute angle.

Denudation. General word for all LANDFORM processes that lead to a lowering of the LANDSCAPE, including WEATHERING, mass movement, EROSION, and transport.

Deposition. Laying down of SEDIMENTS that have been transported by water, WIND, or ice.

Deranged drainage. LANDSCAPE whose integrated drainage network has been destroyed by irregular glacial DEPOSITION, yielding numerous shallow LAKE BASINS.

Derivative maps. MAPS that are prepared or derived by combining information from several other maps.

Desalinization. Process of removing SALT and MINERALS from seawater or from saline water occurring in AQUIFERS beneath the land surface to render it fit for AGRICULTURE or other human use.

Desert climate. Low PRECIPITATION, low HUMIDITY, high daytime TEMPERATURES, and abundant sunlight are characteristics of desert climates. The hot DESERTS of the world generally are located on the western sides of CONTINENTS, at LATITUDES from fifteen to thirty DEGREES north or south of the EQUATOR. One definition, based on precipitation, defines deserts as areas that receive between 0 and 9 inches (0 to 250 millimeters) of precipitation per year. REGIONS receiving more precipitation are considered to have a SEMIDESERT climate, in which some AGRICULTURE is possible.

Desert pavement. Surface covered with smoothed PEBBLES and gravels, found in arid areas where DEFLATION (WIND EROSION) has removed the smaller particles. Called a "gibber plain" in Australia.

Desertification. Increase in DESERT areas worldwide, largely as a result of overgrazing or poor agricultural practices in semiarid and marginal CLIMATES. DEFORESTATION, DROUGHT, and POPULATION increase also contribute to desertification. The REGION of Africa just south of the Sahara Desert, known as the SAHEL, is the largest and most dramatic demonstration of desertification.

Detrital rock. SEDIMENTARY ROCK composed mainly of grains of silicate MINERALS as opposed to grains of calcite or CLAYS.

Devolution. Breaking up of a large COUNTRY into smaller independent political units is the final and most extreme form of devolution. The Soviet Union devolved from one single country into fifteen separate countries in 1991. At an intermediate level, devolution refers to the granting of political autonomy or self-government to a REGION, without a complete split. The reopening of the Scottish Parliament in 1999 and the Northern Ireland parliament in 2000 are examples of devolution; the Parliament of the United Kingdom had previously met only in London and made laws there for all parts of the country. Canada experienced devolution with the creation of the new territory of Nunavut, whose residents elect the members of their own legislative assembly.

Dew point. TEMPERATURE at which an AIR mass becomes saturated and can hold no more moisture. Further cooling leads to CONDENSATION. At ground level, this produces DEW.

Diagenesis. Conversion of unconsolidated SEDIMENT into consolidated ROCK after burial by the processes of compaction, cementation, recrystallization, and replacement.

Diaspora. Dispersion of a group of people from one CULTURE to a variety of other REGIONS or to other lands. A Greek word, used originally to refer to the Jewish diaspora. Jewish people now live in many countries, although they have Israel as a HOMELAND. Similar to this are the diasporas of the Irish and the overseas Chinese.

Diastrophism. Deformation of the earth's CRUST by faulting or FOLDING.

Diatom ooze. Deposit of soft mud on the OCEAN floor consisting of the shells of diatoms, which are microscopic single-celled creatures with SILICA-rich shells. Diatom ooze deposits are located in the southern Pacific around Antarctica and in the northern Pacific. Other PELAGIC, or deep-ocean, SEDIMENTS include CLAYS and calcareous ooze.

Dike (geology). LANDFORM created by IGNEOUS intrusion when MAGMA or molten material within the earth forces its way in a narrow band through overlying ROCK. The dike can be exposed at the surface through EROSION.

Dike (water). Earth wall or DAM built to prevent flooding; an EMBANKMENT or artificial LEVEE. Sometimes specifically associated with structures built in the Netherlands to prevent the entry of seawater. The land behind the dikes was reclaimed for AGRICULTURE; these new fields are called POLDERS.

Distance-decay function. Rate at which an activity diminishes with increasing distance. The effect that distance has as a deterrent on human activity is sometimes described as the FRICTION OF DISTANCE. It occurs because of the time and cost of overcoming distances between people and their desired activity. An example of the distance-decay function is the rate of visitors to a football stadium. The farther people have to travel, the less likely they are to make this journey.

Distributary. STREAM that takes waters away from the main CHANNEL of a RIVER. A DELTA usually comprises many distributaries. Also called distributary channel.

Diurnal range. Difference between the highest and lowest TEMPERATURES registered in one twenty-four-hour period.

Diurnal tide. Having only one high tide and one low tide each lunar DAY; TIDES on some parts of the Gulf of Mexico are diurnal.

Divergent boundary. BOUNDARY that results where two PLATES are moving apart from each other, as is the case along MID-OCEANIC RIDGES.

Divergent margin. Area where the earth's CRUST and LITHOSPHERE form by SEAFLOOR SPREADING.

Doline. Large SINKHOLE or circular depression formed in LIMESTONE areas through the CHEMICAL WEATHERING process of carbonation.

Dolomite. MINERAL consisting of calcium and magnesium carbonate compounds that often forms from PRECIPITATION from seawater; it is abundant in ancient ROCKS.

Downwelling. Sinking of OCEAN water.

Drainage basin. Area of the earth's surface that is drained by a STREAM. Drainage basins vary greatly in size, but each is separated from the next by RIDGES, or drainage DIVIDES. The CATCHMENT of the drainage basin is the WATERSHED.

Drift ice. ARCTIC or ANTARCTIC ice floating in the open SEA.

Drumlin. Low HILL, shaped like half an egg, formed by DEPOSITION by CONTINENTAL GLACIERS. A drumlin is composed of TILL, or mixed-size materials. The wider end faces upstream of the glacier's movement; the tapered end points in the direction of the ice movement. Drumlins usually occur in groups or swarms.

Dust devil. Whirling cloud of DUST and small debris, formed when a small patch of the earth's surface becomes heated, causing hot AIR to rise; cooler air then flows in and begins to spin. The resulting dust devil can grow to heights of 150 feet (50 meters) and reach speeds of 35 miles (60 km.) per hour.

Dust dome. Dome of AIR POLLUTION, composed of industrial gases and particles, covering every large CITY in the world. The pollution sometimes is carried downwind to outlying areas.

Earth pillar. Formation produced when a boulder or caprock prevents EROSION of the material directly beneath it, usually CLAY. The clay is easily eroded away by water during RAINFALL, except where the overlying ROCK protects it. The result is a tall, slender column, as high as 20 feet (6.5 meters) in exceptional cases.

Earth radiation. Portion of the electromagnetic spectrum, from about 4 to 80 microns, in which the earth emits about 99 percent of its RADIATION.

Earth tide. Slight deformation of Earth resulting from the same forces that cause OCEAN TIDES, those that are exerted by the MOON and the SUN.

Earthflow. Term applied to both the process and the LANDFORM characterized by fluid downslope movement of SOIL and ROCK over a discrete plane of failure; the landform has a HUMMOCKY surface and usually terminates in discrete lobes.

Earth's heat budget. Balance between the incoming SOLAR RADIATION and the outgoing terrestrial reradiation.

Eclipse, lunar. Obscuring of all or part of the light of the MOON by the shadow of the earth. A lunar eclipse occurs at the full moon up to three times a year. The surface of the Moon changes from gray to a reddish color, then back to gray. The sequence may last several hours.

Eclipse, solar. At least twice a year, the SUN, MOON, and Earth are aligned in one straight line. At that time, the Moon obscures all the light of the Sun along a narrow band of the earth's surface, causing a total eclipse; in REGIONS of Earth adjoining that area, there is a partial eclipse. A corona (halo of light) can be seen around the Sun at the total eclipse. Viewing a solar eclipse with naked eyes is extremely dangerous and can cause blindness.

Ecliptic, plane of. Imaginary plane that would touch all points in the earth's ORBIT as it moves around the SUN. The angle between the plane of the ecliptic and the earth's AXIS is 66.5 DEGREES.

Edge cities. Forms of suburban downtown in which there are nodal concentrations of office space and shopping facilities. Edge cities are located close to major freeways or highway intersections, on the outer edges of METROPOLITAN AREAS.

Effective temperature. TEMPERATURE of a PLANET based solely on the amount of SOLAR RADIATION that the planet's surface receives; the effective temperature of a planet does not include the GREENHOUSE temperature enhancement effect.

Ejecta. Material ejected from the CRATER made by a meteoric impact.

Ekman layer. REGION of the SEA, from the surface to about 100 meters down, in which the WIND directly affects water movement.

Eluviation. Removal of materials from the upper layers of a SOIL by water. Fine material may be removed by SUSPENSION in the water; other material is removed by SOLUTION. The removal by solution is called LEACHING. Eluviation from an upper layer leads to illuviation in a lower layer.

Enclave. Piece of territory completely surrounded by another COUNTRY. Two examples are Lesotho, which is surrounded by the Republic of South Africa, and the Nagorno-Karabakh REGION, populated by Armenians but surrounded

by Azerbaijan. The term is also used for smaller regions, such as ethnic neighborhoods within larger cities. See also EXCLAVE.

Endemic species. Species confined to a restricted area in a restricted ENVIRONMENT.

Endogenic sediment. SEDIMENT produced within the water column of the body in which it is deposited; for example, calcite precipitated in a LAKE in summer.

Environmental degradation. Situation that occurs in slum areas and SQUATTER SETTLEMENTS because of poverty and inadequate INFRASTRUCTURE. Too-rapid human POPULATION growth can lead to the accumulation of human waste and garbage, the POLLUTION of GROUNDWATER, and DENUDATION of nearby FORESTS. As a result, LIFE EXPECTANCY in such degraded areas is lower than in the RURAL communities from which many of the settlers came. INFANT MORTALITY is particularly high. When people leave an area because of such environmental degradation, that is referred to as ecomigration.

Environmental determinism. Theory that the major influence on human behavior is the physical ENVIRONMENT. Some evidence suggests that TEMPERATURE, PRECIPITATION, sunlight, and TOPOGRAPHY influence human activities. Originally espoused by early German geographers, this theory has led to some extreme stances, however, by authors who have sought to explain the dominance of Europeans as a result of a cool temperate CLIMATE.

Eolian (aeolian). Relating to, or caused by, WIND. In Greek mythology, Aeolus was the ruler of the winds. EROSION, TRANSPORT, and DEPOSITION are common eolian processes that produce LANDFORMS in DESERT REGIONS.

Eolian deposits. Material transported by the WIND.

Eolian erosion. Mechanism of EROSION or CRATER DEGRADATION caused by WIND.

Eon. Largest subdivision of geologic time; the two main eons are the PRECAMBRIAN (c. 4.6 billion years ago to 544 million years ago) and the PHANEROZOIC (c. 544 million years ago to the present).

Ephemeral stream. Watercourse that has water for only a DAY or so.

Epicontinental sea. Shallow SEAS that are located on the CONTINENTAL SHELF, such as the North Sea or Hudson Bay. Also called an EPEIRIC SEA.

Epifauna. Organisms that live on the seafloor.

Epilimnion. Warmer surface layer of water that occurs in a LAKE during summer stratification; during spring, warmer water rises from great depths, and it heats up through the summer SEASON.

Equal-area projection. MAP PROJECTION that maintains the correct area of surfaces on7 a MAP, although shape distortion occurs. The property of such a map is called equivalence.

Erg. Sandy DESERT, sometimes called a SEA of SAND. Erg deserts account for less than 30 percent of the world's deserts. "Erg" is an Arabic word.

Eruption, volcanic. Emergence of MAGMA (molten material) at the earth's surface as LAVA. There are various types of volcanic eruptions, depending on the chemistry of the magma and its viscosity. Scientists refer to effusive and explosive eruptions. Low-viscosity magma generally produces effusive eruptions, where the lava emerges gently, as in Hawaii and Iceland, although explosive events can occur at those SITES as well. Gently sloping SHIELD VOLCANOES are formed by effusive eruptions; FLOODS, such as the Columbia Plateau, can also result. Explosive eruptions are generally associated with SUBDUCTION. Much gas, including steam, is associated with magma formed from OCEANIC CRUST, and the compressed gas helps propel the explosion. COMPOSITE CONES, such as Mount Saint Helens, are created by explosive eruptions.

Escarpment. Steep slope, often almost vertical, formed by faulting. Sometimes called a FAULT SCARP.

Esker. Deposit of coarse gravels that has a sinuous, winding shape. An esker is formed by a STREAM of MELTWATER that flowed through a tunnel it formed under a CONTINENTAL GLACIER. Now that the continental glaciers have melted, eskers can be found exposed at the surface in many PLACES in North America.

Estuarine zone. Area near the COASTLINE that consists of estuaries and coastal saltwater WETLANDS.

Etesian winds. WINDS that blow from the north over the Mediterranean during July and August.

Ethnocentrism. Belief that one's own ETHNIC GROUP and its CULTURE are superior to any other group.

Ethnography. Study of different CULTURES and human societies.

Eustacy. Any change in global SEA LEVEL resulting from a change in the absolute volume of available sea water. Also known as eustatic sea-level change.

Eustatic movement. Changes in SEA LEVEL.

Exclave. Territory that is part of one COUNTRY but separated from the main part of that country by another country. Alaska is an exclave of the United States; Kaliningrad is an exclave of Russia. See also ENCLAVE.

Exfoliation. When GRANITE rocks cooled and solidified, removal of the overlyingrock that was present reduced the pressure on the granite mass, allowing it to expand and causing sheets or layers of rock to break off. An exfoliation DOME, such as Half Dome in Yosemite National Park, is the resultant LANDFORM.

Exotic stream. RIVER that has its source in an area of high RAINFALL and then flows through an arid REGION or DESERT. The Nile River is the most famous exotic STREAM. In the United States, the Colorado River is a good example of an exotic stream.

Expansion-contraction cycles. Processes of wetting-drying, heating-cooling, or freezing-thawing, which affect SOIL particles differently according to their size.

Extrusive rock. Fine-grained, or glassy, ROCK which was formed from a MAGMA that cooled on the surface of the earth.

Fall line. Edge of an area of uplifted land, marked by WATERFALLS where STREAMS flow over the edge.

Fata morgana. Large mirage. Originally, the name given to a multiple mirage phenomenon often observed over the Straits of Messina and supposed to be the work of the fairy ("fata") Morgana. Another famous fata morgana is located in Antarctica.

Fathometer. Instrument that uses sound waves or sonar to determine the depth of water or the depth of an object below the water.

Fault drag. Bending of ROCKS adjacent to a FAULT.

Fault line. Line of breakage on the earth's surface. FAULTS may be quite short, but many are extremely long, even hundreds of miles. The origin of the faulting may lie at a considerable depth below the surface. Movement along the fault line generates EARTHQUAKES.

Fault plane. Angle of a FAULT. When fault blocks move on either side of a fault or fracture, the movement can be vertical, steeply inclined, or sometimes horizontal. In a NORMAL FAULT, the fault plane is steep to almost vertical. In a REVERSE FAULT, one block rides over the other, forming an overhanging FAULT SCARP. The angle of inclination of the fault plane from the horizontal is called the dip. The inclination of a fault plane is generally constant throughout the length of the fault, but there can be local variations in slope. In a STRIKE-SLIP FAULT the movement is horizontal, so no fault scarp is produced, although the FAULT LINE may be seen on the surface.

Fault scarp. FAULTS are produced through breaking or fracture of the surface ROCKS of the earth's CRUST as a result of stresses arising from tectonic movement. A NORMAL FAULT, one in which the earth movement is predominantly vertical, produces a steep fault scarp. A STRIKE-SLIP FAULT does not produce a fault scarp.

Feldspar. Family name for a group of common MINERALS found in such ROCKS as GRANITE and composed of silicates of aluminum together with potassium, sodium, and calcium. Feldspars are the most abundant group of minerals within the earth's CRUST. There are many varieties of feldspar, distinguished by variations in chemistry and crystal structure. Although feldspars have some economic uses, their principal importance lies in their role as rock-forming minerals.

Felsic rocks. IGNEOUS ROCKS rich in potassium, sodium, aluminum, and SILICA, including GRANITES and related rocks.

Fertility rate. DEMOGRAPHIC MEASURE of the average number of children per adult female in any given POPULATION. Religious beliefs, education, and other cultural considerations influence fertility rates.

Fetch. Distance along a large water surface over which a WIND of almost uniform direction and speed blows.

Feudalism. Social and economic system that prevailed in Europe before the INDUSTRIAL REVOLUTION. The land was owned and controlled by a minority comprising noblemen or lords; all other people were peasants or serfs, who worked as agricultural laborers on the lords' land. The peasants were not free to leave, or to do anything without their lord's permission. Other REGIONS such as China and Japan also had a feudal system in the past.

Firn. Intermediate stage between SNOW and glacial ice. Firn has a granular TEXTURE, due to compaction. Also called NÉVÉ.

Fission, nuclear. Splitting of an atomic nucleus into two lighter nuclei, resulting in the release of neutrons and some of the binding ENERGY that held the nucleus together.

Fissure. Fracture or crack in ROCK along which there is a distinct separation.

Flash flood. Sudden rush of water down a STREAM CHANNEL, usually in the DESERT after a short but intense STORM. Other causes, such as a DAM failure, could lead to a flash flood.

Flood control. Attempts by humans to prevent flooding of STREAMS. Humans have consistently settled on FLOODPLAINS and DELTAS because of the fertile SOIL for AGRICULTURE, and attempts at flood control date back thousands of years. In strictly agricultural societies such as ancient Egypt, people built VILLAGES above the FLOOD levels, but transport and industry made riverside LOCATIONS desirabl and engineers devised technological means to try to prevent flood damage. Artificial LEVEES, RESERVOIRS, and DAMS of ever-increasing size were built on

RIVERS, as well as bypass CHANNELS leading to artificial floodplains. In many modern dam construction projects, the production of HYDRO-ELECTRIC POWER was more important than flood control. Despite modern TECHNOLOGY, floods cause the largest loss of human life of all natural disasters, especially in low-income countries such as Bangladesh.

Flood tide. Rising or incoming tide. Most parts of the world experience two flood TIDES in each 24-hour period.

Floodplain. Flat, low-lying land on either side of a STREAM, created by the DEPOSITION of ALLUVIUM from floods. Also called ALLUVIAL PLAIN.

Fluvial. Pertaining to running water; for example, fluvial processes are those in which running water is the dominant agent.

Fog deserts. Coastal DESERTS where FOG is an important source of moisture for plants, animals, and humans. The fog forms because of a cold OCEAN CURRENT close to the SHORE. The Namib Desert of southwestern Africa, the west COAST of California, and the Atacama Desert of Peru are coastal deserts.

Föhn wind. WIND warmed and dried by descent, usually on the LEE side of a MOUNTAIN. In North America, these winds are called the CHINOOK.

Fold mountains. ROCKS in the earth's CRUST can be bent by compression, producing folds. The Swiss Alps are an example of complex FOLDING, accompanied by faulting. Simple upward folds are ANTICLINES, downward folds are SYNCLINES; but subsequent EROSION can produce LANDSCAPES with synclinal MOUNTAINS.

Folding. Bending of ROCKS in the earth's CRUST, caused by compression. The rocks are deformed, sometimes pushed up to form mountain RANGES.

Foliation. TEXTURE or structure in which MINERAL grains are arranged in parallel planes.

Food web. Complex network of FOOD CHAINS. Food chains are interconnected, because many organisms feed on a variety of others, and in turn may be eaten by any of a number of predators.

Forced migration. MIGRATION that occurs when people are moved against their will. The Atlan-tic slave trade is an example of forced migration. People were shipped from Africa to countries in Europe, Asia, and the New World as forced immigrants. Within the United States, some NATIVE AMERICANS were forced by the federal government to migrate to new reservations.

Ford. Short shallow section of a RIVER, where a person can cross easily, usually by walking or riding a horse. To cross a STREAM in such a manner.

Formal region. Cultural REGION in which one trait, or group of traits, is uniform. LANGUAGE might be the basis of delineation of a formal cultural region. For example, the Francophone region of Canada constitutes a formal region based on one single trait. One might also identify a formal Mormon region centered on the STATE of Utah, combining RELIGION and LANDSCAPE as defining traits. Cultural geographers generally identify formal regions using a combination of traits.

Fossil fuel. Deposit rich in hydrocarbons, formed from organic materials compressed in ROCK layers—COAL, OIL, and NATURAL GAS.

Fossil record. Fossil record provides evidence that addresses fundamental questions about the origin and history of life on the earth: When life evolved; how new groups of organisms originated; how major groups of organisms are related. This record is neither complete nor without biases, but as scientists' understanding of the limits and potential of the fossil record grows, the interpretations drawn from it are strengthened.

Fossilization. Processes by which the remains of an organism become preserved in the ROCK record.

Fracture zones. Large, linear zones of the seafloor characterized by steep CLIFFS, irregular TOPOG-RAPHY, and FAULTS; such zones commonly cross and displace oceanic RIDGES by faulting.

Free association. Relationship between sovereign NATIONS in which one nation—invariably the larger—has responsibility for the other nation's defense. The Cook Islands in the South Pacific have such a relationship with New Zealand.

Friction of distance. Distance is of prime importance in social, political, economic, and other relationships. Large distance has a negative effect on human activity. The time and cost of overcoming distance can be a deterrent to various activities. This has been called the friction of distance.

Frigid zone. Coldest of the three CLIMATE zones proposed by the ancient Greeks on the basis of their theories about the earth. There were two frigid zones, one around each POLE. The Greeks believed that human life was possible only in the TEMPERATE ZONE.

Fringing reef. Type of CORAL REEF formed at the SHORELINE, extending out from the land in shallow water. The top of the coral may be exposed at low TIDE.

Frontier Thesis. Thesis first advanced by the U.S. historian Frederick Jackson Turner, who declared that U.S. history and the U.S. character were shaped by the existence of empty, FRONTIER lands that led to exploration and westward expansion and DEVELOPMENT. The closing of the frontier occurred when transcontinental railroads linked the East and West Coasts and SETTLEMENTS spread across the United States. This thesis was used by later historians to explain the history of South Africa, Canada, and Australia. Critics of the Frontier Thesis point out that minorities and women were excluded from this view of history.

Frost wedging. Powerful form of PHYSICAL WEATHERING of ROCK, in which the expansion of water as it freezes in JOINTS or cracks shatters the rock into smaller pieces. Also known as frost shattering.

Fumarole. Crack in the earth's surface from which steam and other gases emerge. Fumaroles are found in volcanic areas and areas of GEOTHERMAL activity, such as Yellowstone National Park.

Fusion energy. Heat derived from the natural or human-induced union of atomic nuclei; in effect, the opposite of FISSION energy.

Gall's projection. MAP PROJECTION constructed by projecting the earth onto a cylinder that intersects the sphere at 45 DEGREES north and 45 degrees south LATITUDE. The resulting map has less distortion of area than the more familiar CYLINDRICAL PROJECTION of Mercator.

Gangue. Apparently worthless ROCK or earth in which valuable gems or MINERALS are found.

Garigue. VEGETATION cover of small shrubs found in Mediterranean areas. Similar to the larger *maquis.*

Genus (plural, genera). Group of closely related species; for example, *Homo* is the genus of humans, and it includes the species *Homo sapiens* (modern humans) and *Homo erectus* (Peking Man, Java Man).

Geochronology. Study of the time SCALE of the earth; it attempts to develop methods that allow the scientist to reconstruct the past by dating events such as the formation of ROCKS.

Geodesy. Branch of applied mathematics that determines the exact positions of points on the earth's surface, the size and shape of the earth, and the variations of terrestrial GRAVITY and MAGNETISM.

Geoid. Figure of the earth considered as a MEAN SEA LEVEL surface extended continuously through the CONTINENTS.

Geologic terrane. Crustal block with a distinct group of ROCKS and structures resulting from a particular geologic history; assemblages of TERRANES form the CONTINENTS.

Geological column. Order of ROCK layers formed during the course of the earth's history.

Geomagnetic elements. MEASUREMENTS that describe the direction and intensity of the earth's MAGNETIC FIELD.

Geomagnetism. External MAGNETIC FIELD generated by forces within the earth; this force attracts materials having similar properties, inducing them to line up (point) along field lines of force.

Geostationary orbit. ORBIT in which a SATELLITE appears to hover over one spot on the PLANET's EQUATOR; this procedure requires that the orbit be high enough that its period matches the planet's rotational period, and have no inclination relative to the equator; for Earth, the ALTITUDE is 22,260 miles (35,903 km.).

Geostrophic. Force that causes directional change because of the earth's ROTATION.

Geotherm. Curve on a TEMPERATURE-depth graph that describes how temperature changes in the subsurface.

Geothermal power. Power having its source in the earth's internal heat.

Glacial erratic. ROCK that has been moved from its original position and transported by becoming incorporated in the ice of a GLACIER. Deposited in a new LOCATION, the rock is noteworthy because its geology is completely different from that of the surrounding rocks. Glacial erratics provide information about the direction of glacial movement and strength of the flow. They can be as small as PEBBLES, but the most interesting erratics are large boulders. Erratics become smoothed and rounded by the transport and EROSION.

Glaciation. This term is used in two senses: first, in reference to the cyclic widespread growth and advance of ICE SHEETS over the polar and high- to mid-LATITUDE REGIONS of the CONTINENTS; second, in reference to the effect of a GLACIER on the TERRAIN it transverses as it advances and recedes.

Global Positioning System (GPS). Group of SATELLITES that ORBIT Earth every twenty-four hours, sending out signals that can be used to locate PLACES on Earth and in near-Earth orbits.

Global warming. Trend of Earth CLIMATES to grow increasingly warm as a result of the GREENHOUSE EFFECT. One of the most dramatic effects of global warming is the melting of the POLAR ICE CAPS and a consequent rise the level of the world's OCEANS.

Gondwanaland. Hypothesized ancient CONTINENT in the SOUTHERN HEMISPHERE that geologists theorize broke into at least two large segments; one segment became India and pushed northward to collide with the Eurasian LANDMASS, while the other, Africa, moved westward.

Graben. Roughly symmetrical crustal depression formed by the lowering of a crustal block between two NORMAL FAULTS that slope toward each other.

Granules. Small grains or pellets.

Gravimeter. Device that measures the attraction of GRAVITY.

Gravitational differentiation. Separation of MINERALS, elements, or both as a result of the influence of a gravitational field wherein heavy phases sink or light phases rise through a melt.

Great circle. Largest circle that goes around a sphere. On the earth, all lines of LONGITUDE are parts of great circles; however, the EQUATOR is the only line of LATITUDE that is a great circle.

Green mud. SOILS that develop under conditions of excess water, or waterlogged soils, can display colors of gray to blue to green, largely because of chemical reactions involving iron. Fine CLAY soils and muds in areas such as BOGS or ESTUARIES can be called green mud. This soil-forming process is called gleization.

Greenhouse effect. Trapping of the SUN's rays within the earth's ATMOSPHERE, with a consequence rise in TEMPERATURES that leads to GLOBAL WARMING.

Greenhouse gas. Atmospheric gas capable of absorbing electromagnetic radiation in the infrared part of the spectrum.

Greenwich mean time. Also known as universal time, the solar mean time on the MERIDIAN running through Greenwich, England—which is used as the basis for calculating time throughout most of the world.

Grid. Pattern of horizontal and vertical lines forming squares of uniform size.

Groundwater movement. Flow of water through the subsurface, known as groundwater movement, obeys set principles that allow hydrologists to predict flow directions and rates.

Groundwater recharge. Water that infiltrates from the surface of the earth downward through SOIL and ROCK pores to the WATER TABLE, causing its level to rise.

Growth pole. LOCATION where high-growth economic activity is deliberately encouraged and promoted. Governments often establish growth poles by creating industrial parks, open cities, special economic zones, new TOWNS, and other incentives. The plan is that the new industries will further stimulate economic growth in a cumulative trend. Automobile plants are a traditional form of growth industry but have been overtaken by high-tech industries and BIOTECHNOLOGY. In France, the term "technopole" is used for a high-tech growth pole. A related concept is SPREAD EFFECTS.

Guyot. Drowned volcanic ISLAND with a flat top caused by WAVE EROSION or coral growth. A type of SEAMOUNT.

Gyre. Large semiclosed circulation patterns of OCEAN CURRENTS in each of the major OCEAN BASINS that move in opposite directions in the Northern and Southern hemispheres.

Haff. Term used for various WETLANDS or LAGOONS located around the southern end of the Baltic Sea, from Latvia to Germany. Offshore BARS of SAND and shingle separate the haffs from the open SEA. One of the largest is the Stettiner Haff, which covers the BORDER REGION between Germany and Poland and is separated from the Baltic by the low-lying ISLAND of Usedom. The Kurisches Haff (in English, the Courtland Lagoon) is located on the Lithuanian border.

Harmonic tremor. Type of EARTHQUAKE activity in which the ground undergoes continuous shaking in response to subsurface movement of MAGMA.

Headland. Elevated land projecting into a body of water.

Headwaters. Source of a RIVER. Also called headstream.

Heat sink. Term applied to Antarctica, whose cold CLIMATE causes warm AIR masses flowing over it to chill quickly and lose ALTITUDE, affecting the entire world's WEATHER.

Heterosphere. Major realm of the ATMOSPHERE in which the gases hydrogen and helium become predominant.

High-frequency seismic waves. EARTHQUAKE WAVES that shake the ROCK through which they travel most rapidly.

Histogram. Bar graph in which vertical bars represent frequency and the horizontal axis represents categories. A POPULATION PYRAMID, or age-sex pyramid, is a histogram, as is a CLIMOGRAPH.

Historical inertia. Term used by economic geographers when heavy industries, such as steelmaking and large manufacture, that require huge capital investments in land and plant continue in operation for long periods, even after they become out of date, uncompetitive, or obsolete.

Hoar frost. Similar to DEW, except that moisture is deposited as ice crystals, not liquid dew, on surfaces such as grass or plant leaves. When moist AIR cools to saturation level at TEMPERATURES below the freezing point, CONDENSATION occurs directly as ice. Technically, hoar frost is not the same as frozen dew, but it is difficult to distinguish between the two.

Hogback. Steeply sloping homoclinal RIDGE, with a slope of 45 DEGREES or more. The angle of the slope is the same as the dip of the ROCK STRATA. These LANDFORMS develop in REGIONS where the underlying rocks, usually SEDIMENTARY, have been folded into anticlinal ridges and synclinal VALLEYS. Differential EROSION causes softer rock layers to wear away more rapidly than the harder layers of rock that form the hogback ridge. A similar feature with a gentler slope is called a CUESTA.

Homosphere. Lower part of the earth's ATMOSPHERE. In this area, 60 miles (100 km.) thick, the component gases are uniformly mixed together, largely through WINDS and turbulent AIR CURRENTS. Above the homosphere is the REGION of the atmosphere called the HETEROSPHERE. There, the individual gases separate out into layers on the basis of their molecular weight. The lighter gases, hydrogen and helium, are at the top of the heterosphere.

Hook. A long, narrow deposit of SAND and SILT that grows outward into the OCEAN from the land is called a SPIT or sandspit. A hook forms when currents or WAVES cause the deposited material to curve back toward the land. Cape Cod is the most famous spit and hook in the United States.

Horse latitudes. Parts of the OCEANS from about 30 to 35 DEGREES north or south of the EQUATOR. In these latitudes, AIR movement is usually light WINDS, or even complete calm, because there are semipermanent high-pressure cells called ANTI-CYCLONES, which are marked by dry subsiding air and fine clear WEATHER. The atmospheric circulation of an anticyclone is divergent and CLOCKWISE in the NORTHERN HEMISPHERE, so to the north of the horse latitudes are the westerly winds and to the south are the northeast TRADE WINDS. In the SOUTHERN HEMISPHERE, the circulation is reversed, producing the easterly winds and the southeast trade winds. It is believed that the name originated because when ships bringing immigrants to the Americas were becalmed for any length of time, horses were thrown overboard because they required too much FRESH WATER. Also called the CALMS OF CANCER.

Horst. FAULT block or piece of land that stands above the surrounding land. A horst usually has been uplifted by tectonic forces, but also could have originated by downward movement or lowering of the adjacent lands. Movement occurs along the parallel faults on either side of a horst. If the land is downthrown instead of uplifted, a VALLEY known as a GRABEN is formed. "Horst" comes from the German word for horse, because the flat-topped feature resembles a vaulting horse used in gymnastics.

Hot spot. PLACE on the earth's surface where heat and MAGMA rise from deep in the interior, perhaps from the lower MANTLE. Erupting VOLCANOES may be present, as in the formation of the Hawaiian Islands. More commonly, the heat from the rising magma causes GROUNDWATER to form HOT SPRINGS, GEYSERS, and other thermal and HYDROTHERMAL features. Yellowstone National Park is located on a hot spot. Also known as a MANTLE PLUME.

Hot spring. SPRING where hot water emerges at the earth's surface. The usual cause is that the GROUNDWATER is heated by MAGMA. A GEYSER is a special type of hot spring at which the water heats under pressure and that periodically spouts hot water and steam. Old Faithful is the best known of many geysers in Yellowstone National Park. In some countries, GEOTHERMAL EN-

ERGY from hot springs is used to generate electricity. Also called thermal spring.

Humus. Uppermost layer of a SOIL, containing decaying and decomposing organic matter such as leaves. This produces nutrients, leading to a fertile soil. Tropical soils are low in humus, because the rate of decay is so rapid. Soils of GRASSLANDS and DECIDUOUS FOREST develop thick layers of humus. In a SOIL PROFILE, the layer containing humus is the O Horizon.

Hydroelectric power. Electricity generated when falling water turns the blades of a turbine that converts the water's potential ENERGY to mechanical energy. Natural WATERFALLS can be used, but most hydroelectric power is generated by water from DAMS, because the flow of water from a dam can be controlled. Hydroelectric generation is a RENEWABLE, clean, cheap way to produce power, but dam construction inundates land, often displacing people, who lose their homes, VILLAGES, and farmland. Aquatic life is altered and disrupted also; for example, Pacific salmon cannot return upstream on the Columbia River to their spawning REGION. In a few coastal PLACES, TIDAL ENERGY is used to generate hydroelectricity; La Rance in France is the oldest successful tidal power plant.

Hydrography. Surveying of underwater features or those parts of the earth that are covered by water, especially OCEAN depths and OCEAN CURRENTS. Hydrographers make MAPS and CHARTS of the ocean floor and COASTLINES, which are used by mariners for NAVIGATION. For centuries, mariners used a leadline, a long rope with a lead weight at the bottom. The line was thrown overboard and the depth of water measured. The unit of MEASUREMENT was FATHOMS (6 feet/1.8 meters), which is one-thousandth of a NAUTICAL MILE. The invention of sonar (underwater echo sounding) has enabled mapping of large areas, and hydrographers currently use both television cameras and SATELLITE data.

Hydrologic cycle. Continuous circulation of the earth's HYDROSPHERE, or waters, through EVAPORATION, CONDENSATION, and PRECIPITATION.

Other parts of the hydrologic cycle include RUN-OFF, INFILTRATION, and TRANSPIRATION.

Hydrostatic pressure. Pressure imposed by the weight of an overlying column of water.

Hydrothermal vents. Areas on the OCEAN floor, typically along FAULT LINES or in the vicinity of undersea VOLCANOES, where water that has percolated into the ROCK reemerges much hotter than the surrounding water; such heated water carries various dissolved MINERALS, including metals and sulfides.

Hyetograph. Chart showing the DISTRIBUTION of RAINFALL over time. Typically, a hyetograph is constructed for a single STORM, showing the amount of total PRECIPITATION accumulating throughout the period. A hyetograph shows how rainfall intensity varies throughout the duration of a storm.

Hygrometer. Instrument for measuring the RELATIVE HUMIDITY of AIR, or the amount of water vapor in the ATMOSPHERE at any time.

Hypsometer. Instrument used for measuring ALTITUDE (height above SEA LEVEL), using boiling water that circulates around a THERMOMETER. Since ATMOSPHERIC PRESSURE falls with increased altitude, the boiling point of water is lower. The hypsometer relies on this difference in boiling point to calculate ELEVATION. A more common instrument for measuring altitude is the ALTIMETER.

Ice blink. Bright, usually yellowish-white glare or reflection on the underside of a CLOUD layer, produced by light reflected from an ice-covered surface such as pack ice. A similar phenomenon of reflection from a snow-covered surface is called snow blink.

Ice-cap climate. Earth's most severe CLIMATE, where the mean monthly TEMPERATURE is never above 32 DEGREES Fahrenheit (0 degrees Celsius). This climate is found in Greenland and Antarctica, which are high PLATEAUS, where KATABATIC WINDS blow strongly and frequently. At these high LATITUDES, INSOLATION (SOLAR ENERGY) is received for a short period in the summer months, but the high reflectivity of the ice and SNOW means that much is reflected back instead of being absorbed by the surface. No VEGETATION can grow, because the LANDSCAPE is permanently covered in ice and snow. Because AIR temperatures are so cold, PRECIPITATION is usually less than 5 inches (13 centimeters) annually. The POLES are REGIONS of stable, high-pressure air, where dry conditions prevail, but strong winds that blow the snow around are common. In the KÖPPEN CLIMATE CLASSIFICATION, the ice-cap climate is signified by the letters *EF.*

Ice sheet. Huge CONTINENTAL GLACIER. The only ice sheets remaining cover most of Antarctica and Greenland. At the peak of the last ICE AGE, around 18,000 years ago, ice covered as much as one-third of the earth's land surfaces. In the NORTHERN HEMISPHERE, there were two great ice sheets—the Laurentide ice sheet, covering North America, and the Scandinavian ice sheet, covering northwestern Europe and Scandinavia.

Ice shelf. Portion of an ICE SHEET extending into the OCEAN.

Ice storm. STORM characterized by a fall of freezing rain, with the formation of glaze on Earth objects.

Icefoot. Long, tapering extension of a GLACIER floating above the seawater where it enters the OCEAN. Eventually, it breaks away and forms an ICEBERG.

Igneous rock. ROCKS formed when molten material or MAGMA cools and crystallizes into solid rock. The type of rock varies with the composition of the magma and, more important, with the rate of cooling. Rocks that cool slowly, far beneath the earth's surface, are igneous INTRUSIVE ROCKS. These have large crystals and coarse grains. GRANITE is the most typical igneous intrusive rock. When cooling is more rapid, usually closer to or at the surface, finer-grained igneous EXTRUSIVE ROCKS such as rhyolite are formed. If the magma flows out to the surface as LAVA, it may cool quickly, forming a glassy rock called obsidian. If there is gas in the lava, rocks full of holes from bubbles of escaping gases form; PUMICE and BASALT are common igneous extrusive rocks.

Impact crater. Generally circular depression formed on the surface of a PLANET by the impact of a high-velocity projectile such as a METEORITE, ASTEROID, or COMET.

Impact volcanism. Process in which major impact events produce huge CRATERS along with MAGMA RESERVOIRS that subsequently produce volcanic activity. Such cratering is clearly visible on the MOON, Mars, Mercury, and probably Venus. It is assumed that Earth had similar craters, but EROSION has erased most of the evidence.

Import substitution. Economic process in which domestic producers manufacture or supply goods or services that were previously imported or purchased from overseas and foreign producers.

Index fossil. Remains of an ancient organism that are useful in establishing the age of ROCKS; index fossils are abundant and have a wide geographic DISTRIBUTION, a narrow stratigraphic RANGE, and a distinctive form.

Indian summer. Short period, usually not more than a week, of unusually warm WEATHER in late October or early November in the NORTHERN HEMISPHERE. Before the Indian summer, TEMPERATURES are cooler and there can be occurrences of FROST. Indian summer DAYS are marked by clear to hazy skies and calm to light WINDS, but nights are cool. The weather pattern is a high-pressure cell or ridge located for a few days over the East Coast of North America. The name originated in New England, referring to the practice of NATIVE AMERICANS gathering foods for winter storage over this brief spell. Similar weather in England is called an Old Wives' summer.

Infant mortality. DEMOGRAPHIC MEASURE calculated as the number of deaths in a year of infants, or children under one year of age, compared with the total number of live births in a COUNTRY for the same year. Low-income countries have high infant mortality rates, more than 100 infant deaths per thousand.

Infauna. Organisms that live in the seafloor.

Infiltration. Movement of water into and through the SOIL.

Initial advantage. In terms of economic DEVELOPMENT, not all LOCATIONS are suited for profitable investment. Some locations offer initial advantages, including an existing skilled labor pool, existing consumer markets, existing plants, and situational advantages. These advantages can also lead to clustering of a number of industries at a particular location and to further economic growth, which will provide the preconditions of initial advantage for further economic development.

Inlier. REGION of old ROCKS that is completely surrounded by younger rocks. These are often PLACES where ORES or MINERALS are found in commercial quantities.

Inner core. The innermost layer of the earth; the inner core is a solid ball with a radius of about 900 miles.

Inselberg. Exposed rocky HILL in a DESERT area, made of resistant ROCKS, rising steeply from the flat surrounding countryside. There are many inselbergs in Africa, but Uluru (Ayers Rock) in Australia is possibly the most famous inselberg. The word is German for "island mountain." A special type of inselberg is a bornhardt.

Insolation. ENERGY received by the earth from the SUN, which heats the earth's surface. The average insolation received at the top of the earth's ATMOSPHERE at an average distance from the Sun is called the SOLAR CONSTANT. Insolation is predominantly shortwave radiation, with wavelengths in the RANGE of 0.39 to 0.76 micrometers, which corresponds to the visible spectrum. Less than half of the incoming SOLAR ENERGY reaches the earth's surface-insolation is reflected back into space by CLOUDS; smaller amounts are reflected back by surfaces, absorbed, or scattered by the atmosphere. Insolation is not distributed evenly over the earth, because of Earth's curved surface. Where the rays are perpendicular, at the SUBSOLAR POINT, insolation is at the maximum. The word is a shortened form of incoming (or intercepted) SOLAR RADIATION.

Insular climate. Island climates are influenced by the fact that no PLACE is far from the SEA. Therefore, both the DIURNAL (daily) TEMPERATURE RANGE and the annual temperature range are small.

Insurgent state. STATE that arises when an uprising or guerrilla movement gains control of part of the territory of a COUNTRY, then establishes its own form of control or government. In effect, the insurgents create a state within a state. In Colombia, for example, the government and armed forces have been unable to control several REGIONS where insurgents have created their own domains. This is generally related to coca growing and the production of cocaine. Civilian farmers are unable to resist the drug-financed "armies."

Interfluve. Higher area between two STREAMS; the surface over which water flows into the stream. These surfaces are subject to RUNOFF and EROSION by RILL action and GULLYING. Over time, interfluves are lowered.

Interlocking spur. STREAM in a hilly or mountainous REGION that winds its way in a sinuous VALLEY between the different RIDGES, slowly eroding the ends of the spurs and straightening its course. The view of interlocking spurs looking upstream is a favorite of artists, as colors change with the receding distance of each interlocking spur.

Intermediate rock. IGNEOUS ROCK that is transitional between a basic and a silicic ROCK, having a SILICA content between 54 and 64 percent.

Internal migration. Movement of people within a COUNTRY, from one REGION to another. Internal MIGRATION in high-income economies is often urban-to-RURAL, such as the migration to the SUN BELT in the United States. In low-income economies, rural-to-URBAN migration is more common.

Intertillage. Mixed planting of different seeds and seedling crops within the same SWIDDEN or cleared patch of agricultural land. Potatoes, yams, corn, rice, and bananas might all be planted. The planting times are staggered throughout the year to increase the variety of crops or nutritional balance available to the subsistence farmer and his or her family.

Intrusive rock. IGNEOUS ROCK which was formed from a MAGMA that cooled below the surface of the earth; it is commonly coarse-grained.

Irredentism. Expansion of one COUNTRY into the territory of a nearby country, based on the residence of nationals in the neighboring country. Hitler used irredentist claims to invade Czechoslovakia, because small groups of German-speakers lived there in the Sudetenland. The term comes from Italian, referring to Italy's claims before World War I that all Italian-speaking territory should become part of Italy.

Isallobar. Imaginary line on a MAP or meteorological chart joining PLACES with an equal change in ATMOSPHERIC PRESSURE over a certain time, often three hours. Isallobars indicate a pressure tendency and are used in WEATHER FORECASTING.

Island arc. Chain of VOLCANOES next to an oceanic TRENCH in the OCEAN BASINS; an oceanic PLATE descends, or subducts, below another oceanic plate at ISLAND arcs.

Isobar. Imaginary line joining PLACES of equal ATMOSPHERIC PRESSURE. WEATHER MAPS show isobars encircling areas of high or low pressure. The spacing between isobars is related to the pressure gradient.

Isobath. Line on a MAP or CHART joining all PLACES where the water depth is the same; a kind of underwater CONTOUR LINE. This kind of map is a BATHYMETRIC CONTOUR.

Isoclinal folding. When the earth's CRUST is folded, the size and shape of the folds vary according to the force of compression and nature of the ROCKS. When the surface is compressed evenly so that the two sides of the fold are parallel, isoclinal folding results. When the sides or slopes of the fold are unequal or dissimilar in shape and angle, this can be an asymmetrical or overturned fold. See also ANTICLINE; SYNCLINE.

Isotherm. Line joining PLACES of equal TEMPERATURE. A world MAP with isotherms of average monthly temperature shows that over the OCEANS, temperature decreases uniformly from the EQUATOR to the POLES, and higher temperatures occur over the CONTINENTS in summer and lower temperatures in winter because of the unequal heating properties of land and water.

Isotropic surface. Hypothetical flat surface or PLAIN, with no variation in any physical attribute. An isotropic surface has uniform ELEVATION, SOIL type, CLIMATE, and VEGETATION. Economic geographic models study behavior on an isotropic surface before applying the results to the real world. For example, in an isotropic model, land value is highest at the CITY center and falls regularly with increasing distance from there. In the real world, land values are affected by elevation, water features, URBAN regulations, and other factors. The von Thuenen model of the Isolated State is based on a uniform plain or isotropic surface.

Isthmian links. Chains of ISLANDS between substantial LANDMASSES.

Isthmus. Narrow strip of land connecting two larger bodies of land. The Isthmus of Panama connects North and South America; the Isthmus of Suez connects Africa and Asia. Both of these have been cut by CANALS to shorten shipping routes.

Jet stream. WINDS that move from west to east in the upper ATMOSPHERE, 23,000 to 33,000 feet (7,000–10,000 meters) above the earth, at about 200 miles (300 km.) per hour. They are narrow bands, elliptical in cross section, traveling in irregular paths. Four jet streams of interest to earth scientists and meteorologists are the polar jet stream and the subtropical jet stream in the Northern and SOUTHERN HEMISPHERES. The polar jet stream is located at the TROPOPAUSE, the BOUNDARY between the TROPOSPHERE and the STRATOSPHERE, along the polar FRONT. There is a complex interaction between surface winds and jet streams. In winter the NORTHERN HEMISPHERE polar front can move as far south as Texas, bringing BLIZZARDS and extreme WEATHER conditions. In summer, the polar jet stream is located over Canada. The subtropical jet stream is located at the tropopause around 30 DEGREES north or south LATITUDE, but it also migrates north or south, depending on the SEASON.

At times, the polar and subtropical jet streams merge for a few DAYS. Aircraft take advantage of the jet stream, or avoid it, depending on the direction of their flight. Upper atmosphere winds are also known as GEOSTROPHIC winds.

Joint. Naturally occurring fine crack in a ROCK, formed by cooling or by other stresses. SEDIMENTARY ROCKS can split along bedding planes; other joints form at right angles to the STRATA, running vertically through the rocks. In IGNEOUS ROCKS such as GRANITE, the stresses of cooling and contraction cause three sets of joints, two vertical and one parallel to the surface, which leads to the formation of distinctive LANDFORMS such as TORS. BASALT often demonstrates columnar jointing, producing tall columns that are mostly hexagonal in section. The presence of joints in BEDROCK hastens WEATHERING, because water can penetrate into the joints. This is particularly obvious in LIMESTONE, where joints are rapidly enlarged by SOLUTION. FROST WEDGING is a type of PHYSICAL WEATHERING that can split large boulders through the expansion when water in a joint freezes to form ice. Compare with FAULTS, which occur through tectonic activity.

Jurassic. Second of the three PERIODS that make up

Kame. Small HILL of gravel or mixed-size deposits, SAND, and gravel. Kames are found in areas previously covered by CONTINENTAL GLACIERS or ICE SHEETS, near what was the outer edge of the ice. They may have formed by materials dropping out of the melting ice, or in a deltalike deposit by a STREAM of MELTWATER. These deposits of which kames are made are called drift. Small LAKES called KETTLES are often found nearby. A closely spaced group of kames is called a kame field.

Karst. LANDSCAPE of SINKHOLES, underground STREAMS and caverns, and associated features created by CHEMICAL WEATHERING, especially SOLUTION, in REGIONS where the BEDROCK is LIMESTONE. The name comes from a region in the southwest of what is now Slovenia, the Krs (Kras) Plateau, but the karst region extends south through the Dinaric Alps bordering the

Adriatic Sea, into Bosnia-Herzegovina and Montenegro. Where limestone is well jointed, RAINFALL penetrates the JOINTS and enters the GROUNDWATER, carrying the MINERALS, especially calcium, away in solution. Most of the famous CAVES and caverns of the world are found in karst areas. The Carlsbad Caverns in New Mexico are a good example. Kentucky, Tennessee, and Florida also have well-known areas of karst. In some tropical countries, a form called tower karst is found. Tall conical or steep-sided HILLS of limestone rise above the flat surrounding landscape. Around 15 percent of the earth's land surface is karst TOPOGRAPHY.

Katabatic wind. GRAVITY DRAINAGE WINDS similar to MOUNTAIN BREEZES but stronger in force and over a larger area than a single VALLEY. Cold AIR collects over an elevated REGION, and the dense cold air flows strongly downslope. The ICE-SHEETS of Antarctica and Greenland produce fierce katabatic winds, but they can occur in smaller regions. The BORA is a strong, cold, squally downslope wind on the Dalmatian COAST of Yugoslavia in winter.

Kettle. Small depression, often a small LAKE, produced as a result of continental GLACIATION. It is formed by an isolated block of ice remaining in the ground MORAINE after a GLACIER has retreated. Deposited material accumulates around the ice, and when it finally melts, a steep hole remains, which often fills with water. Walden Pond, made famous by writer Henry David Thoreau, is a glacial kettle.

Khamsin. Hot, dry, DUST-laden WIND that blows in the eastern Sahara, in Egypt, and in Saudi Arabia, bringing high TEMPERATURES for three or four DAYS. Winds can reach GALE force in intensity. The word Khamsin is Arabic for "fifty" and refers to the period between March and June when the khamsin can occur.

Knickpoint. Abrupt change in gradient of the bed of a RIVER or STREAM. It is marked by a WATER-FALL, which over time is eroded by FLUVIAL action, restoring the smooth profile of the riverbed. The knickpoint acts as a TEMPORARY BASE LEVEL for the upper part of the stream.

Knickpoints can occur where a hard layer of ROCK is slower to erode than the rocks downstream, for example at Niagara Falls. Other knickpoints and waterfalls can develop as a result of tectonic forces. UPLIFT leads to new EROSION by a stream, creating a knickpoint that gradually moves upstream. The bed of a tributary GLACIER is often considerably higher than the VALLEY of the main glacier, so that after the glaciers have melted, a waterfall emerges over this knickpoint from the smaller hanging valley to join the main stream. Yosemite National Park has several such waterfalls.

Köppen climate classification. Commonly used scheme of CLIMATE classification that uses statistics of average monthly TEMPERATURE, average monthly PRECIPITATION, and total annual precipitation. The system was devised by Wladimir Köppen early in the twentieth century.

La Niña. WEATHER phenomenon that is the opposite part of EL NIÑO. When the SURFACE WATER in the eastern Pacific Ocean is cooler than average, the southeast TRADE WINDS blow strongly, bringing heavy rains to countries of the western Pacific. Scientists refer to the whole RANGE of TEMPERATURE, pressure, WIND, and SEA LEVEL changes as the SOUTHERN OSCILLATION (ENSO). The term "El Niño" gained wide currency in the U.S. media after a strong ENSO warm event in 1997–1998. A weak ENSO cold event, or La Niña, followed it in 1998. Means "the little girl" in Spanish. Alternative terms are "El Viejo" and "anti-El Niño."

Laccolith. LANDFORM of INTRUSIVE volcanism formed when viscous MAGMA is forced between overlying sedimentary STRATA, causing the surface to bulge upward in a domelike shape.

Lahar. Type of mass movement in which a MUD-FLOW occurs because of a volcanic explosion or ERUPTION. The usual cause is that the heat from the LAVA or other pyroclastic material melts ice and SNOW at the VOLCANO's SUMMIT, causing a hot mudflow that can move downslope with great speed. The eruption of Mount Saint Helens in 1985 was accompanied by a lahar.

Lake basin. Enclosed depression on the surface of the land in which SURFACE WATERS collect; BASINS are created primarily by glacial activity and tectonic movement.

Lakebed. Floor of a LAKE.

Land bridge. Piece of land connecting two CONTINENTS, which permits the MIGRATION of humans, animals, or plants from one area to another. Many former land bridges are now under water, because of the rise in SEA LEVEL after the last ICE AGE. The Bering Strait connecting Asia and North America was an important land bridge for the latter continent.

Land hemisphere. Because the DISTRIBUTION of land and water surfaces on Earth is quite asymmetrical on either side of the EQUATOR, the NORTHERN HEMISPHERE might well be called the land hemisphere. For many centuries, Europeans refused to believe that there was not an equal area of land in the SOUTHERN HEMISPHERE. Explorers such as James Cook were dispatched to seek such a "Great South Land."

Landmass. Large area of land—an ISLAND or a CONTINENT.

Landsat. Space-exploration project begun in 1972 to MAP the earth continuously with SATELLITE imaging. The satellites have collected data about the earth: its AGRICULTURE, FORESTS, flat lands, MINERALS, waters, and ENVIRONMENT. These were the first satellites to aid in Earth sciences, helping to produce the best maps available and assisting farmers around the world to improve their crop yields.

Language family. Group of related LANGUAGES believed to have originated from a common prehistoric language. English belongs in the Indo-European language family, which includes the languages spoken by half of the world's peoples.

Lapilli. Small ROCK fragments that are ejected during volcanic ERUPTIONS. A lapillus ranges from about the size of a pea to not larger than a walnut. Some lapilli form by accretion of VOLCANIC ASH around moisture droplets, in a manner similar to hailstone formation. Lapilli sometimes form into a textured rock called lapillistone.

Laterite. Bright red CLAY SOIL, rich in iron oxide, that forms in tropical CLIMATES, where both TEMPERATURE and PRECIPITATION are high year-round, as ROCKS weather. It can be used in brick making and is a source of iron. When the soil is rich in aluminum, it is called BAUXITE. When laterite or bauxite forms a hard layer at the surface, it is called duricrust. Australia and sub-Saharan Africa have large areas of duricrust, some of which is thought to have formed under previous conditions during the TRIASSIC period.

Laurasia. Hypothetical SUPERCONTINENT made up of approximately the present CONTINENTS of the NORTHERN HEMISPHERE.

Lava tube. Cavern structure formed by the draining out of liquid LAVA in a pahoehoe flow.

Layered plains. Smooth, flat REGIONS believed to be composed of materials other than sulfur compounds.

Leaching. Removal of nutrients from the upper horizon or layer of a SOIL, especially in the humid TROPICS, because of heavy RAINFALL. The remaining soil is often bright red in color because iron is left behind. Despite their bright color, tropical soils are infertile.

Leeward. Rear or protected side of a MOUNTAIN or RANGE is the leeward side. Compare to WINDWARD.

Legend. Explanation of the different colors and symbols used on a MAP. For example, a map of the world might use different colors for high-income, middle-income, and low-income economies. A historical map might use different colors for countries that were once colonies of Britain, France, or Spain.

Light year. Distance traveled by light in one year; widely used for measuring stellar distances, it is equal to roughly 6 trillion miles (9.5 million km.).

Lignite. Low-grade COAL, often called brown coal. It is mined and used extensively in eastern Germany, Slovakia, and the Moscow Basin.

Liquefaction. Loss in cohesiveness of water-saturated SOIL as a result of ground shaking caused by an EARTHQUAKE.

Lithification. Process whereby loose material is transformed into solid ROCK by compaction or cementation.

Lithology. Description of ROCKS, such as rock type, MINERAL makeup, and fluid in rock pores.

Lithosphere. Solid outermost layer of the earth. It varies in thickness from a few miles to more than 120 miles (200 km.). It is broken into pieces known as TECTONIC PLATES, some of which are extremely large, while others are quite small. The upper layer of the lithosphere is the CRUST, which may be CONTINENTAL CRUST or OCEANIC CRUST. Below the crust is a layer called the ASTHENOSPHERE, which is weaker and plastic, enabling the motion of tectonic plates.

Littoral. Adjacent to or related to a SEA.

Llanos. Grassy REGION in the Orinoco Basin of Venezuela and part of Colombia. SAVANNA VEGETATION gradually gives way to scrub at the outer edges of the *llanos.* The area is relatively undeveloped.

Loam. SOIL TEXTURE classification, indicating a soil that is approximately equal parts of SAND, SILT, and CLAY. Farmers generally consider a sandy loam to be the best soil texture because of its water-retaining qualities and the ease with which it can be cultivated.

Local sea-level change. Change in SEA LEVEL only in one area of the world, usually by land rising or sinking in that specific area.

Lode deposit. Primary deposit, generally a VEIN, formed by the filling of a FISSURE with MINERALS precipitated from a HYDROTHERMAL solution.

Loess. EOLIAN, or wind-blown, deposit of fine, silt-sized, light-colored material. Loess covers about 10 percent of the earth's land surface. The loess PLATEAU of China is good agricultural land, although susceptible to EROSION. Loess has the property of being able to form vertical CLIFFS or BLUFFS, and many people have built dwellings in the steep cliffs above the Huang He (Yellow) River. In the United States, loess deposits are found in the VALLEYS of the Platte, Missouri, Mississippi, and Ohio Rivers, and on the Columbia Plateau. A German word, meaning loose or unconsolidated, which comes from loess deposits along the Rhine River.

Longitudinal bar. Midchannel accumulation of SAND and gravel with its long end oriented roughly parallel to the RIVER flow.

Longshore current. Current in the OCEAN close to the SHORE, in the surf zone, produced by WAVES approaching the COAST at an angle. Also called a LITTORAL current. The longshore current combined with wave action can move large amounts of SAND and other BEACH materials down the coast, a process called LONGSHORE DRIFT.

Longshore drift. The movement of SEDIMENT parallel to the BEACH by a LONGSHORE CURRENT.

Maar. Explosion vent at the earth's surface where a volcanic cone has not formed. A small ring of pyroclastic materials surrounds the maar. Often a LAKE occupies the small CRATER of a maar. A larger form is called a TUFF RING.

Macroburst. Updrafts and downdrafts within a CUMULONIMBUS CLOUD or THUNDERSTORM can cause severe TURBULENCE. A DOWNBURST within a thunderstorm when windspeeds are greater than 130 miles (210 km.) per hour and over areas of 2.5 square miles (5 sq. km.) or more is called a macroburst. See also MICROBURST.

Magnetic poles. Locations on the earth's surface where the earth's MAGNETIC FIELD is perpendicular to the surface. The magnetic poles do not correspond exactly to the geographic North Pole and South Pole, or earth's AXIS; the difference is called magnetic variation or DECLINATION.

Magnetic reversal. Change in the earth's MAGNETIC FIELD from the North Pole to the South MAGNETIC POLE.

Magnetic storm. Rapid changes in the earth's MAGNETIC FIELD as a result of the bombardment of the earth by electrically charged particles from the SUN.

Magnetosphere. REGION surrounding a PLANET where the planet's own MAGNETIC FIELD predominates over magnetic influences from the SUN or other planets.

Mantle convection. Thermally driven flow in the earth's MANTLE thought to be the driving force of PLATE TECTONICS.

Mantle plume. Rising jet of hot MANTLE material that produces tremendous volumes of basaltic LAVA. See also HOT SPOT.

Map projection. Mathematical formula used to transform the curved surface of the earth onto a flat plane or sheet of paper. Projections are divided into three classes: CYLINDRICAL, CONICAL, and AZIMUTHAL.

Marchland. FRONTIER area where boundaries are poorly defined or absent. The marches themselves were a type of BOUNDARY REGION. Marchlands have changed hands frequently throughout history. The name is related to the fact that armies marched across them.

Mass balance. Summation of the net gain and loss of ice and SNOW mass on a GLACIER in a year.

Mass extinction. Die-off of a large percentage of species in a short time.

Mass wasting. Downslope movement of Earth materials under the direct influence of GRAVITY.

Massif. French term used in geology to describe very large, usually IGNEOUS INTRUSIVE bodies.

Meandering river. RIVER confined essentially to a single CHANNEL that transports much of its SEDIMENT load as fine-grained material in SUSPENSION.

Mechanical weathering. Another name for PHYSICAL WEATHERING, or the breaking down of ROCK into smaller pieces.

Mechanization. Replacement of human labor with machines. Mechanization occurred in AGRICULTURE as tractors, reapers, picking machinery, and similar technological inventions took the place of human farm labor. Mechanization in industry was part of the INDUSTRIAL REVOLUTION, as spinning and weaving machines were introduced into the textile industry.

Medical geography. Branch of geography specializing in the study of health and disease, with a particular emphasis on the areal spread or DIFFUSION of disease. The spatial perspective of geography can lead to new medical insights. Geographers working with medical researchers in

Africa have made great contributions to understanding the role of disease on that CONTINENT. John Snow's studies of the origin and spread of cholera in London in 1854 mark the beginnings of medical geography.

Megalopolis. Conurbation formed when large cities coalesce physically into one huge built-up area. Originally coined by the French geographer Jean Gottman in the early 1960s for the northeastern part of the United States, from Boston to Washington, D.C.

Mesa. Flat-topped HILL with steep sides. EROSION removes the surrounding materials, while the mesa is protected by a cap of harder, more resistant ROCK. Usually found in arid REGIONS. A larger LANDFORM of this type is a PLATEAU; a smaller feature is a BUTTE. The Colorado Plateau and Grand Canyon in particular are rich in these landforms. From the Spanish word for table.

Mesosphere. Atmospheric layer above the STRATOSPHERE where TEMPERATURE drops rapidly.

Mestizo. Person of mixed European and Amerindian ancestry, especially in countries of LATIN AMERICA.

Metamorphic rock. Any ROCK whose mineralogy, MINERAL chemistry, or TEXTURE has been altered by heat, pressure, or changes in composition; metamorphic rocks may have IGNEOUS, SEDIMENTARY, or other, older metamorphic rocks as their precursors.

Metamorphic zone. Areas of ROCK affected by the same limited RANGE of TEMPERATURE and pressure conditions, commonly identified by the presence of a key individual MINERAL or group of minerals.

Meteor. METEOROID that enters the ATMOSPHERE of a PLANET and is destroyed through frictional heating as it comes in contact with the various gases present in the atmosphere.

Meteorite. Fragment of an ASTEROID that survives passage through the ATMOSPHERE and strikes the surface of the earth.

Meteoroid. Small planetary body that enters Earth's ATMOSPHERE because its path intersects the earth's ORBIT. Friction caused by the earth's atmosphere on the meteoroid creates a glowing METEOR, or "shooting star." This is a common phenomenon, and most meteors burn away completely. Those that are large enough to reach the ground are called METEORITES.

Microburst. Brief but intense downward WIND, lasting not more than fifteen minutes over an area of 0.6 to 0.9 square mile (1.5–8 sq. km.). Usually associated with THUNDERSTORMS, but are quite unpredictable. The sudden change in wind direction associated with a microburst can create wind shear that causes airplanes to crash, especially if it occurs during takeoff or landing. See also MACROBURST.

Microclimate. CLIMATE of a small area, at or within a few yards of the earth's surface. In this REGION, variations of TEMPERATURE, PRECIPITATION, and moisture can have a pronounced effect on the bioclimate, influencing the growth or well-being of plants and animals, including humans. DEW or FROST, RAIN SHADOW effects, wind-tunneling between tall buildings, and similar phenomena are studied by microclimatologists. Horticulturists know the variations in aspect that affect INSOLATION and temperature, so that certain plants grow best on south-facing walls, for example. The growing of grapes for wine production is a major industry where microclimatology is essential. The study of microclimatology was pioneered by the German meteorologist Rudolf Geiger.

Microcontinent. Independent LITHOSPHERIC PLATE that is smaller than a CONTINENT but possesses continental-type CRUST. Examples include Cuba and Japan.

Microstates. Tiny countries. In 2000, seventeen independent countries each had an area of less than 200 square miles (520 sq. km.). The smallest microstate is Vatican City, with an area of 0.2 square miles (0.5 sq. km.). Most of the world's microstates are island NATIONS, including Nauru, Tuvalu, Marshall Islands, Saint Kitts and Nevis, Seychelles, Maldives, Malta, Grenada, Saint Vincent and the Grenadines, Barbados, Antigua and Barbuda, and Palau.

Mineral species. Mineralogic division in which all the varieties in any one species have the same basic physical and chemical properties.

Monadnock. Isolated HILL far from a STREAM, composed of resistant BEDROCK. Monadnocks are found in humid temperate REGIONS. A similar LANDFORM in an arid region is an INSELBERG.

Monogenetic. Pertaining to a volcanic ERUPTION in which a single vent is used only once.

Moraine. Materials transported by a GLACIER, and often later deposited as a RIDGE of unsorted ROCKS and smaller material. Lateral moraine is found at the side of the glacier; medial moraine occurs when two glaciers join. Other types of moraine include ABLATION moraine, ground moraine, and push, RECESSIONAL, and TERMINAL MORAINE.

Mountain belts. Products of PLATE TECTONICS, produced by the CONVERGENCE of crustal PLATES. Topographic MOUNTAINS are only the surficial expression of processes that profoundly deform and modify the CRUST. Long after the mountains themselves have been worn away, their former existence is recognizable from the structures that mountain building forms within the ROCKS of the crust.

Nappe. Huge sheet of ROCK that was the upper part of an overthrust fold, and which has broken and traveled far from its original position due to the tremendous forces. The Swiss Alps have nappes in many LOCATIONS.

Narrows. STRAIT joining two bodies of water.

Nation-state. Political entity comprising a COUNTRY whose people are a national group occupying the area. The concept originated in eighteenth century France; in practice, such cultural homogeneity is rare today, even in France.

Natural increase, rate of. DEMOGRAPHIC MEASURE of POPULATION growth: the difference between births and deaths per year, expressed as a percentage of the POPULATION. The rate of natural increase for the United States in 2000 was 0.6 percent. In countries where the population is decreasing, the DEATH RATE is greater than the BIRTH RATE.

Natural selection. Main process of biological evolution; the production of the largest number of offspring by individuals with traits that are best adapted to their ENVIRONMENTS.

Nautical mile. Standard MEASUREMENT at SEA, equalling 6,076.12 feet (1.85 km.). The mile used for land measurements is called a statute mile and measures 5,280 feet (1.6 km.).

Neap tide. TIDE with the minimum RANGE, or when the level of the high tide is at its lowest.

Near-polar orbit. Earth ORBIT that lies in a plane that passes close to both the north and south POLES.

Nekton. PELAGIC organisms that can swim freely, without having to rely on OCEAN CURRENTS or WINDS. Nekton includes shrimp; crabs; oysters; MARINE reptiles such as turtles, crocodiles, and snakes; and even sharks; porpoises; and whales.

Net migration. Net balance of a COUNTRY or REGION's IMMIGRATION and EMIGRATION.

Nomadism. Lifestyle in which pastoral people move with grazing animals along a defined route, ensuring adequate pasturage and water for their flocks or herds. This lifestyle has decreased greatly as countries discourage INTERNATIONAL MIGRATION. A more restricted form of nomadism is TRANSHUMANCE.

North geographic pole. Northernmost REGION of the earth, located at the northern point of the PLANET's AXIS of ROTATION.

North magnetic pole. Small, nonstationary area in the Arctic Circle toward which a COMPASS needle points from any LOCATION on the earth.

Notch. Erosional feature found at the base of a SEA CLIFF as a result of undercutting by WAVE EROSION, bioabrasion from MARINE organisms, and dissolution of ROCK by GROUNDWATER seepage. Also known as a nip.

Nuclear energy. ENERGY produced from a naturally occurring isotope of uranium. In the process of nuclear FISSION, the unstable uranium isotope absorbs a neutron and splits to form tin and molybdenum. This releases more neurons, so a chain reaction proceeds, releasing vast amounts of heat energy. Nuclear energy was seen in the 1950s as the energy of the future, but safety fears and the problem of disposal of radioactive nuclear waste have led to public condemnation of nuclear power plants.

Nuée ardente. Hot cloud of ROCK fragments, ASH, and gases that suddenly and explosively erupt from some VOLCANOES and flow rapidly down their slopes.

Nunatak. Isolated MOUNTAIN PEAK or RIDGE that projects through a continental ICE SHEET. Found in Greenland and Antarctica.

Obduction. Tectonic collisional process, opposite in effect to SUBDUCTION, in which heavier OCEANIC CRUST is thrust up over lighter CONTINENTAL CRUST.

Oblate sphere. Flattened shape of the earth that is the result of ROTATION.

Occultation. ECLIPSE of any astronomical object other than the SUN or the MOON caused by the Moon or any PLANET, SATELLITE, or ASTEROID.

Ocean basins. Large worldwide depressions that form the ultimate RESERVOIR for the earth's water supply.

Ocean circulation. Worldwide movement of water in the SEA.

Ocean current. Predictable circulation of water in the OCEAN, caused by a combination of WIND friction, Earth's ROTATION, and differences in TEMPERATURE and density of the waters. The five great oceanic circulations, known as GYRES, are in the North Pacific, North Atlantic, South Pacific, South Atlantic, and Indian Oceans. Because of the CORIOLIS EFFECT, the direction of circulation is CLOCKWISE in the NORTHERN HEMISPHERE and COUNTERCLOCKWISE in the SOUTHERN HEMISPHERE, except in the Indian Ocean, where the direction changes annually with the pattern of winds associated with the Asian MONSOON. Currents flowing toward the EQUATOR are cold currents; those flowing away from the equator are warm currents. An important current is the warm Gulf Stream, which flows north from the Gulf of Mexico along the East Coast of the United States; it crosses the North Atlantic, where it is called the North Atlantic Drift, and brings warmer conditions to the western parts of Europe. The West Coast of the United States is affected by the cool, south-flowing California Current. The cool Humboldt, or Peru, Current, which flows north along the South American coast, is an important indicator of whether there will be an EL NIÑO event. Deep currents, below 300 feet (100 meters), are extremely complicated and difficult to study.

Oceanic crust. Portion of the earth's CRUST under its OCEAN BASINS.

Oceanic island. ISLANDS arising from seafloor volcanic ERUPTIONS, rather than from continental shelves. The Hawaiian Islands are the best-known examples of oceanic islands.

Off-planet. Pertaining to REGIONS off the earth in orbital or planetary space.

Ore deposit. Natural accumulation of MINERAL matter from which the owner expects to extract a metal at a profit.

Orogeny. MOUNTAIN-building episode, or event, that extends over a period usually measured in tens of millions of years; also termed a revolution.

Orographic precipitation. Phenomenon caused when an AIR mass meets a topographic barrier, such as a mountain RANGE, and is forced to rise; the air cools to saturation, and orographic precipitation falls on the WINDWARD side as rain or SNOW. The lee side is a RAIN SHADOW. This effect is noticeable on the West Coast of the United States, which has RAINFOREST on the windward side of the MOUNTAINS and DESERTS on the lee.

Orography. Study of MOUNTAINS that incorporates assessment of how they influence and are affected by WEATHER and other variables.

Oscillatory flow. Flow of fluid with a regular back-and-forth pattern of motion.

Overland flow. Flow of water over the land surface caused by direct PRECIPITATION.

Oxbow lake. LAKE created when floodwaters make a new, shorter CHANNEL and abandon the loop of a MEANDER. Over time, water in the oxbow lake evaporates, leaving a dry, curving, low-lying area known as a meander scar. Oxbow lakes are common on FLOODPLAINS. Another name for this feature is a cut-off.

Ozone hole. Decrease in the abundance of ANTARCTIC OZONE as sunlight returns to the POLE in early springtime

Ozone layer. Narrow band of the STRATOSPHERE situated near 18 miles (30 km.) above the earth's surface, where molecules of OZONE are concentrated. The average concentration is only one in 4 million, but this thin layer protects the earth by absorbing much of the ultraviolet light from the SUN and reradiating it as longer-wavelength radiation. Scientists were disturbed to discover that the ozonosphere was being destroyed by photochemical reaction with CHLOROFLUOROCARBONS (CFCs). The OZONE HOLES over the South and North Poles negatively affect several animal species, as well as humans; skin cancer risk is increasing rapidly as a consequence of depletion of the ozone layer. Stratospheric ozone should not be confused with ozone at lower levels, which is a result of PHOTOCHEMICAL SMOG. Also called the ozonosphere.

P wave. Fastest elastic wave generated by an EARTHQUAKE or artificial ENERGY source; basically an acoustic or shock wave that compresses and stretches solid material in its path.

Pangaea. Name used by Alfred Wegener for the SUPERCONTINENT that broke apart to create the present CONTINENTS.

Parasitic cone. Small volcanic cone that appears on the flank of a larger VOLCANO, or perhaps inside a CALDERA.

Particulate matter. Mixture of small particles that adversely affect human health. The particles may come from smoke and DUST and are in their highest concentrations in large URBAN AREAS, where they contribute to the "DUST DOME." Increased occurrences of illnesses such as asthma and bronchitis, especially in children, are related to high concentrations of particulate matter.

Pastoralism. Type of AGRICULTURE involving the raising of grazing animals, such as cattle, goats, and sheep. Pastoral nomads migrate with their domesticated animals in order to ensure sufficient grass and water for the animals.

Paternoster lakes. Small circular LAKES joined by a STREAM. These lakes are the result of glacial EROSION. The name comes from the resemblance to rosary beads and the accompanying prayer (the Our Father).

Pedestal crater. A CRATER that has assumed the shape of a pedestal as a result of unique shaping processes caused by WIND.

Pedology. Scientific study of SOILS.

Pelagic. Relating to life-forms that live on or in open SEAS, rather than waters close to land.

Peneplain. In the geomorphic CYCLE, or cycle of LANDFORM development, described by W. M. Davis, the final stage of EROSION led to the creation of an extensive land surface with low RELIEF. Davis named this a peneplain, meaning "almost a plain." It is now known that tectonic forces are so frequent that there would be insufficient time for such a cycle to complete all stages required to complete this landform.

Percolation. Downward movement of part of the water that falls on the surface of the earth, through the upper layers of PERMEABLE SOIL and ROCKS under the influence of GRAVITY. Eventually, it accumulates in the zone of SATURATION as GROUNDWATER.

Perforated state. STATE whose territory completely surrounds another state. The classic example of a perforated state is South Africa, within which lies the COUNTRY of Lesotho. Technically, Italy is perforated by the MICROSTATES of San Marino and Vatican City.

Perihelion. Point in Earth's REVOLUTION when it is closest to the SUN (usually on January 3). At perihelion, the distance between the earth and the Sun is 91,500,000 miles (147,255,000 km.). The opposite of APHELION.

Periodicity. The recurrence of related phenomena at regular intervals.

Permafrost. Permanently frozen SUBSOIL. The condition occurs in perennially cold areas such as the ARCTIC. No trees can grow because their roots cannot penetrate the permafrost. The upper portion of the frozen SOIL can thaw briefly in the summer, allowing many smaller plants to thrive in the long daylight. Permafrost occurs in about 25 percent of the earth's land surface, and the condition even hampers construction in REGIONS such as Siberia and ARCTIC Canada.

Perturb. To change the path of an orbiting body by a gravitational force.

Petrochemical. Chemical substance obtained from NATURAL GAS or PETROLEUM.

Petrography. Description and systematic classification of ROCKS.

Photochemical smog. Mixture of gases produced by the interaction of sunlight on the gases emanating from automobile exhausts. The gases include OZONE, nitrogen dioxide, carbon monoxide, and peroxyacetyl nitrates. Many large cities suffer from poor AIR quality because of photochemical smog. Severe health problems arise from continued exposure to photochemical smog.

Photometry. Technique of measuring the brightness of astronomical objects, usually with a photoelectric cell.

Phylogeny. Study of the evolutionary relationships among organisms.

Phylum. Major grouping of organisms, distinguished on the basis of basic body plan, grade of anatomical complexity, and pattern of growth or development.

Physiography. The PHYSICAL GEOGRAPHY of a PLACE—the LANDFORMS, water features, CLIMATE, SOILS, and VEGETATION.

Piedmont glacier. GLACIER formed when several ALPINE GLACIERS join together into a spreading glacier at the base of a MOUNTAIN or RANGE. The Malaspina glacier in Alaska is a good example of a piedmont glacier.

Place. In geographic terms, space that is endowed with physical and human meaning. Geographers study the relationship between people, places, and ENVIRONMENTS. The five themes that geographers use to examine the world are LOCATION, place, human/environment interaction, movement, and REGIONS.

Placer. Accumulation of valuable MINERALS formed when grains of the minerals are physically deposited along with other, nonvaluable mineral grains.

Planetary wind system. Global atmospheric circulation pattern, as in the BELT of prevailing westerly WINDS.

Plantation. Form of AGRICULTURE in which a large area of agricultural land is devoted to the production of a single cash crop, for export. Many plantation crops are tropical, such as bananas, sugarcane, and rubber. Coffee and tea plantations require cooler CLIMATES. Formerly, slave labor was used on most plantations, and the owners were Europeans.

Plate boundary. REGION in which the earth's crustal PLATES meet, as a converging (SUBDUCTION ZONE), diverging (MID-OCEAN RIDGE), TRANSFORM FAULT, or collisional interaction.

Plate tectonics. Theory proposed by German scientist Alfred Wegener in 1910. Based on extensive study of ancient geology, STRATIGRAPHY, and CLIMATE, Wegener concluded that the CONTINENTS were formerly one single enormous LANDMASS, which he named PANGAEA. Over the past 250 million years, Pangaea broke apart, first into LAURASIA and GONDWANALAND, and subsequently into the present continents. Earth scientists now believe that the earth's CRUST is composed of a series of thin, rigid PLATES that are in motion, sometimes diverging, sometimes colliding.

Plinian eruption. Rapid ejection of large volumes of VOLCANIC ASH that is often accompanied by the collapse of the upper part of the VOLCANO. Named either for Pliny the Elder, a Roman naturalist who died while observing the ERUPTION of Mount Vesuvius in 79 CE, or for Pliny the Younger, his nephew, who chronicled the eruption.

Plucking. Term used to describe the way glacial ice can erode large pieces of ROCK as it makes its way downslope. The ice penetrates JOINTS, other openings on the floor, or perhaps the side wall, and freezes around the block of stone, tearing it away and carrying it along, as part of the glacial MORAINE. The rocks contribute greatly to glacial ABRASION, causing deep grooves or STRIATIONS in some places. The jagged torn surface left behind is subject to further plucking. ALPINE GLACIERS can erode steep VALLEYS called glacial TROUGHS.

Plutonic. IGNEOUS ROCKS made of MINERAL grains visible to the naked eye. These igneous rocks have cooled relatively slowly. GRANITE is a good example of a plutonic rock.

Pluvial period. Episode of time during which rains were abundant, especially during the last ICE AGE, from a few million to about 10,000 years ago.

Polar stratospheric clouds. CLOUDS of ice crystals formed at extremely low TEMPERATURES in the polar STRATOSPHERE.

Polder. Lands reclaimed from the SEA by constructing DIKES to hold back the sea and then pumping out the water retained between the dikes and the land. Before AGRICULTURE is possible, the SOIL must be specially treated to remove the SALT. Some polders are used for recreational land; cities also have been built on polders. The largest polders are in the Netherlands, where the northern part, known as the Low Netherlands, covers almost half of the total area of this COUNTRY.

Polygenetic. Pertaining to volcanism from several physically distinct vents or repeated ERUPTIONS from a single vent punctuated by long periods of quiescence.

Polygonal ground. Distinctive geological formation caused by the repetitive freezing and thawing of PERMAFROST.

Possibilism. Concept that arose among French geographers who rejected the concept of ENVIRONMENTAL DETERMINISM, instead asserting that the relationship between human beings and the ENVIRONMENT is interactive.

Potable water. FRESH WATER that is being used for domestic consumption.

Potholes. Circular depressions formed in the bed of a RIVER when the STREAM flows over BEDROCK. The scouring of PEBBLES as a result of water TURBULENCE wears away the sides of the depression, deepening it vertically and producing a smooth, rounded pothole. (In modern parlance, the term is also applied to holes in public roads.)

Primary minerals. MINERALS formed when MAGMA crystallizes.

Primary wave. Compressional type of EARTHQUAKE wave, which can travel in any medium and is the fastest wave.

Primate city. CITY that is at least twice as large as the next-largest city in that COUNTRY. The "law of the primate city" was developed by U.S. geographer Mark Jefferson, to analyze the phenomenon of countries where one huge city dominates the political, economic, and cultural life of that country. The size and dominance of a primate city is a PULL FACTOR and ensures its continuing dominance.

Principal parallels. The most important lines of LATITUDE. PARALLELS are imaginary lines, parallel to the EQUATOR. The principal parallels are the equator at zero DEGREES, the tropic of CANCER at 23.5 degrees North, the tropic of CAPRICORN at 23.5 degrees south, the Arctic Circle at 66.5 degrees north, and the Antarctic Circle at 66.5 degrees south.

Protectorate. COUNTRY that is a political DEPENDENCY of another NATION; similar to a COLONY, but usually having a less restrictive relationship with its overseeing power.

Proterozoic eon. Interval between 2.5 billion and 544 million years ago. During this PERIOD in the GEOLOGIC RECORD, processes presently active on Earth first appeared, notably the first clear evidence for PLATE TECTONICS. ROCKS of the Proterozoic eon also document changes in conditions on Earth, particularly an apparent increase in atmospheric oxygen.

Pull factors. Forces that attract immigrants to a new COUNTRY or LOCATION as permanent settlers. They include economic opportunities, educational facilities, land ownership, gold rushes, CLIMATE conditions, democracy, and similar factors of attraction.

Push factors. Forces that encourage people to migrate permanently from their HOMELANDS to settle in a new destination. They include war, persecution for religious or political reasons, hunger, and similar negative factors.

Pyroclasts. Materials that are ejected from a VOLCANO into the AIR. Pyroclastic materials return to Earth at greater or lesser distances, depending on their size and the height to which they are thrown by the explosion of the volcano. The largest pyroclasts are volcanic bombs. Smaller pieces are volcanic blocks and scoria. These generally fall back onto the volcano and roll down the sides. Even smaller pyroclasts are LAPILLI, cinders, and VOLCANIC ASH. The finest pyroclastic materials may be carried by WINDS for great distances, even completely around the earth, as was the case with DUST from the Krakatoa explosion in 1883 and the early 1990s explosions of Mount Pinatubo in the Philippines.

Qanat. Method used in arid REGIONS to bring GROUNDWATER from mountainous regions to lower and flatter agricultural land. A qanat is a long tunnel or series of tunnels, perhaps more than a mile long. The word *qanat* is Arabic, but the first qanats are thought to have been constructed in Farsi-speaking Persia more than 2,000 years ago. Qanats are still used there, as well as in Afghanistan and Morocco.

Quaternary sector. Economic activity that involves the collection and processing of information. The rapid spread of computers and the Internet caused a major increase in the importance of employment in the quaternary sector.

Radar imaging. Technique of transmitting radar toward an object and then receiving the reflected radiation so that time-of-flight MEASUREMENTS provide information about surface TOPOGRAPHY of the object under study.

Radial drainage. The pattern of STREAM courses often reveals the underlying geology or structure of a REGION. In a radial drainage pattern, streams radiate outward from a center, like spokes on a wheel, because they flow down the slopes of a VOLCANO.

Radioactive minerals. MINERALS combining uranium, thorium, and radium with other elements. Useful for nuclear TECHNOLOGY, these minerals furnish the basic isotopes necessary not only for nuclear reactors but also for advanced medical treatments, metallurgical analysis, and chemicophysical research.

Rain gauge. Instrument for measuring RAINFALL, usually consisting of a cylindrical container open to the sky.

Rain shadow. Area of low PRECIPITATION located on the LEEWARD side of a topographic barrier such as a mountain RANGE. Moisture-laden WINDS are forced to rise, so they cool ADIABATICALLY, leading to CONDENSATION and precipitation on the WINDWARD side of the barrier. When the AIR descends on the other side of the MOUNTAIN, it is dry and relatively warm. The area to the east of the Rocky Mountains is in a rain shadow.

Range, mountain. Linear series of MOUNTAINS close together, formed in an OROGENY, or mountain-building episode. Tall mountain ranges such as the Rocky Mountains are geologically much younger than older mountain ranges such as the Appalachians.

Rapids. Stretches of RIVERS where the water flow is swift and turbulent because of a steep and rocky CHANNEL. The turbulent conditions are called WHITE WATER. If the change in ELEVATION is greater, as for small WATERFALLS, they are called CATARACTS.

Recessional moraine. Type of TERMINAL MORAINE that marks a position of shrinkage or wasting or a GLACIER. Continued forward flow of ice is maintained so that the debris that forms the moraine continues to accumulate. Recessional moraines occur behind the terminal moraine.

Recumbent fold. Overturned fold in which the upper part of the fold is almost horizontal, lying on top of the nearest adjacent surface.

Reef (geology). VEIN of ORE, for example, a reef of gold.

Reef (marine). Underwater ridge made up of sand, rocks, or coral that rises near to the water's surface.

Refraction of waves. Bending of waves, which can occur in all kinds of waves. When OCEAN WAVES approach a COAST, they start to break as they approach the SHORE because the depth decreases.

The wave speed is retarded and the WAVE CREST seems to bend as the wavelength decreases. If waves are approaching a coast at an oblique angle, the crest line bends near the shore until it is almost parallel. If waves are approaching a BAY, the crests are refracted to fit the curve of the bay.

Regression. Retreat of the SEA from the land; it allows land EROSION to occur on material formerly below the sea surface.

Relative humidity. Measure of the HUMIDITY, or amount of moisture, in the ATMOSPHERE at any time and place compared with the total amount of moisture that same AIR could theoretically hold at that TEMPERATURE. Relative humidity is a ratio that is expressed as a percentage. When the air is saturated, the relative humidity reaches 100 percent and rain occurs. When there is little moisture in the air, the relative humidity is low, perhaps 20 percent. Relative humidity varies inversely with temperature, because warm air can hold more moisture than cooler air. Therefore, when temperatures fall overnight, the air often becomes saturated and DEW appears on grass and other surfaces. The human COMFORT INDEX is related to the relative humidity. Hot temperatures are more bearable when relative humidity is low. Media announcers frequently use the term "humidity" when they mean relative humidity.

Replacement rate. The rate at which females must reproduce to maintain the size of the POPULATION. It corresponds to a FERTILITY RATE of 2.1.

Reservoir rock. Geologic ROCK layer in which OIL and gas often accumulate; often SANDSTONE or LIMESTONE.

Retrograde orbit. ORBIT of a SATELLITE around a PLANET that is in the opposite sense (direction) in which the planet rotates.

Retrograde rotation. ROTATION of a PLANET in a direction opposite to that of its REVOLUTION.

Reverse fault. Feature produced by compression of the earth's CRUST, leading to crustal shortening. The UPTHROWN BLOCK overhangs the downthrown block, producing a FAULT SCARP where the overhang is prone to LANDSLIDES. When the movement is mostly horizontal, along a low angle FAULT, an overthrust fault is formed. This is commonly associated with extreme FOLDING.

Reverse polarity. Orientation of the earth's MAG-NETIC FIELD so that a COMPASS needle points to the SOUTHERN HEMISPHERE.

Ria coast. Ria is a long narrow ESTUARY or RIVER MOUTH. COASTS where there are many rias show the effects of SUBMERGENCE of the land, with the SEA now occupying former RIVER VALLEYS. Generally, there are MOUNTAINS running at an angle to the coast, with river valleys between each RANGE, so that the ria coast is a succession of estuaries and promontories. The submergence can result from a rising SEA LEVEL, which is common since the melting of the PLEISTOCENE GLACIERS, or it can be the result of SUBSIDENCE of the land. There is often a great TIDAL RANGE in rias, and in some, a tidal BORE occurs with each TIDE. The eastern coast of the United States, from New York to South Carolina, is a ria coast. The southwest coast of Ireland is another. The name comes from Spain, where rias occur in the south.

Richter scale. SCALE used to measure the magnitude of EARTHQUAKES; named after U.S. physicist Charles Richter, who, together with Beno Gutenberg, developed the scale in 1935. The scale is a quantitative measure that replaced the older MERCALLI SCALE, which was a descriptive scale. Numbers range from zero to nine, although there is no upper limit. Each whole number increase represents an order of magnitude, or an increase by a factor of ten. The actual MEASUREMENT was logarithm to base 10 of the maximum SEISMIC WAVE amplitude (in thousandths of a millimeter) recorded on a standard SEISMO-GRAPH at a distance of 60 miles (100 km.) from the earthquake EPICENTER.

Rift valley. Long, low REGION of the earth's surface; a VALLEY or TROUGH with FAULTS on either side. Unlike valleys produced by EROSION, rift valleys are produced by tectonic forces that have caused the faults or fractures to develop in the ROCKS of Earth's CRUST. TENSION can lead to the block of land between two faults dropping in ELEVATION compared to the surrounding blocks, thus forming the rift valley. A small LANDFORM produced

in this way is called a GRABEN. A rift valley is a much larger feature. In Africa, the Great Rift Valley is partially occupied by Lake Malawi and Lake Tanganyika, as well as by the Red Sea.

Ring dike. Volcanic LANDFORM created when MAGMA is intruded into a series of concentric FAULTS. Later EROSION of the surrounding material may reveal the ring dike as a vertical feature of thick BASALT rising above the surroundings.

Ring of Fire. Zone of volcanic activity and associated EARTHQUAKES that marks the edges of various TECTONIC PLATES around the Pacific Ocean, especially those where SUBDUCTION is occurring.

Riparian. Term meaning related to the BANKS of a STREAM or RIVER. Riparian VEGETATION is generally trees, because of the availability of moisture. RIPARIAN RIGHTS allow owners of land adjacent to a river to use water from the river.

River terraces. LANDFORMS created when a RIVER first produces a FLOODPLAIN, by DEPOSITION of ALLUVIUM over a wide area, and then begins downcutting into that alluvium toward a lower BASE LEVEL. The renewed EROSION is generally because of a fall in SEA LEVEL, but can result from tectonic UPLIFT or a change in CLIMATE pattern due to increased PRECIPITATION. On either side of the river, there is a step up from the new VALLEY to the former alluvium-covered floodplain surface, which is now one of a pair of river terraces. This process may occur more than once, creating as many as three sets of terraces. These are called depositional terraces, because the terrace is cut into river deposits. Erosional terraces, in contrast, are formed by lateral migration of a river, from one part of the valley to another, as the river creates a floodplain. These terraces are cut into BEDROCK, with only a thin layer of alluvium from the point BAR deposits, and they do not occur in matching pairs.

River valleys. VALLEYS in which STREAMS flow are produced by those streams through long-term EROSION and DEPOSITION. The LANDFORMS produced by FLUVIAL action are quite diverse, ranging from spectacular CANYONS to wide, gently sloping valleys. The patterns formed by stream networks are complex and generally reflect the BEDROCK geology and TERRAIN characteristics.

Rock avalanche. Extreme case of a rockfall. It occurs when a large mass of ROCK moves rapidly down a steeply sloping surface, taking everything that lies in its path. It can be started by an EARTHQUAKE, rock-blasting operations, or vibrations from thunder or artillery fire.

Rock cycle. Cycle by which ROCKS are formed and reformed, changing from one type to another over long PERIODS of geologic time. IGNEOUS ROCKS are formed by cooling from molten MAGMA. Once exposed at the surface, they are subject to WEATHERING and EROSION. The products of erosion are compacted and cemented to form SEDIMENTARY ROCKS. The heat and pressure accompanying a volcanic intrusion causes adjacent rocks to be altered into METAMORPHIC ROCKS.

Rock slide. Event that occurs when water lubricates an unconsolidated mass of weathered ROCK on a steep slope, causing rapid downslope movement. In a RIVER VALLEY where there are steep SCREE slopes being constantly carried away by a swiftly flowing STREAM, the undercutting at the base can lead to constant rockslides of the surface layer of rock. A large rockslide is a ROCK AVALANCHE.

S waves. Type of SEISMIC disturbance of the earth when an EARTHQUAKE occurs. In an S wave, particles move about at right angles to the direction in which the wave is traveling. S waves cannot pass through the earth's CORE, which is why scientists believe the INNER CORE is liquid. Also called transverse wave, shear wave, or secondary wave.

Sahel. Southern edge of the Sahara Desert; a great stretch of semiarid land extending from the Atlantic Ocean in Senegal and Mauritania through Mali, Burkina Faso, Nigeria, Niger, Chad, and Sudan. Northern Ethiopia, Eritrea, Djibouti, and Somalia usually are included also. This transition zone between the hot DESERT and the tropical SAVANNA has low summer RAINFALL of less than 8 inches (200 millimeters) and a natural VEGETATION of low grasses with some small shrubs. The REGION traditionally has been used for PASTORALISM, raising goats, camels, and occasionally sheep. Since a prolonged DROUGHT in the 1970s, DESERTIFICATION, SOIL EROSION, and FAMINE have plagued the Sahel. The narrow band between the northern Sahara and the Mediterranean North African COAST is also called Sahel. "Sahel" is the Arabic word for edge.

Saline lake. LAKE with elevated levels of dissolved solids, primarily resulting from evaporative concentration of SALTS; saline lakes lack an outlet to the SEA. Well-known examples include Utah's Great Salt Lake, California's Mono Lake and Salton Sea, and the Dead Sea in the Middle East.

Salinization. Accumulation of SALT in SOIL. When IRRIGATION is used to grow crops in semiarid to arid REGIONS, salinization is frequently a problem. Because EVAPORATION is high, water is drawn upward through the soil, depositing dissolved salts at or near the surface. Over years, salinization can build up until the soil is no longer suitable for AGRICULTURE. The solution is to maintain a plentiful flow of water while ensuring that the water flows through the soil and is drained away.

Salt domes. Formations created when deeply buried salt layers are forced upwards. SALT under pressure is a plastic material, one that can flow or move slowly upward, because it is lighter than surrounding SEDIMENTARY ROCKS. The salt forms into a plug more than a half mile (1 km.) wide and as much as 5 miles (8 km.) deep, which passes through overlying sedimentary rock layers, pushing them up into a dome shape as it passes. Some salt domes emerge at the earth's surface; others are close to the surface and are easy to mine for ROCK SALT. OIL and NATURAL GAS often accumulate against the walls of a salt dome. Salt domes are numerous around the COAST of the Gulf of Mexico, in the North Sea REGION, and in Iran and Iraq, all of which are major oil-producing regions.

Sand dunes. Accumulations of SAND in the shape of mounds or RIDGES. They occur on some COASTS and in arid REGIONS. Coastal dunes are formed when the prevailing WINDS blow strongly on-shore, piling up sand into dunes, which may become stabilized when grasses grow on them. DESERT sand dunes are a product of DEFLATION, or wind EROSION removing fine materials to leave a DESERT PAVEMENT in one region and sand deposits in another. Sand dunes are classified by their shape into barchans, or crescent-shaped dunes; seifs or LONGITUDINAL DUNES; TRANSVERSE DUNES; star dunes; and sand drifts or sand sheets.

Sapping. Natural process of EROSION at the bases of HILL slopes or CLIFFS whereby support is removed by undercutting, thereby allowing overlying layers to collapse; SPRING SAPPING is the facilitation of this process by concentrated GROUNDWATER flow, generally at the heads of VALLEYS.

Saturation, zone of. Underground REGION below the zone of AERATION, where all pore space is filled with water. This water is called GROUNDWATER; the upper surface of the zone of saturation is the WATER TABLE.

Scale. Relationship between a distance on a MAP or diagram and the same distance on the earth. Scale can be represented in three ways. A linear, or graphic, scale uses a straight line, marked off in equally spaced intervals, to show how much of the map represents a mile or a kilometer. A representative fraction (RF) gives this scale as a ratio. A verbal scale uses words to explain the relationship between map size and actual size. For example, the RF 1:63,360 is the same as saying "one inch to the mile."

Scarp. Short version of the word "ESCARPMENT," a short steep slope, as at the edge of a PLATEAU. EARTHQUAKES lead to the formation of FAULT SCARPS.

Schist. METAMORPHIC ROCK that can be split easily into layers. Schist is commonly produced from the action of heat and pressure on SHALE or SLATE. The rock looks flaky in appearance. Mica-schists are shiny because of the development of visible mica. Other schists include talc-schist, which contains a large amount of talc, and hornblende-schist, which develops from basaltic rocks.

Scree. Broken, loose ROCK material at the base of a slope or CLIFF. It is often the result of FROST WEDGING of BEDROCK cliffs, causing rockfall. Another name for scree is TALUS.

Sedimentary rocks. ROCKS formed from SEDIMENTS that are compressed and cemented together in a process called LITHIFICATION. Sedimentary rocks cover two-thirds of the earth's land surface but are only a small proportion of the earth's CRUST. SANDSTONE is a common sedimentary rock. Sedimentary rocks form STRATA, or layers, and sometimes contain FOSSILS.

Seif dunes. Long, narrow RIDGES of SAND, built up by WINDS blowing at different times of year from two different directions. Seif dunes occur in parallel lines of sand over large areas, running for hundreds of miles in the Sahara, Iran, and central Australia. Another name for seif dunes is LONGITUDINAL DUNES. The Arabic word means sword.

Seismic activity. Movements within the earth's CRUST that often cause various other geological phenomena to occur; the activity is measured by SEISMOGRAPHS.

Seismology. The scientific study of EARTHQUAKES. It is a branch of GEOPHYSICS. The study of SEISMIC WAVES has provided a great deal of knowledge about the composition of the earth's interior.

Shadow zone. When an EARTHQUAKE occurs at one LOCATION, its waves travel through the earth and are detected by SEISMOGRAPHS around the world. Every earthquake has a shadow zone, a band where neither P nor S WAVES from the earthquake will be detected. This shadow zone leads scientists to draw conclusions about the size, density, and composition of the earth's CORE.

Shale oil. SEDIMENTARY ROCK containing sufficient amounts of hydrocarbons that can be extracted by slow distillation to yield OIL.

Shallow-focus earthquakes. EARTHQUAKES having a focus less than 35 miles (60 km.) below the surface.

Shantytown. URBAN SQUATTER SETTLEMENT, usually housing poor newcomers.

Shield. Large part of the earth's CONTINENTAL CRUST, comprising very old ROCKS that have been eroded to REGIONS of low RELIEF. Each CONTINENT has a shield area. In North America, the Canadian Shield extends from north of the Great Lakes to the Arctic Ocean. Sometimes known as a CONTINENTAL SHIELD.

Shield volcano. VOLCANO created when the LAVA is quite viscous or fluid and highly basaltic. Such lava spreads out in a thin sheet of great radius but comparatively low height. As flows continue to build up the volcano, a low DOME shape is created. The greatest shield volcanoes on Earth are the ISLANDS of Hawaii, which rise to a height of almost 30,000 feet (10,000 meters) above SEA LEVEL.

Shock city. CITY that typifies disturbing changes in social and cultural conditions or in economic conditions. In the nineteenth century, the shock city of the United States was Chicago.

Sierra. Spanish word for a mountain RANGE with a serrated crest. In California, the Sierra Nevada is an important range, containing Mount Whitney, the highest PEAK in the continental United States.

Sill. Feature formed by INTRUSIVE volcanic activity. When LAVA is forced between two layers of ROCK, it can form a narrow horizontal layer of BASALT, parallel with the adjacent beds. Although it resembles a windowsill in its flatness, a sill may be hundreds of miles long and can range in thickness from a few centimeters to considerable thickness.

Siltation. Build-up of SILT and SAND in creeks and waterways as a result of SOIL EROSION, clogging water courses and creating DELTAS at RIVER MOUTHS. Siltation often results from DEFORESTATION or removal of tree cover. Such ENVIRONMENTAL DEGRADATION causes loss of agricultural productivity, worsening of water supply, and other problems.

Sima. Abbreviation for SILICA and *ma*gnesium. These are the two principal constituents of heavy ROCKS such as BASALT, which forms much

of the OCEAN floor. Lighter, more abundant rock is SIAL.

Sinkhole. Circular depression in the ground surface, caused by WEATHERING of LIMESTONE, mainly through the effects of SOLUTION on JOINTS in the ROCK. If a STREAM flows above ground and then disappears down a sinkhole, the feature is called a swallow hole. In everyday language, many events that cause the surface to collapse are called sinkholes, even though they are rarely in limestone and rarely caused by weathering.

Sinking stream. STREAM or RIVER that loses part or all of its water to pathways dissolved underground in the BEDROCK.

Situation. Relationship between a PLACE, such as a TOWN or CITY, and its RELATIVE LOCATION within a REGION. A situation on the COAST is desirable in terms of overseas TRADE.

Slip-face. LEEWARD side of a SAND DUNE. As the WIND piles up sand on the WINDWARD side, it then slips down the rear or slip-face. The angle of the slip-face is gentler than the angle of the windward slope.

Slump. Type of LANDSLIDE in which the material moves downslope with a rotational motion, along a curved slip surface.

Snout. Terminal end of a GLACIER.

Snow line. The height or ELEVATION at which snow remains throughout the year, without melting away. Near the EQUATOR, the snow line is more than 15,000 feet (almost 5,000 meters); at higher LATITUDES, the snow line is correspondingly lower, reaching SEA LEVEL at the POLES. The actual snow line varies with the time of year, retreating in summer and coming lower in winter.

Soil horizon. SOIL consists of a series of layers called horizons. The uppermost layer, the O horizon, contains organic materials such as decayed leaves that have been changed into HUMUS. Beneath this is the A horizon, the TOPSOIL, where farmers plow and plant seeds. The B HORIZON often contains MINERALS that have been washed downwards from the A horizon, such as calcium, iron, and aluminum. The A and B horizons together comprise a solum, or true soil. The C horizon is weathered BEDROCK, which contains pieces of the original ROCK from which the soil formed. Another name for the C horizon is REGOLITH. Beneath this is the R horizon, or bedrock.

Soil moisture. Water contained in the unsaturated zone above the WATER TABLE.

Soil profile. Vertical section of a SOIL, extending through its horizon into the unweathered parent material.

Soil stabilization. Engineering measures designed to minimize the opportunity and/or ability of EXPANSIVE SOILS to shrink and swell.

Solar energy. One of the forms of ALTERNATIVE or RENEWABLE ENERGY. In the late 1990s, the world's largest solar power generating plant was located at Kramer Junction, California. There, solar energy heats huge OIL-filled containers with a parabolic shape, which produces steam to drive generating turbines. An alternative is the production of energy through photovoltaic cells, a TECHNOLOGY that was first developed for space exploration. Many individual homes, especially in isolated areas, use this technology.

Solar system. SUN and all the bodies that ORBIT it, including the PLANETS and their SATELLITES, plus numerous COMETS, ASTEROIDS, and METEOROIDS.

Solar wind. Gases from the SUN's ATMOSPHERE, expanding at high speeds as streams of charged particles.

Solifluction. Word meaning flowing SOIL. In some REGIONS of PERMAFROST, where the ground is permanently frozen, the uppermost layer thaws during the summer, creating a saturated layer of soil and REGOLITH above the hard layer of frozen ground. On slopes, the material can flow slowly downhill, creating a wavy appearance along the hillslope.

Solution. Form of CHEMICAL WEATHERING in which MINERALS in a ROCK are dissolved in water. Most substances are soluble, but the combination of water with CARBON DIOXIDE from the ATMOSPHERE means that RAINFALL is slightly acidic, so

that the chemical reaction is often a combination of solution and carbonation.

Sound. Long expanse of the SEA, close to the COAST, such as a large ESTUARY. It can also be the expanse of sea between the mainland and an ISLAND.

Source rock. ROCK unit or bed that contains sufficient organic carbon and has the proper thermal history to generate OIL or gas.

Spatial diffusion. Notion that things spread through space and over time. An understanding of geographic change depends on this concept. Spatial diffusion can occur in various ways. Geographers distinguish between expansion diffusion, relocation diffusion, and hierarchical diffusion.

Spheroidal weathering. Form of ROCK WEATHERING in which layers of rock break off parallel to the surface, producing a rounded shape. It results from a combination of physical and CHEMICAL WEATHERING. Spheroidal weathering is especially common in GRANITE, leading to the creation of TORS and similar rounded features. Onion-skin weathering is a term sometimes used, especially when this is seen on small rocks.

Spring tide. TIDE of maximum RANGE, occurring when lunar and solar tides reinforce each other, a few DAYS after the full and new MOONS.

Squall line. Line of vigorous THUNDERSTORMS created by a cold downdraft that spreads out ahead of a fast-moving COLD FRONT.

Stacks. Pieces of ROCK surrounded by SEA water, which were once part of the mainland. WAVE EROSION has caused them to be isolated. Also called sea stacks.

Stalactite. Long, tapering piece of calcium carbonate hanging from the roof of a LIMESTONE CAVE or cavern. Stalactites are formed as water containing the MINERAL in solution drips downward. The water evaporates, depositing the dissolved minerals.

Stalagmite. Column of calcium carbonate growing upward from the floor of a LIMESTONE CAVE or cavern.

Steppe. Huge REGION of GRASSLANDS in the midlatitudes of Eurasia, extending from central

Europe to northeast China. The region is not uniform in ELEVATION; most of it is rolling PLAINS, but some mountain RANGES also occur. These have not been a barrier to the migratory lifestyle of the herders who have occupied the steppe for many centuries. The Asian steppe is colder than the European steppe, because of greater elevation and greater continentality. The best-known rulers from the steppe were the Mongols, whose empire flourished in the thirteenth and fourteenth centuries. Geographers speak of a steppe CLIMATE, a semiarid climate where the EVAPORATION rate is double that of PRECIPITATION. South of the steppe are great DESERTS; to the north are midlatitude mixed FORESTS. In terms of climate and VEGETATION, the steppe is like the short-grass PRAIRIE vegetation west of the Mississippi River. Also called steppes.

Storm surge. General rise above normal water level, resulting from a HURRICANE or other severe coastal STORM.

Strait. Relatively narrow body of water, part of an OCEAN or SEA, separating two pieces of land. The world's busiest SEAWAY is the Johore Strait between the Malay Peninsula and the island of Sumatra.

Strata. Layers of SEDIMENT deposited at different times, and therefore of different composition and TEXTURE. When the sediments are laid down, strata are horizontal, but subsequent tectonic processes can lead to tilting, FOLDING, or faulting. Not all SEDIMENTARY ROCKS are stratified. Singular form of the word is stratum.

Stratified drift. Material deposited by glacial MELTWATERS; the water separates the material according to size, creating layers.

Stratigraphy. Study of sedimentary STRATA, which includes the concept of time, possible correlation of the ROCK units, and characteristics of the rocks themselves.

Stratovolcano. Type of VOLCANO in which the ERUPTIONS are of different types and produce different LAVAS. Sometimes an eruption ejects cinder and ASH; at other times, viscous lava flows down the sides. The materials flow, settle, and fall to produce a beautiful symmetrical LANDFORM with a broad circular base and concave slopes tapering upward to a small circular CRATER. Mount Rainier, Mount Saint Helens, and Mount Fuji are stratovolcanoes. Also known as a COMPOSITE CONE.

Streambed. Channel through which a STREAM flows. Dry streambeds are variously known as ARROYOS, DONGAS, WASHES, and WADIS.

Strike. Term used when earth scientists study tilted or inclined beds of SEDIMENTARY ROCK. The strike of the inclined bed is the direction of a horizontal line along a bedding plane. The strike is at right angles to the dip of the rocks.

Strike-slip fault. In a strike-slip fault, the surface on either side of the fault moves in a horizontal plane. There is no vertical displacement to form a FAULT SCARP, as there is with other types of faults. The San Andreas Fault is a strike-slip fault. Also called a transcurrent fault.

Subduction zone. CONVERGENT PLATE BOUNDARY where an oceanic PLATE is being thrust below another plate.

Sublimation. Process by which water changes directly from solid (ice) to vapor, or vapor to solid, without passing through a liquid stage.

Subsolar point. Point on the earth's surface where the SUN is directly overhead, making the Sun's rays perpendicular to the surface. The subsolar point receives maximum INSOLATION, compared with other PLACES, where the Sun's rays are oblique.

Sunspots. REGIONS of intense magnetic disturbances that appear as dark spots on the solar surface; they occur approximately every eleven years.

Supercontinent. Vast LANDMASS of the remote geologic past formed by the collision and amalgamation of crustal PLATES. Hypothesized supercontinents include PANGAEA, GONDWANALAND, and LAURASIA.

Supersaturation. State in which the AIR's RELATIVE HUMIDITY exceeds 100 percent, the condition necessary for vapor to begin transformation to a liquid state.

Supratidal. Referring to the SHORE area marginal to shallow OCEANS that are just above high-tide level.

Swamp. WETLAND where trees grow in wet to water-logged conditions. Swamps are common close to the RIVER on FLOODPLAINS, as well as in some coastal areas.

Swidden. Area of land that has been cleared for SUBSISTENCE AGRICULTURE by a farmer using the technique of slash-and-burn. A variety of crops is planted, partly to reduce the risk of crop failure. Yields are low from a swidden because SOIL fertility is low and only human labor is used for CLEARING, planting, and harvesting. See also INTERTILLAGE.

Symbolic landscapes. LANDSCAPES centered on buildings or structures that are so visually emblematic that they represent an entire CITY.

Syncline. Downfold or TROUGH shape that is formed through compression of ROCKS. An upfold is an ANTICLINE.

Tableland. Large area of land with a mostly flat surface, surrounded by steeply sloping sides, or ESCARPMENTS. A small PLATEAU.

Taiga. Russian name for the vast BOREAL FORESTS that cover Siberia. The marshy ground supports a tree VEGETATION in which the trees are CONIFEROUS, comprising mostly pine, fir, and larch.

Talus. Broken and jagged pieces of ROCK, produced by WEATHERING of steep slopes, that fall to the base of the slope and accumulate as a talus cone. In high MOUNTAINS, a ROCK GLACIER may form in the talus. See also SCREE.

Tarn. Small circular LAKE, formed in a CIRQUE, which was previously occupied by a GLACIER.

Tectonism. The formation of MOUNTAINS because of the deformation of the CRUST of the earth on a large scale.

Temporary base level. STREAMS or RIVERS erode their beds down toward a BASE LEVEL—in most cases, SEA LEVEL. A section of hard ROCK may slow EROSION and act as a temporary, or local, base level. Erosion slows upstream of the temporary base level. A DAM is an artificially constructed temporary base level.

Tension. Type of stress that produces a stretching and thinning or pulling apart of the earth's CRUST. If the surface breaks, a NORMAL FAULT is created, with one side of the surface higher than the other.

Tephra. General term for volcanic materials that are ejected from a vent during an ERUPTION and transported through the AIR, including ASH (volcanic), BLOCKS (volcanic), cinders, LAPILLI, scoria, and PUMICE.

Terminal moraine. RIDGE of unsorted debris deposited by a GLACIER. When a glacier erodes it moves downslope, carrying ROCK debris and creating a ground MORAINE of material of various sizes, ranging from big angular blocks or boulders down to fine CLAY. At the terminus of the glacier, where the ice is melting, the ground moraine is deposited, building the ridge of unsorted debris called a terminal moraine.

Terrain. Physical features of a REGION, as in a description of rugged terrain. It should not be confused with TERRANE.

Terrane. Piece of CONTINENTAL CRUST that has broken off from one PLATE and subsequently been joined to a different plate. The terrane has quite different composition and structure from the adjacent continental materials. Alaska is composed mostly of terranes that have accreted, or joined, the North American plate.

Terrestrial planet. Any of the solid, rocky-surfaced bodies of the inner SOLAR SYSTEM, including the PLANETS Mercury, Venus, Earth, and Mars and Earth's SATELLITE, the MOON.

Terrigenous. Originating from the WEATHERING and EROSION of MOUNTAINS and other land formations.

Texture. One of the properties of SOILS. The three textures are SAND, SILT, and CLAY. Texture is measured by shaking the dried soil through a series of sieves with mesh of reducing diameters. A mixture of sand, silt, and clay gives a LOAM soil.

Thermal equator. Imaginary line connecting all PLACES on Earth with the highest mean daily TEMPERATURE. The thermal equator moves south of the EQUATOR in the SOUTHERN HEMISPHERE summer, especially over the CONTINENTS of South America, Africa, and Australia. In the northern summer, the thermal equator moves far into Asia, northern Africa, and North America.

Thermal pollution. Disruption of the ECOSYSTEM caused when hot water is discharged, usually as a thermal PLUME, into a relatively cooler body of water. The TEMPERATURE change affects the aquatic ecosystem, even if the water is chemically pure. Nuclear power-generating plants use large volumes of water in the process and are important sources of thermal pollution.

Thermocline. Depth interval at which the TEMPERATURE of OCEAN water changes abruptly, separating warm SURFACE WATER from cold, deep water.

Thermodynamics. Area of science that deals with the transformation of ENERGY and the laws that govern these changes; equilibrium thermodynamics is especially concerned with the reversible conversion of heat into other forms of energy.

Thermopause. Outer limit of the earth's ATMOSPHERE.

Thermosphere. Atmospheric zone beyond the MESOSPHERE in which TEMPERATURE rises rapidly with increasing distance from the earth's surface.

Thrust belt. Linear BELT of ROCKS that have been deformed by THRUST FAULTS.

Thrust fault. FAULT formed when extreme compression of the earth's CRUST pushes the surface into folds so closely spaced that they overturn and the ROCK then fractures along a fault.

Tidal force. Gravitational force whose strength and direction vary over a body and thus act to deform the body.

Tidal range. Difference in height between high TIDE and low tide at a given point.

Tidal wave. Incorrect name for a TSUNAMI.

Till. Mass of unsorted and unstratified SEDIMENTS deposited by a GLACIER. Boulders and smaller rounded ROCKS are mixed with CLAY-sized materials.

Timberline. Another term for tree line, the BOUNDARY of tree growth on MOUNTAIN slopes. Above the timberline, TEMPERATURES are too cold for tree growth.

Tombolo. Strip of SAND or other SEDIMENT that connects an ISLAND or SEA stack to the mainland. Mont-Saint-Michel is linked to the French mainland by a tombolo.

Topography. Description of the natural LANDSCAPE, including LANDFORMS, RIVERS and other waters, and VEGETATION cover.

Topological space. Space defined in terms of the connectivity between LOCATIONS in that space. The nature and frequency of the connections are measured, while distance between locations is not considered an important factor. An example of topological space is a transport network diagram, such as a bus route or a MAP of an underground rail system. Networks are most concerned with flows, and therefore with connectivity.

Toponyms. PLACE names. Sometimes, names of features and SETTLEMENTS reveal a good deal about the history of a REGION. For example, the many names starting with "San" or "Santa" in the Southwest of the United States recall the fact that Spain once controlled that area. The scientific study of place names is toponymics.

Tor. Rocky outcrop of blocks of ROCK, or corestones, exposed and rounded by WEATHERING. Tors frequently form in GRANITE, where three series of JOINTS often developed as the rock originally cooled when it was formed.

Transform faults. FAULTS that occur along DIVERGENT PLATE boundaries, or SEAFLOOR SPREADING zones. The faults run perpendicular to the spreading center, sometimes for hundreds of miles, some for more than five hundred miles. The motion along a transform fault is lateral or STRIKE-SLIP.

Transgression. Flooding of a large land area by the SEA, either by a regional downwarping of continental surface or by a global rise in SEA LEVEL.

Transmigration. Policy of the government of Indonesia to encourage people to move from the densely overcrowded ISLAND of Java to the sparsely populated other islands.

Transverse bar. Flat-topped body of SAND or gravel oriented transverse to the RIVER flow.

Trophic level. Different types of food relations that are found within an ECOSYSTEM. Organisms that derive food and ENERGY through PHOTOSYNTHE-SIS are called autotrophs (self-feeders) or producers. Organisms that rely on producers as their source of energy are called heterotrophs (feeders on others) or consumers. A third trophic level is represented by the organisms known as decomposers, which recycle organic waste.

Tropical cyclone. STORM that forms over tropical OCEANS and is characterized by extreme amounts of rain, a central area of calm AIR, and spinning WINDS that attain speeds of up to 180 miles (300 km.) per hour.

Tropical depression. STORM with WIND speeds up to 38 miles (64 km.) per hour.

Tropopause. BOUNDARY layer between the TROPO-SPHERE and the STRATOSPHERE.

Troposphere. Lowest and densest of Earth's atmospheric layers, marked by considerable TURBU-LENCE and a decrease in TEMPERATURE with increasing ALTITUDE.

Tsunami. SEISMIC SEA WAVE caused by a disturbance of the OCEAN floor, usually an EARTH-QUAKE, although undersea LANDSLIDES or volcanic ERUPTIONS can also trigger tsunami.

Tufa. LIMESTONE or calcium carbonate deposit formed by PRECIPITATION from an alkaline LAKE. Mono Lake is famous for the dramatic tufa towers exposed by the lowering of the level of lake water. Also known as TRAVERTINE.

Tumescence. Local swelling of the ground that commonly occurs when MAGMA rises toward the surface.

Tunnel vent. Central tube in a volcanic structure through which material from the earth's interior travels.

U-shaped valley. Steep-sided VALLEY carved out by a GLACIER. Also called a glacial TROUGH.

Ubac slope. Shady side of a MOUNTAIN, where local or microclimatic conditions permit lower TIMBERLINES and lower SNOW LINES than occur on a sunny side.

Ultimate base level. Level to which a STREAM can erode its bed. For most RIVERS, this is SEA LEVEL. For streams that flow into a LAKE, the ultimate base level is the level of the lakebed.

Unconfined aquifer. AQUIFER whose upper BOUND-ARY is the WATER TABLE; it is also called a water table aquifer.

Underfit stream. STREAM that appears to be too small to have eroded the VALLEY in which it flows. A RIVER flowing in a glaciated valley is a good example of underfit.

Uniformitarianism. Theory introduced in the early nineteenth century to explain geologic processes. It used to be believed that the earth was only a few thousand years old, so the creation of LANDFORMS would have been rapid, even catastrophic. This theory, called CATASTROPHISM, explained most landforms as the result of the Great Flood of the Bible, when Noah, his family, and animals survived the deluge. Uniformitarian- ism, in contrast, stated that the processes in operation today are slow, so the earth must be immensely older than a mere few thousand years.

Universal time (UT). See GREENWICH MEAN TIME.

Universal Transverse Mercator. Projection in which the earth is divided into sixty zones, each six DEGREES of LONGITUDE wide. In a traditional Mercator projection, the earth is seen as a sphere with a cylinder wrapped around the EQUATOR. UTM can be visualized as a series of six-degree side strips running transverse, or north-south.

Unstable air. Condition that occurs when the AIR above rising air is unusually cool so that the rising air is warmer and accelerates upward.

Upthrown block. When EARTHQUAKE motion produces a FAULT, the block of land on one side is displaced vertically relative to the other. The higher is the upthrown block; the lower is the downthrown block.

Upwelling. OCEAN phenomenon in which warm SURFACE WATERS are pushed away from the

COAST and are replaced by cold waters that carry more nutrients up from depth.

Urban heat island. Cities experience a different MICROCLIMATE from surrounding REGIONS. The CITY TEMPERATURE is typically higher by a few DEGREES, both DAY and night, because of factors such as surfaces with higher heat absorption, decreased WIND strength, human heat-producing activities such as power generation, and the layer of AIR POLLUTION (DUST DOME).

Vadose zone. The part of the SOIL also known as the zone of AERATION, located above the WATER TABLE, where space between particles contains AIR.

Valley train. Fan-shaped deposit of glacial MORAINE that has been moved down-valley and redeposited by MELTWATER from the GLACIER.

Van Allen radiation belts. Bands of highly energetic, charged particles trapped in Earth's MAGNETIC FIELD. The particles that make up the inner BELT are energetic protons, while the outer belt consists mainly of electrons and is subject to DAY-night variations.

Varnish, desert. Shiny black coating often found over the surface of ROCKS in arid REGIONS. This is a form of OXIDATION or CHEMICAL WEATHERING, in which a coating of manganese oxides has formed over the exposed surface of the rock.

Varve. Pair of contrasting layers of SEDIMENT deposited over one year's time; the summer layer is light, and the winter layer is dark.

Ventifacts. PEBBLES on which one or more sides have been smoothed and faceted by ABRASION as the WIND has blown SAND particles.

Volcanic island arc. Curving or linear group of volcanic ISLANDS associated with a SUBDUCTION ZONE.

Volcanic rock. Type of IGNEOUS ROCK that is erupted at the surface of the earth; volcanic rocks are usually composed of larger crystals inside a fine-grained matrix of very small crystals and glass.

Volcanic tremor. Continuous vibration of long duration, detected only at active VOLCANOES.

Volcanology. Scientific study of VOLCANOES.

Voluntary migration. Movement of people who decide freely to move their place of permanent residence. It results from PULL FACTORS at the chosen destination, together with PUSH FACTORS in the home situation.

Warm temperate glacier. GLACIER that is at the melting TEMPERATURE throughout.

Water power. Generally means the generation of electricity using the ENERGY of falling water. Usually a DAM is constructed on a RIVER to provide the necessary height difference. The potential energy of the falling water is converted by a water turbine into mechanical energy. This is used to power a generator, which produces electricity. Also called HYDROELECTRIC POWER. Another form of water power is tidal power, which uses the force of the incoming and outgoing TIDE as its source of energy.

Water table. The depth below the surface where the zone of AERATION meets the zone of SATURATION. Above the water table, there may be some SOIL MOISTURE, but most of the pore space is filled with air. Below the water table, pore space of the ROCKS is occupied by water that has percolated down through the overlying earth material. This water is called GROUNDWATER. In practice, the water table is rarely as flat as a table, but curved, being far below the surface in some PLACES and even intersecting the surface in others. When GROUNDWATER emerges at the surface, because it intersects the water table, this is called a SPRING. The depth of the water table varies from SEASON to season, and with pumping of water from an AQUIFER.

Watershed. The whole surface area of land from which RAINFALL flows downslope into a STREAM. The watershed comprises the STREAMBED or CHANNEL, together with the VALLEY sides, extending up to the crest or INTERFLUVE, which separates that watershed from its neighbor. Each watershed is separated from the next by the drainage DIVIDE. Also called a DRAINAGE BASIN.

Waterspout. TORNADO that forms over water, or a tornado formed over land which then moves over water. The typical FUNNEL CLOUD, which reaches down from a CUMULONIMBUS CLOUD, is a narrow rotating STORM, with WIND speeds reaching hundreds of miles per hour.

Wave crest. Top of a WAVE.

Wave-cut platform. As SEA CLIFFS are eroded and worn back by WAVE attack, a wave-cut platform is created at the base of the cliffs. ABRASION by ROCK debris from the cliffs scours the platform further, as waves wash to and fro and TIDES ebb and flow. The upper part of the wave-cut platform is exposed at high tide. These areas contain rockpools, which are rich in interesting MARINE life-forms. Offshore beyond the platform, a wave-built TERRACE is formed by DEPOSITION.

Wave height. Vertical distance between one WAVE CREST and the adjacent WAVE TROUGH.

Wave length. Distance between two successive WAVE CRESTS or two successive WAVE TROUGHS.

Wave trough. The low part of a WAVE, between two WAVE CRESTS.

Weather analogue. Approach to WEATHER FORECASTING that uses the WEATHER behavior of the past to predict what a current weather pattern will do in the future.

Weather forecasting. Attempt to predict WEATHER patterns by analysis of current and past data.

Wilson cycle. Creation and destruction of an OCEAN BASIN through the process of SEAFLOOR SPREADING and SUBDUCTION of existing ocean basins.

Wind gap. Abandoned WATER GAP. The Appalachian Mountains contain both wind gaps and water gaps.

Windbreak. Barrier constructed at right angles to the prevailing WIND direction to prevent damage to crops or to shelter buildings. Generally, a row of trees or shrubs is planted to form a windbreak. The feature is also called a shelter belt.

Windchill. MEASUREMENT of apparent TEMPERATURE that quantifies the effects of ambient WIND and temperature on the rate of cooling of the human body.

World Aeronautical Chart. International project undertaken to map the entire world, begun during World War II.

World city. CITY in which an extremely large part of the world's economic, political, and cultural activity occurs. In the year 2018, the top ten world cities were London, New York City, Tokyo, Paris, Singapore, Amsterdam, Seoul, Berlin, Hong Kong, and Sydney.

Xenolith. Smaller piece of ROCK that has become embedded in an IGNEOUS ROCK during its formation. It is a piece of older rock that was incorporated into the fluid MAGMA.

Xeric. Description of SOILS in REGIONS with a MEDITERRANEAN CLIMATE, with moist cool winters and long, warm, dry summers. Since summer is the time when most plants grow, the lack of SOIL MOISTURE is a limiting factor on plant growth in a xeric ENVIRONMENT.

Xerophytic plants. Plants adapted to arid conditions with low PRECIPITATION. Adaptations include storage of moisture in tissue, as with cactus plants; long taproots reaching down to the WATER TABLE, as with DESERT shrubs; or tiny leaves that restrict TRANSPIRATION.

Yardangs. Small LANDFORMS produced by WIND EROSION. They are a series of sharp RIDGES, aligned in the direction of the wind.

Yazoo stream. TRIBUTARY that flows parallel to the main STREAM across the FLOODPLAIN for a considerable distance before joining that stream. This occurs because the main stream has built up NATURAL LEVEES through flooding, and because RELIEF is low on the floodplain. The yazoo stream flows in a low-lying wet area called backswamps. Named after the Yazoo River, a tributary of the Mississippi.

Zero population growth. Phenomenon that occurs when the number of deaths plus EMIGRATION is matched by the number of births plus IMMIGRATION. Some European countries have reached zero population growth.

BIBLIOGRAPHY

THE NATURE OF GEOGRAPHY

Adams, Simon, Anita Ganeri, and Ann Kay. *Geography of the World*. London: DK, 2010. Print.

Harley, J. B., and David Woodward, eds. *The History of Cartography: Cartography in the Traditional Islamic and South Asian Societies*. Vol. 2, book 1. Chicago: University of Chicago Press, 1992. Offers a critical look at maps, mapping, and mapmakers in the Islamic world and South Asia.

_____, eds. *The History of Cartography: Cartography in the Traditional East and Southeast Asian Societies*. Vol. 2, book 2. Chicago: University of Chicago Press, 1994. Similar in thrust and breadth to volume 2, book 1.

Marshall, Tim, and John Scarlett. *Prisoners of Geography: Ten Maps That Tell You Everything You Need to Know About Global Politics*. London : Elliott and Thompson Limited, 2016. Print

Nijman, Jan. *Geography: Realms, Regions, and Concepts*. Hoboken, NJ : Wiley, 2020. Print.

Snow, Peter, Simon Mumford, and Peter Frances. *History of the World Map by Map*. New York: DK Smithsonian, 2018.

Woodward, David, et al., eds. *The History of Cartography: Cartography in the Traditional African, American, Arctic, Australian, and Pacific Societies*. Vol. 2, book 3. Chicago: University of Chicago Press, 1998. Investigates the roles that maps have played in the wayfinding, politics, and religions of diverse societies such as those in the Andes, the Trobriand Islanders of Papua-New Guinea, the Luba of central Africa, and the Mixtecs of Central America.

PHYSICAL GEOGRAPHY

Christopherson, Robert W, and Ginger H. Birkeland. *Elemental Geosystems*. Hoboken, NJ : Wiley, 2016. Print.

Lutgens, Frederick K., and Edward J. Tarbuck. *Foundations of Earth Science*. Upper Saddle River, N.J.: Prentice-Hall, 2017. Undergraduate text for an introductory course in earth science, consisting of seven units covering basic principles in geology, oceanography, meteorology, and astronomy, for those with little background in science.

McKnight, Tom. Physical Geography: A Landscape Appreciation. 12th ed. New York: Prentice Hall, 2017. Now-classic college textbook that has become popular because of its illustrations, clarity, and wit. Comes with a CD-ROM that takes readers on virtual-reality field trips.

Robinson, Andrew. *Earth Shock: Climate Complexity and the Force of Nature*. New York: W. W. Norton, 1993. Describes, illustrates, and analyzes the forces of nature responsible for earthquakes, volcanoes, hurricanes, floods, glaciers, deserts, and drought. Also recounts

how humans have perceived their relationship with these phenomena throughout history.

Weigel, Marlene. *UxL Encyclopedia of Biomes*. Farmington Hills, Mich.: Gale Group, 1999. This three-volume set should meet the needs of seventh grade classes for research. Covers all biomes such as the forest, grasslands, and desert. Each biome includes sections on development of that particular biome, type, and climate, geography, and plant and animal life.

Woodward, Susan L. *Biomes of Earth*. Westport, CT: Greenwood Press, 2003. Print.

HUMAN GEOGRAPHY

Blum, Richard C, and Thomas C. Hayes. *An Accident of Geography: Compassion, Innovation, and the Fight against Poverty*. Austin, TX: Greenleaf Book Group, 2016. Print.

Dartnell, Lewis. *Origins: How the Earth Made Us*. New York: Hachette Book Group, 2019. Print.

Glantz, Michael H. *Currents of Change: El Niño's Impact on Climate and Society*. New York: Cambridge University Press, 1996. Aids readers in understanding the complexities of the earth's weather pattern, how it relates to El Niño, and the impact upon people around the globe.

Morland, Paul. *The Human Tide: How Population Shaped the Modern World*. New York: PublicAffairs, 2019. Print.

Novaresio, Paolo. *The Explorers: From the Ancient World to the Present*. New York: Stewart, Tabori and Chang, 1996. Describes amazing journeys and exhilarating discoveries from the earliest days of seafaring to the first landing on the moon and beyond.

Rosin, Christopher J, Paul Stock, and Hugh Campbell. *Food Systems Failure: The Global Food Crisis and the Future of Agriculture*. New York: Routledge, 2014. Print.

ECONOMIC GEOGRAPHY

Diamond, Jared M. *Guns, Germs, and Steel: The Fates of Human Societies*. New York: Norton, 2011. Print.

Esping-Andersen, Gosta. *Social Foundations of Postindustrial Economies*. New York: Cambridge University Press, 1999. Examines such topics as social risks and welfare states, the structural bases of postindustrial employment, and recasting welfare regimes for a postindustrial era.

Michaelides, Efstathios E. S. *Alternative Energy Sources*. Berlin: Springer Berlin, 2014. Print. This book offers a clear view of the role each form of alternative energy may play in supplying energy needs in the near future. It details the most common renewable energy sources as well as examines nuclear energy by fission and fusion energy.

Robertson, Noel, and Kenneth Blaxter. *From Dearth to Plenty: The Modern Revolution in Food Production*. New York: Cambridge University Press, 1995. Tells a story of scientific discovery and its exploitation for technological advance in agriculture. It encapsulates the history of an important period, 1936-86, when government policy sought to aid the competitiveness of the agricultural industry through fiscal measures and by encouraging scientific and technical innovation.

REGIONAL GEOGRAPHY

Biger, Gideon, ed. *The Encyclopedia of International Boundaries*. New York: Facts on File, 1995. Entries for approximately 200 countries are arranged alphabetically, each beginning with introductory information describing demographics, political structure, and political and cultural history. The boundaries of each state are then described with details of the geographical setting, historical background, and present political situation, including unresolved claims and disputes.

Leinen, Jo, Andreas Bummel, and Ray Cunningham. *A World Parliament: Governance and Democracy in the 21st Century*. Berlin Democracy Without Borders, 2018. Print.

Pitts, Jennifer. *Boundaries of the International: Law and Empire*. Cambridge, Mass: Harvard University Press, 2018. Print.

Index

O

P

F L A G S O F T H E W O R L D

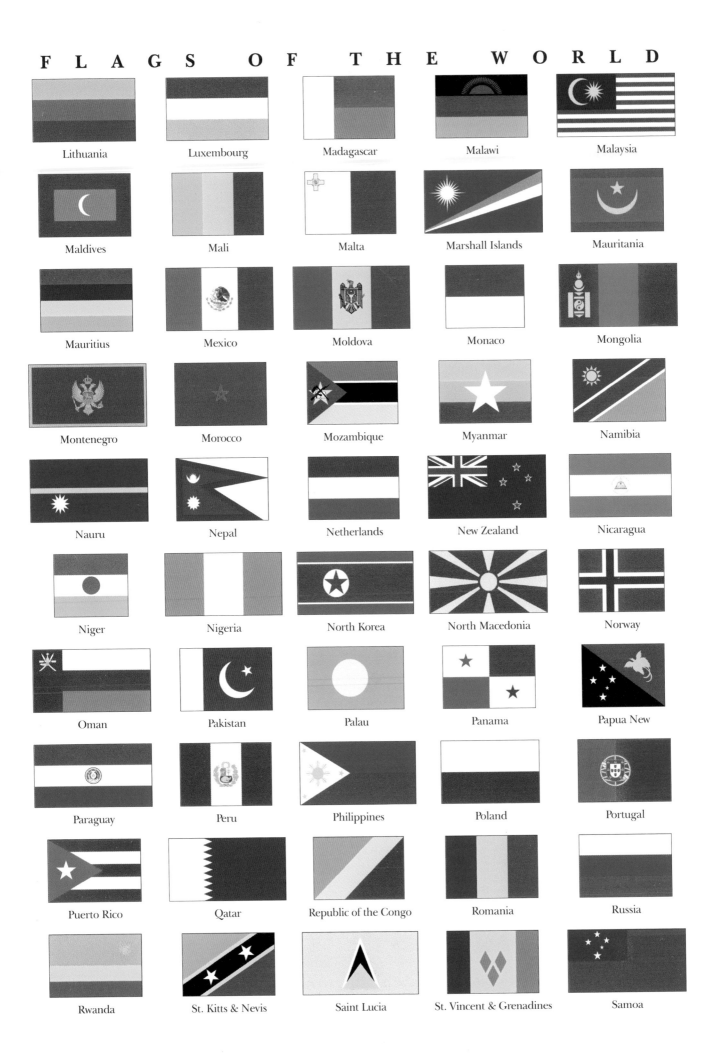

Lithuania

Luxembourg

Madagascar

Malawi

Malaysia

Maldives

Mali

Malta

Marshall Islands

Mauritania

Mauritius

Mexico

Moldova

Monaco

Mongolia

Montenegro

Morocco

Mozambique

Myanmar

Namibia

Nauru

Nepal

Netherlands

New Zealand

Nicaragua

Niger

Nigeria

North Korea

North Macedonia

Norway

Oman

Pakistan

Palau

Panama

Papua New

Paraguay

Peru

Philippines

Poland

Portugal

Puerto Rico

Qatar

Republic of the Congo

Romania

Russia

Rwanda

St. Kitts & Nevis

Saint Lucia

St. Vincent & Grenadines

Samoa